Diego Gonzalez-Aguilera, Fabio Remondino,
Pablo Rodríguez-Gonzálvez and Erica Nocerino
(Eds.)

Remote Sensed Data and Processing Methodologies for 3D Virtual Reconstruction and Visualization of Complex Architectures

MDPI

This book is a reprint of the Special Issue that appeared in the online, open access journal, *Remote Sensing* (ISSN 2072-4292) from 2015–2016 (available at: http://www.mdpi.com/journal/remotesensing/special_issues/3D-ARCH).

Guest Editors

Diego Gonzalez-Aguilera
University of Salamanca
Spain

Fabio Remondino
Bruno Kessler Foundation (FBK)
Italy

Pablo Rodríguez-Gonzálvez
Department of Cartographic and
Land Engineering
University of Salamanca
Spain

Erica Nocerino
Bruno Kessler Foundation (FBK)
Italy

Editorial Office
MDPI AG
St. Alban-Anlage 66
Basel, Switzerland

Publisher
Shu-Kun Lin

Managing Editor
Xuanxuan Guan

1. Edition 2016

MDPI • Basel • Beijing • Wuhan • Barcelona

ISBN 978-3-03842-237-2 (Hbk)
ISBN 978-3-03842-238-9 (PDF)

Table of Contents

Chapter 1: Aerial Photogrammetry

Chapter 2: Terrestrial Photogrammetry

Chapter 3: Multi-Sensor Fusion

Chapter 4: 3D Modeling

Chapter 5: Structural Analysis

List of Contributors

Mingyao Ai School of Remote Sensing and Information Engineering, Wuhan University, No.129, Luoyu Road, Wuhan 430079, China.

Fabrizio I. Apollonio Department of Architecture, University of Bologna, Bologna 40136, Italy.

Andrea Ballabeni Department of Architecture, University of Bologna, Bologna 40136, Italy.

Marco Giorgio Bevilacqua Energy, Systems, Territory and Construction Engineering Department (DESTeC), University of Pisa, Largo Lucio Lazzarino 1, 56122 Pisa, Italy.

Roland Billen Department of Geography, Geomatics Unit, University of Liège, Liège, B4000, Belgium.

Jan Boehm Civil, Environmental & Geomatic Engineering, University College London, Gower Street, London WC1E 6BT, UK.

Christian Boucheny Electricité De France Lab, 1 Avenue du Général de Gaulle, 92140 Clamart, France.

Manuel Cabaleiro Department of Material Engineering, Applied Mechanics and Construction, School of Industrial Engineering, University of Vigo, 36208 Vigo, Spain.

Mariana B. Campos Department of Cartography, Univ. Estadual Paulista (UNESP), Presidente Prudente, SP 19060-900, Brazil.

Gabriella Caroti Civil and Industrial Engineering Department (DICI), University of Pisa, Largo Lucio Lazzarino 1, 56122 Pisa, Italy.

Borja Conde Department of Material Engineering, Applied Mechanics and Construction, School of Industrial Engineering, University of Vigo, 36208 Vigo, Spain.

Fabien Dory Electricité De France Lab, 1 Avenue du Général de Gaulle, 92140 Clamart, France.

Beatriz Felipe-García Institute for Regional Development (IDR), Albacete, University of Castilla La Mancha, 02071 Albacete, Spain.

Caiwu Fu Institute of National Culture Development, Wuhan University, Luojiashan, Wuhan 430070, China.

Marco Gaiani Department of Architecture, University of Bologna, Bologna 40136, Italy.

Jesús García-Gago Department of Land and Cartographic Engineering, High Polytechnic School of Avila, University of Salamanca, Hornos Caleros, 50, 05003 Avila, Spain.

Diego Gonzalez-Aguilera Department of Land and Cartographic Engineering, High Polytechnic School of Avila, University of Salamanca, Hornos Caleros, 50, 05003 Avila, Spain

Pierre Grussenmeyer ICube Laboratory UMR 7357, Photogrammetry and Geomatics Group, National Institute of Applied Sciences (INSA Strasbourg), 24, Boulevard de la Victoire, 67084 Strasbourg Cedex, France.

Bingxuan Guo State Key Laboratory of Information Engineering in Surveying, Mapping and Remote Sensing, Wuhan University, Wuhan 430079, China; Collaborative Innovation Center for Geospatial Technology, Wuhan 430079, China.

Ayman Habib Lyles School of Civil Engineering, Purdue University, West Lafayette, IN 47907, USA.

David Hernández-López Institute for Regional Development (IDR), Albacete, University of Castilla La Mancha, 02071 Albacete, Spain.

Jesús Herrero-Pascual Department of Land and Cartographic Engineering, High Polytechnic School of Avila, University of Salamanca, Hornos Caleros, 50, 05003 Avila, Spain.

Qingwu Hu School of Remote Sensing and Information Engineering, Wuhan University, No.129, Luoyu Road, Wuhan 430079, China.

Jingyi Huang School of Remote Sensing and Information Engineering, Wuhan University, Wuhan 430079, China

Shan Huang School of Remote Sensing and Information Engineering, Wuhan University, No.129 Luoyu Road, Wuhan 430079, China

Xu Huang School of Remote Sensing and Information Engineering, Wuhan University, Wuhan 430079, China.

Jean-François Hullo Electricité De France Lab, 1 Avenue du Général de Gaulle, 92140 Clamart, France.

Ivana Ivánová Department of Cartography, Univ. Estadual Paulista (UNESP), Presidente Prudente, SP 19060-900, Brazil.

Tao Ke School of Remote Sensing and Information Engineering, Wuhan University, No.129 Luoyu Road, Wuhan 430079, China.

Ulrich Krispel Fraunhofer Austria Research GmbH, Visual Computing & Technische Universität Graz, Inffeldgasse 16c, 8010 Graz, Austria.

Elise Lachat ICube Laboratory UMR 7357, Photogrammetry and Geomatics Group, National Institute of Applied Sciences (INSA Strasbourg), 24, Boulevard de la Victoire, 67084 Strasbourg Cedex, France.

Susana Lagüela Department of Land and Cartographic Engineering, H igh Polytechnic School of Avila, University of Salamanca, Hornos Caleros, 50, 05003 Avila, Spain; Applied Geotechnologies Research Group, University of Vigo. Rúa Maxwell s/n, Campus Lagoas-Marcosende, Vigo 36310, Spain.

Tania Landes ICube Laboratory UMR 7357, Photogrammetry and Geomatics Group, National Institute of Applied Sciences (INSA Strasbourg), 24, Boulevard de la Victoire, 67084 Strasbourg Cedex, France.

Deren Li State Key Laboratory of Information Engineering in Surveying, Mapping and Remote Sensing, Wuhan University, Wuhan 430079, China; Collaborative Innovation Center for Geospatial Technology, Wuhan 430079, China.

Linhui Li State Key Laboratory of Information Engineering in Surveying, Mapping and Remote Sensing, Wuhan University, Wuhan 430079, China.

Qian Li School of Remote Sensing and Information Engineering, Wuhan University, Wuhan 430079, China.

Yun-Jou Lin Lyles School of Civil Engineering, Purdue University, West Lafayette, IN 47907, USA.

Jianchen Liu School of Remote Sensing and Information Engineering, Wuhan University, Wuhan 430079, China.

Xinyi Liu School of Remote Sensing and Information Engineering, Wuhan University, Wuhan 430079, China.

Luis López-Fernández Department of Land and Cartographic Engineering, High Polytechnic School of Avila, University of Salamanca, Hornos Caleros, 50, 05003 Avila, Spain.

Hongshu Lu Electronic Science and Engineering, National University of Defence Technology, Changsha 410000, China.

Hélène Macher ICube Laboratory UMR 7357, Photogrammetry and Geomatics Group, National Institute of Applied Sciences (INSA Strasbourg), 24, Boulevard de la Victoire, 67084 Strasbourg Cedex, France.

Jakub Stefan Markiewicz Department of Photogrammetry, Remote Sensing and Spatial Information Systems, Faculty of Geodesy and Cartography, Warsaw University of Technology, sqr. Politechniki 1, Warsaw 00-661, Poland.

Isabel Martínez-Espejo Zaragoza Civil and Industrial Engineering Department (DICI), University of Pisa, Largo Lucio Lazzarino 1, 56122 Pisa, Italy.

Arnaud Mas Electricité De France Lab, 1 Avenue du Général de Gaulle, 92140 Clamart, France.

Erica Nocerino 3D Optical Metrology (3DOM) Unit, Bruno Kessler Foundation (FBK), Trento 38123, Italy.

Zhe Peng State Key Laboratory of Information Engineering in Surveying, Mapping and Remote Sensing, Wuhan University, Wuhan 430079, China; Collaborative Innovation Center for Geospatial Technology, 129 Luoyu Road, Wuhan 430079, China.

Inmaculada Picón Department of Land and Cartographic Engineering, High Polytechnic School of Avila, University of Salamanca, Hornos Caleros, 50, 05003 Avila, Spain.

Andrea Piemonte Civil and Industrial Engineering Department (DICI), University of Pisa, Largo Lucio Lazzarino 1, 56122 Pisa, Italy.

Piotr Podlasiak Department of Photogrammetry, Remote Sensing and Spatial Information Systems, Faculty of Geodesy and Cartography, Warsaw University of Technology, sqr. Politechniki 1, Warsaw 00-661, Poland.

Susana Del Pozo Department of Land and Cartographic Engineering, High Polytechnic School of Avila, University of Salamanca, Hornos Caleros, 50, 05003 Avila, Spain.

Fabio Remondino 3D Optical Metrology (3DOM) Unit, Bruno Kessler Foundation (FBK), Trento 38123, Italy.

Pablo Rodríguez-Gonzálvez Department of Land and Cartographic Engineering, High Polytechnic School of Avila, University of Salamanca, Hornos Caleros, 50, 05003 Avila, Spain.

Luis Javier Sánchez-Aparicio Department of Land and Cartographic Engineering, High Polytechnic School of Avila, University of Salamanca, Hornos Caleros, 50, 05003 Avila, Spain.

Marcello Seddaiu Dipartimento di Storia, Scienze dell' Uomo e della Formazione, Università degli Studi di Sassari, Sassari, Italy.

Christoph Schinko Fraunhofer Austria Research GmbH, Visual Computing & Technische Universität Graz, Inffeldgasse 16c, 8010 Graz, Austria.

Zhenfeng Shao State Key Laboratory of Information Engineering in Surveying, Mapping and Remote Sensing, Wuhan University, Wuhan 430079, China; Collaborative Innovation Center for Geospatial Technology, 129 Luoyu Road, Wuhan 430079, China.

Chao Song School of Remote Sensing and Information Engineering, Wuhan University, Wuhan 430079, China.

Min Tang School of Remote Sensing and Information Engineering, Wuhan University, No.129 Luoyu Road, Wuhan 430079, China.

Guillaume Thibault Electricité De France Lab, 1 Avenue du Général de Gaulle, 92140 Clamart, France.

Charles Thomson Civil, Environmental & Geomatic Engineering, University College London, Gower Street, London WC1E 6BT, UK.

Antonio M. G. Tommaselli Department of Cartography, Univ. Estadual Paulista (UNESP), Presidente Prudente, SP 19060-900, Brazil.

Jose Alberto Torres-Martínez Department of Land and Cartographic Engineering, High Polytechnic School of Avila, University of Salamanca, Hornos Caleros, 50, 05003 Avila, Spain.

Torsten Ullrich Fraunhofer Austria Research GmbH, Visual Computing & Technische Universität Graz, Inffeldgasse 16c, 8010 Graz, Austria.

Alberto Villarino Department of Land and Cartographic Engineering, High Polytechnic School of Avila, University of Salamanca, Hornos Caleros, 50, 05003 Avila, Spain.

Shaohua Wang School of International Software, Wuhan University, No.129, Luoyu Road, Wuhan 430079, China.

Wende Wang Institute of National Culture Development, Wuhan University, Luojiashan, Wuhan 430070, China.

Xiongwu Xiao State Key Laboratory of Information Engineering in Surveying, Mapping and Remote Sensing, Wuhan University, Wuhan 430079, China; Collaborative Innovation Center for Geospatial Technology, 129 Luoyu Road, Wuhan 430079, China.

Xuan Xu School of Remote Sensing and Information Engineering, Wuhan University, No.129 Luoyu Road, Wuhan 430079, China.

Nan Yang State Key Laboratory of Information Engineering in Surveying, Mapping and Remote Sensing, Wuhan University, Wuhan 430079, China; Collaborative Innovation Center for Geospatial Technology, 129 Luoyu Road, Wuhan 430079, China.

Dengbo Yu School of Remote Sensing and Information Engineering, Wuhan University, No.129, Luoyu Road, Wuhan 430079, China.

Dorota Zawieska Department of Photogrammetry, Remote Sensing and Spatial Information Systems, Faculty of Geodesy and Cartography, Warsaw University of Technology, sqr. Politechniki 1, Warsaw 00-661, Poland.

Lei Zhang State Key Laboratory of Information Engineering in Surveying, Mapping and Remote Sensing, Wuhan University, Wuhan 430079, China; Collaborative Innovation Center for Geospatial Technology, 129 Luoyu Road, Wuhan 430079, China.

Peng Zhang Institute of Hydroecology, Ministry of Water Resource and Chinese Academy of Sciences, Wuhan 430079, China.

Yongjun Zhang School of Remote Sensing and Information Engineering, Wuhan University, Wuhan 430079, China

Zuxun Zhang School of Remote Sensing and Information Engineering, Wuhan University, No.129 Luoyu Road, Wuhan 430079, China.

About the Guest Editors

Diego González-Aguilera received a B.Sc. degree in surveying engineering and a M.Sc. degree in geodesy and cartography engineering from Salamanca University, Spain, in 1999 and 2001, respectively. He was a Research Assistant at INRIA, Grenoble, France, in the Institute of Computer Vision and Robotics, where he also performed his Ph.D. research on 3D reconstruction from a single view, graduating in 2005. He has authored more than 100 research articles in international journals and conference proceedings. Based on his thesis results, he has received four international awards from the International Societies of Photogrammetry and Remote Sensing (ISPRS/ASPRS). In 2005, he founded the research group TIDOP (Geomatics technologies for the 3D digitalization and modeling of complex objects) at the University of Salamanca (http://tidop.usal.es/). Dr. González-Aguilera is an Assistant Professor at the University of Salamanca (with accreditation for Full Professor). He is a member of the ISPRS, where he serves as Secretary of the Working Group on Terrestrial 3D Modelling: Algorithms and Methods and regularly as a Program Committee Member for conferences and a reviewer for related journals.

Fabio Remondino received a Ph.D. in Photogrammetry in 2006 from ETH Zurich, Switzerland and now leads the 3D Optical Metrology unit (http://3dom.fbk.eu) at the Bruno Kessler Foundation (http://www.fbk.eu), a public research center in Trento, Italy. His research interests include geospatial data collection and processing, heritage documentation, 3D modeling, and sensor and data integration. He is the author of over 150 scientific publications in journals and at international conferences, five books, and editor of seven special issues in journals. He has received 10 awards for best papers at conferences, and organized 26 scientific events and 29 summer schools and tutorials. He was President of the ISPRS Technical Commission V (2012–2016), President of EuroSDR Commission I, and Vice-President of CIPA Heritage Documentation.

Pablo Rodríguez-Gonzálvez received a B.Sc. degree in surveying engineering in 2004 from Oviedo University and a M.Sc. degree in geodesy and cartography in 2006 from Salamanca University. He received for both degrees the First National End of Degree Award. He obtained his Ph.D. in 2011, from Salamanca University, for which received an Extraordinary Ph.D. award. His main research line is photogrammetry application to different fields (automotive, forensic sciences, etc.). He is involved in research on UAS and LiDAR, and applications of TLS and gaming sensors to engineering and architecture. He is also interested in radiometric and geometric calibration of different Geomatic sensors.

Erica Nocerino received a Ph.D. in "Aerospatial Engineering, Naval Architecture, and Quality Control" from the Federico II University of Naples, and in March 2015 she received a second Ph.D. degree in "Geomatics, Navigation, and Geodesy" from the Parthenope University of Naples. Since 2012, she has been working as a research scientist in the 3D Optical Metrology unit of the Bruno Kessler Foundation (http://www.fbk.eu). Her research interests are in the area of photogrammetry (close range aerial, terrestrial, and underwater), laser scanning, 3D recording and reverse engineering, automation, and image analysis. She has received four awards at international conferences and is involved in teaching and tutoring activities in international and national workshops and summer schools.

Preface to "Remote Sensed Data and Processing Methodologies for 3D Virtual Reconstruction and Visualization of Complex Architectures"

The topic *"Remote Sensed Data and Processing Methodologies for 3D Virtual Reconstruction and Visualization of Complex Architectures"* is a growing subject that involves many different disciplines, such as architecture, cultural heritage, engineering, archaeology, and virtual reality. The increase is driven by the current availability of new sensors for remote acquisition along with open and big data sources. Despite availability and advances at sensor- and data-level, there is a need for reliable, affordable, and powerful methods and tools for realizing photo-realistic, metric, re-usable, and semantic-aware 3D products. In this challenging and inspiring transition, the scientific community is putting great effort to design, research, develop and validate novel easy-to-use, ease-to-learn and low-cost frameworks for 3D modelling and further understanding of real environments.

This book originates from the ISPRS/CIPA 3D-ARCH workshop *"3D Virtual Reconstruction and Visualization of Complex Architectures"* which was held in Ávila, Spain in February 2015. The workshop brought scientists, developers and advanced users in photogrammetry, computer vision, 3D modelling, and related topics to present and share their latest advancements and achievements.

This book brings together 19 peer-reviewed contributions from various authors, including extended papers presented at the 3D-ARCH workshop, covering topics related to the 3D virtual reconstruction and visualization of complex scenarios. Due to the high transversality of 3D and intra-sectoral applications, the key topics addressed in the book are: **aerial photogrammetry**, especially focused on the low-altitude remote sensing images which take advantage of classical platforms, stressing the automated processing stages of data segmentation and dense point cloud generation; **terrestrial photogrammetry**, as a fundamental technique for architectural heritage documentation, with all latest improvements in the automated 3D reconstruction pipeline; **multi-sensor fusion**, as a solution for a broad spectrum of applications and solutions, encompassing pathological mapping, urban-scale analysis, and optimization of orthoimage generation, solar panel installation, and surveying of complex architectures; **3D modelling**, which covers the modelling procedures involved when using gaming sensors, while producing building information models (BIM), or for procedural building modelling; **structural analysis**, focused on reverse engineering and finite element modelling (FEM) for stability evaluation and conservation measures.

We would like to thank everyone involved in the editing of this book: the invaluable contribution of the almost 80 authors, along with the excellent help of reviewers and the MDPI *Remote Sensing* journal team.

Diego González-Aguilera, Fabio Remondino,
Pablo Rodriguez-Gonzalvez and Erica Nocerino
Guest Editors

Chapter 1:
Aerial Photogrammetry

Multi-View Stereo Matching Based on Self-Adaptive Patch and Image Grouping for Multiple Unmanned Aerial Vehicle Imagery

Xiongwu Xiao, Bingxuan Guo, Deren Li, Linhui Li, Nan Yang, Jianchen Liu, Peng Zhang and Zhe Peng

Abstract: Robust and rapid image dense matching is the key to large-scale three-dimensional (3D) reconstruction for multiple Unmanned Aerial Vehicle (UAV) images. However, the following problems must be addressed: (1) the amount of UAV image data is very large, but ordinary computer memory is limited; (2) the patch-based multi-view stereo-matching algorithm (PMVS) does not work well for narrow-baseline cases, and its computing efficiency is relatively low, and thus, it is difficult to meet the UAV photogrammetry's requirements of convenience and speed. This paper proposes an Image-grouping and Self-Adaptive Patch-based Multi-View Stereo-matching algorithm (IG-SAPMVS) for multiple UAV imagery. First, multiple UAV images were grouped reasonably by a certain grouping strategy. Second, image dense matching was performed in each group and included three processes. (1) Initial feature-matching consists of two steps: The first was feature point detection and matching, which made some improvements to PMVS, according to the characteristics of UAV imagery. The second was edge point detection and matching, which aimed to control matching propagation during the expansion process; (2) The second process was matching propagation based on the self-adaptive patch. Initial patches were built that were centered by the obtained 3D seed points, and these were repeatedly expanded. The patches were prevented from crossing the discontinuous terrain by using the edge constraint, and the extent size and shape of the patches could automatically adapt to the terrain relief; (3) The third process was filtering the erroneous matching points. Taken the overlap problem between each group of 3D dense point clouds into account, the matching results were merged into a whole. Experiments conducted on three sets of typical UAV images with different texture features demonstrate that the proposed algorithm can address a large amount of UAV image data almost without computer memory restrictions, and the processing efficiency is significantly better than that of the PMVS algorithm and the matching accuracy is equal to that of the state-of-the-art PMVS algorithm.

Reprinted from *Remote Sens*. Cite as: Xiao, X.; Guo, B.; Li, D.; Li, L.; Yang, N.; Liu, J.; Zhang, P.; Peng, Z. Multi-View Stereo Matching Based on Self-Adaptive Patch and Image Grouping for Multiple Unmanned Aerial Vehicle Imagery. *Remote Sens*. **2016**, *8*, 89.

1. Introduction

The image sequences of UAV low-altitude photogrammetry are characterized by large scale, high resolution and rich texture information, which make it suitable for three-dimensional (3D) observation and its role as a primary source of fine 3D data [1,2]. UAV photogrammetry systems consist of airborne sensors, airborne Global Navigation Satellite Systems (GNSS) (for example, Global Positioning Systems, GPS) and Inertial Navigation Systems (INS), flight control systems and other components, which can provide aerial images and position and pose (POS) data [3,4]. As light and small low-altitude remote sensing aircraft, UAVs offer advantages such as flexibility, ease of operation, convenience, safety and reliability, and low costs [3,5,6]. They can be widely used in many applications such as large-scale mapping [7], true orthophoto generation [8], environmental surveying [9], archaeology and cultural heritage [10], traffic monitoring [11], 3D city modeling [12], and especially emergency response [13]; each field contributes to the rapid development of the technology and offers extensive markets [2,14].

Reconstructing 3D models of objects based on large-scale and high-resolution image sequences obtained by UAV low-altitude photogrammetry demands rapid modeling speeds, high automaticity and low costs. These attributes rely upon the technology in the digital photogrammetry and computer vision fields, and image dense matching is exactly the key to this problem. However, because of their small size, UAVs are vulnerable to airflow, resulting in instability in flight attitudes, which leads to images with large tilt angles and irregular tilt directions [15,16]. Until now, most UAV photogrammetry systems have used non-metric cameras, which generate a large number of images with small picture formats, resulting in a small base-to-height ratio [5]. The characteristics described above present many difficulties and challenges for robust and rapid image matching. Thus, the research on and implementation of UAV multi-view stereo-matching are of great practical significance and scientific value.

The goal of the multi-view stereo is to reconstruct a complete 3D object model from a collection of images taken from known camera viewpoints [17]. Over the last decade, a number of high-quality algorithms have been developed, and the state of the art is improving rapidly. According to [18], multi-view stereo algorithms can be roughly categorized into four classes: (1) Voxel-based approaches [19–24] require knowing a bounding box that contains the scene, and their accuracy is limited by the resolution of the voxel grid. A simple example of this approach is the graph cut algorithm [22,25,26], which transforms 3D reconstruction into finding the minimum cut of the constructed graph; (2) Algorithms based on deformable polygonal meshes [27–29] demand a good starting point—for example, a visual hull model [30,31]—to initialize the corresponding optimization process, which limits their applicability. The spacing curve [29] first extracts the outline of the

4

object, establishing a rough visual hull, and then photo consistency constraints are adopted to carve the visual hull and finally recover the surface model. Voxel-based or polygonal mesh–based methods are often limited to object data sets (scene data sets or crowd scene data sets are hard to handle), and they are not flexible; (3) Approaches based on multiple depth maps [32–35] are more flexible, but the depth maps tend to be noisy and highly redundant, leading to wasted computational effort. Therefore, these algorithms typically require additional post-processing steps to clean up and merge the depth maps [36]. The Semi-Global Matching (SGM) algorithm [35] and its acceleration algorithms [37] are widely used in many applications [38,39]. The study in [40] enhanced the SGM approach with the capacity to search pixel correspondences using dynamic disparity search ranges, and introduced a correspondence linking technique for disparity map fusion (disparity maps are generated for each reference view and its two adjacent views) in a sequence of images, which is most similar to [41]; (4) patch-based methods [42,43] represent scene surfaces by collections of small patches (or surfels). They use matching propagation to achieve dense matching. Typical algorithms include the patch propagation algorithm [18], belief propagation algorithm [44], and triangle constrained image matching propagation [45,46]. Patch-based matching in the scene space is much more reasonable than rectangular window matching in the image space [22,26,33] because it adds the surface normal and position information. Furukawa [18] generates a sparse set of patches corresponding to the salient image features, and then spreads the initial matches to nearby pixels and filters incorrect matches to maintain surface accuracy and completeness. This algorithm can handle a variety of data sets and allows outliers or obstacles in the images. Furthermore, it does not require any assumption on the topology of an object or a scene and does not need any initialization, for example a visual hull model, a bounding box, or valid depth ranges that are required in most other competing approaches, but it can take advantage of such information when available. The state-of-the-art algorithm achieves extremely high performance on a great deal of MVS datasets [47] and is suitable for large-scale high-resolution multi-view stereo [48], but does not work well for narrow-baseline cases [18]. To improve the processing efficiency of PMVS, Mingyao Ai [16] feeds the PMVS software with matched points (as seed points) to obtain a dense point cloud.

This paper proposes a multi-view stereo-matching method for low-altitude UAV data, which is characterized by a large number of images, a small base-to-height ratio, large tilt angles and irregular tilt directions. The proposed method is based on an image-grouping strategy and some control strategies suitable for UAV image matching and a self-adaptive patch-matching propagation method. It is used to improve upon the state-of-the-art PMVS algorithm in terms of the processing capacity and efficiency. Practical applications indicate that the proposed method greatly improves processing capacity and efficiency, while the matching precision is equal to that of the PMVS algorithm.

The paper is organized as follows. Section 2 describes the issues and countermeasures for UAV image matching and the improved multi-view stereo-matching method based on the PMVS algorithm for UAV data. In Section 3, based on experiments using three typical sets of UAV data with different texture features, the processing efficiency and matching accuracy of the proposed multi-view stereo-matching method for UAV data are analyzed and discussed. Conclusions are presented in the last section.

2. Methodology

2.1. The Issues and Countermeasures Related to UAV Image Matching

Because of the low flight altitude of UAVs, UAV images have high resolution and rich features. However, we cannot avoid mismatching because of the impact of deformation, occlusion, discontinuity and repetitive texture. For image deformation problems, we can establish a general affine transformation model [18,49,50]. For occlusion problems, because the occluded part of an image maybe visible in other images, the multi-view redundancy matching strategy was generally used [18,50]. For discontinuity problems, local smooth constraints were introduced to match the sparse texture areas, and edges were used to control smooth constraints [51]. For repetitive texture problems, the epipolar constraint is a good choice. However, it is ineffective for ambiguous matches when the texture and epipolar line have similar directions. This paper used the matching method based on patch, which can solve the problem.

2.2. PMVS Algorithm

We will briefly introduce the Patch-based Multi-View Stereo (PMVS) algorithm [18]; then we will employ it and make improvements in the field of UAV image dense matching. PMVS [18,52] is a multi-view stereo software that uses a set of images as well as the camera parameters as inputs and then reconstructs the 3D structure of an object or a scene that is visible in the images. The software outputs both the 3D coordinate and the surface normal at each oriented point. The algorithm consists of three procedures: (1) initial matching where sparse (3D) seed points are generated; (2) expansion where the initial matches are spread to nearby pixels and dense matches are obtained; (3) filtering where visibility constraints are used to eliminate incorrect matches. After the first step (generating seed points), the next two steps need to cycle three times.

First, use the Difference-of-Gaussian (DOG) [53] and Harris [54] operators to detect blob and corner features. To ensure uniform coverage, lay over each image a regular grid of 32×32 pixel blocks and return as features the four local maxima with the strongest responses in each block for each operator. Consider each image as reference image $R(p)$ in turn and other images that meet the geometric constraint conditions as search images $I(p)$. For each feature f detected in $R(p)$, collect in $I(p)$ the set F of features f' of the same type (Harris or DOG) that lie within two pixels from the corresponding epipolar lines in $I(p)$, and triangulate the 3D points associated with the pairs (f, f'). Sort these points in order of increasing distance from the optical center of the corresponding camera. Initial a patch from these points one by one and also initial corresponding image sets $V(p)$, $V^*(p)$ (images in $V(p)$ satisfy the angle constraint, and images in $V^*(p)$ satisfy the correlation coefficient constraint). Then, use a conjugate-gradient method [55] to refine the center and normal vector of the patch and update $V(p)$ and $V^*(p)$. If $|V^*(p)| \geqslant 3$, the patch generation is deemed a success, and the patch is stored in the corresponding cells of the visible images. To speed up the computation, once a patch has been reconstructed and stored in a cell, all the features in the cell are removed and no longer used.

Second, repeat taking existing patches and generating new ones in nearby empty spaces. The expansion is unnecessary if a patch has already been reconstructed there or if there is segmentation information (depth discontinuity) when viewed from the camera. The new patch's normal vector is the same as that of the seed patch. The new patch's center is the intersection of the light through the neighborhood image cell of the image point f and the seed patch plane. The rest is similar to the procedure for generating the seed patch: Refine and verify the new patch, update $V(p)$ and $V^*(p)$, and if $|V^*(p)| \geqslant 3$, accept the new patch as a success. The new patches are also participating in the expansion as seed patches. The goal of the expansion step is to reconstruct at least one patch in each image cell.

Third, remove erroneous patches using three filters that rely on visibility consistency, a weak form of regularization, and clustering constraint.

2.3. The Design and Implementation of IG-SAPMVS

The proposed IG-SAPMVS mainly processes multiple UAV images with known orientation elements and outputs a dense colored point cloud. First, given that the number of images may be too large and considering the memory limit of an ordinary computer, we need to group the images, which will be described in Section 2.3.4. Then, we process each group in turn, which is partitioned into three parts: (1) multi-view initial feature-matching; (2) matching propagation based on the self-adaptive patch; and (3) filtering the erroneous matching points. Finally, we need to merge the 3D point cloud results of all the groups into a whole, which will be described in Section 2.3.5.

7

2.3.1. Multi-View Initial Feature-Matching

This procedure had two steps: The first was feature point detection and matching, which made some improvements to PMVS, according to the characteristics of UAV imagery. The second was edge detection and matching, which aimed to control matching propagation during the expansion process.

The proposed method followed PMVS by setting up regular grids on all of the images, used Harris and DOG operators to detect feature points, and then matched the feature points. For each matched feature point f in the reference image, when finding candidate matching points on the corresponding epipolar line in the search image, there is an improvement compared to PMVS. In general cases, the rough ground elevation scope of the region photographed by UAV is known. Denoting the rough ground elevation scope by (Z_{min}, Z_{max}), there is a corresponding scope denoted by (P_{min}, P_{max}) in the epipolar line of the search images; thus, we can simply seek the corresponding image points in that scope. Taking into account that the orientation elements of the reference image may not be very accurate, the search range (the red box in Figure 1) was expanded to two pixels around the epipolar line. Then, we calculated the correlation coefficient between each potential candidate matching point in that scope and the matched feature point f. Because of the unstable flight attitudes of the UAV, the image deformation is large. It is not advisable to use the traditional correlation coefficient calculation method that assumes the relevant window may be simply along the direction of the image. Thus, we designed the relevant window along the direction of the epipolar line; that is to say, the edge of the relevant window was parallel with the epipolar line (Figure 1).

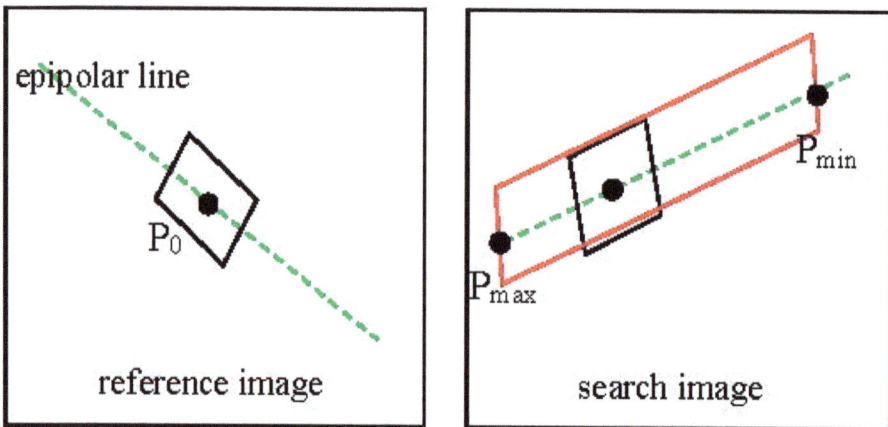

Figure 1. The relevant window (black window) is consistent with the direction of the epipolar line (green dotted line).

The potential candidate feature point whose correlation coefficient is greater than γ_1 (in this paper, we set γ_1 to be 0.7) was used as candidate match point f' and added to set F. The set F was sorted according to the value of the correlation coefficient from big to small, and then, the 3D points associated with the pairs (f, f') were triangulated. According to [18], we considered these 3D points as potential patch centers. Next, we refined and verified each patch candidate associated with the potential patch center in turn according to the PMVS method. Finally, we obtained the accurate 3D seed points with normal vectors.

Edge feature detection and matching, taking the method of Li Zhang [51] as a reference, first used the Canny operator [56] to detect edge points on every image, and then performed a match for the dominant points and well-distributed points on the edges. The only difference was that after obtaining the 3D edge points using the method of [51], we used the PMVS method to refine and verify each 3D edge point.

2.3.2. Matching Propagation Based on Self-Adaptive Patch

In PMVS, the 3D seed points are very sparse, and dense matching mainly depends on the expansion step; moreover, it is time-consuming. However, such speed makes it difficult to meet the requirements of real-time and convenient UAV photogrammetry. Thus, this paper presented matching propagation based on the self-adaptive patch method to improve processing efficiency. The basic concept was to build initial patches centered by the 3D seed points that had already been obtained. The extents and shapes of the patches could adapt to the terrain relief automatically: When the surface was smooth, the size of the patch would become bigger to cover the entire smooth area; if the terrain was very rough, the size of the patch would become smaller to describe the details of the surface (Figure 2a). In Figure 2, different sizes and shapes of patches are conformed to the different terrain features. There are more spread points on the larger extent patches.

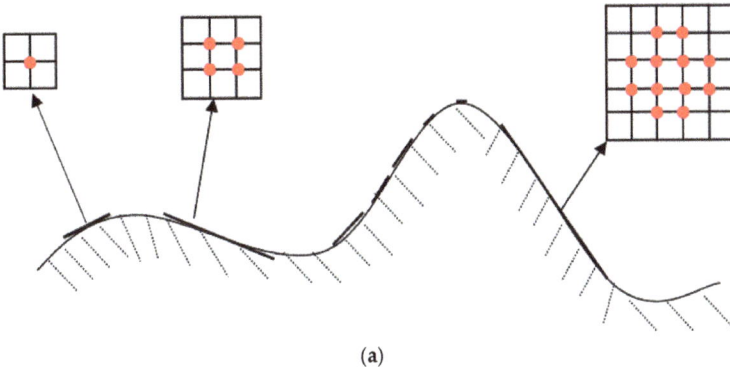

(a)

Figure 2. *Cont.*

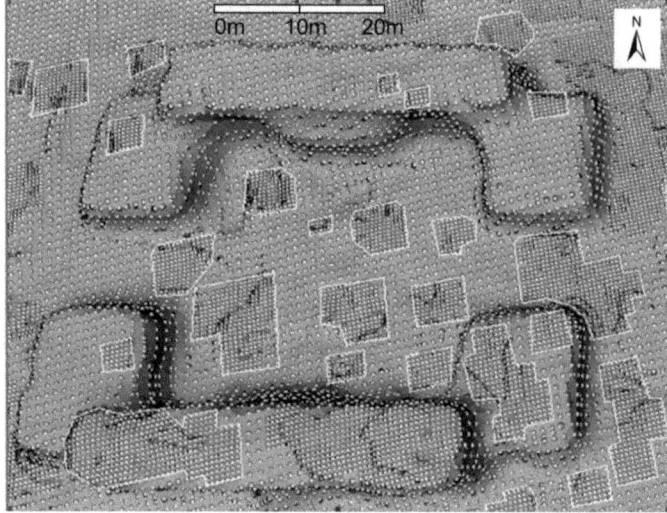

(b)

Figure 2. (**a**) Self-adaptive patch; (**b**) Schematic diagram of self-adaptive patches on the DSM.

The initial size of a patch must ensure that the projection of the patch onto the reference image can cover the size of $m_2 \times m_2$ square pixels (in this paper, m_2 is also denoted as the length of the patch, and the initial value of m_2 is $3 \times 7 = 21$, meaning it is three times the initial patch size of PMVS). There was an affine transformation between the patch and its projection onto the image, and the affine transformation parameters could be determined by the center coordinate and normal vector of the patch. Taking the projection of the patch onto the image as a relevant window, the correlation coefficient between the reference image I_0 and each search image I_j donated by $ncc(I_0, I_j, patch)$ was calculated. Because UAV imagery is generally taken by an ordinary non-metric digital camera, the imagery usually has three channels: R, G, B. Thus, when calculating the correlation coefficient $ncc(I_0, I_j, patch)$, we made full use of the information of the three channels. The formula is as follows:

$$
\begin{aligned}
&ncc(I_0, I_j, patch) \\
&= \sum_{i=1}^{m_2^2} \frac{(R_0[i] - R_{ave0})(R_j[i] - R_{avej}) + (G_0[i] - G_{ave0})(G_j[i] - G_{avej}) + (B_0[i] - B_{ave0})(B_j[i] - B_{avej})}{\sigma_0 \cdot \sigma_j}
\end{aligned}
\qquad (1)
$$

where

$$
\sigma_0 = \sqrt{\frac{\sum_{i=1}^{m_2^2} [(R_0[i] - R_{ave0})^2 + (G_0[i] - G_{ave0})^2 + (B_0[i] - B_{ave0})^2]}{m_2 \times 3}}
$$

10

$$\sigma_j = \sqrt{\frac{\sum\limits_{i=1}^{m_2^2}\left[(R_j[i] - R_{avej})^2 + (G_j[i] - G_{avej})^2 + (B_j[i] - B_{avej})^2\right]}{m_2 \times 3}}$$

$$R_{ave0} = \frac{\sum\limits_{i=1}^{m_2^2} R_0[i]}{m_2^2} \qquad\qquad R_{avej} = \frac{\sum\limits_{i=1}^{m_2^2} R_j[i]}{m_2^2}$$

Then, we computed the average value of the correlation coefficients as follows:

$$NCC_{patch} = \frac{1}{n}\sum_{j=1}^{n} ncc\left(I_0, I_j, patch\right) \qquad\qquad (2)$$

where n is the number of search images.

If the NCC_{patch} is greater than the threshold γ_2 ($\gamma_2 = 0.8$), the surface area covered by the patch is smooth and can be similarly treated as a plane, so that some new 3D points (near the patch center) in the patch plane can be directly generated. The normal vectors of the new 3D points are the same as the normal vector $n(p)$ of the patch. According to the properties of affine transformation, a plane π through affine transformation becomes another plane $\pi*$, and the affine transformation parameters for each point on the plane π are the same [18,57]; thus, we can compute the corresponding coordinates of the 3D new points in the patch by using the center point coordinate $c(p)$ and normal vector $n(p)$ of this patch plane. The calculation process is as follows:

(1) Calculate the xyz-plane coordinate system of the patch, that is the x-axis is $\mathbf{p_x}$ (p_{x1}, p_{x2}, p_{x3}) and the y-axis is $\mathbf{p_y}$ (p_{y1}, p_{y2}, p_{y3}), and the normal vector $n(p)$ of the patch is seen as the z-axis (Figure 3). The patch center p is considered the origin of the xyz-plane coordinate system of the patch. The y-axis $\mathbf{p_y}$ is the vector that is perpendicular to the normal vector $n(p)$ and the X_c-axis of the image space coordinate system; thus, $\mathbf{p_y} = n(p) \times X_c$. The x-axis $\mathbf{p_x}$ is the vector that is perpendicular to the y-axis $\mathbf{p_y}$ and the normal vector $n(p)$; thus, $\mathbf{p_x} = \mathbf{p_y} \times n(p)$. Then, $\mathbf{p_x}$ and $\mathbf{p_y}$ are normalized to the unit vector.

(2) Calculate the ground resolution of the image as follows:

$$d = \frac{d_{pc}}{f_c \cdot \cos\theta} \qquad\qquad (3)$$

where d_{pc} represents the distance between the patch center p and the projection center of the image, f_c represents the focal length, θ represents the angle between the light through the patch center p and the normal vector $n(p)$ of the patch, and d actually represents the corresponding distance in the direction of $n(p)$ for one pixel in the image.

(3) Suppose a new point's plane coordinate in the patch plane coordinate system is $(\Delta x, \Delta y)$. The plane coordinates of the new points in the patch plane coordinate system are as shown in Figure 4. To ensure matching accuracy, the spread size is in the range of $\left(-\dfrac{m_2 - 1}{4}, \dfrac{m_2 - 1}{4} \right)$ (the patch size is $m_2 \times m_2$) when the patch center p is considered the origin of the patch plane coordinate system.

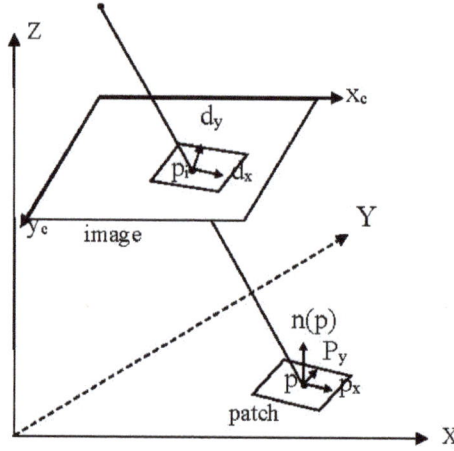

Figure 3. The xyz-plane coordinate system of the patch.

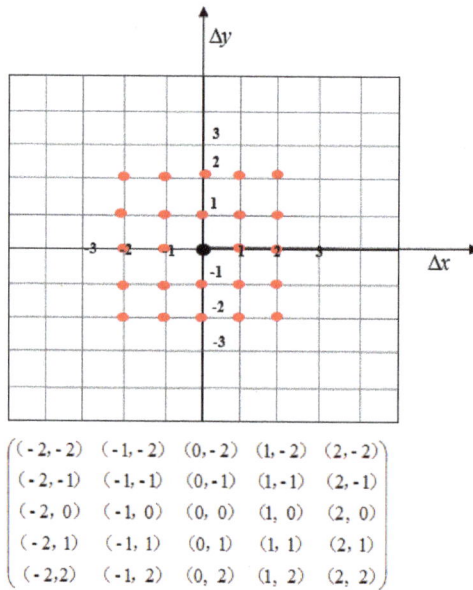

$$
\begin{pmatrix}
(-2,-2) & (-1,-2) & (0,-2) & (1,-2) & (2,-2) \\
(-2,-1) & (-1,-1) & (0,-1) & (1,-1) & (2,-1) \\
(-2, 0) & (-1, 0) & (0, 0) & (1, 0) & (2, 0) \\
(-2, 1) & (-1, 1) & (0, 1) & (1, 1) & (2, 1) \\
(-2, 2) & (-1, 2) & (0, 2) & (1, 2) & (2, 2)
\end{pmatrix}
$$

Figure 4. The plane coordinates of the new points (red points) in the patch plane.

(4) Calculate the XYZ-coordinate of the new point in the object space coordinate system. Suppose a (3D) new point is $P = (X_P, Y_P, Z_P)$ and the patch center p is (X_c, Y_c, Z_c). We calculate the XYZ-coordinate of the new point as follows:

$$\mathbf{P} = \mathbf{p} + d \cdot \Delta x \cdot \mathbf{p_x} + d \cdot \Delta y \cdot \mathbf{p_y}$$

$$\begin{pmatrix} X_P \\ Y_P \\ Z_P \end{pmatrix} = \begin{pmatrix} X_c \\ Y_c \\ Z_c \end{pmatrix} + d \cdot \Delta x \cdot \begin{pmatrix} p_{x1} \\ p_{x2} \\ p_{x3} \end{pmatrix} + d \cdot \Delta y \cdot \begin{pmatrix} p_{y1} \\ p_{y2} \\ p_{y3} \end{pmatrix} \tag{4}$$

After obtaining the XYZ-coordinate of a new point in the object space coordinate system, the new point is projected onto the reference image and search images so that we can obtain the corresponding image points.

As a result, new 3D points are spread by the 3D seed point, as shown in Figure 5.

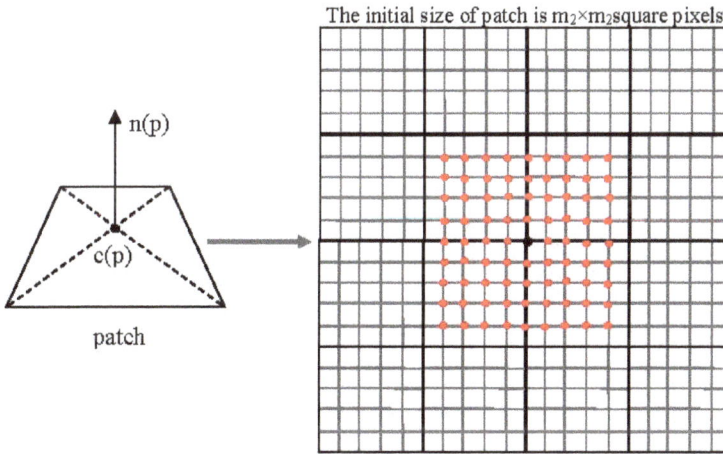

Figure 5. The new 3D points (red points) are propagated by the seed point (black point).

In our method, the size and shape of the patch can self-adapt to the texture feature: (1) For big and very smooth areas, the generated new 3D points will be directly added to the seed point set, and the newly added seed points are also in the original patch plane and they can spread further so that it enlarges the size of the original patch. Thus, the shape formed by all the spread points in the original patch would not be a regular square (Figure 2); (2) For discontinuous terrain, the length of the patch m_2 would be shortened to avoid crossing the edge, resulting in a smaller spread size (the spread size is in the range of $\left(-\dfrac{m_2 - 1}{4}, \dfrac{m_2 - 1}{4} \right)$) and a smaller size of the patch ($m_2 \times m_2$); (3) For a large relief terrain, the size and shape of the patch would be automatically shrunk from different directions, resulting in a smaller spread size. For such cases, the generated new 3D points would be processed following the method of PMVS. First, they are refined and verified one by one, and then the 3D points that are successfully constructed are added to the 3D seed point set. The newly added seed points are not in the original patch plane, and, thus, they will not enlarge the size of the original patch.

2.3.3. The Strategy of Matching Propagation

In Self-Adaptive Patch-based Multi-View Stereo-matching (SAPMVS), the size and shape of the patch should adapt to the texture feature: increasing the patch size in smooth areas and decreasing the patch size in undulating terrain. In addition, the patch should avoid crossing discontinuous terrain. The strategy of matching propagation based on the self-adaptive patch is as follows (Figure 6):

(1)   Get a 3D seed point from the initial feature-matching set $SEED$, and build a patch by using that point as the patch center. The initial size of the patch should ensure that the projection of the patch onto the reference image is of size $m_2 \times m_2$ square pixels. In this paper, m_2 is also denoted as the length of the patch, and the initial value of m_2 is 21. However, if the set $SEED$ becomes empty, stop the whole matching propagation process.

(2)   If the patch does not contain any edge points, go to step 3. Otherwise, adjust the size and shape of the patch so that the edge (we used the edge points to determine the edge) is not crossed. In Figure 7, the patch is partitioned into two parts by an edge. We need to build a new patch from the part that contains the center point. The center point of the new patch is unchanged, and the shape is a square. One edge of the new patch is parallel to the edge. Thus, the length of the patch m_2 decreased and the shape also changed. Finally, go to step 3.

(3)   Take the projection of the patch onto each of the images as the relevant window, and use Equation (1) to calculate the correlation coefficient $ncc(I_0, I_j, patch)$ between the reference image and each search image. Then, use Equation (2) to calculate the average value of the correlation coefficients NCC_{patch}.

(4)   If $NCC_{patch} < \gamma_2$ $(\gamma_2 = 0.8)$, the size of the patch needs to be adjusted; thus, go to step 5, or else generate some new 3D points near the patch center in the range of $\left(-\dfrac{m_2 - 1}{4}, \dfrac{m_2 - 1}{4} \right)$ on the patch plane. Details are presented in Section 2.3.2. Here, we need to judge whether another point has already been generated in that place. The judging method is as follows: project the new point onto the target image; if there is another point in the image pixel of the new point, give up the newly generated point. Directly add the remaining new points to the set $SEED$, and go to step 1.

14

(5) From one direction (e.g., the right), shrink the patch once by two pixels and calculate the NCC_{patch} in the meantime. If the value of NCC_{patch} is increased, continue to shrink the size of the patch in the same direction; otherwise, change the direction (e.g., left, up and down) to shrink the patch. The process above continues until $NCC_{patch} > \gamma_2$. However, if the process continues until $m_2 = 1$, go to step 1. After finishing the size and shape adjustment process, if the length of the patch m_2 is greater than λ ($\lambda = 14$) pixels, go to step 4, or else go to step 6).

(6) Generate new 3D points near the patch center in the range of $\left(-\dfrac{m_2 - 1}{4}, \dfrac{m_2 - 1}{4} \right)$ on the patch plane (if $\dfrac{m_2 - 1}{4} < 1$, the scope becomes $(-1, 1)$). Then, refine and verify them one by one. Add the points that are constructed successfully to the set $SEED$, and go to step 1.

In the end, filter the incorrect points following the PMVS filtering method.

2.3.4. Image-Grouping

Generally, a UAV flight varies between one and three hours. The number of images can reach from 1000 to 2000. It demands a high-performance computer and, in particular, a memory with large capacity. Thus, we should divide the whole region into small regions and process each separately. Finally, combine the matching result of each image group into a whole point cloud. When we divide the whole region, we must not only consider the memory constraint but also process more images at one time. Suppose the number of images in each group is no more than n_{max} (the value of n_{max} is related to the computer memory and the size of the image; in this paper, $n_{max} = 6$). The process for grouping images is as follows:

(1) Calculate the position and the size of the associated area (footprint) of every image, that is the corresponding ground points' XY-plane coordinates for the four corner points of the image.

Firstly, compute the size of the footprint (the length and width of the ground region) as follows:

$$width = imgWidth \times \frac{H}{f_c} \quad \text{and} \quad length = imgHeight \times \frac{H}{f_c} \tag{5}$$

where $imgWidth$ and $imgHeight$ are the width and height of the image. H is the flight height relative to the ground, and f_c is the focal length of the image.

Then, compute the four XY-plane coordinates (X_i, Y_i) of the footprint as follows:

$$
\begin{aligned}
X_i &= \cos\left(kappa\right) \times \frac{width}{2} - \sin\left(kappa\right) \times \frac{length}{2} + x_0 \,, \; (i = 1,2,3,4) \\
Y_i &= \sin\left(kappa\right) \times \frac{width}{2} + \cos\left(kappa\right) \times \frac{length}{2} + y_0 \,, \; (i = 1,2,3,4)
\end{aligned}
\tag{6}
$$

where *kappa* is the rotation angle of the image, and (x_0, y_0) is the coordinate of the projection center.

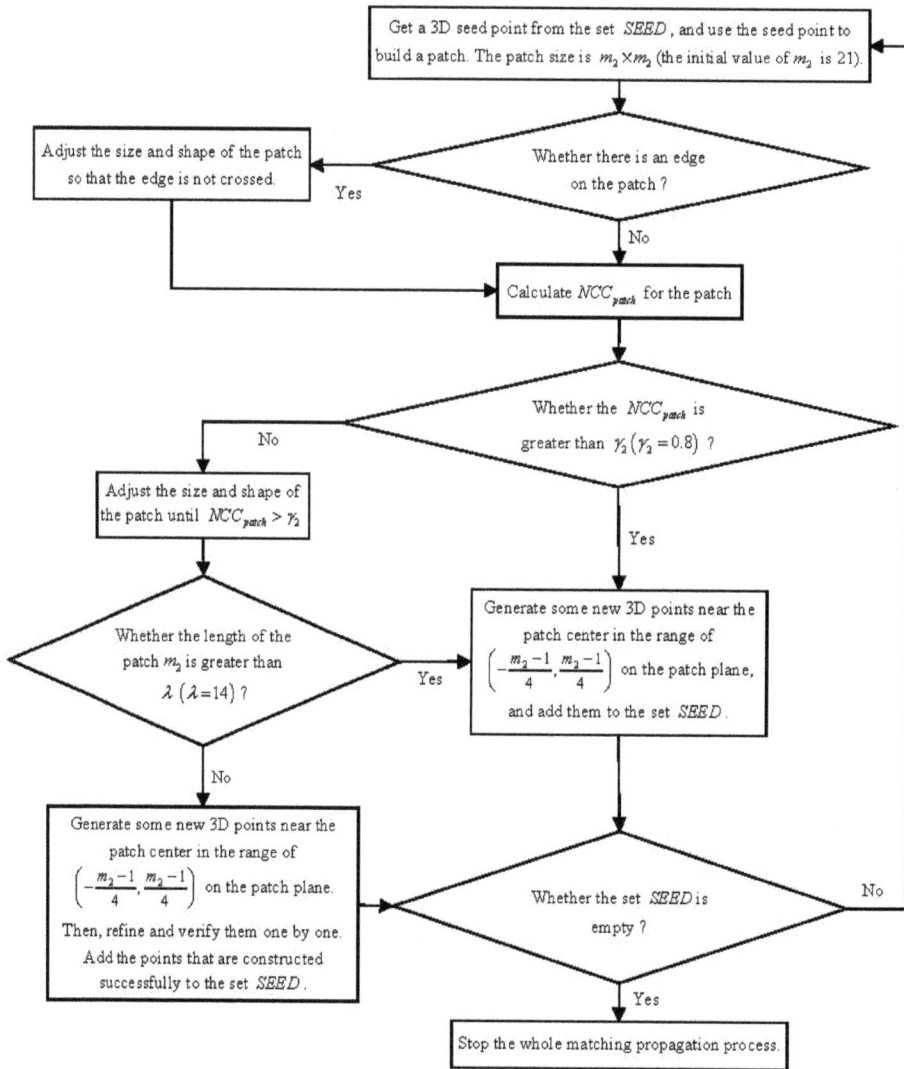

Figure 6. The flow chart of the matching propagation based on self-adaptive patch.

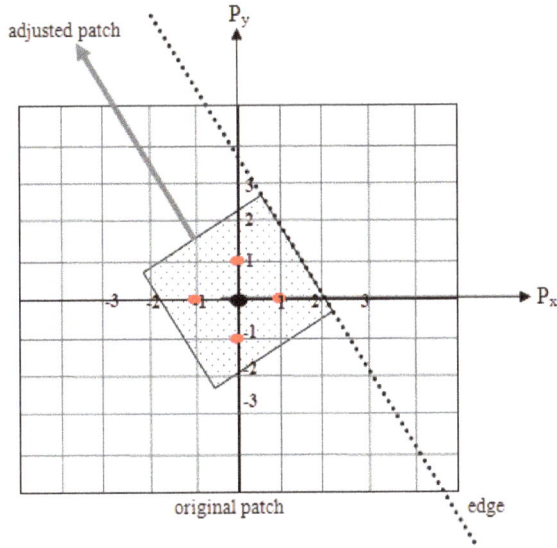

Figure 7. Adjusting patch according to the edge.

(2) Compute the minimum enclosing rectangle of the entire photographed area according to the footprints of all the images.

(3) Divide the minimum enclosing rectangle into $N \times M$ blocks to ensure the number of images that are completely within each block is no more than but close to n_{max} (Figure 8). Compute the footprint of each block, and enlarge the block. In Figure 8, the black dotted box represents the enlarged block, which is denoted by bigBlock.

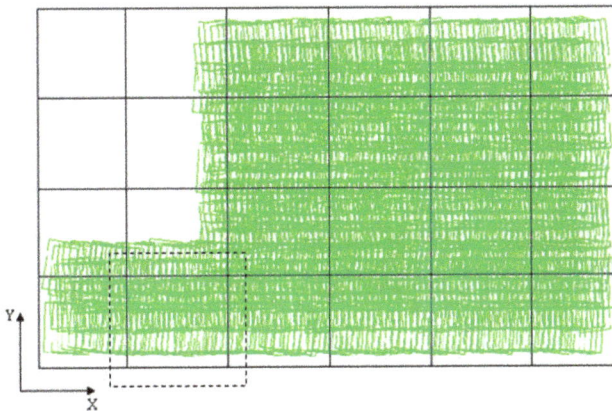

Figure 8. To set up $N \times M$ regular grids in a minimum enclosing rectangle.

(4)   According to the footprint of every image and the footprint of every bigBlock, take all the images belonging to one bigBlock as a group.

2.3.5. Group Matching and Merging the Results

Process each image group in turn using the proposed Self-Adaptive Patch-based Multi-View Stereo-matching algorithm (SAPMVS). SAPMVS is partitioned into three parts: (1) multi-view initial feature-matching, which is introduced in Section 2.3.1; (2) matching propagation based on the self-adaptive patch, which is introduced in Section 2.3.3; (3) filtering the erroneous matching points following the PMVS filtering method. As a result, we obtain the group matching results.

Finally, we need to merge all group matching results into a whole. The merge process is as follows: (1) Remove the redundant points in bigBlock; that is, retain only the points in the block and abandon the points that exceed the extent of the block; (2) Obtain the final point cloud of the entire photographed area by merging the point clouds of every block.

To avoid gaps that appear on the edges of the blocks in the final result, in step (1) we should preserve the (3D) edge points of each block and, in step (2) remove the (3D) repetitive points at the block edges according to the corresponding image points of those 3D points. If there is a corresponding image point that belongs to two 3D points, we preserve only one of the two 3D points (or refine and verify the 3D repetitive points following the PMVS method).

3. Experiments and Results

3.1. Evaluation Index and Method

Currently, there is no unified approach to evaluating multiple image dense matching algorithms. Usually, we can compare them in the following aspects [58]: (1) accuracy, which indicates the degree of correct matching quantitatively; (2) reliability, which represents the degree of precluding overall classification error; (3) versatility, the ability to apply the algorithm to different image scenes; (4) complexity, the cost of equipment and calculation.

In the field of computer vision, to evaluate the accuracy of a multi-view dense matching algorithm, 3D reconstruction of the scene is carried out by means of the dense matching algorithm, and then the 3D reconstruction model is compared with the high-accuracy real surface model (real data are obtained by laser), and the performance of the dense matching algorithm is evaluated in terms of accuracy and completeness [17]. In the field of digital photogrammetry, we can use the corresponding image points obtained by the multi-view stereo image matching to obtain their corresponding object points by means of forward intersection, and we can then generate the Digital Surface Model (DSM) and Digital Elevation Model

(DEM). Therefore, we can evaluate the accuracy of DSM and DEM and thereby evaluate the accuracy of the matching algorithm indirectly; the accuracy of the DEM and DSM can be evaluated in relation to higher-precision reference data, such as laser point cloud data and manual measurement control point data [46,51].

As the output result of our algorithm is the 3D point cloud and the computational cost is large, we will evaluate our algorithm from three aspects: visual inspection, quantitative description and complexity. Visual inspection ensures that the shape of the 3D point cloud is consistent with the actual terrain. Quantitative description is necessary to compare the 3D point cloud with the high-precision reference data, such as laser point cloud data and artificial measurement control point data. Complexity mainly refers to the requirements of devices and the cost in time of computing. On the other hand, the main input data of our algorithm are the images that have lens distortion removed [59–61] and their orientation elements, while the accuracy of the orientation elements can directly affect the accuracy of the imaging geometric model, which may lead to errors in the forward intersection.

3.2. Experiments and Analysis

To evaluate the performance of our multi-view stereo-matching algorithm for multiple UAV imagery in a more in-depth and comprehensive way, we will use three typical sets of UAV data with different texture features viewed from three perspectives: visual inspection, quantitative description and complexity.

3.2.1. The Experimental Platform

The proposed algorithm is implemented in Visual C++ and a PC with Intel® Core™ i7 CPU 920 2.67 GHz processors, 3.25 GB RAM, and Microsoft Windows Xp Sp3 64.

3.2.2. The First Dataset: Northwest University Campus, China

This group of experiments uses UAV imagery data taken at the Northwest University campus in Shaanxi, China, by a Cannon EOS 400D. The photography flying height is 700 m, and the ground resolution of the imagery is approximately 0.166 m. The shooting lasted 40 min, and there are a total of 67 images. The specific parameters of the photography can be seen in Table 1.

This set of data is provided by Xi'an Dadi Surveying and Mapping Corporation. We used a commercial UAV photogrammetric processing software called GodWork, which was developed by Wuhan University, to perform automatic aerotriangulation to obtain the precise orientation elements of the images (we can also use the structure-from-motion software "VisualSFM" by Changchang Wu [62,63] to estimate the precise camera pose). The accuracy of aerotriangulation was as follows: the value of the unit-weight mean square error (Sigma0) was 0.49 pixels, and the average residual of the image point was 0.23 pixels. Because the data had no manual control point information, the bundle adjustment method and the orientation elements were under freenet. Figure 9 is the tracking map of the 67 images under freenet. We used GodWork software to remove the lens distortion of the 67 images. Table 2 shows part of the corrected images' external orientation elements.

Table 1. The parameters of the UAV photography in Northwest University.

Camera Name	CCD Size (mm × mm)	Image Resolution (pixels × pixels)	Pixel Size (μm)	Focal Length (mm)	Flying Height (m)	Ground Resolution (m)	Number of Images
Canon EOS 400D	22.16 × 14.77	3888 × 2592	5.7	24	700	0.166	67

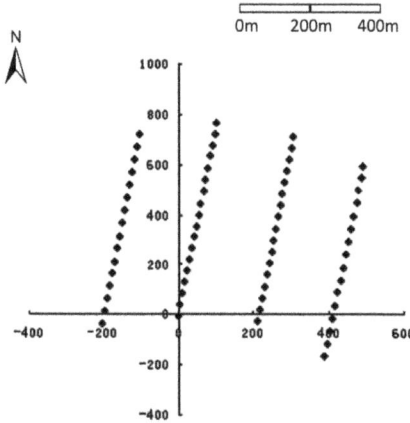

Figure 9. The tracking map of the 67 images under freenet.

Table 2. The external orientation elements of a portion of the corrected images.

Image Name	X (m)	Y (m)	Z (m)	φ (Degree)	ω (Degree)	κ (Degree)
IMG_0555	−201.736	−31.7532	−1.35375	−1.0234	−0.44255	166.7002
IMG_0554	−194.482	17.90618	−1.22801	−0.71193	−0.1569	166.2114
IMG_0553	−187.641	67.63342	−0.87626	0.320481	−0.03621	166.3385
IMG_0552	−180.965	116.052	−0.61774	0.518773	−0.87317	166.7509
IMG_0551	−174.434	166.2001	−0.59227	0.454139	−0.86679	167.2784
IMG_0550	−168.264	214.3878	−0.79246	−0.57551	−0.67894	167.5912
IMG_0549	−162.096	265.1268	−0.70428	−1.08565	−0.7456	167.166
IMG_0548	−156.002	314.2426	−0.53393	−1.40444	−0.84598	166.5402
IMG_0547	−148.987	367.3003	−0.30983	−0.77136	−0.86104	166.5097
IMG_0546	−142.152	417.2658	−0.04708	−0.38776	−0.89084	166.5219
IMG_0545	−135.102	466.7365	0.507349	−0.08542	−0.09929	166.4838
IMG_0544	−128.322	520.032	1.246216	−0.2938	−0.41415	166.7157
IMG_0543	−121.833	569.8722	1.862483	−0.20825	−0.21687	167.1119

First, we used the proposed UAV multiple image–grouping strategy to divide this set of 67 images into 12 groups. The serial number of each image group is shown in Table 3.

Table 3. Image-grouping result of the Northwest University data.

Group	Image Number	Corresponding Image Name	Number of Images
0	0 1 2 3 4 5	IMG_1093~IMG_1089	6
1	6 7 8 9 10 11	IMG_ 1087~IMG_1082	6
2	12 13 14 15	IMG_1081~IMG_1078	4
3	16 17 18 19 20 21	IMG_0102~IMG_0107	6
4	22 23 24 25 26 27	IMG_0108~IMG_0113	6
5	28 29 30 31 32	IMG_0114~IMG_0118	5
6	33 34 35 36 37 38	IMG_0641~IMG_0646	6
7	39 40 41 42 43 44	IMG_0647~IMG_0652	6
8	45 46 47 48 49 50	IMG_0653~IMG_0658	6
9	51 52 53 54 55 56	IMG_0555~IMG_0550	6
10	57 58 59 60 61 62	IMG_0549~IMG_0544	6
11	63 64 65 66	IMG_0543~IMG_0540	4

After image-grouping, we used the proposed Self-Adaptive Patch-based Multi-View Stereo-matching algorithm (SAPMVS) to address each image group, and obtained the 3D dense point cloud data of each image group. Then, we merged the 3D dense point clouds of each group; the merged 3D dense point cloud is shown in Figure 10 (the small black areas in the figures are water). We found that the merged 3D dense point cloud has 8,526,192 points, and the point density is approximately three points per square meter; thus, the ground resolution is approximately 0.3 m.

Figure 10. The merged 3D dense point cloud of Northwest University.

21

Because of the lack of control point data or high-precision reference data in the data set, such as the laser point cloud, we use the visual inspection method to evaluate the results of the proposed algorithm, *i.e.*, whether the shape of the 3D point cloud is consistent with the actual terrain. We compared the 3D dense point cloud and the corresponding corrected images that had the lens distortion removed, as shown in Figure 11. By comparing the point clouds and images in Figure 11, it can be seen that the 3D dense point clouds of the proposed algorithm accurately described the terrain features of the Northwest University campus as well as the shape and distribution of physical objects (such as roads and buildings).

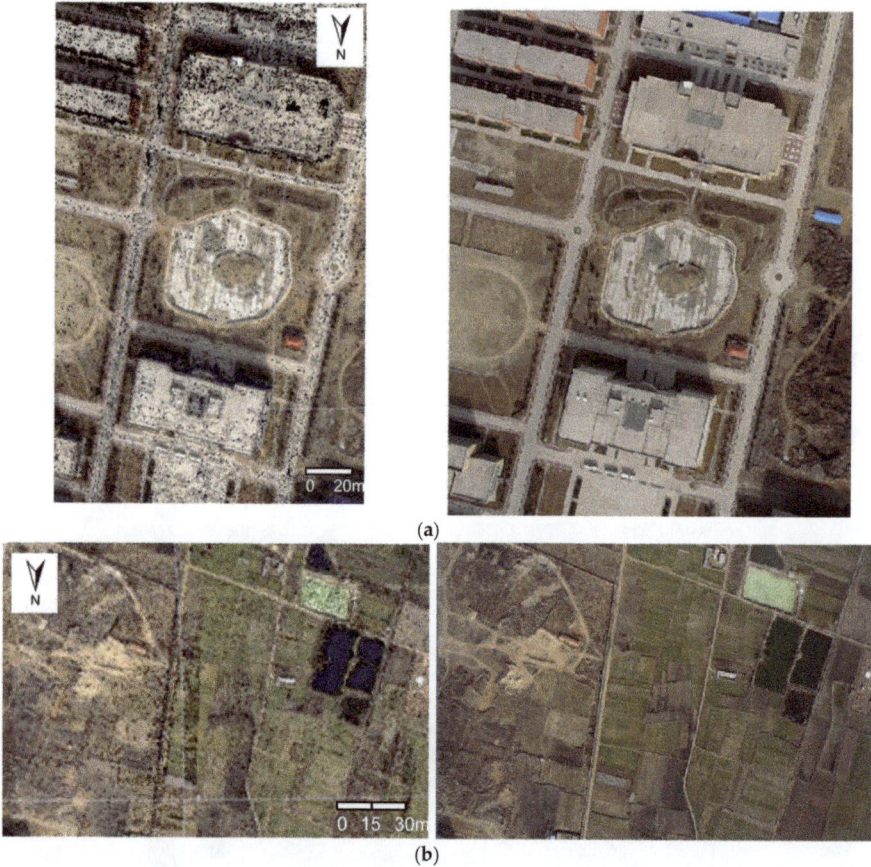

(a)

(b)

Figure 11. Comparison of the same areas in the 3D dense point clouds and the corresponding images. (**a**) The building area; (**b**) the flat area.

For further analysis of the accuracy and efficiency of the proposed algorithm, we used the proposed IG-SAPMVS algorithm and PMVS algorithm [18,52], respectively, to process this set of data, and recorded the processing time and the 3D point cloud

22

results. Table 4 shows the statistics for these two algorithms with respect to the processing time and the point number of 3D dense point clouds. Figure 12 shows the final 3D dense point cloud results.

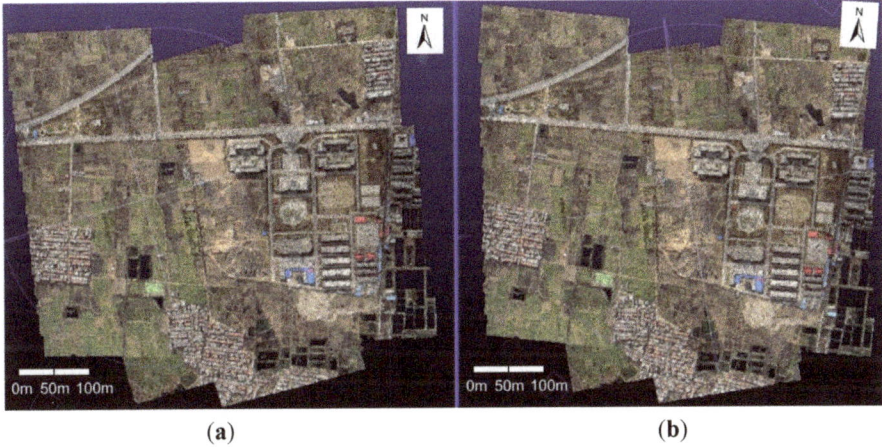

(a)   (b)

Figure 12. Final 3D point cloud results of the Northwest University campus using IG-SAPMVS (a) and PMVS (b).

From Table 4, it can be seen that the processing time of the proposed IG-SAPMVS algorithm is approximately 0.5 times that of the PMVS algorithm; thus, the calculation efficiency of the proposed IG-SAPMVS algorithm is significantly higher than that of the PMVS algorithm. Because the terrain relief of the test area is not large and there are many flat square grounds in the test area, the proposed Self-Adaptive Patch-based Multi-View Stereo-matching algorithm (SAPMVS) can spread more quickly than the PMVS algorithm in the matching propagation process. On the other hand, based on Table 4, it can be seen that the point number of the 3D dense point cloud by the proposed IG-SAPMVS algorithm is 1.15 times that of the PMVS algorithm; Figure 12 illustrates that the 3D dense point cloud result of the proposed IG-SAPMVS algorithm is almost the same as that of the PMVS algorithm based on visual inspection. In general, the proposed IG-SAPMVS algorithm outperforms the PMVS algorithm in computing efficiency and the quantity of 3D dense point clouds.

Table 4. Statistics for the proposed IG-SAPMVS algorithm and PMVS algorithm.

Algorithm	RunTime (h:min:s)	Point Cloud Amount	Number of Images
IG-SAPMVS	2:41:38	8526192	67
PMVS	4:8:30	7428720	67

3.2.3. The Second Dataset: Remote Mountains

This group of experiments uses UAV imagery data of remote mountains characterized by large relief, heavy vegetation and a small amount of physical objects, such as roads and buildings, in China; they were also taken by a Cannon EOS 400D with a focus of 24 mm. The photography flying height is approximately 1900 m and the ground resolution of the imagery is approximately 0.451 m. There are a total of 125 images. Figure 13 is the GPS tracking map under the geodetic control network. We also used GodWork software to remove the lens distortion of the 125 images.

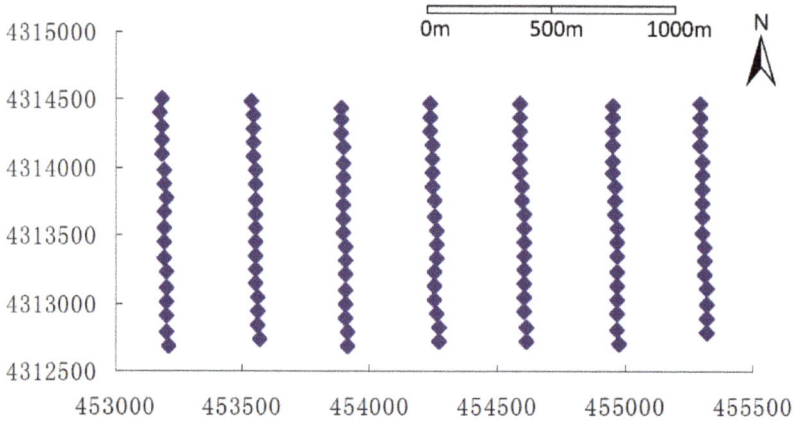

Figure 13. The GPS tracking map of the UAV images taken in remote mountains under geodetic control network.

This set of data is also provided by Xi'an Dadi Surveying and Mapping Corporation. Because of the low accuracy of the airborne GPS/IMU data, it cannot meet the requirements of the proposed multi-view stereo-matching algorithm. We used the commercial UAV photogrammetric processing software GodWork, which was developed by Wuhan University, to perform automatic aerotriangulation and obtain the images' precise exterior orientation elements. The accuracy of aerotriangulation was as follows: the value of the unit-weight mean square error (Sigma0) was 0.77 pixels, and the average residual of the image point was 0.36 pixels. Table 5 shows part of the corrected images' external orientation elements.

First, we also used the UAV multiple image-grouping strategy to divide this set of 125 images into 21 groups. After image-grouping, we used the proposed Self-Adaptive Patch-based Multi-View Stereo-matching algorithm (SAPMVS) to address each image group, and obtained the 3D dense point cloud data of each image group. Then, we merged the 3D dense point clouds of each group, and the merged 3D dense point cloud is shown in Figure 14 (the small black areas in the figures are water). We found that the merged 3D dense point cloud has 12,509,202 points, and the point density is approximately one point per square meter; thus, the ground resolution is approximately 1 m.

Table 5. The external orientation elements of part of the corrected images (the remote mountains).

Image Name	X (m)	Y (m)	Z (m)	φ (Degree)	ω (Degree)	κ (Degree)
IMG_0250	453208.9	4312689	1926.36	4.251745	−2.30691	10.23632
IMG_0251	453205.6	4312796	1926.462	2.852207	−6.04494	10.94959
IMG_0252	453200.6	4312900	1921.71	0.681977	−7.93856	11.60872
IMG_0253	453198.6	4313007	1920.499	−0.01863	−3.6135	8.714555
IMG_0254	453198.7	4313117	1915.972	3.277204	−6.38877	8.599633
IMG_0255	453199.3	4313230	1910.448	6.64432	−6.29777	8.965715
IMG_0256	453196.7	4313341	1904.105	3.33602	−6.5811	10.72329
IMG_0257	453194.1	4313450	1903.386	1.594531	−4.486	10.71933
IMG_0258	453192.8	4313559	1902.593	−4.04339	−5.57166	8.112464
IMG_0259	453194.9	4313667	1899.551	3.77633	−2.96418	7.066141
IMG_0260	453198.2	4313776	1899.075	7.240338	−5.29839	8.254281
IMG_0261	453196.6	4313882	1898.725	3.41796	−1.02482	10.90049
IMG_0262	453193.1	4313986	1895.928	3.000324	−5.9483	12.35882
IMG_0263	453185.4	4314095	1896.413	1.532879	−4.00544	11.27756
IMG_0264	453184	4314196	1895.204	4.379199	−2.4512	11.95034

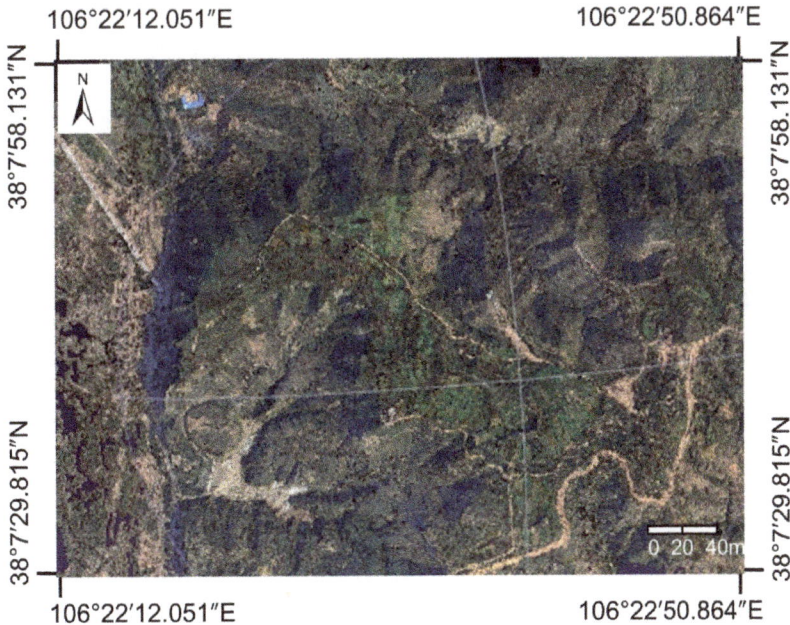

(a)

Figure 14. *Cont.*

25

(b)

Figure 14. The merged 3D dense point cloud of the remote mountains. (**a**) The plan view of the merged 3D dense point cloud; (**b**) The side views of the merged 3D point-cloud.

For further analysis of the accuracy and efficiency of the proposed algorithm, we used the proposed IG-SAPMVS algorithm and PMVS algorithm [18,52], respectively, to process this set of remote mountain data and recorded the processing time and the 3D point cloud results. Table 6 shows the statistics of these two algorithms with respect to the processing time and the point number of the 3D dense point clouds. Figure 15 shows the final results of the 3D dense point cloud.

Table 6. The statistics of the proposed IG-SAPMVS algorithm and PMVS algorithm.

Algorithm	RunTime (h:min:s)	Point Cloud Amount	Number of Images
IG-SAPMVS	4:36:5	12509202	125
PMVS	15:48:11	8953228	125

From Table 6, it can be seen that the processing time of the proposed IG-SAPMVS algorithm is about one-third of that of the PMVS algorithm; thus, the efficiency of the proposed IG-SAPMVS algorithm is significantly higher than that of the PMVS algorithm. Obviously, even in the remote mountain field with complex terrain, the proposed Self-Adaptive Patch-based Multi-View Stereo-matching algorithm (SAPMVS) can spread more quickly than the PMVS algorithm in the matching

26

propagation process. On the other hand, based on Table 6, it can be seen that the point number of the 3D dense point cloud by the proposed IG-SAPMVS algorithm is 1.40 times that of the PMVS algorithm, and Figure 15 illustrates that the 3D dense point cloud result of the proposed IG-SAPMVS algorithm is nearly the same as the PMVS algorithm based on visual inspection. In general, the proposed IG-SAPMVS algorithm significantly outperforms the PMVS algorithm in computing efficiency and the quantity of 3D dense point clouds.

(a)   (b)

Figure 15. Final 3D point cloud results of the remote mountains using IG-SAPMVS (**a**) and PMVS (**b**).

3.2.4. The Third Dataset: Vaihingen, Germany

The third dataset was captured over Vaihingen, Germany, by the German Society for Photogrammetry, Remote Sensing and Geoinformation (DGPF) [64]. It consists of three test areas of various object classes (three yellow areas in Figure 16).

- **Area 1 "Inner City":** This test area is situated in the center of the city of Vaihingen. It is characterized by dense development consisting of historic buildings with rather complex shapes, but there are also some trees (Figure 17a).
- **Area 2 "High Riser":** This area is characterized by a few high-rise residential buildings that are surrounded by trees (Figure 17b).
- **Area 3 "Residential Area":** This is a purely residential area with small detached houses (Figure 17c).

Figure 16. The Vaihingen test areas.

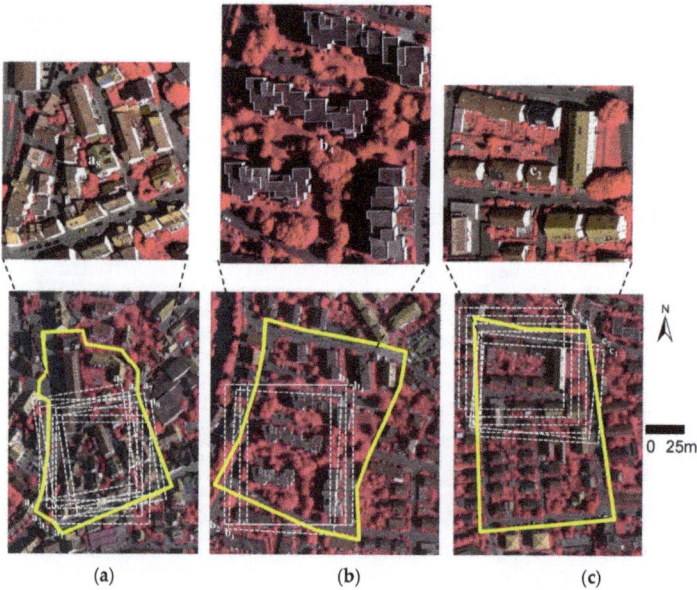

Figure 17. The three test sites in Vaihingen. (**a**) a_1-a_8: the eight cut images of the "Inner City" from the original images: 10030061.jpg, 10030062.jpg, 10040083.jpg, 10040084.jpg, 10050105.jpg, 10050106.jpg, 10250131.jpg, 10250132.jpg, respectively; (**b**) b_1-b_4: the four cut images of the "High Riser" from the original images: 10040082.jpg, 10040083.jpg, 10050104.jpg, 10050105.jpg, respectively; (**c**) c_1-c_6: the six cut images of the "Residential Area" from the original images: 10250134.jpg, 10250133.jpg, 10040083.jpg, 10040084.jpg, 10050105.jpg, 10050106.jpg, respectively.

The data include high-resolution digital aerial images and orientation parameters and airborne laser scanner data (available in [65]).

Digital Aerial Images and Orientation Parameters: The images are a part of the Intergraph/ZI DMC block with 8 cm ground resolution [64]. Each area is visible in multiple images from several strips. The orientation parameters are distributed together with the images. The accuracy of aerotriangulation is as follows: the value of unit-weight mean square error (Sigma0) is about 0.25 pixels. Table 7 shows the external orientation elements of the images in the test region.

Airborne Laser Scanner Data: The test area was covered by 10 strips captured with a Leica ALS50 system. Inside an individual strip, the average point density is 4 points /m^2 [66]. The airborne laser scanner data of the test region are shown in Figure 18.

Table 7. The external orientation elements of the experimental images (Vaihingen data).

Image Name	X (m)	Y (m)	Z (m)	ω (Degree)	φ (Degree)	κ (Degree)
10030060.tif	496803.043	5420298.566	1163.983	2.50674	0.73802	199.32970
10030061.tif	497049.238	5420301.525	1163.806	2.05968	0.67409	199.23470
10030062.tif	497294.288	5420301.839	1163.759	1.97825	0.51201	198.84290
10030063.tif	497539.821	5420299.469	1164.423	1.40457	0.38326	198.88310
10040081.tif	496558.488	5419884.008	1181.985	−0.87093	0.36520	−199.20110
10040082.tif	496804.479	5419882.183	1183.373	−0.26935	−0.63812	−198.97290
10040083.tif	497048.699	5419882.847	1184.616	0.34834	−0.40178	−199.44720
10040084.tif	497296.587	5419884.550	1185.010	0.81501	−0.53024	−199.35600
10040085.tif	497540.779	5419886.806	1184.876	1.38534	−0.46333	−199.85010
10050103.tif	496573.389	5419477.807	1161.431	−0.48280	−0.03105	−0.23869
10050104.tif	496817.972	5419476.832	1161.406	−0.65210	−0.06311	−0.17326
10050105.tif	497064.985	5419476.630	1159.940	−0.74655	0.11683	−0.09710
10050106.tif	497312.996	5419477.065	1158.888	−0.53451	−0.19025	−0.13489
10050107.tif	497555.389	5419477.724	1158.655	−0.55312	−0.12844	−0.13636
10250130.tif	497622.784	5420189.950	1180.494	0.09448	3.41227	−101.14170
10250131.tif	497630.734	5419944.364	1181.015	0.61065	2.54420	−97.84478
10250132.tif	497633.024	5419698.973	1179.964	1.27053	1.62793	−97.23292
10250133.tif	497628.317	5419452.807	1179.237	0.90688	0.83308	−98.72504
10250134.tif	497620.954	5419207.621	1178.201	0.17675	1.27920	−101.86160
10250135.tif	497617.307	5418960.618	1176.629	0.22019	1.47729	−101.55860

Because this dataset's image pixel resolution (7680 × 13824 square pixels) is large, it often exhausted the computer memory in the experiment when processing the original images. In addition, the imaging of any of the three experimental areas is only a small part of each image. Therefore, we can cut out the three experimental areas in each of the original images separately (Figure 17). We used the proposed

Self-Adaptive Patch-based Multi-View Stereo-matching algorithm (SAPMVS) to address the three sets of cut images separately, and obtained the 3D dense point cloud data of each dataset. The 3D dense point cloud data of each dataset are shown in Figure 19. The statistics of the results by the proposed SAPMVS algorithm are shown in Table 8.

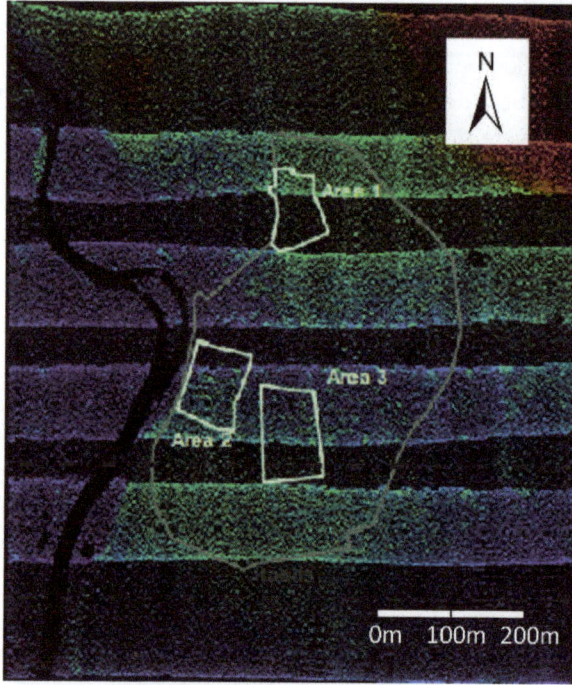

Figure 18. The airborne laser scanner data of the experimental region (Vaihingen).

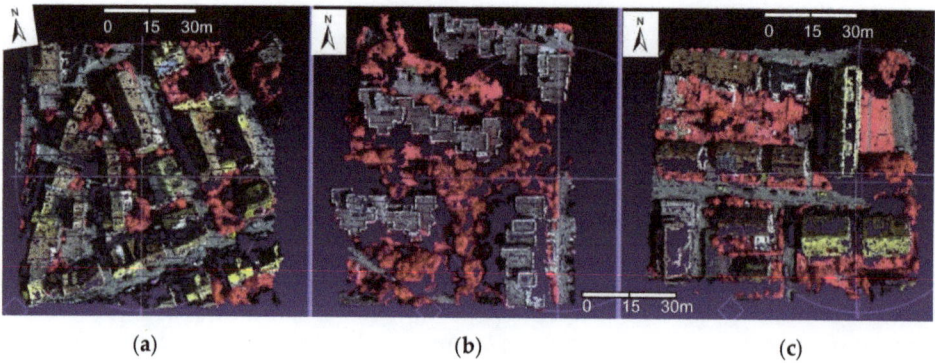

(a)   (b)   (c)

Figure 19. Final 3D point cloud results for the three sets of cut images. (**a**) Area 1: "Inner City"; (**b**) Area 2: "High Riser"; (**c**) Area 3: "Residential Area".

Table 8. The statistics for the results of the Vaihingen data by the proposed SAPMVS algorithm.

Experiment Area	Number of Images (Image Pixel Resolution)	Point Amount	Average Distance between Points	RunTime (min:s)
Area 1: "Inner City"	8 (1200*1200)	253125	16 cm	11:33
Area 2: "High Riser"	4 (1200*1600)	220073	16 cm	6:30
Area 3: "Residential Area"	6 (1400*1300)	259637	16 cm	9:46

From Table 8 and Figure 19, it can be seen that the computational efficiency of the proposed SAPMVS algorithm is high and the ground resolution of the obtained 3D dense point cloud is approximately 0.16 m.

To quantitatively describe the accuracy of the proposed algorithm, we can compare the obtained 3D point cloud results by the PMVS and the proposed algorithm with the high-precision airborne laser scanner data, respectively. The specific evaluation method is performed as follows: For each 3D point of the obtained point-cloud result that was assumed to be P_i ($i \in [1,n]$), we determine all the laser points near the point P_i (the XY-plane distance between point P_i and the laser point should be smaller than the threshold value d, which is related to the average point density of the airborne laser scanner data, in our experiment $d = 0.25m$) in the laser point cloud data; we then calculate the average elevation Za_i of these nearby laser points as the reference elevation of the point P_i [36]. Finally, we compare the elevation Z_i of the point P_i with its reference elevation Za_i and calculate the root mean square error ($RMSE$) and the maximum error (Max) of the obtained 3D point cloud [67]. The calculation formulas are as follows:

$$
RMSE = \sqrt{\frac{\sum\limits_{i=1}^{n} (Z_i - Za_i)^2}{n}} \tag{7}
$$

$$
Max = \max |Z_i - Za_i|, \ i \in [1,n] \tag{8}
$$

where n represents the point number of the obtained 3D dense point cloud result. It should be noted that using this method to evaluate the accuracy of the obtained 3D dense point cloud has one drawback: At the edge or the fracture line, the value of $|Z_i - Za_i|$ may become a large value, which is inconsistent with the actual situation, resulting in a pseudo-error. That is to say, if a 3D point P_i is on the edge (or at one side of the edge) and the nearby laser points are located outside of the edge (or on the other side of the edge), the value of $|Z_i - Za_i|$ may become very large, but the large error is not real (in fact, in that case, the maximum error Max is not meaningful). The quantitative evaluation results if we do not remove the pseudo-errors (from statistics, the frequency of pseudo-errors is relatively small) are shown in Table 9.

Table 9. The quantitative evaluation accuracy of the obtained 3D dense point cloud without removing pseudo-errors. (**a**) PMVS; (**b**) The proposed algorithm.

(a)				
Experiment Area	Checkpoint Amount	RMSE(m)	Max(m)	Percentage of Errors within 1 m
Area 1: "Inner City"	245752	2.180527	20.410379	66.7%
Area 2: "High Riser"	213679	4.032463	30.742815	46.1%
Area 3: "Residential Area"	252568	2.349705	18.903685	74.1%

(b)				
Experiment Area	Checkpoint Amount	RMSE(m)	Max(m)	Percentage of Errors within 1 m
Area 1: "Inner City"	253125	2.164632	20.307628	66.8%
Area 2: "High Riser"	220073	3.950138	29.812880	46.4%
Area 3: "Residential Area"	259637	2.328481	18.713413	74.2%

Because there are a certain number of pseudo-errors that may be very large, when evaluating the accuracy of the obtained 3D dense point cloud, we mainly focus on the percentage of errors within 1 m (for errors within 1 m, the vast majority should be a true error; this has valuable reference meaning) and then the *RMSE* value. However, the value of *Max* is a maximum pseudo-error and is not meaningful, and if there is no such pseudo-error, the actual *RMSE* value will be much smaller.

Table 9a shows that the percentages of errors within 1 m by the PMVS algorithm for the three experiment areas ("Area 1", "Area 2", and "Area 3") are 66.7%, 46.1% and 74.1%, respectively, and the corresponding *RMSE* values are 2.180527 m, 4.032463 m and 2.349705 m, respectively. Table 9b shows that the percentages of errors within 1 m by the proposed algorithm for the three experiment areas ("Area 1", "Area 2", and "Area 3") are 66.8%, 46.4% and 74.2%, respectively, and the corresponding *RMSE* values are 2.164632 m, 3.950138 m and 2.328481 m, respectively. It can be seen that the accuracy of the proposed algorithm is slightly higher than that of the PMVS algorithm for the three experiment areas. Additionally, it can be seen that the matching accuracy from high to low is "Area 3", "Area 1", and "Area 2". Such an experimental result is reasonable because there are mainly low residential buildings in "Area 3", the buildings in "Area 1" are more complex, and the buildings in "Area 2" are very tall; thus, the matching difficulty is gradually increased.

To evaluate the actual accuracy of the PMVS and the proposed algorithm more precisely, we need to delete some of the large pseudo-errors when calculating the *RMSE* value. We can take a simple approach inspired by [68–71]: For "Area 3" and

32

"Area 1", if the value $|Z_i - Za_i|$ of a 3D point is greater than the Elevation Error Threshold 1 ($EET1$), we consider the value of $|Z_i - Za_i|$ as a pseudo-error and delete the 3D point (do not use it as a checkpoint), while for "Area 2", if the value $|Z_i - Za_i|$ of a 3D point is greater than the Elevation Error Threshold 2 ($EET2$), we consider the value of $|Z_i - Za_i|$ as a pseudo-error and delete the 3D point (do not use it as a checkpoint). The actual and more precise quantitative evaluation results of the two algorithms are shown in Tables 10 and 11.

Table 10. The actual quantitative evaluation accuracy of the obtained 3D dense-point cloud by removing the large pseudo-errors ($EET1 = 6m$, $EET2 = 9m$). (a) PMVS; (b) The proposed method.

(a)				
Experiment Area	Point Amount	Checkpoint Amount	RMSE(m)	Percentage of Errors within 1 m
Area 1: "Inner City"	245752	243179	1.312648	67.5%
Area 2: "High Riser"	213679	212151	3.402339	46.5%
Area 3: "Residential Area"	252568	250463	1.587426	74.7%

(b)				
Experiment Area	Point Amount	Checkpoint Amount	RMSE(m)	Percentage of Errors within 1 m
Area 1: "Inner City"	253125	250488	1.301095	67.5%
Area 2: "High Riser"	220073	218508	3.352631	46.7%
Area 3: "Residential Area"	259637	257470	1.571167	74.8%

Table 10a shows that after we deleted the large pseudo-errors ($EET1 = 6m$, $EET2 = 9m$), the $RMSE$ values of the PMVS algorithm for the three experiment areas ("Area 1", "Area 2", and "Area 3") are 1.312648 m, 3.402339 m and 1.587426 m, respectively. Table 10b shows that after we deleted the large pseudo-errors ($EET1 = 6m$, $EET2 = 9m$), the $RMSE$ values of the proposed algorithm for the three experiment areas ("Area 1", "Area 2", and "Area 3") are 1.301095 m, 3.352631 m and 1.571167 m, respectively. It can be seen that the accuracy of the proposed algorithm is almost equal to that of the PMVS algorithm for the three experiment areas. In fact, the pseudo-errors that still remain act as a kind of constraint on the precision of the two algorithms.

Table 11. The actual quantitative evaluation accuracy of the obtained 3D dense point cloud by removing the pseudo-errors ($EET1 = 2m$, $EET2 = 3m$). (**a**) PMVS; (**b**) The proposed method.

(a)				
Experiment Area	**Point Amount**	**Checkpoint Amount**	**RMSE (m)**	**Percentage of Errors within 1 m**
Area 1: "Inner City"	245752	240826	0.880695	68.2%
Area 2: "High Riser"	213679	173487	1.351428	57.1%
Area 3: "Residential Area"	252568	248728	0.898527	75.3%

(b)				
Experiment Area	**Point Amount**	**Checkpoint Amount**	**RMSE (m)**	**Percentage of Errors within 1 m**
Area 1: "Inner City"	253125	248113	0.870425	68.1%
Area 2: "High Riser"	220073	178565	1.316283	57.2%
Area 3: "Residential Area"	259637	255674	0.886161	75.3%

Table 11a shows that after we deleted almost all the pseudo-errors ($EET1 = 2m$, $EET2 = 3m$), the $RMSE$ values by the PMVS algorithm for the three experiment areas ("Area 1", "Area 2", and "Area 3") are 0.880695 m, 1.351428 m and 0.898527 m, respectively. Table 11b shows that after we deleted almost all the pseudo-errors ($EET1 = 2m$, $EET2 = 3m$), the $RMSE$ values by the proposed algorithm for the three experiment areas ("Area 1", "Area 2", and "Area 3") are 0.870425 m, 1.316283 m and 0.886161 m, respectively. It can be seen that the accuracy of the proposed algorithm is almost equal to that of the PMVS algorithm for the three experiment areas. In fact, because the building coverage of the three experimental regions is high, the overall accuracy of the dense matching algorithm is bound to decrease. In addition, the precision of the image orientation elements may act as a small type of constraint on the precision of the proposed algorithm.

3.2.5. Discussion

Based on experiments on three typical sets of UAV data with different texture features, the experimental conclusions are as follows:

(1) The proposed multi-view stereo-matching algorithm based on matching control strategies suitable for multiple UAV imagery and the self-adaptive patch can address the UAV image data with different texture features effectively. The obtained dense point cloud has a realistic effect, and the precision is equal to that of the PMVS algorithm.

(2)   Due to the image-grouping strategy, the proposed algorithm can handle a large amount of data on a typical computer and is, to a large degree, not restricted by the memory of the computer.

(3)   The proposed matching propagation method based on the self-adaptive patch is superior to that of the state-of-the-art PMVS algorithm in terms of the processing efficiency, and has, to some extent, improved the accuracy of the multi-view stereo-matching algorithm by the self-adaptive spread patch sizes according to terrain relief, e.g., a small spread patch size for large relief terrain, and avoiding crossing the terrain edge, *i.e.*, the fracture line.

However, there are several limitations with respect to the proposed approach. The following aspects should be addressed in future work:

(1)   The accuracy and completeness of the proposed algorithm.

In cases of special UAV data characterized by serious terrain discontinuity or other difficult conditions (low texture and so on), the accuracy and completeness of the proposed algorithm is not very high [18]. It needs to be further optimized (for example, after multi-view stereo-matching, we can post-process the 3D dense point cloud to fill in possible holes and obtain a complete mesh model [18,72]).

(2)   The accuracy evaluation method of the proposed algorithm.

In the experimental section, the quantitative description method for the precision of the proposed algorithm may have certain shortcomings that act as constraints on the evaluation precision of the proposed algorithm. We have implemented a simple and effective solution; however, it can be replaced or further improved upon. There are primarily two approaches that can help avoid the problem of pseudo-error effectively: (1) First, the reference model was aligned to its image set using an iterative optimization approach (an Iterative Closest Point alignment, ICP) that minimizes the photo-consistency function between the reference mesh and the images. The alignment parameters consist of a translation, rotation, and uniform scale. Second, we compare the elevation between the dense point cloud obtained from the images and the reference mesh obtained from the high precision laser points [17,73]; (2) We can compare the elevation between the high precision Ground Control Points (GCPs) and the DSM (volume or mesh) or DEM obtained from the dense point cloud for the images [74–77].

4. Conclusions

Multi-view dense image-matching is a hot topic in the field of digital photogrammetry and computer vision. In this paper, according to the characteristics of UAV imagery, we proposed a multi-view stereo image-matching method for UAV images based on image-grouping and the self-adaptive patch. This algorithm mainly processed multiple UAV images with the known orientation elements and could output a colorful 3D dense point cloud. The main processing procedures were as follows: First, the UAV images were divided into groups automatically. Then, each image group was processed in turn, and the processing flow was partitioned into three parts: (1) multi-view initial feature-matching; (2) matching propagation based on the self-adaptive patch; and (3) filtering the erroneous matching points. Finally, considering the overlap problem between groups, the matching results were merged into a whole.

The innovations of this paper were as follows: (1) An image-grouping strategy for multi-view UAV image-matching was designed; as a result, the proposed algorithm can address a large number of UAV image data without the restriction of computer memory; (2) According to the characteristics of UAV imagery, some matching control strategies were proposed for multiple UAV image-matching, which could improve the efficiency of the initial feature-matching process; (3) A new matching propagation method was designed based on the self-adaptive patch. In the matching propagation process, the sizes and shapes of the patches could adapt to the terrain relief automatically, and the patches were prevented from crossing the terrain edge, *i.e.*, the fracture line. Compared with the matching propagation method of the PMVS algorithm, the proposed self-adaptive patch-based matching propagation method not only reduced computing time markedly, but also enhanced integrity to some extent.

In sum, many practices indicate that the proposed method can address a large amount of UAV image data with almost no computer memory restrictions and significantly surpasses the PMVS algorithm in processing efficiency, while the matching precision is equal to that of the PMVS algorithm.

Acknowledgments: This work was supported by the National Basic Research Program of China (Grant Nos. 2012CB719905, 2012CB721317), the National Natural Science Foundation of China (Grant No. 41127901), the Scientific Research Program of China in Public Interest of Surveying, Mapping and Geographic Information (Grant No. 201412015) and the Open Foundation of State Key Laboratory of Information Engineering in Surveying, Mapping and Remote Sensing (Grant No. (13) Key Project). We are thankful for Yasutaka Furukawa and Jean Ponce for providing the open source code of PMVS algorithm.

Author Contributions: Xiongwu Xiao, Linhui Li and Zhe Peng designed the multi-view stereo-matching algorithm for multiple UAV images based on image-grouping and self-adaptive patch; Xiongwu Xiao and Linhui Li wrote the manuscript; BingxuanGuo and Deren Li helped to design the multi-view stereo-matching algorithm for multiple UAV images based on image-grouping and self-adaptive patch and helped to outline the manuscript structure; Jianchen Liu and Nan Yang helped to remove lens distortion of the original images to get the corrected images and calculated the orientation elements of the corrected images by GodWork software; Peng Zhang prepared the UAV image data; Peng Zhang and Nan Yang helped to prepare the manuscript. All authors read and approved the final manuscript.

Conflicts of Interest: The authors declare no conflict of interest.

References

1. Lin, Z.; Su, G.; Xie, F. UAV borne low altitude photogrammetry system. *Int. Arch. Photogramm. Remote Sens. Spat. Inf. Sci.* **2012**, *XXXIX-B1*, 415–423.

2. Tong, X.; Liu, X.; Chen, P.; Liu, S.; Luan, K.; Li, L.; Liu, S.; Liu, X.; Xie, H.; Jin, Y.; *et al.* Integration of UAV-based photogrammetry and terrestrial laser scanning for the three-dimensional mapping and monitoring of open-pit mine areas. *Remote Sens.* **2015**, *7*, 6635–6662.

3. Colomina, I.; Molina, P. Unmanned aerial systems for photogrammetry and remote sensing: A review. *ISPRS J. Photogramm. Remote Sens.* **2014**, *92*, 79–97.

4. Ahmad, A. Digital mapping using low altitude UAV. *Pertan. J. Sci. Technol.* **2011**, *19*, 51–58.

5. Eisenbeiss, H. UAV Photogrammetry. Ph.D. Thesis, Swiss Federal Institute of Technology Zurich, Zurich, Switzerland, 2009.

6. Udin, W.S.; Ahmad, A. Large scale mapping using digital aerial imagery of unmanned aerial vehicle. *Int. J. Sci. Eng. Res.* **2012**, *3*, 1–6.

7. Remondino, F.; Barazzetti, L.; Nex, F.; Scaioni, M.; Sarazzi, D. UAV photogrammetry for mapping and 3d modeling–current status and future perspectives. In Proceedings of the International Conference on Unmanned Aerial Vehicle in Geomatics (UAV-g), Zurich, Switzerland, 14–16 September 2011; pp. 1–7.

8. Bachmann, F.; Herbst, R.; Gebbers, R.; Hafner, V.V. Micro UAV based georeferenced orthophoto generation in VIS+ NIR for precision agriculture. *Int. Arch. Photogramm. Remote Sens. Spat. Inf. Sci.* **2013**, *XL-1/W2*, 11–16.

9. Lin, Y.; Saripalli, S. Road detection from aerial imagery. In Proceedings of the 2012 IEEE International Conference on Robotics and Automation (ICRA), Saint Paul, MN, USA, 14–18 May 2012; pp. 3588–3593.

10. Bendea, H.; Chiabrando, F.; Tonolo, F.G.; Marenchino, D. Mapping of archaeological areas using a low-cost UAV. In Proceedings of the XXI International CIPA Symposium, Athens, Greece, 1–6 October 2007.

11. Coifman, B.; McCord, M.; Mishalani, R.G.; Iswalt, M.; Ji, Y. Roadway traffic monitoring from an unmanned aerial vehicle. *IEE Proc. Intell. Trans. Syst.* **2006**, *153*, 11–20.

12. Wang, J.; Lin, Z.; Li, C. Reconstruction of buildings from a single UAV image. In Proceedings of the International Society for Photogrammetry and Remote Sensing Congress, Istanbul, Turkey, 12–23 July 2004; pp. 100–103.

13. Quaritsch, M.; Kuschnig, R.; Hellwagner, H.; Rinner, B.; Adria, A.; Klagenfurt, U. Fast aerial image acquisition and mosaicking for emergency response operations by collaborative UAVs. In Proceedings of the International ISCRAM Conference, Lisbon, Portugal, 8–11 May 2011; pp. 1–5.

14. Lin, Z. UAV for mapping—Low altitude photogrammetry survey. *Int. Arch. Photogramm. Remote Sens. Spat. Inf. Sci.* **2008**, *XXXVII*, 1183–1186.

15. Zhang, Y.; Xiong, J.; Hao, L. Photogrammetric processing of low-altitude images acquired by unpiloted aerial vehicles. *Photogramm. Rec.* **2011**, *26*, 190–211.

16. Ai, M.; Hu, Q.; Li, J.; Wang, M.; Yuan, H.; Wang, S. A robust photogrammetric processing method of low-altitude UAV images. *Remote Sens.* **2015**, *7*, 2302–2333.

17. Seitz, S.M.; Curless, B.; Diebel, J.; Scharstein, D.; Szeliski, R. A comparison and evaluation of multi-view stereo reconstruction algorithms. In Proceedings of the 2006 IEEE Computer Society Conference on Computer Vision and Pattern Recognition (CVPR'06), New York, NY, USA, 17–22 June 2006; pp. 519–528.

18. Furukawa, Y.; Ponce, J. Accurate, dense, and robust multi-view stereopsis. *IEEE Trans. Pattern Anal. Mach. Intell.* **2010**, *32*, 1362–1376.

19. Faugeras, O.; Keriven, R. Variational principles, surface evolution, pdes, level set methods, and the stereo problem. *IEEE Trans. Image Process.* **1998**, *7*, 336–344.

20. Pons, J.P.; Keriven, R.; Faugeras, O. Multi-view stereo reconstruction and scene flow estimation with a global image-based matching score. *Int. J. Comput. Vis.* **2007**, *72*, 179–193.

21. Tran, S.; Davis, L. 3d surface reconstruction using graph cuts with surface constraints. In Proceedings of the European Conf. Computer Vision (ECCV 2006), Graz, Austria, 7–13 May 2006; pp. 219–231.

22. Vogiatzis, G.; Torr, P.H.S.; Cipolla, R. Multi-view stereo via volumetric graph-cuts. In Proceedings of the 2005 IEEE Computer Society Conference on Computer Vision and Pattern Recognition (CVPR 2005), San Diego, CA, USA, 20–25 June 2005; pp. 391–398.

23. Hornung, A.; Kobbelt, L. Hierarchical volumetric multi-view stereo reconstruction of manifold surfaces based on dual graph embedding. In Proceedings of the 2006 IEEE Computer Society Conference on Computer Vision and Pattern Recognition (CVPR 2006), New York, NY, USA, 17–22 June 2006; pp. 503–510.

24. Sinha, S.N.; Mordohai, P.; Pollefeys, M. Multi-view stereo via graph cuts on the dual of an adaptive tetrahedral mesh. In Proceedings of the 11th IEEE International Conference on Computer Vision (ICCV 2007), Rio de Janeiro, Brazil, 14–21 October 2007; pp. 1–8.

25. Kolmogorov, V.; Zabih, R. Multi-camera scene reconstruction via graph cuts. In Proceedings of the 7th European Conference on Computer Vision (ECCV 2002), Copenhagen, Denmark, 28–31 May 2002; pp. 82–96.

26. Vogiatzis, G.; Hernandez, C.; Torr, P.H.S.; Cipolla, R. Multi-view stereo via volumetric graph-cuts and occlusion robust photo-consistency. *IEEE Trans. Pattern Anal. Mach. Intell.* **2007**, *29*, 2241–2246.

27. Esteban, C.H.; Schmitt, F. Silhouette and stereo fusion for 3d object modeling. In Proceedings of the Fourth International Conference on 3-D Digital Imaging and Modeling (3DIM 2003), Banff, AB, Canada, 6–10 October 2003; pp. 46–53.

28. Zaharescu, A.; Boyer, E.; Horaud, R. Transformesh: A topology-adaptive mesh-based approach to surface evolution. In Proceedings of the 8th Asian Conference on Computer Vision (ACCV 2007), Tokyo, Japan, 18–22 November 2007; pp. 166–175.

29. Furukawa, Y.; Ponce, J. Carved visual hulls for image-based modeling. *Int. J. Comput. Vis.* **2009**, *81*, 53–67.

30. Baumgart, B.G. Geometric Modeling for Computer Vision. Ph.D. Thesis, Stanford University, Stanford, CA, USA, 1974.

31. Laurentini, A. The visual hull concept for silhouette-based image understanding. *IEEE Trans. Pattern Anal. Mach. Intell.* **1994**, *16*, 150–162.

32. Strecha, C.; Fransens, R.; Gool, L.V. Combined depth and outlier estimation in multi-view stereo. In Proceedings of the 2006 IEEE Computer Society Conference on Computer Vision and Pattern Recognition (CVPR 2006), New York, NY, USA, 17–22 June 2006; pp. 2394–2401.

33. Goesele, M.; Curless, B.; Seitz, S.M. Multi-view stereo revisited. In Proceedings of the IEEE Conference Computer Vision and Pattern Recognition (CVPR 2006), New York, NY, USA, 17–22 June 2006; pp. 2402–2409.

34. Bradley, D.; Boubekeur, T.; Heidrich, W. Accurate multi-view reconstruction using robust binocular stereo and surface meshing. In Proceedings of the IEEE Conference Computer Vision and Pattern Recognition (CVPR 2008), Anchorage, AK, USA, 23–28 June 2008; pp. 1–8.

35. Hirschmuller, H. Stereo processing by semi-global matching and mutual information. *IEEE Trans. Pattern Anal. Mach. Intell.* **2008**, *30*, 328–341.

36. Furukawa, Y.; Curless, B.; Seitz, S.M.; Szeliski, R. Towards internet-scale multi-view stereo. In Proceedings of the 2010 IEEE Conference on Computer Vision and Pattern Recognition (CVPR 2010), San Francisco, CA, USA, 13–18 June 2010; pp. 1434–1441.

37. Hirschmüller, H. Semi-global matching–Motivation, developments and applications. In Proceedings of the Photogrammetric Week, Stuttgart, Germany, 5–9 September 2011; pp. 173–184.

38. Haala, N.; Rothermel, M. Dense multi-stereo matching for high quality digital elevation models. *Photogramm. Fernerkund. Geoinf.* **2012**, *2012*, 331–343.

39. Rothermel, M.; Haala, N. Potential of dense matching for the generation of high quality digital elevation models. *Int. Arch. Photogramm. Remote Sens. Spat. Inf. Sci.* **2011**, *XXXVIII-4-W19*, 1–6.

40. Rothermel, M.; Wenzel, K.; Fritsch, D.; Haala, N. Sure: Photogrammetric surface reconstruction from imagery. In Proceedings of the LC3D Workshop, Berlin, Germany, 4–5 December 2012.

41. Koch, R.; Pollefeys, M.; Gool, L.V. Multi viewpoint stereo from uncalibrated video sequences. In Proceedings of the 5th European Conference on Computer Vision (ECCV'98), Freiburg, Germany, 2–6 June 1998; pp. 55–71.

42. Lhuillier, M.; Long, Q. A quasi-dense approach to surface reconstruction from uncalibrated images. *IEEE Trans. Pattern Anal. Mach. Intell.* **2005**, *27*, 418–433.

43. Habbecke, M.; Kobbelt, L. Iterative multi-view plane fitting. In Proceedings of the 11th Fall Workshop Vision, Modeling, and Visualization, Aachen, Germany, 22–24 November 2006; pp. 73–80.

44. Sun, J.; Zheng, N.N.; Shum, H.Y. Stereo matching using belief propagation. *IEEE Trans. Pattern Anal. Mach. Intell.* **2003**, *25*, 787–800.

45. Zhu, Q.; Zhang, Y.; Wu, B.; Zhang, Y. Multiple close-range image matching based on a self-adaptive triangle constraint. *Photogramm. Rec.* **2010**, *25*, 437–453.

46. Wu, B. A Reliable Image Matching Method Based on Self-Adaptive Triangle Constraint. Ph.D. Thesis, Wuhan University, Wuhan, China, 2006.

47. Liu, Y.; Cao, X.; Dai, Q.; Xu, W. Continuous depth estimation for multi-view stereo. In Proceedings of the 2009 IEEE Conference on Computer Vision and Pattern Recognition (CVPR 2009), Miami, FL, USA, 20–25 June 2009; pp. 2121–2128.

48. Hiep, V.H.; Keriven, R.; Labatut, P.; Pons, J.P. Towards high-resolution large-scale multi-view stereo. In Proceedings of the 2009 IEEE Conference on Computer Vision and Pattern Recognition (CVPR 2009), Miami, FL, USA, 20–25 June 2009; pp. 1430–1437.

49. Zhang, Z.; Zhang, J. *Digital Photogrammetry*; Wuhan University Press: Wuhan, China, 1997.

50. Jiang, W. Multiple Aerial Image Matching and Automatic Building Detection. Ph.D. Thesis, Wuhan University, Wuhan, China, 2004.

51. Zhang, L. Automatic Digital Surface Model (Dsm) Generation from Linear Array Images. Ph.D. Thesis, Swiss Federal Institute of Technology (ETH), Zurich, Switzerland, 2005.

52. Furukawa, Y.; Ponce, J. Patch-Based Multi-View Stereo Software. Available online: http://www.di.ens.fr/pmvs/ (accessed on 14 November 2015).

53. David, G.L. Distinctive image features from scale-invariant keypoints. *Int. J. Comput. Vis.* **2004**, *60*, 91–110.

54. Harris, C.G.; Stephens, M.J. A combined corner and edge detector. In Proceedings of the Fourth Alvey Vision Conference, Manchester, UK, 31 August–2 September 1988; pp. 147–151.

55. Naylor, W.; Chapman, B. Free Software Which You Can Download. Available online: http://www.willnaylor.com/wnlib.html (accessed on 14 November 2015).

56. Canny, J. A computational approach to edge detection. *IEEE Trans. Pattern Anal. Mach. Intell.* **1986**, *PAMI-8*, 679–698.

57. Zhou, X. *Higher Geometry*; Science Press: Beijing, China, 2003.

58. Liu, Z. Research on Stereo Matching of Computer Vision. Ph.D. Thesis, Nanjing University of Science and Technology, Nanjing, China, 2005.

59. Fraser, C.S. Digital camera self-calibration. *ISPRS J. Photogramm. Remote Sens.* **1997**, *52*, 149–159.

60. Image Coordinate Correction Function in Australis. Available online: http://www.photometrix.com.au/downloads/australis/Image%20Correction%20Model.pdf (accessed on 14 November 2015).

61. Liu, C.; Jia, Y.; Cai, W.; Wang, T.; Song, Y.; Sun, X.; Jundong, Z. Camera calibration optimization technique based on genetic algorithms. *J. Chem. Pharm. Res.* **2014**, *6*, 97–103.

62. Wu, C. Visualsfm: A Visual Structure from Motion System. Available online: http://ccwu.me/vsfm/ (accessed on 14 November 2015).

63. Snavely, N. Bundler: Structure from Motion (SFM) for Unordered Image Collections. Available online: http://www.cs.cornell.edu/~snavely/bundler/ (accessed on 14 November 2015).

64. Cramer, M. The dgpf-test on digital airborne camera evaluation—Overview and test design. *Photogramm. Fernerkund. Geoinf.* **2010**, *2010*, 73–82.

65. Isprs Test Project on Urban Classification, 3d Building Reconstruction and Semantic Labeling. Available online: http://www2.isprs.org/commissions/comm3/wg4/tests.html (accessed on 14 November 2015).

66. Haala, N.; Hastedt, H.; Wolf, K.; Ressl, C.; Baltrusch, S. Digital photogrammetric camera evaluation—Generation of digital elevation models. *Photogramm. Fernerkund. Geoinf.* **2010**, *2010*, 99–115.

67. Xiao, X.; Guo, B.; Pan, F.; Shi, Y. Stereo matching with weighted feature constraints for aerial images. In Proceedings of the seventh International Conference on Image and Graphics (ICIG 2013), Qingdao, China, 26–28 July 2013; pp. 562–567.

68. Höhle, J. The eurosdr project "automated checking and improving of digital terrain models". In Proceedings of the ASPRS 2007 Annual Conference, Tampa, FL, USA, 7–11 May 2007.

69. Wenzel, K.; Rothermel, M.; Fritsch, D.; Haala, N. Image acquisition and model selection for multi-view stereo. *Int. Arch. Photogramm. Remote Sens. Spat. Inf. Sci.* **2013**, *XL-5/W1*, 251–258.

70. Hobi, M.L.; Ginzler, C. Accuracy assessment of digital surface models based on Worldview-2 and ADS80 stereo remote sensing data. *Sensors* **2012**, *12*, 6347–6368.

71. Ginzler, C.; Hobi, M. Countrywide stereo-image matching for updating digital surface models in the framework of the Swiss national forest inventory. *Remote Sens.* **2015**, *7*, 4343–4370.

72. Kazhdan, M.; Hoppe, H. Screened Poisson Surface Reconstruction. Available online: http://www.cs.jhu.edu/~misha/Code/PoissonRecon/ (accessed on 15 November 2015).

73. Strecha, C.; Von Hansen, W.; Gool, L.V.; Fua, P.; Thoennessen, U. On benchmarking camera calibration and multi-view stereo for high resolution imagery. In Proceedings of the 2008 IEEE Conference on Computer Vision and Pattern Recognition (CVPR 2008), Anchorage, AK, USA, 23–28 June 2008; pp. 1–8.

74. Pix4d White Paper—How Accurate Are UAV Surveying Methods? Available online: https://support.pix4d.com/hc/en-us/article_attachments/200932859/Pix4D_White_paper_How_accurate_are_UAV_surveying_methods.pdf (accessed on15 November 2015).

75. Küng, O.; Strecha, C.; Beyeler, A.; Zufferey, J.C.; Floreano, D.; Fua, P.; Gervaix, F. The accuracy of automatic photogrammetric techniques on ultra-light UAV imagery. In Proceedings of the International Conference on Unmanned Aerial Vehicle in Geomatics (UAV-g), Zurich, Switzerland, 14–16 September 2011; pp. 14–16.

76. Cryderman, C.; Mah, S.B.; Shufletoski, A. Evaluation of UAV photogrammetric accuracy for mapping and earthworks computations. *Geomatica* **2014**, *68*, 309–317.

77. Rock, G.; Ries, J.B.; Udelhoven, T. Sensitivity analysis of UAV-photogrammetry for creating digital elevation models (DEM). *Int. Arch. Photogramm. Remote Sens. Spat. Inf. Sci.* **2011**, *XXXVIII-1/C22*, 1–5.

Multi-Class Simultaneous Adaptive Segmentation and Quality Control of Point Cloud Data

Ayman Habib and Yun-Jou Lin

Abstract: 3D modeling of a given site is an important activity for a wide range of applications including urban planning, as-built mapping of industrial sites, heritage documentation, military simulation, and outdoor/indoor analysis of airflow. Point clouds, which could be either derived from passive or active imaging systems, are an important source for 3D modeling. Such point clouds need to undergo a sequence of data processing steps to derive the necessary information for the 3D modeling process. Segmentation is usually the first step in the data processing chain. This paper presents a region-growing multi-class simultaneous segmentation procedure, where planar, pole-like, and rough regions are identified while considering the internal characteristics (*i.e.*, local point density/spacing and noise level) of the point cloud in question. The segmentation starts with point cloud organization into a kd-tree data structure and characterization process to estimate the local point density/spacing. Then, proceeding from randomly-distributed seed points, a set of seed regions is derived through distance-based region growing, which is followed by modeling of such seed regions into planar and pole-like features. Starting from optimally-selected seed regions, planar and pole-like features are then segmented. The paper also introduces a list of hypothesized artifacts/problems that might take place during the region-growing process. Finally, a quality control process is devised to detect, quantify, and mitigate instances of partially/fully misclassified planar and pole-like features. Experimental results from airborne and terrestrial laser scanning as well as image-based point clouds are presented to illustrate the performance of the proposed segmentation and quality control framework.

Reprinted from *Remote Sens.* Cite as: Habib, A.; Lin, Y.-J. Multi-Class Simultaneous Adaptive Segmentation and Quality Control of Point Cloud Data. *Remote Sens.* **2016**, *8*, 104.

1. Introduction

Urban planning, heritage documentation, military simulation, airflow analysis, transportation management, and Building Information Modeling (BIM) are among the applications that need accurate 3D models of the sites in question. Optical imaging and laser scanning systems are the two leading data acquisition modalities for 3D model generation. Acquired images can be manipulated to produce a point cloud along the visible surface within the field of view of the camera stations. Laser scanning systems, on the other hand, are capable of directly providing accurate point clouds at high density. To allow for the derivation of semantic information, image and laser-based point clouds need to undergo a sequence of data processing steps to meet the demands of Digital Building Model—DBM—generation, urban planning [1], as-built mapping of industrial sites, transportation infrastructure systems [2], cultural heritage documentation [3], and change detection. Point cloud segmentation according to pre-defined criteria is one of the initial steps in the data processing chain. More specifically, the segmentation of planar, pole-like, and rough regions from a given point cloud is quite important for ensuring the validity and reliability of the generated 3D models.

As mentioned earlier, optical imagery and laser scanners are two major sources for indirectly or directly deriving point clouds, which can meet the demands of the intended 3D modeling applications. Electro-Optical (EO) sensors onboard space borne, airborne, and terrestrial platforms are capable of acquiring imagery with high resolution, which could be used for point cloud generation. Identification of conjugate points in overlapping images is a key prerequisite for image-based point cloud generation. Within the photogrammetric community, area-based and feature-based matching techniques have been used [4]. Area-based image matching is performed by comparing the gray values within a defined template in one image to those within a larger search window in an overlapping image to identify the location that exhibits the highest similarity. Pratt [5] proposed the Normalized Cross-Correlation (NCC) measure, which compensates for local brightness and contrast variations between the gray values within the template and search windows. Feature-based matching, on the other hand, compares the attributes of extracted features (e.g., points, lines, and regions) from overlapping images. Scale Invariant Feature Transform (SIFT) detector and descriptor can be used to identify and provide the attributes for key image points (Lowe, 2004). The SIFT descriptor can be then used to identify conjugate point features in overlapping images. Alternatively, Canny edge detection and linking can be used to derive linear features from imagery [6]. Then, Generalized Hough Transform can be used to identify conjugate points along detected edges [7]. Area and feature-based image matching techniques are not capable of providing dense point clouds, which are needed for 3D object modeling (*i.e.*, they are mainly used for automated recovery of image orientation). Recently developed dense image matching techniques can generate point clouds that exhibit high level of detail [8–10].

In contrast to imaging sensors, laser scanners can directly derive dense point clouds. Depending on the used platform, a laser scanner can be categorized either as an Airborne Laser Scanner (ALS), a Terrestrial Laser Scanner (TLS), or a Mobile Terrestrial Laser Scanner (MTLS). TLS systems provide point clouds that are referred to the laser-unit coordinate system. For ALS and MTLS systems, the onboard direct geo-referencing unit allows for the derivation of the point cloud coordinates relative to a global reference frame (e.g., WGS84). ALS systems are used for collecting relatively coarse-scale elevation data. Due to the pulse repetition rate, flying height, and speed of available systems/platforms, the Local Point Density (LPD) within ALS point clouds is lower, when compared with TLS and MTLS point clouds. The point density within an ALS-based point cloud can range from 1 to 40 pts/m^2 [11]. Such point density is suitable for Digital Terrain Model (DTM) generation [12,13] and Digital Building Model (DBM) generation at a low level of detail [14,15]. However, ALS cannot provide point clouds, which are useful for modeling building façades, above-ground pole-like features such as light poles, and trees. As a result of their proximity to the objects of interest, TLS and MTLS systems can deliver dense point clouds for the extraction and accurate modeling of transportation corridors, building façades, and trees/bushes. El-Halawany *et al.* [16] utilized MTLS point clouds to identify ground/non-ground points and extract road curbs for transportation management applications. TLS and MTLS point clouds have been also used for 3D pipeline modeling, which is valuable for plant maintenance and operation [17,18] and building façade modeling [19].

Point-cloud-based object modeling usually starts with a segmentation process to categorize the data into subgroups that share similar characteristics. Segmentation approaches can be generally classified as being either spatial or parameter domain. For the spatial-domain approach, e.g., region-growing based segmentation, the point cloud is segmented into subgroups according to the spatial proximity and similarity of local attributes of its constituents [20]. More specifically, starting from seed points/regions, the region-growing process augments neighboring points using a pre-defined similarity measure. The spatial proximity and local attribute determination depends on whether the point cloud is represented as raster, Triangular Irregular Network (TIN), or un-structured set. Rottensteiner and Briese [21] interpolated non-organized point clouds to generate a Digital Surface Model (DSM), which is then used to detect building regions through height and region-growing analysis of the DSM-based binary image. The region-growing process is terminated whenever the Root Mean Square Error (RMSE) of a plane-fitting process exceeds a pre-set threshold. Forlani *et al.* [22] used a region-growing process to segment raster elevation data, where the height gradient between neighboring cells is used as the stopping criterion. For TIN-based point clouds, the spatial neighborhood among the generated triangles and the similarity of the respective surface normals have been used for the segmentation process [23]. For non-organized point clouds, data structuring approaches (e.g., Kd-trees or Octree data structures) are used to identify local neighborhoods and derive the respective attributes [24,25]. Yang and Dong [26] classified point clouds using Support Vector Machines (SVMs) into planar, linear, and spherical local neighborhoods. Then, region growing is implemented by checking the similarity of derived attributes such as principal direction, normal vector, and intensity. Region-growing segmentation approaches are usually preferred due to their computational efficiency. However, their performance is quite sensitive to noise level within the point cloud in question as well as the selected seed-points/regions [27–29].

For the parameter-domain approach, a feature vector is first defined for the individual points using their local neighborhoods. Then, the feature vectors are incorporated in an attribute space/accumulator array where peak-detection techniques are used to identify clusters—i.e., points sharing similar feature vectors. Filin and Pfeifer [30] used a slope-adaptive neighborhood to derive the local surface normal for the individual points. Then, they defined a feature vector that encompasses the position of the point and the normal vector to the tangent plane at that point. Then, a mode-seeking algorithm is used to identify clusters in the resulting attribute space [31]. Biosca and Lerma [32] utilized three attributes—namely, normal distance to the fitted plane through a local neighborhood from a defined origin, normal vector to the fitted plane, and normal distance between the point in question and the fitted plane—to define a feature vector. Then, an unsupervised fuzzy clustering approach is implemented to identify peaks in the attribute space. Lari and Habib [29] introduced an approach where the individual points have been classified as either belonging to planar or linear/cylindrical local neighborhoods using Principal Component Analysis (PCA). Then, the attributes of the classified features are stored in different accumulator arrays where peaks are identified without the need for tessellating such array to detect planar and pole-like features. Parameter-domain segmentation techniques do not depend on seed points. However, the identification of peaks in the constructed attribute space is a time-consuming process, whose complexity depends on the dimensionality of the involved feature vector [27]. Moreover, spatially-disconnected segments that share the same attributes will be erroneously grouped together. In general, existing spatial-domain and parameter-domain segmentation techniques do not deal with simultaneous segmentation of planar, pole-like, and rough regions in a given point cloud.

The outcome of a segmentation process usually suffers from some artifacts [33]. The traditional approach for Quality Control (QC) of the segmentation result is based on having reference data, which is manually generated, and deriving correctness and completeness measures [34,35]. The correctness measure evaluates the percentage of correctly-segmented constituents of regions in a given class relative the total size of that class in the segmentation outcome. The completeness measure, on the other hand, represents the percentage of correctly-segmented constituents of regions in a given class relative to the total size of that class in the reference data. The reliance on reference data to evaluate the correctness and completeness measures is a major disadvantage of such QC measures. Therefore, prior research has addressed the possibility of deriving QC measures that are not based on reference data. More specifically, Belton, Nurunnabi *et al.*, and Lari and Habib [36–38] developed QC measures that make hypotheses regarding possible segmentation problems, propose procedures for detecting instances of such problems, and develop mitigation approaches to fix such problems without the need for having reference data. Over-segmentation—where a single planar/pole-like feature is segmented into more than one region, and under segmentation—where multiple planar/pole-like features are segmented as one region are key segmentation problems that have been considered by prior literature. More specifically, problems associated with planar and pole-like feature segmentation are independently addressed. However, segmentation problems arising from possible competition among neighboring planar and pole-like features have not been addressed by prior research.

In this paper, we present a region-growing and quality-control framework for the segmentation of planar, pole-like, and rough features. The main characteristics of the proposed procedure are as follows:

1. Planar and varying-radii pole-like features are simultaneously segmented,
2. ALS, TLS, MTLS, and image-based point clouds can be manipulated by the proposed segmentation procedure,
3. The region-growing process starts from optimally-selected seed regions to reduce the sensitivity of the segmentation outcome to the choice of the seed location,
4. The region-growing process considers variations in the local characteristics of the point cloud (*i.e.*, local point density/spacing and noise level),
5. The QC process considers possible competition among neighboring planar and pole-like features for the same points,
6. The QC procedure considers possible artifacts arising from the sequence of the region growing process, and
7. The QC process considers the possibility of having partially or fully misclassified planar and pole-like features.

The paper starts with a presentation of the proposed segmentation and quality control procedures. Then, comprehensive results from ALS, TLS, and image-based point clouds are discussed to illustrate the feasibility of the proposed procedure. Finally, the paper concludes with a summary of the main characteristics, as well as the limitations of the proposed methodology/framework together with recommendations for future research.

2. Proposed Methodology

As can be seen in Figure 1, the proposed methodology proceeds according to the following steps: (1) structuring and characterization of the point cloud; (2) distance-based region growing starting from randomly-selected seed points to define seed regions with pre-defined size; (3) PCA-based classification and feature modeling of generated seed regions; (4) Sequential region-growing according to the quality of fit between neighboring points and the fitted-model through the constituents of the seed regions; (5) PCA-based classification, model-fitting, and region growing of non-segmented points; (6) distance-based region growing for the segmentation of rough points; and (7) quality control of the segmentation outcome. The following subsections introduce the technical details of these steps.

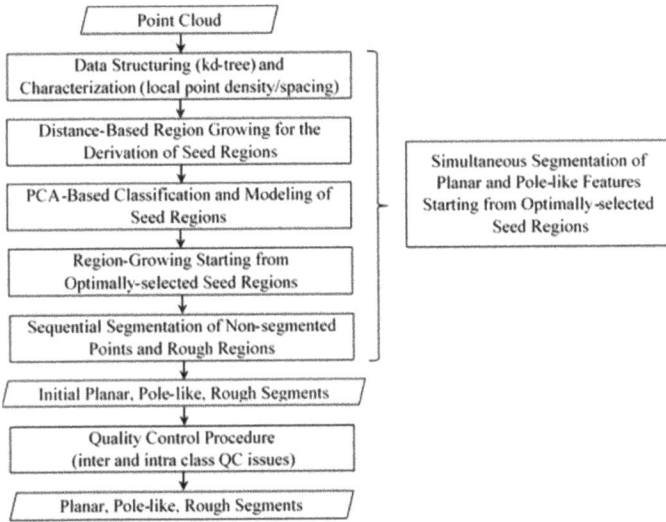

Figure 1. Framework for the multi-class segmentation and quality control procedure.

2.1. Simultaneous Segmentation of Planar and Pole-Like Features Starting from Optimally-Selected Seed Regions

In this subsection, we introduce the conceptual basis and implementation details for the first four steps of the processing framework in Figure 1 (*i.e.*, data structuring and characterization, establishing seed regions, PCA-based classification and modeling of the seed regions, and sequential region-growing from optimally-selected seed regions).

2.1.1. Data Structuring and Characterization

For non-organized point clouds, it is important to re-organize such data to facilitate the identification of the nearest neighbor or nearest *n*-neighbors for a given point. TIN, grid, voxel, Octree, and kd-tree data structures are possible alternatives for facilitating the search within a non-organized point cloud [39–44]. In this research, the kd-tree data structure is utilized for sorting and organizing a set of points since it leads to a balanced tree—*i.e.*, a binary tree with the minimum depth—which improves the efficiency of the neighborhood-search process. The kd-tree data structure is established by recursive sequential subdivision of the three-dimensional space along the X, Y, and Z directions starting with the one that has the longest extent. The splitting plane is defined to be perpendicular to the direction in question and passes through the point with the median coordinate along that direction. The 3D recursive splitting proceeds until all the points are inserted in the kd-tree.

The outcome of any region-growing segmentation approach depends on the search radius, which is used to identify neighboring points that satisfy a predefined similarity criterion. This search radius should be based on the Local Point Density/Spacing (LPD/LPS) for the point under consideration. For either laser-based or image-based point clouds, the LPD/LPS will change depending on the utilized sensor and/or platform as well as the sensor-to-object distance. For image-based point clouds, the LPD/LPS can be also affected by object texture or illumination conditions. Therefore, we need to estimate a unique LPD/LPS for every point within the dataset in question. More specifically, for every point, we establish a local neighborhood that contains a pre-specified number of points. As stated in Lari and Habib [36], the evaluation of the LPD/LPS requires the identification of the nature of the local surface at the vicinity of the query point (*i.e.*, LPD/LPS evaluation depends on whether the local surface is defined by a planar, thin linear, cylindrical, or rough feature—please, refer to the reported statistics in Table 1). The number of used points to define the local surface should be large enough to ensure that the local surface is correctly identified for valid estimation of the LPD/LPS. In this research, a total of 70 points have been used to define the local neighborhood for a given point. Then, a PCA procedure is used to identify the nature of the defined local neighborhood—*i.e.*, determine whether it is part of a planar, pole-like, or rough region [45]. Depending on the identified class, the corresponding LPD—*pnts/m* for thin pole-like features, *pnts/m^2* for planar and cylindrical features, and *pnts/m^3* for rough regions—and the corresponding LPS are estimated according to the established measures in Lari and Habib [38].

2.1.2. Distance-Based Region Growing for the Derivation of Seed Regions

This step starts by forming a set of seed points that are randomly distributed within the point cloud in question. Rather than directly defining seed regions, which are centered at the randomly-established seed points, we define the seed regions through a distance-based region growing. More specifically, starting from a user-defined percentage of randomly-selected seed points, we perform a distance-based region growing (*i.e.*, the spatial closeness of the points to the seed point in question as determined by the LPS is the only used criterion). The distance-based region growing continues until pre-specified region size is attained.

This approach for seed-region definition will ensure that the seed region is large enough, while avoiding the risk of having the seed region comprised of points from two or more different classes. Therefore, when dealing with different features that are spatially close to each other, we ensure that the seed regions belong to the individual objects as long as the spatial separation between those features is larger than the LPS. Having larger seed regions that belong to individual objects will lead to better identification of the respective models associated with those neighborhoods, which in turn will increase the reliability of the segmentation procedure.

Table 1. LPD statistics for the different datasets.

	ALS	TLS1	TLS2	TLS3	DIM
Number of Points	812,980	170,296	201,846	455,167	230,434
Max. Planar LPD (pts/m^2)	4.518	1549	324.54	73,443	404
Min. Planar LPD (pts/m^2)	0.058	1.687	≈0.000	17.351	1.234
Mean Planar LPD (pts/m^2)	2.596	781	27.305	17,685	104
Max. Linear LPD (pts/m)	10.197	140	8.473	1,186	0
Min. Linear LPD (pts/m)	7.708	16.777	8.473	55.093	0
Mean Linear LPD (pts/m)	8.960	62.396	8.473	276	0
Max. Cylindrical LPD (pts/m^2)	3.204	2,423	999	34,337	375
Min. Cylindrical LPD (pts/m^2)	2.059	6.132	1.707	10.063	3.066
Mean Cylindrical LPD (pts/m^2)	2.990	313	41.216	6,055	46.045
Max. Rough LPD (pts/m^3)	1.329	9,267	2,120	1,954,807	1217
Min. Rough LPD (pts/m^3)	0.001	1.980	≈0.000	11.688	0.053
Mean Rough LPD (pts/m^3)	0.363	1818	26.238	230,796	145

2.1.3. PCA-Based Classification and Modeling of Seed Regions

Now that we defined the seed regions, we use PCA to identify whether they belong to planar, pole-like, or rough neighborhoods. More specifically, the relationships among the normalized Eigen values of the dispersion matrix of the points within a seed region relative to its centroid are used to identify planar seed regions (*i.e.*, where two of the normalized Eigen values are significantly larger than the third one), pole-like seed regions (*i.e.*, where one of the normalized Eigen values is significantly larger the other two), and rough seed regions (*i.e.*, where the three normalized Eigen values are of similar magnitude). For planar and pole-like seed regions, a Least Squares Adjustment (LSA) model-fitting procedure is used to derive the plane/pole-like parameters together with the quality of fit between the points within the seed region and the defined model as represented by the respective a-posteriori variance factor (this a-posteriori variance factor will be used as an indication of the local noise level within the seed region). For a planar seed region, the LSA estimates the three plane parameters—a, b, and c—using either Equation (1),

(2), or (3) (the choice of the appropriate plane equation depends on the orientation of the Eigen vector corresponding to the smallest Eigen value—*i.e.*, the one defining the normal to the plane)—refer to Figure 2. For a pole-like feature, the LSA estimates its radius together with four parameters that define the coordinates of a point along the axis and the axis orientation—p, q, a, and b—using either Equation (4), (5), or (6) (the choice of the appropriate equation depends on the orientation of the Eigen vector corresponding to the largest Eigen value—*i.e.*, the one defining the axis orientation of the pole-like feature)—refer to Figure 3. One should note that the variable t in Equations (4)–(6), depends on the distance between the projection of any point onto the axis of the pole-like feature and the utilized point along the axis—*i.e.*, (p,q,0) for the axis defined by Equation (4), (p,0,q) for the axis defined by Equation (5), or (0,p,q) for the axis defined by Equation (6) (refer to Figure 3).

$$z = ax + by + c \tag{1}$$

$$y = ax + bz + c \tag{2}$$

$$x = ay + bz + c \tag{3}$$

$$\begin{aligned} x &= p + t\,a \\ y &= q + t\,b \\ z &= t \end{aligned} \tag{4}$$

$$\begin{aligned} x &= p + t\,a \\ y &= t \\ z &= q + t\,b \end{aligned} \tag{5}$$

$$\begin{aligned} x &= t \\ y &= p + t\,a \\ z &= q + t\,b \end{aligned} \tag{6}$$

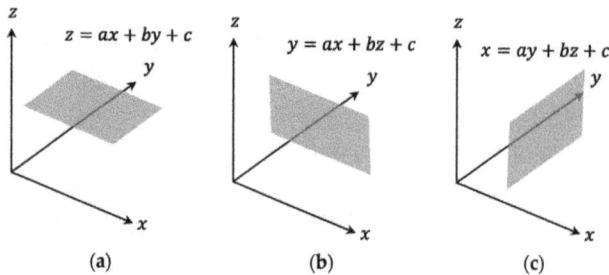

Figure 2. Representation scheme for 3D planar features; planes that are almost parallel to the $xy - plane$ (**a**); planes that are almost parallel to the $xz - plane$ (**b**); and planes that are almost parallel to the $yz - plane$ (**c**).

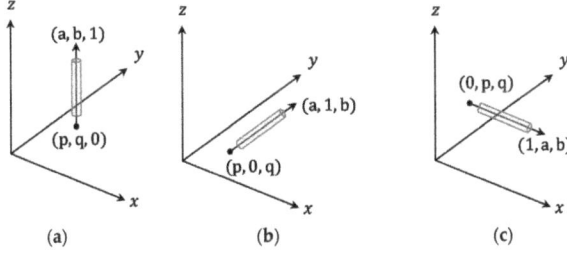

Figure 3. Representation scheme for 3D pole-like features; pole-like features that are almost parallel to the $z-axis$ (**a**); pole-like features that are almost parallel to the $y-axis$ (**b**); and pole-like features that are almost parallel to the $x-axis$ (**c**).

2.1.4. Region-Growing Starting from Optimally-Selected Seed Regions

In this research, the seed regions representing planar and pole-like features are sorted according to an ascending order for the evaluated a-posteriori variance factor in the previous step. One should note that such a-posteriori variance factor is an indication of the normal distances between the points within the seed region and the best-fitted model—*i.e.*, it is an indication of the noise level in the dataset as well as the compatibility of the physical surface and the underlying mathematical model. Starting with the seed region that has the minimum a-posteriori variance factor, a region-growing process is implemented while considering the spatial proximity as defined by the LPS and the normal distance to the defined model through the seed region as the similarity criteria. Throughout the region-growing process, the model parameters and a-posteriori variance factor are sequentially updated. For a given seed region, the growing process will proceed until no more points could be added. The sequential region growing according to the established quality of fit—*i.e.*, a-posteriori variance factor—will ensure that seed regions showing better fit to the planar or pole-like feature model are considered first. Thus, rather than starting the region growing from randomly established seed points, we start the growing from locations that exhibit good fit with the pre-defined models for planar and pole-like features.

2.1.5. Sequential Segmentation of Non-Segmented Points and Rough Regions

Depending on the user-defined percentage of seed points, one should expect that some points might not be segmented or considered since they are not within the immediate vicinity of seed points that belong to the same class or they happen to be at the neighborhood of rough seed regions. To consider such situations, we implement a sequential region-growing process by going through the points within the kd-tree data structure starting from its root and identifying the points that have not been segmented/classified so far. Whenever a non-segmented/non-classified point within the kd-tree data structure is encountered, the following region-growing procedure is implemented:

1. Starting from a non-segmented/non-classified point, a distance-based region growing is implemented, according the established LPS, until a pre-defined seed-region size is achieved.
2. For the established seed region, PCA is used to decide whether the seed region represents planar, pole-like, or rough neighborhood. If the seed region is deemed as being part of a planar or pole-like feature, the parameters of the respective model are estimated through a LSA procedure.
3. A region-growing process is carried out using the LPS and quality of fit with the established model in the previous step as the similarity measures. Throughout the region-growing process, the model parameters and the respective a-posteriori variance factor are sequentially updated.
4. Steps 1–3 are repeated until all the non-segmented/non-classified nodes within the kd-tree data structure are considered.

The last step of the segmentation process, is grouping neighboring points that belong to rough regions. This is carried out according to the following steps:

1. For the seed regions, which have been classified as being part of rough neighborhoods during the first or the second stages of the segmentation procedure, we conduct a distance-based region-growing of non-segmented points.
2. Finally, we inspect the kd-tree starting from its root node to identify non-segmented/non-classified nodes, which are utilized as seed points for a distance-based segmentation of rough regions.

At this stage, the constituents of a point cloud have been classified and segmented into planar, pole-like, and rough segments. For planar and pole-like features, we have also established the respective model parameters and a-posteriori variance factor, which describes the average normal distance between the constituents of a region and the best-fit model.

2.2. Quality Control of the Segmentation Outcome

In spite of the facts that, (1) the proposed region-growing segmentation strategy has been designed to optimally-select seed regions that exhibit the best quality of fit to the LSA-based planar/pole-like models; and (2) the region growing is based on the established LPS for the individual points, one cannot guarantee that the segmentation outcome will be perfect (*i.e.*, the segmentation outcome might still exhibit artifacts). For example, one should expect that segmented regions at an earlier stage might invade segmented regions at a later stage. Additionally, due to the location of the randomly-established seed points and the nature of the objects within the point cloud, there might be instances where seed regions are wrongly classified (e.g., a portion of a planar feature is wrongly classified as a pole-like feature or a set of contiguous pole-like features are identified as a planar segment). As has been mentioned in the Introduction, prior research has dealt with the detection and mitigation of over-segmentation and under-segmentation problems. However, prior research does not consider potential artifacts that might arise when simultaneously segmenting planar, pole-like, and rough regions. The proposed quality control framework proceeds according to the following three stages; namely, (1) developing a list of hypothesized artifacts/problems that might take place during the segmentation process; (2) developing procedures for the detection of instances of such artifacts/problems without the need for having reference data; and (3) developing approaches to mitigate such problems whenever detected. The following list provides a summary of hypothesized problems that might take place within a multi-class simultaneous segmentation of planar and pole-like features; Figure 4a–f is a schematic illustration of such problems—in sub-figures *a*, *b*, *c*, *e*, and *f* classified planar regions are displayed in light blue while classified pole-like features are displayed in light green:

1. Misclassified planar features: Depending on the LPD/LPS and pre-set size for the seed regions, a pole-like feature might be wrongly classified as a planar region. This situation might be manifested in one of the following scenarios:

   a   Single pole-like feature is wrongly classified as a planar region (Figure 4a), and

   b   Multiple contiguous pole-like features are classified as a single planar region (Figure 4b).

2. Misclassified linear features: depending on the location of the randomly-established seed points, a portion of a planar region might be classified as a single pole-like feature (Figure 4c).

3. Partially misclassified planar and pole-like features: Depending on the order of the region growing process, segmented planar/pole-like features at the earlier stage of the segmentation process might invade neighboring planar/pole-like features. This situation might be manifested in one of the following scenarios:

   a   Earlier-segmented planar regions invade neighboring planar features (Figure 4d),

   b   Earlier-segmented planar regions fully or partially invade neighboring pole-like features (Figure 4e, where a planar region partially invade a neighboring pole-like feature), and

   c   Earlier-segmented pole-like features invade neighboring planar features (Figure 4f).

53

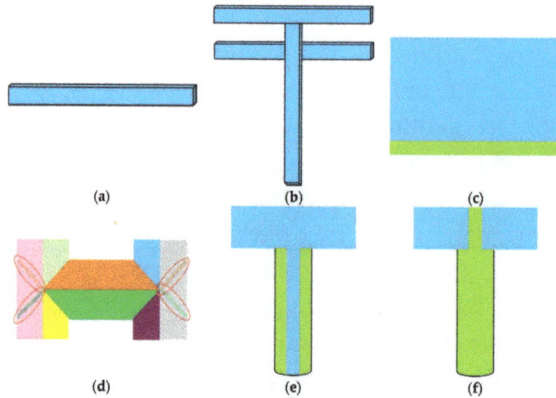

Figure 4. Possible segmentation artifacts; misclassified planar features (**a,b**); misclassified pole-like feature (**c**); partially misclassified planar features (**d,e**); and partially misclassified pole-like feature (**f**)—planar and pole-like features are displayed in light blue and light green, respectively, in subfigures (**a**), (**b**), (**c**), (**e**), (**f**).

The above problems can be categorized as follows: (1) Interclass competition for neighboring points; (2) Intraclass competition for neighboring points; and (3) Fully/partially-misclassified planar and pole-like features. To deal with such segmentation problems, we introduce the following procedure to detect and mitigate instances of such problems:

1. itial mitigation of interclass competition for neighboring points: A key problem in region-growing segmentation is that derived regions at an early stage might invade neighboring features of the same or different class, which are derived at a later stage. In this QC category, we consider potential invasion among features that belong to different classes. Specifically, for segmented features in a given class (*i.e.*, planar or pole-like features), features in the other classes (including rough regions) will be considered as potential candidates that could be incorporated into the constituent regions of the former class. For example, the constituents of pole-like features and rough regions will be considered as potential candidates that could be incorporated into planar features. In this case, if a planar feature has potential candidates, which are spatially close as indicated by the established LPS, and the normal distance between those potential candidates and the LSA-based model through that planar feature is within the respective a-posteriori variance factor, those potential candidates will be incorporated into the planar feature in question. The same procedure is applied for pole-like features, while considering planar and rough regions as potential candidates. In this regard, the respective QC measure—$QC_{interclass\ competition}$—is evaluated according to Equation (7), where $n_{incorporated}$ represents the number

54

of incorporated points from other classes and $n_{potential\ candidates}$ represents the number of potential candidates for this class. For that QC measure, lower percentage indicates lower instances of points that have been incorporated from other classes.

$$QC_{interclass\ competition} = n_{incorporated}/n_{potential\ candidates} \qquad (7)$$

2. Mitigation of intraclass competition for neighboring points: This problem takes place whenever a feature, which has been derived at the earlier stage of the region growing, invades other features from the same class that have been segmented at a later stage. One can argue that intraclass competition for pole-like features is quite limited (this is mainly due to the narrow spread of pole-like features across its axis). Therefore, for this QC measure, we only consider intraclass completion for planar features (as can be seen in Figure 4d, where the middle planar regions invade the left and right planar features with the invading portions highlighted by red ellipses). Detection and mitigation of such problem starts by deriving the inner and outer boundaries of the segmented planar regions (Figure 5 illustrates an example of inner and outer boundaries for a given segment). The inner and outer boundaries can be derived using the minimum convex hull and inter-point-maximum-angle procedures presented by Sampath and Shan [44] and Lari and Habib [45], respectively. Then, for each of the planar regions, we check if some of their constituents are located within the boundaries of neighboring regions and at the same time the normal distances between such constituents and the fitted model through the neighboring regions are within their respective a-posteriori variance factor. In such a case, the individual points that satisfy these conditions are transformed from the invading planar feature to the invaded one. For such QC category, the respective measure is determined according to Equation (8), where $n_{invading}$ represents the number of invading planar points that have been transformed from the invading to the invaded segments and $n_{plane\ total}$ represents the total number of originally-segmented planar points. In this case, lower percentage indicates lower instances of such problem.

$$QC_{intraclass\ competition} = n_{invading}/n_{plane\ total} \qquad (8)$$

3. Single pole-like feature wrongly classified as a planar one: To detect such instances (Figure 4a is a schematic illustration of such situation), we perform PCA of the constituents of the individual planar features. For such segmentation problem, the PCA-based normalized Eigen values will indicate 1-D spread of such regions. Whenever such scenario is encountered, the LSA-based parameters of the fitted cylinder through this feature together with the

respective a-posteriori variance factor are derived. The planar feature will be reclassified as a pole-like one if the latter's a-posteriori variance factor is almost equivalent to the planar-based one. For this case, the respective QC measure—$QC_{reclassified\ linear\ feaure}$—is represented by Equation (9), where $n_{reclassified\ lines}$ is the number of points within reclassified linear features and $n_{plane\ total}$ is the total number of points within the originally-segmented planar features. In this case, lower percentage indicates fewer instances of such a problem.

$$QC_{reclassified\ linear\ features} = n_{reclassified\ lines}/n_{plane\ total} \qquad (9)$$

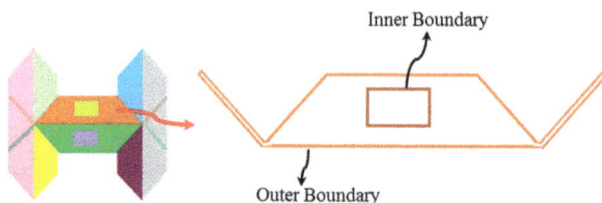

Figure 5. Inner and outer boundary derivation for the identification of intraclass competition for neighboring points.

4. Mitigation of fully or partially misclassified pole-like features: For this problem (as illustrated by Figure 4c,f), we identify pole-like features or portions of pole-like features that are encompassed within neighboring planar features. The process starts with identifying neighboring pole-like and planar features where the axis of the pole-like feature is perpendicular to the planar-feature normal. Then, the constituents of the pole-like feature are projected onto the plane defined by the planar feature. Instances, where the pole-like feature is encompassed—either fully or partially—within the planar feature, are identified by slicing the pole-like feature in the across direction to its axis. For each of the slices, we determine the closest planar point(s) that does (do) not belong to the pole-like feature in question (e.g., point *a* in Figure 6a or points *a and b* in Figure 6b). If the closest point(s) happen to be immediate neighbor(s) of the constituents of that slice (as defined by the established LPS), then one can suspect that the portion of the pole-like feature at the vicinity of that slice might be encompassed within the neighboring planar region (*i.e.*, that portion of the pole-like feature might be invading the planar region). To confirm or reject this suspicion, we evaluate the normal distances between the constituents of the slice and the neighboring planar region. If these normal distances are within the respective a-posteriori variance factor for the planar region, we

confirm that the slice is encompassed within the planar region. Whenever the pole-like feature is fully encompassed within the planar region (Figure 6a), all the slices will have immediate neighbors from that planar region while having minimal normal distances. Consequently, the entire pole-like feature will be reassigned to the planar region. On the other hand, whenever the linear feature is partially encompassed within the planar region, we identify the slices where the closest neighbors to such slices are not immediate neighbors (Figure 6b). The portion of the pole-like feature, which is defined by such slices, will be retained while the other portion will be reassigned to the planar region. The QC measure in this case is defined by Equation (10), where $n_{encompassed\ line\ points}$ represents the number of points within the pole-like feature that are encompassed within the planar feature and $n_{line\ total}$ is the total number of points within the originally-segmented linear features. Lower percentage indicates fewer instances of such problem.

$$QC_{partially/fully\ misclassified\ pole-like\ features} = n_{encompassed\ line\ points}/n_{line\ total} \qquad (10)$$

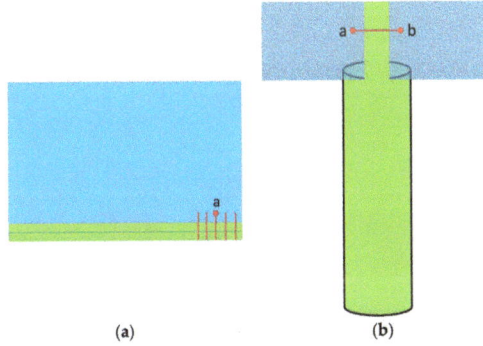

Figure 6. Slicing and immediate-neighbors concept for the identification of fully/partially misclassified pole-like features (**a**)/(**b**).

5. Mitigation of fully or partially misclassified planar features: The conceptual basis of the implemented procedure to detect instances of such problem (as illustrated by Figure 4b,e) is that whenever planar features are either fully (Figure 4b) or partially (Figure 4e) misclassified, a significant portion of the encompassing Minimum Bounding Rectangle (MBR) will not be occupied by those features (refer to Figure 7a). In this regard, one should note that the MBR denotes the smallest area rectangle that encompasses the identified boundary of the planar region in question [46]. Therefore, to detect instances of such problem, we start by defining the MBR for the individual planar regions. Then,

57

we evaluate the ration between the area of the planar region in question and the area of the encompassing MBR. Whenever this area is below a pre-defined threshold, we suspect that the planar feature in question might contain pole-like features, which will take the form of tentacles to the original planar region (as can be seen in Figure 7a). To identify such features, we perform a 2D-linear feature segmentation procedure, which is similar to the one proposed earlier with the exception that it is conducted in 2D rather than 3D (*i.e.*, the line parameters would include slope, intercept, and width)—refer to Figure 7b. More specifically, pre-defined percentage of seed points are established. Then, a distance-based region growing is carried out to define seed regions with pre-set size. A 2D-PCA and line fitting procedure is conducted to identify seed regions that represent 2D lines. Those seed regions are then incorporated within a region-growing process that considers both the spatial closeness of the points and their normal distance to the fitted 2D lines. Following the 2D-line segmentation, an over-segmentation quality control is carried out to identify single linear features that have been identified as multiple ones. Moreover, the conducted QC in the previous step is implemented to identify partially misclassified linear features—*i.e.*, the invading portion of the linear feature(s) (refer to Figure 7c). The QC measure for such problem is evaluated according to Equation (11), where $n_{misclassified\ plane\ points}$ represents the number of points within the planar feature that belong to 2D lines and $n_{plane\ total}$ is the total number of originally-segmented planar points. Lower percentage indicates fewer instances of such problem.

$$QC_{partially/fully\ misclassified\ planar\ features} = n_{misclassified\ plane\ points}/n_{plane\ total} \qquad (11)$$

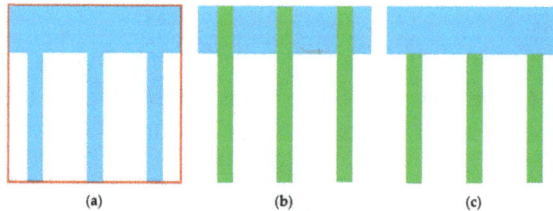

Figure 7. Segmented planar feature—in light blue—and the encompassing MBR—in red (**a**); segmented linear features—in green (**b**); and final segmentation after the identification of partially-misclassified linear features (**c**).

3. Experimental Results

To illustrate the performance of the segmentation and quality control procedure, this section provides the segmentation and quality control results using ALS, TLS, and image-based point clouds. The main objectives of the conducted experiments are as follows:

1. Prove the feasibility of the proposed segmentation procedure in handling data with significant variation in LPD/LPS as well as inherent noise level,
2. Prove the feasibility of the proposed segmentation procedure in handling data with different distribution and concentration of planar, pole-like, and rough regions,
3. Prove the capability of the proposed QC procedure in detecting and quantifying instances of the hypothesized segmentation problems, and
4. Prove the capability of the proposed QC procedure in mitigating instances of the hypothesized segmentation problems.

The following subsections provide the datasets description, segmentation results, and the outcome of the quality control procedure.

3.1. Datasets Description

Airborne Laser Scanner Dataset—ALS: This dataset is captured by an Optech ALTM 3100 over an urban area that includes planar roofs, roads, and trees/bushes. The extent of the covered area is roughly 0.5 km × 0.5 km. Figure 8a shows a perspective view of the ALS point cloud, where the color is based on the height of the different points.

First Terrestrial Laser Scanner Dataset—TLS1: This dataset is captured by a FARO Focus3D X330 scanner. The effective scan distance for this scanner ranges from 0.6 m up to 330 m. The ranging error is ±2 mm. The scanner is positioned at the vicinity of a building façade with planar and cylindrical features whose radii is almost 0.6 m. The extent of the covered area is approximately 35 m × 20 m × 10 m. Figure 8b illustrates the perspective view of this dataset with the colors derived from the scanner-mounted camera.

Second Terrestrial Laser Scanner Dataset—TLS2: This dataset is captured by Leica HDS 3000 scanner. The effective san distance for this unit ranges up to 300 m with ±6 mm position accuracy at 50 m. The covered area includes a planar building façade, some light poles, and trees/bushes. The extent of the covered area is almost 250 m × 200 m × 26 m. A perspective view of this dataset is illustrated in Figure 8c.

Third Terrestrial Laser Scanner Dataset—TLS3: This dataset covers an electrical substation and is captured by a FARO Focus3D X130 scanner. The effective scan distance ranges from 0.6 m up to 130 m. The ranging error is ±2 mm. The dataset is mainly comprised of pole-like features with relatively small radii. The extent of the covered area is roughly 12 m × 10 m × 6 m. A perspective view of this dataset is provided in Figure 8d with the colors derived from the scanner-mounted camera.

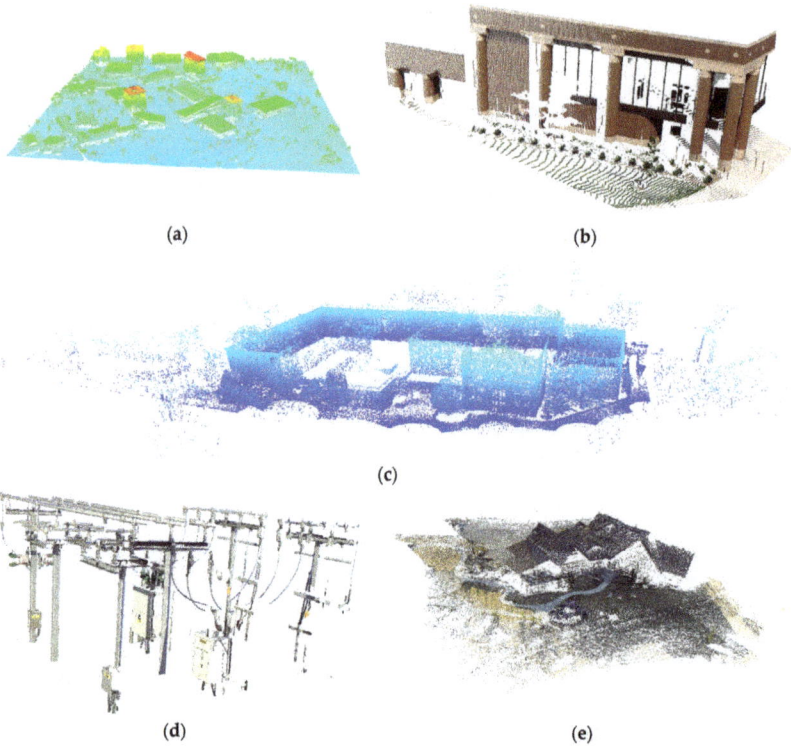

Figure 8. Perspective views of the point clouds from the ALS (**a**); TLS1 (**b**); TLS2 (**c**); TLS3 (**d**); and DIM (**e**) datasets.

Dense Image Matching Dataset—DIM: This dataset, which is shown in Figure 8e, is derived from a block of 28 images captured by a GoPro 3 camera onboard a DJI Phantom 2 UAV platform over a building with complex roof structure. The extent of the covered area is approximately 100 m × 130 m × 17 m. A Structure from Motion (SfM) approach developed by He and Habib [46] is adopted for automated determination of the frame camera EOPs as well as sparse point cloud representing the imaged area relative to an arbitrarily-defined local reference frame. Then, a semi-global dense matching is used to derive a dense point cloud from the involved images [9].

The processing framework starts with data structuring as well as deriving the LPD/LPS for the point clouds in the different datasets. For LPD/LPS estimation, the closest 70 points have been used. Ratios among the PCA-based Eigen values are used to classify the local neighborhoods into planar, pole-like, and rough regions. For Planar regions, the smallest normalized Eigen value should be less than 0.03, while the ration between the other two should be larger than 0.6. For pole-like neighborhoods, on the other hand, the largest normalized Eigen value should be larger than 0.7. The number of the involved points and the statistics of the LPD for the different datasets are listed in Table 1, where one can observe the significant variations in the derived LPD values.

3.2. Segmentation Results

This section provides the segmentation results for planar, pole-like, and rough regions from the different datasets. Before discussing the segmentation results, we introduce the different thresholds, the rationale for setting them up, and the utilized numerical values. The proposed region-growing segmentation methodology involves three thresholds: (1) Percentage of randomly-selected seed points relative to the total number of available points within the dataset—For the above datasets, this percentage is set to 10%. One should note that using larger percentage value did not make a significant impact on the segmentation results; (2) Pre-set size of the seed regions—This size should be set-up in a way to ensure that the seed region is large enough for reliable estimation of the model parameters associated with that region. For the conducted tests, the pre-set region size is set to 100; (3) Normal distance threshold—In general, the normal distance threshold for the region-growing process is based on the derived a-posteriori variance factor from the LSA parameter estimation procedure. However, we set upper threshold values that depend on the sensor specifications (*i.e.*, the normal distance thresholds are not allowed to go beyond these values). For the conducted experiments, the ALS-based region-growing normal distance is set to 0.2 m. For the TLS and DIM datasets, the normal distance threshold is set to 0.05 m. The proposed methodology is implemented in C#. The experiments are conducted using a computer with 16 GB RAM and Intel(R) Core(TM) i7-4790 CPU @3.60 GHz. The time performance of the proposed data structuring, characterization, and segmentation is listed in Table 2.

Table 2. Time performance of the proposed segmentation.

	ALS	TLS1	TLS2	TLS3	DIM
Number of Points	812,980	170,296	201,846	455,167	230,434
Data Structuring and Characterization (mm:ss)	08:15	01:48	02:02	06:57	02:34
Segmentation Time (mm:ss)	11:40	02:55	01:33	06:37	03:56
Total Time (mm:ss)	19:55	04:43	03:35	13:34	06:30

Figures 9 and 10 present the feature classification and segmentation results, respectively. For the classification results in Figure 9, planar, pole-like, and rough regions are shown in blue, green, and red, respectively. As can be seen in Figure 9, ALS, TLS2, and DIM datasets are mainly comprised of planar and rough regions. TLS1 and TLS3, on the other hand, mainly include planar and pole-like features, where large-radii pole-like features are present in TLS1 and the majority of TLS3 is comprised of small-radii cylinders. In Figure 10, the segmented planar, pole-like, and rough regions are shown in different colors. Visual inspection of the results in Figure 10 indicates that a good segmentation has been achieved. To quantitatively evaluate the quality of such segmentation, the previously-discussed QC measures are used to denote the frequency of detected artifacts.

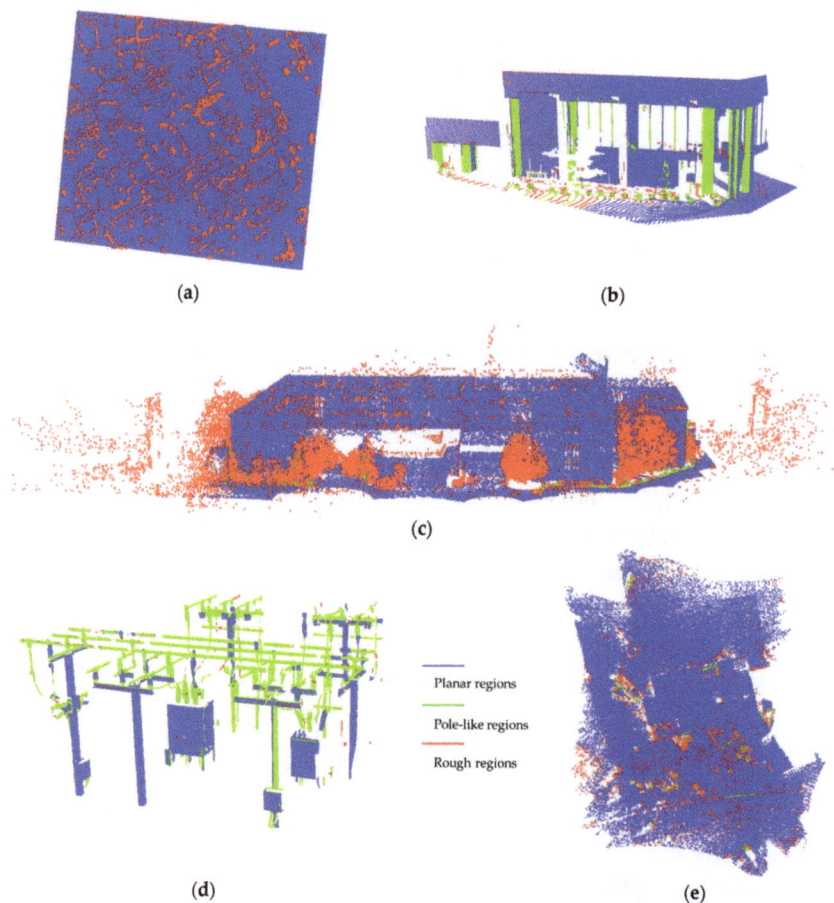

Figure 9. Perspective views of the classified point clouds for the ALS (**a**); TLS1 (**b**); TLS2 (**c**); TLS3 (**d**); and DIM (**e**) datasets (planar, pole-like, and rough regions are shown in blue, green, and red, respectively).

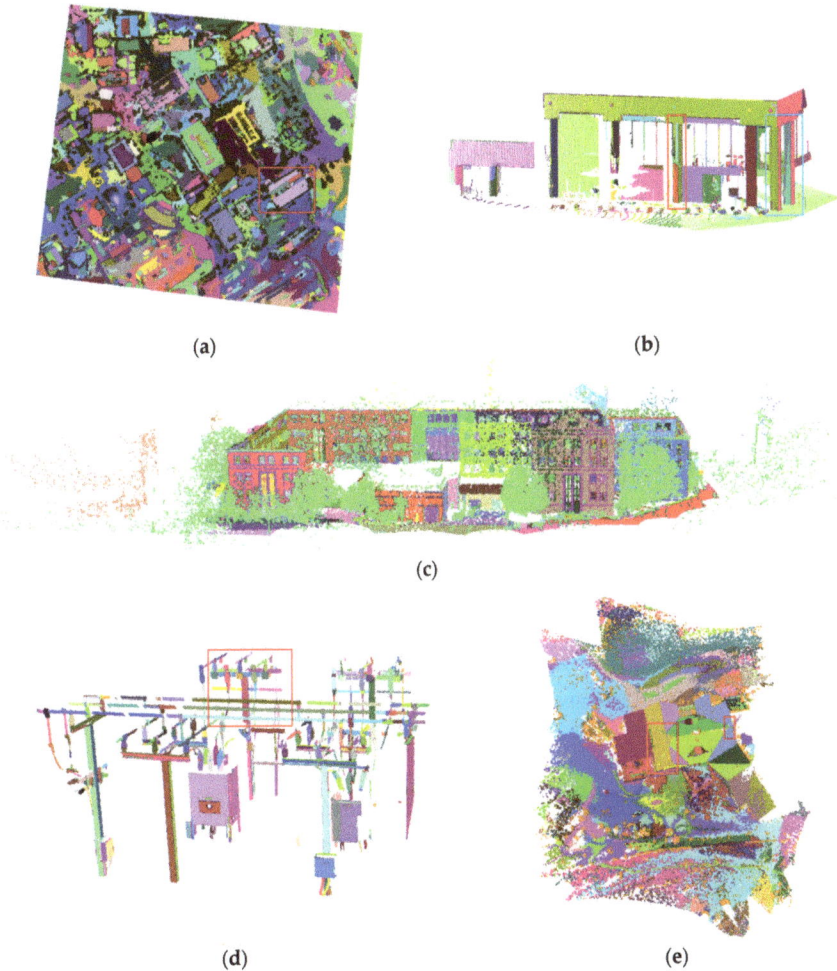

Figure 10. Perspective views of the segmented point clouds for the ALS (**a**); TLS1 (**b**); TLS2 (**c**); TLS3 (**d**); and DIM (**e**) datasets (different segments are shown in different colors).

3.3. Quality Control Outcome

The quality control procedure has been implemented according to the following sequence: QC1) detection and mitigation of single pole-like features that have been misclassified as planar ones, QC2) initial mitigation of interclass competition for neighboring points, QC3) detection and mitigation of over-segmentation problems, QC4) detection and mitigation of intraclass competition for neighboring points, QC5) detection and mitigation of fully/partially misclassified pole-like features, and finally QC6) detection and mitigation of fully/partially misclassified planar features. One

63

should note that for QC3, the respective over-segmentation measure is evaluated as the ration between the merged segments in a given class relative to the total number of segments in that class. Figure 11, presents the segmentation results following these QC procedures. For TLS1 and TLS3, segmentation results for planar and pole-like features are presented separately since those datasets have significant portions that pertain to such classes. Figure 12 illustrated examples of the detected/mitigated problems through the different QC measures. More specifically, Figure 12a shows portions of a cylindrical column, as highlighted by the red rectangle, that have been originally classified as planar regions and after QC1, they have been correctly reclassified as pole-like features. Figure 12b shows examples of points from other classes, in red, that have been incorporated into planar and pole-like features, in yellow, after implementing QC2. An example of corrected over-segmentation of pole like features after QC3 is illustrated in Figure 12c (compare the segmentation results in Figures 10b and 12c). Detection and mitigation of intraclass competition for neighboring points after QC4 is shown in Figure 12d (refer to the highlighted regions within the red rectangles before and after QC4). The results of mitigating fully/partially misclassified linear regions after QC5 are shown in Figure 12e (refer to the results after the over-segmentation in Figure 12c and those in Figure 12e, where one can see the correct mitigation of partially-misclassified pole-like features). Finally, Figure 12f shows an example of the segmentation results after applying QC6 that identifies/corrects partially/fully misclassified planar features (compare the results in Figures 10b and 12f). The proposed QC procedures provide quantitative measures that indicate the frequency of the segmentation problems. Such quantitative measures are presented in Table 3, where closer investigation reveals the following:

1. For TLS1 and TLS3, which include significant number of pole-like features, a higher percentage of misclassified single pole-like features (QC1) is observed. TLS1 has pole-like features with larger radii. Therefore, there is higher probability that seed regions along cylindrical features with high point density are misclassified as planar ones. For TLS3, misclassified pole-like features are caused by having several thin beams in the dataset.

2. For interclass competition for neighboring points (QC2), airborne datasets with predominance of planar features have higher percentage of $QC_{interclass\ competition}\ (planar)$—refer to the results for the ALS and DIM datasets. On the other hand, $QC_{interclass\ competition}\ (pole-like)$ has higher percentages in datasets that have significant portions belonging to cylindrical features (*i.e.*, TLS1 and TLS3).

3. Due to inherent noise in the datasets as well as the strict normal distance thresholds as defined by the derived a-posteriori variance factor, over-segmentation problems (QC4) are present. In this regard, one should note

that over-segmentation problems are easier to handle than under-segmentation ones, which could arise from relaxed normal-distance thresholds.

4. Intraclass competition for neighboring points (QC4) are quite minimal. This is evident by the reported low percentages for this category.

5. For partially/misclassified pole-like features, higher percentages of QC5 when dealing with low number of points in such classes is not an indication of a major issue in the segmentation procedure (e.g., QC5 for ALS and DIM where the percentages of the points that belong to pole-like feature are almost 0% and 4%, respectively).

6. For partially/misclassified planar features, higher percentages of QC6 should be expected when dealing with datasets that have pole-like features with large radii or several interconnected linear features that are almost coplanar (this is the case for TLS1 and TLS3, respectively).

(a)

(b)

(c)

(d)

(e)

Figure 11. *Cont.*

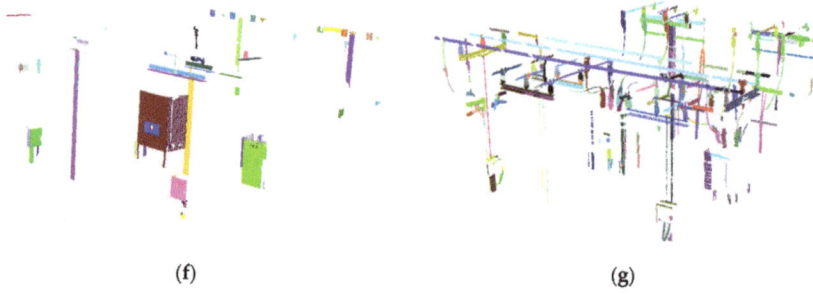

(f)   **(g)**

Figure 11. Perspective views of the segmented point clouds after the quality control procedure for the ALS–planar (**a**); DIM–planar (**b**); TLS1–planar (**c**); TLS1–pole-like (**d**); TLS2–planar (**e**); TLS3–planar (**f**); and TLS3–pole-like (**g**) datasets—different segments are shown in different colors.

Table 3. QC measures for the different datasets.

		ALS	TLS1	TLS2	TLS3	DIM
QC1	$n_{reclassified\ lines}$ / $n_{plane\ total}$ / $QC_{reclassified\ linear\ feaure}$	101/ 716,628/ ≈0.000	5,439/ 123,370/ 0.044	402/ 126,193/ 0.003	25,484/ 224,635/ 0.113	71/ 211,553/ ≈0.000
QC2 Planar	$n_{incorporated}$ / $n_{potential\ candidates}$ / $QC_{interclass\ competition}$	31,700/ 96,453/ 0.328	5,457/ 52,365/ 0.104	2,788/ 76,055/ 0.036	24,469/ 256,016/ 0.095	3,991/ 18,952/ 0.210
QC2 Pole-like	$n_{incorporated}$ / $n_{potential\ candidates}$ / $QC_{interclass\ competition}$	0/ 812,879/ 0	22,193/ 123,042/ 0.180	4,379/ 194,014/ 0.022	29,340/ 208,937/ 0.140	5,198/ 227,486/ 0.022
QC3 Planar	$n_{merged\ planar}$ / $n_{segmented\ planar}$ / $QC_{over\ segmentation}$	618/ 801/ 0.771	23/ 59/ 0.389	278/ 367/ 0.757	8/ 86/ 0.093	163/ 195/ 0.835
QC3 Pole-like	$n_{merged\ linear}$ / $n_{segmented\ linear}$ / $QC_{over\ segmentation}$	0/ 4/ 0	21/ 113/ 0.185	8/ 144/ 0.055	152/ 430/ 0.353	38/ 55/ 0.69
QC4	$n_{invading}$ / $n_{plane\ total}$ / $QC_{intraclass\ competition}$	21,690/ 748,227/ 0.028	857/ 123,388/ 0.006	4,521/ 128,579/ 0.035	5,427/ 223,620/ 0.024	3,381/ 215,473/ 0.015
QC5	$n_{encompassed\ line\ points}$ / $n_{line\ total}$ / $QC_{misclassified\ pole-like}$	101/ 101/ 1	5,841/ 69,447/ 0.084	1,866/ 12,211/ 0.152	9,748/ 275,570/ 0.035	3,427/ 8,146/ 0.420
QC6	$n_{misclassified\ planar}$ / $n_{plane\ total}$ / $QC_{misclassified\ planar}$	N/A	29,647/ 123,388/ 0.240	N/A	73,187/ 223,620/ 0.327	N/A

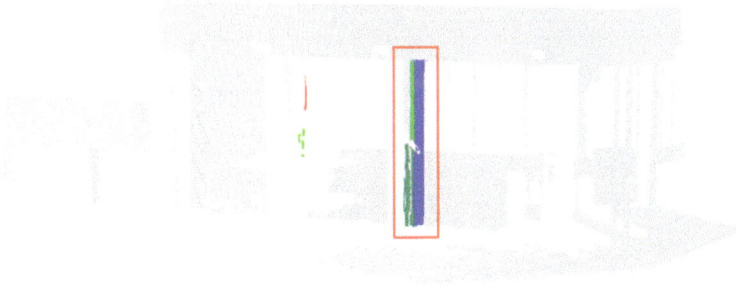

(a)

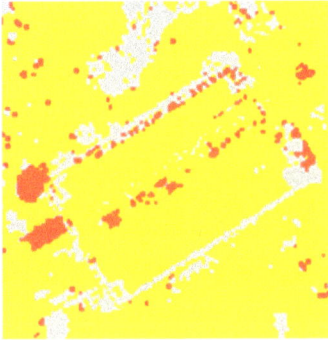 

(b)

(c)

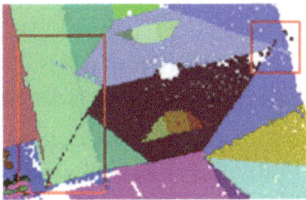

(d)

Figure 12. *Cont.*

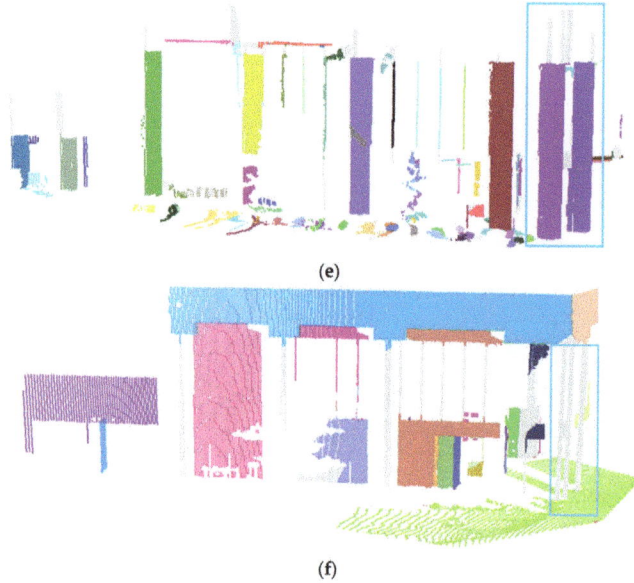

(e)

(f)

Figure 12. Examples of improved segmentation quality by the different QC measures. (**a**) After QC1: reclassified pole-like features; (**b**) After QC2: Interclass competition (planar and pole-like); (**c**) After QC3: Over-segmentation (pole-like); (**d**) Before and after QC4: Intraclass competition (planar); (**e**) After QC5: Misclassified pole-like; (**f**) After QC6: Misclassified planar.

4. Conclusions and Recommendations for Future Work

Segmentation of point clouds into planar, pole-like, and rough regions is the first step in the data-processing chain for object modeling. This paper presents a region-growing segmentation procedure that simultaneously identify planar, pole-like, and rough features in point clouds while considering variations in LPS and noise level. In addition to these characteristics, the proposed region-growing segmentation starts from optimally-selected seed regions that are sorted according to their quality of fit to the LSA-based parametric representation of pole-like and planar regions. Given that segmentation artifacts cannot be avoided, a QC methodology is introduced to consider possible problems arising from the sequential-segmentation procedure (*i.e.*, possible invasion of earlier-segmented regions to later-segmented ones) and possible competition between the different segments for neighboring points. The main advantages of the proposed QC procedure include: (1) It does not need reference data; (2) It provides quantitative estimate of the frequency of detected instances of hypothesized problems; and (3) It encompasses a mitigation mechanism that eliminate instances of such problems. In summary, the proposed processing framework tries to optimize the segmentation procedure and at the same time, potential artifacts are detected, quantified, and mitigated.

To illustrate the performance of the segmentation and quality control procedures, we conducted experimental results using real datasets from airborne and terrestrial laser scanners as well as image-based point clouds. The segmentation results have been proven to be quite reliable while relying on few thresholds that could be easily established. Moreover, the QC procedure has been successful in detecting and eliminating possible problems that could be present in the segmentation results.

Future research will be focusing on establishing additional constraints to ensure even more reliable selection of seed regions. In addition, color/intensity information after accurate geometric and radiometric sensor calibration will be used to improve the segmentation results. We will be also considering other segmentation problems that could be mitigated through improved QC procedures. Finally, the outcome from the segmentation and QC procedures will be used to make hypotheses regarding the generated segments (e.g., building rooftops, building façades, light poles, road surfaces, trees, and bushes).

Acknowledgments: The authors acknowledge the financial support of the Lyles School of Civil Engineering and Faculty of Engineering at Purdue University of this research work. We are also thankful to the members of the Digital Photogrammetry Research Group at Purdue University who helped in the data collection for the experimental results.

Author Contributions: A. Habib and Y.J. Lin conceived and designed the proposed methodology and conducted experiments. Y.J. Lin implemented the methodology and analyzed the experiments under A. Habib's supervision. The manuscript was written by A. Habib.

Conflicts of Interest: The authors declare no conflicts of interest.

References

1. Gonzalez-Aguilera, D.; Crespo-Matellan, E.; Hernandez-Lopez, D.; Rodriguez-Gonzalvez, P. Automated urban analysis based on LiDAR-derived building models. *IEEE Trans. Geosci. Remote Sens.* **2013**, *51*, 1844–1851.

2. Golparvar-Fard, M.; Balali, V.; de la Garza, J.M. Segmentation and recognition of highway assets using image-based 3D point clouds and semantic Texton forests. *J. Comput. Civ. Eng.* **2012**, *29*, 04014023.

3. Alshawabkeh, Y.; Haala, N. Integration of digital photogrammetry and laser scanning for heritage documentation. *Int. Arch. Photogramm. Remote Sens.* **2004**, *35*, B5.

4. Gruen, A. Development and status of image matching in photogrammetry. *Photogramm. Rec.* **2012**, *27*, 36–57.

5. Pratt, W.K. *Digital Image Processing*; Wiley: New, York, NY, USA, 1978.

6. Canny, J. A computational approach to edge detection. *IEEE Trans. Pattern Anal. Mach. Intell.* **1986**, *PAMI-8*, 679–698.

7. Ballard, D.H. Generalizing the Hough transform to detect arbitrary shapes. *Pattern Recognit.* **1981**, *13*, 111–122.

8. Hirschmüller, H. Accurate and efficient stereo processing by semi-global matching and mutual information. In Proceedings of the 2005 IEEE Computer Society Conference on Computer Vision and Pattern Recognition (CVPR '05), San Diego, CA, USA, 20–25 June 2005; pp. 807–814.

9. Hirschmüller, H. Stereo processing by semiglobal matching and mutual information. *IEEE Trans. Pattern Anal. Mach. Intell.* **2008**, *30*, 328–341. PubMed]

10. Haala, N. The landscape of dense image matching algorithms. In *Photogrammetric Week' 13*; Fritsch, D., Ed.; Wichmann: Stuttgart, Germany, 2013; pp. 271–284.

11. Hyyppä, J.; Wagner, W.; Hollaus, M.; Hyyppä, H. Airborne laser scanning. In *SAGE Handbook of Remote Sensing*; SAGE: New York, NY, USA, 2009; pp. 199–211.

12. Sithole, G.; Vosselman, G. Experimental comparison of filter algorithms for bare-Earth extraction from airborne laser scanning point clouds. *ISPRS J. Photogramm. Remote Sens.* **2004**, *59*, 85–101.

13. Liu, X. Airborne LiDAR for DEM generation: Some critical issues. *Prog. Phys. Geogr.* **2008**, *32*, 31–49.

14. Habib, A.F.; Zhai, R.; Kim, C. Generation of complex polyhedral building models by integrating stereo-aerial imagery and lidar data. *Photogramm. Eng. Remote Sens.* **2010**, *76*, 609–623.

15. Kwak, E.; Habib, A. Automatic representation and reconstruction of DBM from LiDAR data using Recursive Minimum Bounding Rectangle. *ISPRS J. Photogramm. Remote Sens.* **2014**, *93*, 171–191.

16. El-Halawany, S.; Moussa, A.; Lichti, D.D.; El-Sheimy, N. Detection of road curb from mobile terrestrial laser scanner point cloud. In Proceedings of the ISPRS Workshop on Laserscanning, Calgary, AB, Canada, 29–31 August 2011.

17. Qiu, R.; Zhou, Q.-Y.; Neumann, U. Pipe-Run Extraction and Reconstruction from Point Clouds. In *Computer Vision–ECCV 2014*; Springer: Cham, Switzerland, 2014; pp. 17–30.

18. Son, H.; Kim, C.; Kim, C. Fully automated as-built 3D pipeline extraction method from laser-scanned data based on curvature computation. *J. Comput. Civ. Eng.* **2014**, *29*, B4014003.

19. Becker, S.; Haala, N. Grammar supported facade reconstruction from mobile lidar mapping. In Proceedings of the ISPRS Workshop, CMRT09-City Models, Roads and Traffic, Paris, France, 3–4 September 2009.

20. Dimitrov, A.; Golparvar-Fard, M. Segmentation of building point cloud models including detailed architectural/structural features and MEP systems. *Autom. Constr.* **2015**, *51*, 32–45.

21. Rottensteiner, F.; Briese, C. A new method for building extraction in urban areas from high-resolution LIDAR data. *Int. Arch. Photogramm. Remote Sens. Spat. Inf. Sci.* **2002**, *34*, 295–301.

22. Forlani, G.; Nardinocchi, C.; Scaioni, M.; Zingaretti, P. Complete classification of raw LIDAR data and 3D reconstruction of buildings. *Pattern Anal. Appl.* **2006**, *8*, 357–374.

23. Vosselman, G.; Gorte, B.G.; Sithole, G.; Rabbani, T. Recognising structure in laser scanner point clouds. *Int. Arch. Photogramm. Remote Sens. Spat. Inf. Sci.* **2004**, *46*, 33–38.

24. Rabbani, T.; van den Heuvel, F.; Vosselmann, G. Segmentation of point clouds using smoothness constraint. *Int. Arch. Photogramm. Remote Sens. Spat. Inf. Sci.* **2006**, *36*, 248–253.

25. Al-Durgham, M.; Habib, A. A framework for the registration and segmentation of heterogeneous lidar data. *Photogramm. Eng. Remote Sens.* **2013**, *79*, 135–145.

26. Yang, B.; Dong, Z. A shape-based segmentation method for mobile laser scanning point clouds. *ISPRS J. Photogramm. Remote Sens.* **2013**, *81*, 19–30.

27. Wang, J.; Shan, J. Segmentation of LiDAR point clouds for building extraction. In Proceedings of the American Society for Photogrammetry and Remote Sensing Annual Conference, Baltimore, MD, USA, 9–13 March 2009; pp. 9–13.

28. Awwad, T.M.; Zhu, Q.; Du, Z.; Zhang, Y. An improved segmentation approach for planar surfaces from unstructured 3D point clouds. *Photogramm. Rec.* **2010**, *25*, 5–23.

29. Lari, Z.; Habib, A. An adaptive approach for the segmentation and extraction of planar and linear/cylindrical features from laser scanning data. *ISPRS J. Photogramm. Remote Sens.* **2014**, *93*, 192–212.

30. Filin, S.; Pfeifer, N. Segmentation of airborne laser scanning data using a slope adaptive neighborhood. *ISPRS J. Photogramm. Remote Sens.* **2006**, *60*, 71–80.

31. Haralock, R.M.; Shapiro, L.G. *Computer and Robot Vision*; Addison-Wesley Longman Publishing Co., Inc.: Boston, MA, USA, 1991.

32. Biosca, J.M.; Lerma, J.L. Unsupervised robust planar segmentation of terrestrial laser scanner point clouds based on fuzzy clustering methods. *ISPRS J. Photogramm. Remote Sens.* **2008**, *63*, 84–98.

33. Pu, S.; Vosselman, G. Automatic extraction of building features from terrestrial laser scanning. *Int. Arch. Photogramm. Remote Sens. Spat. Inf. Sci.* **2006**, *36*, 25–27.

34. Heipke, C.; Mayer, H.; Wiedemann, C.; Jamet, O. Evaluation of automatic road extraction. *Int. Arch. Photogramm. Remote Sens.* **1997**, *32*, 151–160.

35. Rutzinger, M.; Rottensteiner, F.; Pfeifer, N. A comparison of evaluation techniques for building extraction from airborne laser scanning. *IEEE J. Sel. Top. Appl. Earth Obs. Remote Sens.* **2009**, *2*, 11–20.

36. Belton, D. Classification and Segmentation of 3D Terrestrial Laser Scanner Point Clouds. Ph.D. Thesis, Curtin University of Technology, Bentley, Australia, 2008.

37. Nurunnabi, A.; Belton, D.; West, G. Robust segmentation for multiple planar surface extraction in laser scanning 3D point cloud data. In Proceedings of the IEEE 2012 21st International Conference on Pattern Recognition (ICPR), Tsukuba, Japan, 11–15 November 2012; pp. 1367–1370.

38. Lari, Z.; Habib, A. New approaches for estimating the local point density and its impact on LiDAR data segmentation. *Photogramm. Eng. Remote Sens.* **2013**, *79*, 195–207.

39. Shewchuk, J.R. Triangle: Engineering a 2D quality mesh generator and Delaunay triangulator. In *Applied Computational Geometry towards Geometric Engineering*; Springer: Berlin, Germany; Heidelberg, Germany, 1996; pp. 203–222.

40.   Priestnall, G.; Jaafar, J.; Duncan, A. Extracting urban features from LiDAR digital surface models. *Comput. Environ. Urban Syst.* **2000**, *24*, 65–78.

41.   Reitberger, J.; Schnörr, C.; Krzystek, P.; Stilla, U. 3D segmentation of single trees exploiting full waveform LIDAR data. *ISPRS J. Photogramm. Remote Sens.* **2009**, *64*, 561–574.

42.   Bentley, J.L. Multidimensional binary search trees used for associative searching. *Commun. ACM* **1975**, *18*, 509–517.

43.   Moore, I.D.; Grayson, R.B.; Ladson, A.R. Digital terrain modelling: A review of hydrological, geomorphological, and biological applications. *Hydrol. Process.* **1991**, *5*, 3–30.

44.   Pulli, K.; Duchamp, T.; Hoppe, H.; McDonald, J.; Shapiro, L.; Stuetzle, W. Robust meshes from multiple range maps. In Proceedings of the IEEE International Conference on Recent Advances in 3-D Digital Imaging and Modeling, Ottawa, ON, Canada, 12–15 May 1997; p. 205.

45.   Shaw, P.J. *Multivariate Statistics for the Environmental Sciences*; Arnold: London, UK, 2003.

46.   He, F.; Habib, A. Linear Approach for Initial Recovery of the Exterior Orientation Parameters of Randomly Captured Images by Low-Cost Mobile Mapping Systems. *ISPRS Int. Arch. Photogramm. Remote Sens. Spat. Inf. Sci.* **2014**, *1*, 149–154.

A Multi-View Dense Point Cloud Generation Algorithm Based on Low-Altitude Remote Sensing Images

Zhenfeng Shao, Nan Yang, Xiongwu Xiao, Lei Zhang and Zhe Peng

Abstract: This paper presents a novel multi-view dense point cloud generation algorithm based on low-altitude remote sensing images. The proposed method was designed to be especially effective in enhancing the density of point clouds generated by Multi-View Stereo (MVS) algorithms. To overcome the limitations of MVS and dense matching algorithms, an expanded patch was set up for each point in the point cloud. Then, a patch-based Multiphoto Geometrically Constrained Matching (MPGC) was employed to optimize points on the patch based on least square adjustment, the space geometry relationship, and epipolar line constraint. The major advantages of this approach are twofold: (1) compared with the MVS method, the proposed algorithm can achieve denser three-dimensional (3D) point cloud data; and (2) compared with the epipolar-based dense matching method, the proposed method utilizes redundant measurements to weaken the influence of occlusion and noise on matching results. Comparison studies and experimental results have validated the accuracy of the proposed algorithm in low-altitude remote sensing image dense point cloud generation.

Reprinted from *Remote Sens.* Cite as: Shao, Z.; Yang, N.; Xiao, X.; Zhang, L.; Peng, Z. A Multi-View Dense Point Cloud Generation Algorithm Based on Low-Altitude Remote Sensing Images. *Remote Sens.* **2016**, *8*, 381.

1. Introduction

With the development of laser scanning and image matching technology, three-dimensional (3D) information has increasingly attracted researchers' attention. Applications of 3D information have extended from digital elevation model (DEM) and digital surface model (DSM) generation to many other fields including archeology [1,2], topographic monitoring [3,4], facial geometry and dynamic capture [5,6], cultural heritage protection [7,8], forest and agriculture modeling [9,10], and medical treatment [11]. Since laser scanning can produce highly accurate, reliable, dense, and more integrated 3D point clouds of objects [12], it has been utilized as the preferred technology for 3D modeling over the last two decades. In recent years, with the significant progress of photogrammetry and computer vision technology, image-based 3D reconstruction stands as a major competitor against laser scanning [13]. Compared with laser scanning, the advantages of image-based 3D reconstruction are that:

- Images can be accepted from any type of camera [14], including calibrated or uncalibrated images, images taken from smartphones or tablets [15], images captured from digital cameras or frames intercepted from video streams [16];
- It is low in cost;
- Point cloud data contains color information; and
- Theoretically, it may produce much denser point clouds [17].

In numerous photographic platforms, low-altitude remote sensing images have been considered a popular data source for large-scale 3D modeling [18]. In addition to sub-decimeter high-resolution imagery [19], a low-altitude remote sensing platform also has several advantages including: flexibility, low cost, simplicity of operation, and ease of maintenance [20].

This paper proposes a multi-view dense point cloud generation algorithm based on low-altitude remote sensing images. The proposed method exploited Patch-based Multi View Stereo (PMVS) [21] results as a seed point cloud. It took advantage of pixels in image windows and object points on patches to expand the seed point cloud. Then, it utilized multi-image projection relationships to improve the accuracy of the point cloud. In summary, the purpose of this paper is a new approach that takes advantage of redundant measurements of multi-images and generates a much denser point cloud than MVS.

The remainder of the paper is structured as follows: related works are presented and compared with each other in Section 2; in Section 3, the proposed method is introduced in detail; in Section 4, experiments are conducted to verify the feasibility of the proposed algorithm in terms of reliability and matching accuracy; and finally, conclusions are stated in Section 5.

2. Related Works

The theory of stereo matching was first investigated in the early mid-1970s [22] and underwent extensive development in the 1990s [17]. During those 10 years, a large number of high accuracy matching applications and commercial photogrammetric systems appeared for digital surface model (DSM) and digital terrain model (DTM) generation from aerial images. In the last decade, image-based 3D reconstruction approaches have been further advanced by recent developments in computer vision and photogrammetry. Additionally, the data source of images has been extended from satellite aerial images to generic photos, such as those taken on mobile photos.

2.1. Two-Frame Dense Matching in Photogrammetry

Since the advent of stereo matching, the derivation of ground object point coordinates from corresponding image pixels has become one of the most key issues in the domain of photogrammetry and remote sensing [23,24]. With the advances of hardware and innovative image matching algorithms, photogrammetry-based 3D modeling can deliver results in a reasonable amount of time. Some researchers have focused on how to utilize photogrammetry technology to produce relatively sparse seed points [25,26], while others have sought to take advantage of the corresponding epipolar lines between two corresponding images to perform pixel-wise dense matching [27–29]. In 2002, Scharstein and Szeliski [27] introduced a taxonomy and evaluation of two-frame stereo dense matching algorithms, dividing it into four primary steps:

- Matching cost computation;
- Cost (support) aggregation;
- Disparity computation/optimization; and
- Disparity refinement.

Based on the implementation employed in the cost (support) aggregation step, dense matching can be divided into two categories: local algorithms and global algorithms. Local algorithms connect the matching costs within a local neighborhood and select the lowest matching cost as a disparity [30], that is "winner takes all". Global algorithms typically define a global energy function which includes a data term and a smoothness term acting on the whole image instead of local cost aggregations [31]. Since the local algorithm uses a part of the local neighborhood for calculations, the processing speed of the local algorithm is faster, and due to the global algorithms taking into account the whole image in processing, the matching accuracy of the global algorithms is greater. Hirschmüller [32] employed Semi-Global Matching (SGM) which integrates the advantages of local and global algorithms and further improved the efficiency and accuracy of dense matching. Despite these advantages, the two-frame method could not evade the key problem that without the redundant measurements, two-frame dense matching was not robust to the noise and occlusion, and the accuracy of the point cloud reconstructed by two-frame matching is inferior to that of multi-view stereo [33].

2.2. Multi-View Stereo in Computer Vision

With the development of a number of different low-cost and open-source software systems, the multi-view stereo method is becoming one of the most popular subjects in computer vision. Multi-view stereo can use redundant information to weaken the influence of occlusion and noise. From the Middlebury evaluation supplied by Seitz *et al.* [34], for a single object or small-scale sense reconstruction, multi-view reconstruction can provide a first-rate result which is comparable to the point cloud obtained from laser scanning. Since the Structure from Motion (SFM) method makes it possible for disordered image calibration, multi-view stereo quickly extends from photogrammetric images to generic photos, even those downloaded from the Internet or captured from mobile phones [35]. Recently, the challenges of multi-view stereo have focused on the following aspects:

- dynamic capture;
- 3D reconstruction from video streams; and
- 3D reconstruction for large-scale scenes.

For large-scale scene reconstruction, although there are plenty of efforts devoted to making point cloud data denser and more accurate, the density and accuracy of the result cannot substitute the laser scanning point cloud. Since low-altitude remote sensing images have many advantages such as flying under the cloud, low cost and fast response, *etc.*, this article focuses on how to apply low-altitude remote sensing images to reconstruct large-scale scenes.

3. Method

The proposed method can be divided into four steps: (1) a PMVS point cloud generation; (2) patch-based point cloud expansion; (3) point cloud optimization; and (4) an outliers filter. In this section, details of the proposed method are introduced. The principle of this algorithm is illustrated in Figure 1. As shown in Figure 1, the proposed method derives from a technique where growing regions start from a set of seed points or patches [36]. The result of PMVS is a set of patches, and the geometric significance of the patch is a local tangent plane of the object. The proposed algorithm utilizes these results as seed points and takes advantage of projection rules between image pixels and patches to segment the generated patches to expand denser patches. Then, a patch-based Multiphoto Geometrically Constrained Matching (MPGC) algorithm is used to optimize the expanded patches to obtain a more accurate result. Finally, a density constraint [37] is employed to filter the outliers.

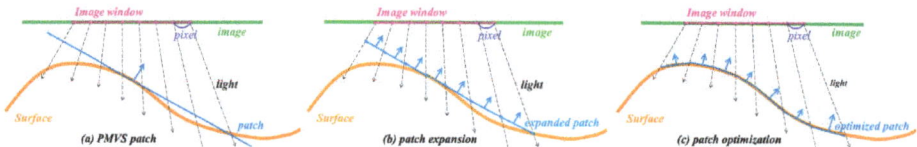

Figure 1. Diagrammatic sketch of the multi-view dense point cloud generation algorithm. (**a**) The result of the seed patch generated from PMVS; (**b**) The expanded patch from the PMVS patch; (**c**) The optimized patch to improve accuracy.

3.1. PMVS Point Cloud Generation

In recent years, many researchers have focused on using MVS to reconstruct large-scale 3D scenes. PMVS is accepted as one of the most popular MVS algorithms due to its accuracy and completeness [8]. By utilizing (1) initial feature matching; (2) patch expansion; and (3) patch filtering, PMVS generates and propagates a semi-dense set of patches [38]. In contrast to a feature-based algorithm, the seed points generated by PMVS have three advantages:

- Much denser: seed points obtained in feature-based matching are expanded in the second step of PMVS;
- Evenly distributed: the PMVS algorithm attempted to reconstruct at least one patch in each image cell with $\beta \times \beta$ pixels;
- More accurate: a Nelder-Mead method [39] was utilized in the PMVS algorithm to refine each patch in the reconstruction model and filter outliers in the last step.

3.2. Patch-Based Point Cloud Expansion

The goal of the expansion step is to expand the seed patch and increase the point cloud density. PMVS attempted to grow a patch starting from a seed matching pixels, and expanding to the neighbor image cells in the visible images until each corresponding image cell reconstructed at least one point. The proposed method utilizes the projection rules to segment the patches into small pieces. Each piece contains one center point, the seed point is growing on the patch and the point cloud is denser.

The result of PMVS records each point in the point cloud with its coordinates (X_c, Y_c, Z_c), color (R, G, B) and normal vector (a, b, c). By projecting the object point $P(X_c, Y_c, Z_c)$ on each image, the image point coordinate $p_i(x_i, y_i)$ (i is the image index) is calculated. Since the distance between the image point and the origin of the image coordinate system is shorter, the projection distortion is smaller, and the proposed method supposes image $I(R)$ as a reference image when the image $I(R)$ is satisfied by:

$$\sqrt{x_R^2 + y_R^2} \leqslant \sqrt{x_i^2 + y_i^2}(i = 1, 2...n, \quad i \neq R) \tag{1}$$

Supposing (X_c, Y_c, Z_c) is the center of the patch, and (a, b, c) is the normal vector, the local tangent plane (*patch* in PMVS) at $P(X_c, Y_c, Z_c)$ is:

$$P: \quad a(X - X_c) + b(Y - Y_c) + c(Z - Z_c) = 0 \tag{2}$$

As illustrated in Figure 2a, the image point $p_i(x_i, y_i)$ is the center of the image window, where the window size is $\mu \times \mu$ pixels. By projecting the image window onto the patch, $\mu \times \mu$ object points are obtained. Theoretically, the density of the point cloud could expand $\mu \times \mu$ times. The overall algorithm description for this step is given in Figure 2b. The result patch P' consists of the coordinates (X, Y, Z), normal vector (a, b, c) and reference image index R.

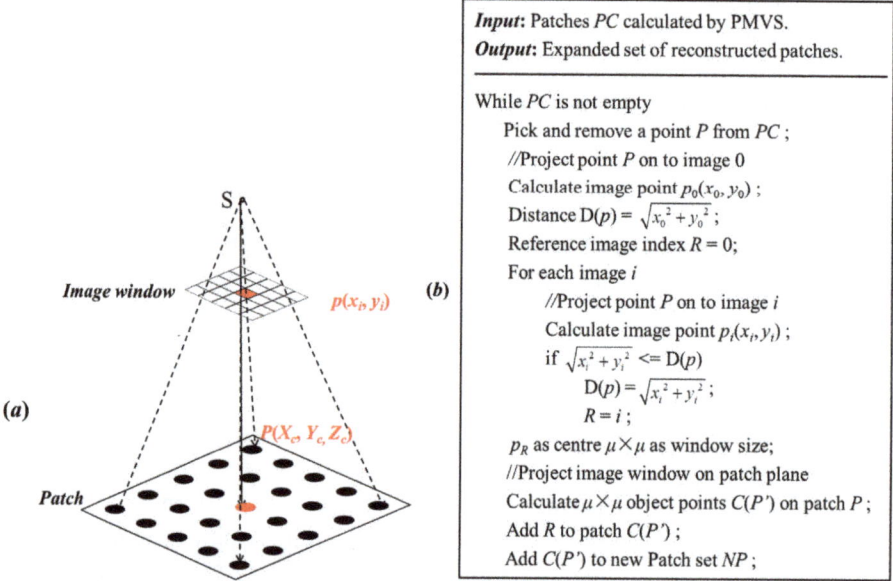

Input: Patches *PC* calculated by PMVS.
Output: Expanded set of reconstructed patches.

While *PC* is not empty
    Pick and remove a point *P* from *PC* ;
    //Project point *P* on to image 0
    Calculate image point $p_0(x_0, y_0)$;
    Distance $D(p) = \sqrt{x_0^2 + y_0^2}$;
    Reference image index $R = 0$;
    For each image *i*
        //Project point *P* on to image *i*
        Calculate image point $p_i(x_i, y_i)$;
        if $\sqrt{x_i^2 + y_i^2} <= D(p)$
            $D(p) = \sqrt{x_i^2 + y_i^2}$;
            $R = i$;
    p_R as centre $\mu \times \mu$ as window size;
    //Project image window on patch plane
    Calculate $\mu \times \mu$ object points $C(P')$ on patch *P* ;
    Add *R* to patch $C(P')$;
    Add $C(P')$ to new Patch set *NP* ;

Figure 2. (**a**) Projection relationship between pixels (grids) in image window and object points (dots) in patch; (**b**) Process of patch-based point cloud expansion algorithm.

3.3. Patch-Based MPGC to Optimize the Point Cloud

PMVS utilized the projection relationship between the patch and the corresponding images to build a function to find the optimal matching pixel:

$$f(z, \alpha, \beta) = \frac{1}{n} \sum_{i=1}^{n} (1 - f_i) \tag{3}$$

In the function above, *i* is the index of the visible images (in PMVS, if patch *p* is visible in image *i*, *i* is considered as a visible image of *p*); *n* is the number of the visible images; f_i is a function that denotes the Normalized Cross-correlation Coefficient (NCC) between corresponding image windows which is obtained by the patch projecting to the reference image (I_0) and visible images (I_i);

$$f_i(z, \alpha, \beta) = NCC(I_0, I_i) \tag{4}$$

78

z is the distance of the patch center moving along the ray; (α, β) are the direct angle of the normal vector (a, b, c). The optimization process employed the Nelder-Mead method [39] to calculate the minimum value of Function (3). From the result of the calculation, the optimal patch (denoted by its center point P' and normal vector (a, b, c)) is obtained:

$$P' = P + z \cdot norm(\overrightarrow{OP}) \tag{5}$$

$$\begin{aligned} a &= \cos \alpha \cos \beta \\ b &= \sin \alpha \cos \beta \\ c &= \sin \beta \end{aligned} \tag{6}$$

As with the optimization method in PMVS, the proposed method also introduces a patch in the optimization step to obtain a better initial value of the optimization function. In the 1990s, Baltsavias [40,41] introduced epipolar line constraints (collinear equation) to Least Square Image Matching (LSM) [42,43] and proposed an extremely useful application named Multi-photo Geometrically Constrained Matching (MPGC). This approach simultaneously derives the accurate coordinates of corresponding object points in the object space coordinate system during the image matching process. It has been widely applied to refine matching results in a three-dimensional reconstruction [25,26,44,45]. The proposed method utilizes a modified MPGC algorithm to optimize the point cloud.

In the traditional LSM method, each pixel in the matching image window is used to build an error equation:

$$v = dh_{0i} + g_i (x_i, y_i) \cdot dh_{1i} + h_{1i} \left(\frac{\partial g_i}{\partial x_i} dx_i + \frac{\partial g_i}{\partial y_i} dy_i \right) - (g_0 (x_0, y_0) - h_{0i} - h_{1i} \cdot g_i (x_i, y_i)) \tag{7}$$

In the error equation above, v is the projection error; h_{0i} and h_{1i} are the radiation distortion coefficients between the reference image and search image i. In the experiments, the initial values of h_{0i} and h_{1i} are usually 0 and 1, respectively. Further, dh_{0i} and dh_{1i} are corrections of parameter h_{0i} and h_{1i}; $g_0(x_0, y_0)$ is the pixel intensity values in the image window of the reference image; $g_i(x_i, y_i)$ is the pixel intensity values of image points (x_i, y_i) in the search image window; ($\partial g_i / \partial x_i$, $\partial g_i / \partial y_i$) is the derivative values of pixel intensity in the x and y directions; (dx_i, dy_i) is the correction values of the image points (x_i, y_i). Therefore, in a matching of the $\mu \times \mu$ pixels image window, the $\mu \times \mu$ error equations can be listed; if $\mu \times \mu$ is larger than the unknown, using least square adjustment, the corresponding pixels (x_i, y_i) can be calculated.

MPGC applied epipolar line constraints to the LSM method, and the coordinates of (x_i, y_i) can be denoted by the interior (x_s, y_s, f) and exterior parameters (projection center $S(X_s, Y_s, Z_s)$, rotation matrix $(a_1, a_2, a_3; b_1, b_2, b_3; c_1, c_2, c_3))$ of image i and the corresponding object point (X, Y, Z):

$$x_i - x_s = -f\frac{a_1(X-X_s)+b_1(Y-Y_s)+c_1(Z-Z_s)}{a_3(X-X_s)+b_3(Y-Y_s)+c_3(Z-Z_s)}$$
$$y_i - y_s = -f\frac{a_2(X-X_s)+b_2(Y-Y_s)+c_2(Z-Z_s)}{a_3(X-X_s)+b_3(Y-Y_s)+c_3(Z-Z_s)} \tag{8}$$

Applying the collinear Equation (8) to the LSM error Equation (7), the optimal object point coordinate can be directly obtained during the process of least square adjustment.

However, despite the fact that MPGC performs well in matching refinement, how to select the initial matching window is still a challenge that has yet to be overcome, because either the accuracy of the result or the efficiency of the process is reliant on the quality of the initial value. The proposed method introduces the patch to MPGC to refine the point cloud. By using the patch set obtained in Section 3.2 as an initial value and projecting each patch onto the visible images to get the initial matching image windows, these initial matching windows have two superior qualities:

- All pixels which are located at the same place in the image matching window between the reference and search images are approximate corresponding pixels.
- Normal vectors in PMVS results as initial normal vectors of the patch plane, by projecting the patch points onto the images which can significantly decrease the projection deformation.

As with PMVS, the optimization algorithm in the proposed method is based on an individual patch, and each patch P' is optimized separately in the following steps: (1) a matching window is selected in reference image R; (2) the matching window is projected onto the patch plane to calculate the corresponding object points $V(P')$ on patch P'; (3) $V(P')$ is projected onto each image except image R to obtain the corresponding points $w(p_i')$ on the search images; (4) if the matching window $w(p_i')$ is located in the range of image I and the Normalized Cross-correlation Coefficient (NCC) is larger than 0.6, then image i is collected into image set $I(p')$; (5) an error equation is built for each corresponding point in the image window between reference image R and search image set $I(p')$; (6) a least square adjustment is applied to calculate the optimal solution. The overall algorithm description for this step is illustrated in Figure 3.

Input: New Patch set NP.

Output: Optimized set of reconstructed patches.

While NP is not empty

    Pick and remove a point P' from NP ;

    //Project point P' on to image R

    Calculate image point $p_R(x_R, y_R)$;

    p_R as centre $v \times v$ as window size $v(p_R)$;

    //Project image window on patch plane

    Calculate $v \times v$ object points $V(P')$ on patch P'

    For each image $i \mathrel{!=} R$

        //Project point $V(P_i')$ on to image i

        Calculate image point set $w(p_i')$;

        if $w(p_i')$ in image range && $NCC(v(p_R), w(p_i')) >= 0.6$

            Add image i to image set $I(p')$;

            Build Error Equation by $I(p')$;

    LSM adjustment;

    if LSM adjustment succeed

        Add P' to refined point cloud RP ;

Figure 3. Process of point cloud optimization algorithm.

The proposed method uses this patch-based MPGC algorithm to optimize the point cloud instead of the PMVS optimization method for the following reasons:

- Epipolar line constraint is the most strict constraint for a single-center projection, especially when the camera parameters are known;
- Least square adjustment can utilize redundant pixels to decrease the influence of the noise, and has a faster speed in the iterative convergence;
- Radiation distortion is taken into account.

3.4. Outliers Filter

To improve the accuracy and reduce the number of outliers in the point cloud, an erroneous point filter step is a prerequisite. The proposed method makes use of a density constraint [37] in the outliers filter step. A radius of one meter is used to compute the local neighborhood of each point. If the number of neighbor points around a center point is lower than a fixed threshold ε, the center point is considered as an outlier that should be removed. In the method of [37], ε was defined as half of the average neighbor number.

4. Experiments and Discussion

4.1. Input Data Sets

In order to evaluate the performance of the proposed method, three sets of low-altitude images were selected. Each image data set consists of five images. The data sets were captured from Northwestern University (a university in Shaanxi Province, China), Yangjiang (a city in Guangdong Province, China) and Hainan (a province in China), respectively. The parameters of the cameras (parameters of the K-matrix) were acquired from laboratory camera calibration and bundle adjustment. Commercial low altitude photogrammetric processing software called GodWork, which was developed by Wuhan University, was used to perform automatic aero-triangulation to acquire external orientation elements (parameters of the C-matrix and R-matrix) of the images. Detailed parameters of the input data sets are provided in Tables 1–3 and the sample input images used in the experiments are shown in Figure 4.

Table 1. The parameters of the photography from Northwest University (unmanned aerial vehicle images).

Camera Name	Area Size (m × m)	CCD Size (mm)	Image Size (pixel)	Pixel Size (μm)	Focal Length (mm)	Flying Height (m)	Ground Resolution (m)	Number of Images
Canon EOS 400D	415.8 × 339.5	22.16 × 14.77	3888 × 2592	5.7	24	600	0.118	5

Table 2. The parameters of the photography from Yangjiang (aerial image captured at nadir).

Camera Name	Area Size (m × m)	CCD Size (mm)	Image Size (pixel)	Pixel Size (μm)	Focal Length (mm)	Flying Height (m)	Ground Resolution (m)	Number of Images
SWDC-5	417 × 426	49.24 × 36.47	8206 × 6078	6	82	800	0.058	5

Table 3. The parameters of the photography from Hainan (unmanned aerial vehicle images).

Camera Name	Area Size (m × m)	CCD Size (mm)	Image Size (pixel)	Pixel Size (μm)	Focal Length (mm)	Flying Height (m)	Ground Resolution (m)	Number of Images
Canon EOS 5D	981.3 × 1004.4	36 × 24	5616 × 3744	6.4	24	650	0.174	5

| (a) Northwestern University | (b) Yangjiang | (c) Hainan |

Figure 4. Sample input images of all the data sets used in the experiments.
(**a**) Northwestern University; (**b**) Yangjiang; (**c**) Hainan.

4.2. Reconstructed Point Cloud

In the expansion step, expanded patch size μ is the only parameter which has to be set up, because the PMVS algorithm attempts to reconstruct at least one patch in each image cell with $\beta \times \beta$ pixels, where μ is usually less than β. From 1 to β, the density of the point cloud result is increased. Taking into account visualization and running speed, our experiments project an image window with 17×17 pixels on the PMVS patch and one pixel as the project interval. The comparison experiments compared the point cloud reconstructed by PMVS, SURE, Pixel4D and the proposed method. Each input data set experimented in the four comparison methods is exactly the same (same images, same camera parameters and same image parameters). The reconstructed point cloud and details are shown in Figures 5–7.

As illustrated in the figures, due to the proposed method's utilization of the PMVS result as a seed patch, the completeness of the point cloud reconstructed by PMVS and that of the proposed method are almost same. The point cloud reconstructed by the Pix4D software program has a better completeness; the point cloud reconstructed by the SURE software program was the poorest. Although SURE failed in the reconstruction of images with complex texture (*i.e.*, the Yangjiang and Hainan data sets), for relatively simple images (the Northwestern University data set) the density of the point cloud was extremely high. From the cut figures on the right of the figure cells, it can be seen that when compared with the other three methods, the point cloud generated by the proposed method is much denser and contains more details. For instance, much plainer silhouettes and roads in the Northwestern University point cloud, cars parked on the side of the basketball court in the Yangjiang point cloud and much more meticulous roofs in the Hainan point cloud data are extracted. Detailed information of the reconstructed result is illustrated in Tables 4 and 5.

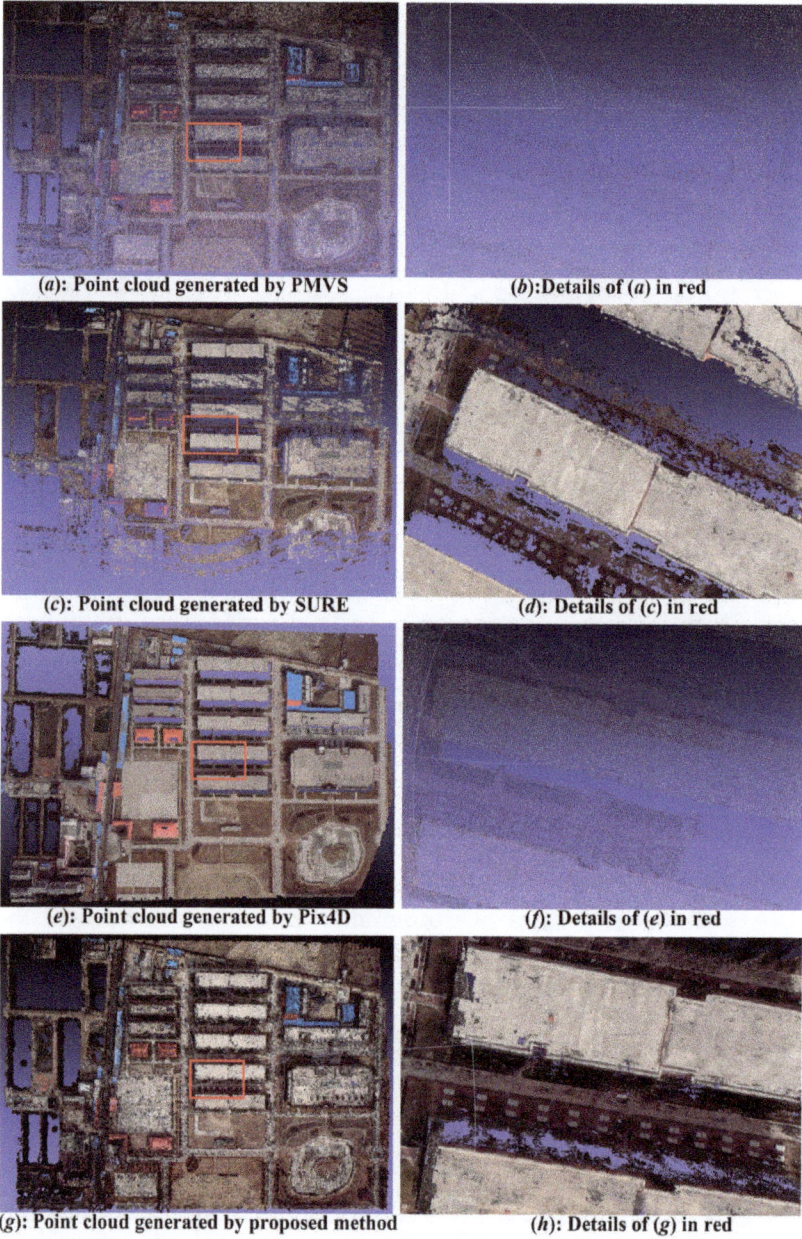

(*a*): Point cloud generated by PMVS
(*b*):Details of (*a*) in red
(*c*): Point cloud generated by SURE
(*d*): Details of (*c*) in red
(*e*): Point cloud generated by Pix4D
(*f*): Details of (*e*) in red
(*g*): Point cloud generated by proposed method
(*h*): Details of (*g*) in red

Figure 5. Examples of reconstructed point cloud with Northwestern University images illustrated by software MeshLab. (**a**) Point cloud generated by PMVS; (**b**) Details of (**a**) in red; (**c**) Point cloud generated by SURE; (**d**) Details of (**c**) in red; (**e**) Point cloud generated by Pix4D; (**f**) Details of (**e**) in red; (**g**) Point cloud generated by proposed method; (**h**) Details of (**g**) in red.

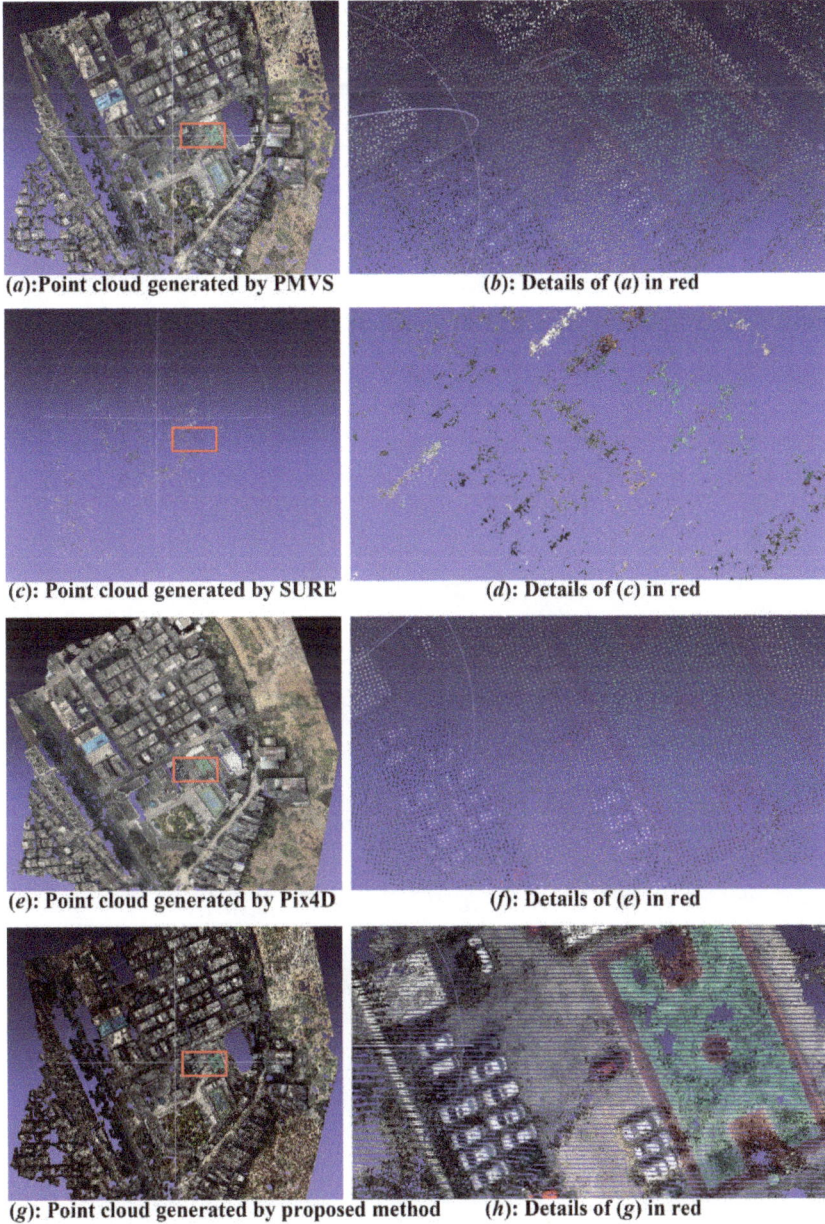

(a):Point cloud generated by PMVS

(b): Details of (a) in red

(c): Point cloud generated by SURE

(d): Details of (c) in red

(e): Point cloud generated by Pix4D

(f): Details of (e) in red

(g): Point cloud generated by proposed method

(h): Details of (g) in red

Figure 6. Examples of reconstructed point cloud with Yangjiang images illustrated by software MeshLab. (**a**) Point cloud generated by PMVS; (**b**) Details of (a) in red; (**c**) Point cloud generated by SURE; (**d**) Details of (c) in red; (**e**) Point cloud generated by Pix4D; (**f**) Details of (e) in red; (**g**) Point cloud generated by proposed method; (**h**) Details of (g) in red.

Figure 7. Examples of reconstructed point cloud with Hainan images illustrated by software MeshLab. (**a**) Point cloud generated by PMVS; (**b**) Details of (**a**) in red; (**c**) Point cloud generated by SURE; (**d**) Details of (**c**) in red; (**e**) Point cloud generated by Pix4D; (**f**) Details of (**e**) in red; (**g**) Point cloud generated by proposed method; (**h**) Details of (**g**) in red.

Table 4. Performance of dense point cloud generated by the proposed method.

Study Area	Seed Patch Number	Expanded Patch Size	Patch Number (after Expand)	Patch Number (after Filter)	Density (patches/m^2)	Times (min)
Northwestern University	107514	17 × 17 (step: 2)	7890775	7802802	55.275	175
Yangjiang	324072	17 × 17 (step: 2)	24369048	24003611	135.122	627
Hainan	178317	17 × 17 (step: 2)	8481032	8474530	8.598	253

Table 5. Comparison of the point cloud performance.

Experimental Method	Northwestern University		Yangjiang		Hainan	
	Point Number	Density (points/m^2)	Point Number	Density (points/m^2)	Point Number	Density (points/m^2)
PMVS	107514	0.762	324072	1.824	178317	0.181
SURE	2053708	14.410	638032	3.592	770993	0.782
Pix4D	525402	3.686	2126320	11.970	1123166	1.140
The proposed method	7802802	55.275	24003611	135.122	8474530	8.598

The third column in Table 4 represents the experiments which used a 17 × 17 image window, and each other pixel in the image window was projected onto the patch. The computational times are recorded in the last column. All timings were obtained on a PC with Intel Core(TM) i7 3.60 GHz processors, 8 GB RAM and a 1 TB SCSI disk device for data storage, and the Microsoft Windows 7 operating system. All the processes were performed offline. From the comparison experiment results in Table 5, it can be noted that the proposed method achieves more than 40 times denser points per m^2 than PMVS and a more than eight times denser point cloud per m^2 than Pix4D. According to the image parameters and the reconstructed results, it can be seen that the density of the point cloud depends on the ground resolution of the input images. As long as the ground resolution is high enough, the proposed method can obtain much denser point clouds than laser scanning [4], such as the point cloud from Yangjiang.

4.3. Point Cloud Accuracy Evaluation

To evaluate the accuracy, each set of point clouds produced by the proposed method were registered into the PMVS model. A relative Euclidean distance (error) comparison between a point from the point cloud and the surface of the PMVS model where this point is supposed to be located is measured.

The accuracy evaluation is based on the method raised by Dai *et al.* [46]. Supposing m_j is the number of points, it should belong to the jth surface of the PMVS model which is denoted as $a_j X + b_j Y + c_j Z + d_j = 0$. The ith point coordinate in point set m_j is denoted as $(X_i{}^j, Y_i{}^j, Z_i{}^j)$; n is the number of surfaces. The average error of the point cloud can be calculated as:

$$\text{error} = \frac{1}{\sum_{j=1}^{n} m_j} \sum_{j=1}^{n} \sum_{i=1}^{m_j} \frac{\left| a_j X_i^j + b_j Y_i^j + c_j Z_i^j + d_j \right|}{\sqrt{a_j^2 + b_j^2 + c_j^2}} \tag{9}$$

Note that if a point's distance to the surface is far beyond the average value, it will be deemed as an outlier and removed from the point cloud set. Details of the accuracy evaluation are listed in Table 6.

Table 6. Evaluation of accuracy.

Study Area	Point Cloud Number	Outlier Number	Outliers/Point Cloud	Average Error (m)
Northwestern University Campus	7802802	1780	$2.281/10^4$	0.332
Yangjiang region	24003611	919	$3.827/10^5$	0.166
Hainan urban district	8474530	8217	$9.695/10^4$	0.480

As illustrated in Table 6, it can be seen that the point clouds generated by the proposed method achieved exceptional results. Specifically, the Yangjiang point cloud data contains less than four outliers in 10^5 points, and the other two data sets contains less than 10 outliers in 10,000 points. The average errors of the point cloud data registered into the PMVS model are all less than 0.5 m. For 3D reconstruction from low-altitude remote sensing images, the accuracy of the point cloud data is reliable. From comparison experiments of image ground resolution and accuracy between these three study areas, it can be noted that the study images which had the highest ground resolution (Yangjiang region) had the most accurate point cloud. With a decrease in ground resolution, the precision was also reduced. It should be noted that parts of the images with weak texture do not be reconstruct well under the proposed method (e.g., flat farmland in the Northwestern University data sets) because feature or seed points to expand these regions are not found. In the three data sets, topographic relief of the Northwestern University model (nearly 30 m) is lower than topographic relief of Yangjiang and Hainan models, which are almost same (nearly 50 m). The Yangjiang point cloud achieved higher accuracy than Northwestern University, which illustrates that, compared with the topographic relief, the influence of the ground resolution and remote sensing platform stability on the accuracy is greater.

5. Conclusions

In this study, a novel algorithm is presented for improving the density of point clouds generated from low-altitude remote sensing images. The proposed algorithm builds an expanded patch for each point in a PMVS point cloud. The method integrates the advantages of Multi-View Stereo and epipolar-based dense matching methods and generates a denser point cloud with more details.

The matching results have illustrated that the proposed approach can achieve a far denser point cloud than PMVS, and the matching accuracy of the proposed method is reliable when using low-altitude remote sensing images. It is important to note that the precision of the image orientation parameter can directly affect the results of the PMVS seed and MPGC refining. Thus, the proposed approach is more suitable for 3D reconstruction using calibrated images with high accuracy. From this work, two potential areas of future research are proposed: (1) raise the efficiency of image matching to extend this method to 3D reconstructions of larger scenes; and (2) improve the PMVS result in areas with little or no texture.

Acknowledgments: This work was supported by National Science & Technology Specific Projects (No.2012YQ1601850 & No.2013BAH42F03), Program for New Century Excellent Talents in University (No.NCET-12-0426) and innovative project of Wuhan University (042016kf0179).

Author Contributions: Nan Yang conceived and designed the experiments; Nan Yang and Zhe Peng performed the experiments; Zhenfeng Shao and Nan Yang and Xiongwu Xiao and Lei Zhang analyzed the data; Zhenfeng Shao and Xiongwu Xiao contributed reagents/materials/analysis tools; Zhenfeng Shao and Nan Yang wrote the paper; Zhenfeng Shao and Lei Zhang helped to prepare the manuscript. All authors read and approved the final manuscript.

Conflicts of Interest: The authors declare no conflict of interest.

Abbreviations

The following abbreviations are used in this manuscript:

MVS	Multi-View Stereo
MPGC	Multiphoto Geometrically Constrained Matching
DEM	Digital Elevation Model
DSM	Digital Surface Model
DTM	Digital Terrain Model
PMVS	Patch-based Multi-View Stereo
SGM	Semi-Global Matching
SFM	Structure from Motion
NCC	Normalized Cross-correlation Coefficient
LSM	Least Square Image Matching

References

1. De Reu, J.; Plets, G.; Verhoeven, G.; De Smedt, P.; Bats, M.; Cherretté, B.; De Maeyer, W.; Deconynck, J.; Herremans, D.; Laloo, P. Towards a three-dimensional cost-effective registration of the archaeological heritage. *J. Archaeol. Sci.* **2013**, *40*, 1108–1121.

2. Capra, A.; Dubbini, M.; Bertacchini, E.; Castagnetti, C.; Mancini, F. 3D reconstruction of an underwater archaelogical site: Comparison between low cost cameras. *ISPRS-Int. Arch. Photogramm. Remote Sens. Spat. Inf. Sci.* **2015**, *1*, 67–72.

3. Gonçalves, J.; Henriques, R. UAV photogrammetry for topographic monitoring of coastal areas. *ISPRS J. Photogramm. Remote Sens.* **2015**, *104*, 101–111.

4. Molina, J.-L.; Rodríguez-Gonzálvez, P.; Molina, M.C.; González-Aguilera, D.; Espejo, F. Geomatic methods at the service of water resources modeling. *J. Hydrol.* **2014**, *509*, 150–162.

5. Beeler, T.; Bickel, B.; Beardsley, P.; Sumner, B.; Gross, M. High-quality single-shot capture of facial geometry. *ACM Trans. Graphics (TOG)* **2010**, *29*, 40.

6. Tung, T.; Matsuyama, T. Geodesic mapping for dynamic surface alignment. *IEEE Trans. Pattern Anal. Mach. Intell.* **2014**, *36*, 901–913.

7. Remondino, F. Heritage recording and 3d modeling with photogrammetry and 3d scanning. *Remote Sens.* **2011**, *3*, 1104–1138.

8. Xu, Z.; Wu, L.; Shen, Y.; Li, F.; Wang, Q.; Wang, R. Tridimensional reconstruction applied to cultural heritage with the use of camera-equipped uav and terrestrial laser scanner. *Remote Sens.* **2014**, *6*, 10413–10434.

9. Rose, J.C.; Paulus, S.; Kuhlmann, H. Accuracy analysis of a multi-view stereo approach for phenotyping of tomato plants at the organ level. *Sensors* **2015**, *15*, 9651–9665.

10. Tao, W. Multi-view dense match for forest area. *ISPRS-Int. Archives Photogramm. Remote. Sens. Spat. Inf. Sci.* **2014**, *1*, 397–400.

11. Lin, B.; Sun, Y.; Qian, X.; Goldgof, D.; Gitlin, R.; You, Y. Video-based 3d reconstruction, laparoscope localization and deformation recovery for abdominal minimally invasive surgery: A survey. *Int. J. Med. Robotics Computer Assist. Surg.* **2015**.

12. Rau, J.-Y.; Jhan, J.-P.; Hsu, Y.-C. Analysis of oblique aerial images for land cover and point cloud classification in an urban environment. *IEEE Trans. Geosci. Remote Sens.* **2015**, *53*, 1304–1319.

13. Vosselman, G. Automated planimetric quality control in high accuracy airborne laser scanning surveys. *ISPRS J. Photogramm.* **2012**, *74*, 90–100.

14. García-Gago, J.; González-Aguilera, D.; Gómez-Lahoz, J.; San José-Alonso, J.I. A photogrammetric and computer vision-based approach for automated 3D architectural modeling and its typological analysis. *Remote Sens.* **2014**, *6*, 5671–5691.

15. Tanskanen, P.; Kolev, K.; Meier, L.; Camposeco, F.; Saurer, O.; Pollefeys, M. Live Metric 3D Reconstruction on Mobile Phones. In Proceedings of the 2013 IEEE International Conference on Computer Vision (ICCV), Sydney, NSW, Australia, 1–8 December 2013; pp. 65–72.

16. Furukawa, Y.; Ponce, J. Dense 3D motion capture from synchronized video streams. In Proceeding of the IEEE Computer Society Conference on Computer Vision and Pattern Recognition, 2008. (CVPR 2008), Anchorage, AK, USA, 23–28 June 2008; pp. 193–211.

17. Remondino, F.; Spera, M.G.; Nocerino, E.; Menna, F.; Nex, F. State of the art in high density image matching. *Photogramm. Rec.* **2014**, *29*, 144–166.

18. Harwin, S.; Lucieer, A. Assessing the accuracy of georeferenced point clouds produced via multi-view stereopsis from unmanned aerial vehicle (uav) imagery. *Remote Sens.* **2012**, *4*, 1573–1599.

19. Turner, D.; Lucieer, A.; Watson, C. An automated technique for generating georectified mosaics from ultra-high resolution unmanned aerial vehicle (uav) imagery, based on structure from motion (sfm) point clouds. *Remote Sens.* **2012**, *4*, 1392–1410.

20. Ai, M.; Hu, Q.; Li, J.; Wang, M.; Yuan, H.; Wang, S. A robust photogrammetric processing method of low-altitude uav images. *Remote Sens.* **2015**, *7*, 2302–2333.

21. Furukawa, Y.; Ponce, J. Accurate, dense, and robust multiview stereopsis. *IEEE Trans. Pattern Anal. Mach. Intell.* **2010**, *32*, 1362–1376.

22. Marr, D.; Poggio, T. Cooperative computation of stereo disparity. *Science* **1976**, *194*, 283–287.

23. Büyüksalih, G.; Koçak, G.; Oruç, M.; Akçin, H.; Jacobsen, K. Accuracy analysis, dem generation and validation using russian tk-350 stereo-images. *Photogramm. Rec.* **2004**, *19*, 200–218.

24. Vassilopoulou, S.; Hurni, L.; Dietrich, V.; Baltsavias, E.; Pateraki, M.; Lagios, E.; Parcharidis, I. Orthophoto generation using ikonos imagery and high-resolution dem: A case study on volcanic hazard monitoring of nisyros island (greece). *ISPRS J. Photogramm.* **2002**, *57*, 24–38.

25. Remondino, F.; El-Hakim, S.; Gruen, A.; Zhang, L. Turning images into 3-D models. *IEEE Signal Process Mag.* **2008**, *25*, 55–65.

26. Goesele, M.; Snavely, N.; Curless, B.; Hoppe, H.; Seitz, S.M. Multi-View Stereo for Community Photo Collections. In Proceeding of the IEEE 11th International Conference on Computer Vision, 2007 (ICCV 2007), Rio de Janeiro, Brazil, 14–21 October 2007; pp. 1–8.

27. Scharstein, D.; Szeliski, R. A taxonomy and evaluation of dense two-frame stereo correspondence algorithms. *Int. J. Computer Vis.* **2002**, *47*, 7–42.

28. Rottensteiner, F.; Sohn, G.; Gerke, M.; Wegner, J.D.; Breitkopf, U.; Jung, J. Results of the isprs benchmark on urban object detection and 3D building reconstruction. *ISPRS J. Photogramm.* **2014**, *93*, 256–271.

29. Shahbazi, M.; Sohn, G.; Théau, J.; Ménard, P. UAV-based point cloud generation for open-pit mine modeling. *Int. Arch. Photogramm. Remote Sens. Spat. Inf. Sci.* **2015**, *40*, 313.

30. Yoon, K.-J.; Kweon, I.S. Adaptive support-weight approach for correspondence search. *IEEE Trans. Pattern Anal. Mach. Intell.* **2006**, *4*, 650–656.

31. Lei, C.; Selzer, J.; Yang, Y.-H. Region-tree based stereo using dynamic programming optimization. In Proceeding of IEEE Computer Society Conference on Computer Vision and Pattern Recognition, New York, NY, USA, 2006; pp. 2378–2385.

32. Hirschmüller, H. Stereo processing by semiglobal matching and mutual information. *IEEE Trans. Pattern Anal. Mach. Intell.* **2008**, *30*, 328–341.

33. Ahmadabadian, A.H.; Robson, S.; Boehm, J.; Shortis, M.; Wenzel, K.; Fritsch, D. A comparison of dense matching algorithms for scaled surface reconstruction using stereo camera rigs. *ISPRS J. Photogramm.* **2013**, *78*, 157–167.

34. Yang, A.; Li, X.; Xie, J.; Wei, Y. Three-dimensional panoramic terrain reconstruction from aerial imagery. *J. Appl. Remote Sens.* **2013**, *7*, 073497.

35. Snavely, N.; Seitz, S.M.; Szeliski, R. Modeling the world from internet photo collections. *Int. J. Comput. Vis.* **2008**, *80*, 189–210.

36. Habbecke, M.; Kobbelt, L. A surface-growing approach to multi-view stereo reconstruction. In Proceeding of the IEEE Conference on Computer Vision and Pattern Recognition (CVPR'07), Minneapolis, MN, USA, 17–22 June 2007; pp. 1–8.

37. Liu, Y.; Dai, Q.; Xu, W. A point-cloud-based multiview stereo algorithm for free-viewpoint video. *IEEE Trans. Visual Comput. Graphics* **2010**, *16*, 407–418.

38. Hiep, V.H.; Keriven, R.; Labatut, P.; Pons, J.-P. Towards high-resolution large-scale multi-view stereo. In Proceeding of the IEEE Conference on Computer Vision and Pattern Recognition, 2009 (CVPR 2009), Miami, FL, USA, 20–25 June 2009; pp. 1430–1437.

39. Nelder, J.A.; Mead, R. A simplex method for function minimization. *Comput. J.* **1965**, *7*, 308–313.

40. Baltsavias, E.P. Multiphoto Geometrically Constrained Matching. Doctoral Dissertation, ETH Zürich, Nr. 9561, Zürich, Switzerland, 1991.

41. Baltsavias, E.P. Digital ortho-images—A powerful tool for the extraction of spatial-and geo-information. *ISPRS J. Photogramm.* **1996**, *51*, 63–77.

42. Ackermann, F. High precision digital image correlation. In Proceedings of the 39th Photogrammetric Week, Stuttgart, Germany, 19–24 September 1983; pp. 231–243.

43. Ackermann, F. Digital image correlation: Performance and potential application in photogrammetry. *Photogramm. Rec.* **1984**, *11*, 429–439.

44. Zhang, L.; Gruen, A. Multi-image matching for dsm generation from ikonos imagery. *ISPRS J. Photogramm.* **2006**, *60*, 195–211.

45. Baltsavias, E.; Gruen, A.; Eisenbeiss, H.; Zhang, L.; Waser, L. High-quality image matching and automated generation of 3D tree models. *Int. J. Remote Sens.* **2008**, *29*, 1243–1259.

46. Dai, F.; Rashidi, A.; Brilakis, I.; Vela, P. Comparison of image-based and time-of-flight-based technologies for three-dimensional reconstruction of infrastructure. *J. Constr. Eng. Manag.* **2012**, *1*, 69–79.

Chapter 2:
Terrestrial Photogrammetry

An Advanced Pre-Processing Pipeline to Improve Automated Photogrammetric Reconstructions of Architectural Scenes

Marco Gaiani, Fabio Remondino, Fabrizio I. Apollonio and Andrea Ballabeni

Abstract: Automated image-based 3D reconstruction methods are more and more flooding our 3D modeling applications. Fully automated solutions give the impression that from a sample of randomly acquired images we can derive quite impressive visual 3D models. Although the level of automation is reaching very high standards, image quality is a fundamental pre-requisite to produce successful and photo-realistic 3D products, in particular when dealing with large datasets of images. This article presents an efficient pipeline based on color enhancement, image denoising, color-to-gray conversion and image content enrichment. The pipeline stems from an analysis of various state-of-the-art algorithms and aims to adjust the most promising methods, giving solutions to typical failure causes. The assessment evaluation proves how an effective image pre-processing, which considers the entire image dataset, can improve the automated orientation procedure and dense 3D point cloud reconstruction, even in the case of poor texture scenarios.

Reprinted from *Remote Sens.* Cite as: Gaiani, M.; Remondino, F.; Apollonio, F.I.; Ballabeni, A. An Advanced Pre-Processing Pipeline to Improve Automated Photogrammetric Reconstructions of Architectural Scenes. *Remote Sens.* **2016**, *8*, 178.

1. Introduction

In the last years, the image-based pipeline for 3D reconstruction purposes has received large interest leading to fully automated methodologies able to process large image datasets and deliver 3D products with a level of detail and precision variable according to the applications [1–3] (Figure 1). Certainly, the integration of automated computer vision algorithms with reliable and precise photogrammetric methods is nowadays producing successful (commercial and open) solutions (often called Structure from Motion (SfM)) for automated 3D reconstructions from large image datasets [4–6].

For terrestrial applications, the level of automation is reaching very high standards and it is increasing the impression that few randomly acquired images (or even found on the Internet) and a black-box tool (or mobile app) are sufficient to produce a metrically precise 3D point cloud or textured 3D model. Such tools are able to ingest and process large quantities of images almost always delivering an apparently successful solution, which is often a local minimum and not the fully correct one (Figure 2). However, non-expert users might not be able to spot such small errors or divergences in the bundle adjustment due to the fact that only a message of successful image orientation is provided, without statistical analyses.

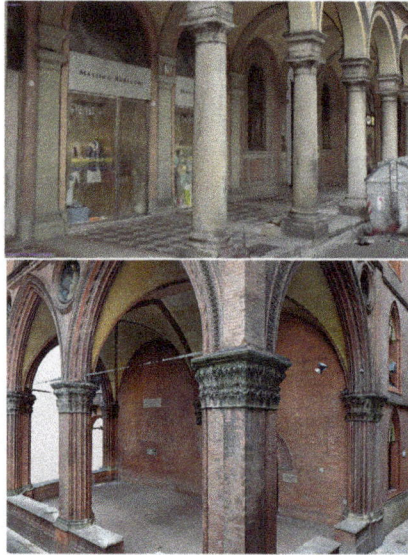

Figure 1. Examples of image-based 3D reconstruction results of complex architectural scenarios achieved with the presented methodology.

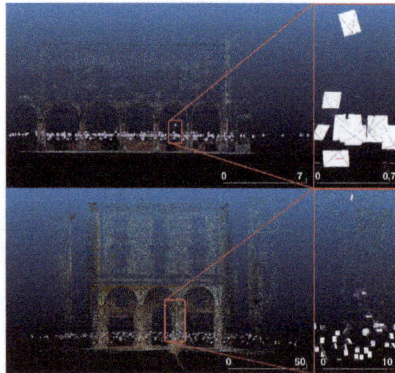

Figure 2. Typical examples of SfM results where, despite a message of successful image orientation and a very small re-projection error, there are some cameras wrongly oriented. A non-expert user could only spot such errors with difficultly and would proceed to the successive processing stages, negatively affecting the final results.

Motion blur, sensor noise and jpeg artifacts are just some of the possible image problems that are negatively affecting automated 3D reconstruction methods. These problems are then coupled with lack of texture scenarios, repeated patterns, illumination changes, *etc*. Therefore, image pre-processing methods are fundamental to improve the image quality for successful photogrammetric processing. Indeed, as the image processing is fully automated, the quality of the input images, in terms of radiometric quality as well as network geometry, is fundamental for a successful 3D reconstruction.

This paper presents an efficient image pre-processing methodology developed to increase the processing performances of the two central steps of the photogrammetric pipeline, *i.e.*, image orientation and dense image matching. The main idea is to minimize typical failure caused by Scale-Invariant Feature Transform (SIFT)-like algorithms [7] due to changes in the illumination conditions or low contrast blobs areas and to improve the performances of dense image matching methods [8]. The methodology tries to: (i) increase the number of correct image correspondences, particularly in textureless areas; (ii) track image features along the largest number of images to increase the reliability of the computed 3D coordinates; (iii) correctly orient the largest number of images; (iv) deliver sub-pixel accuracy at the end of the bundle adjustment procedure; and (v) provide dense, complete and noise-free 3D point clouds (Figure 3). The work investigated various state-of-the-art algorithms aiming to adapt the most promising methods and give solutions at the aforementioned specific problems, thus creating a powerful solution to radiometrically improve the image quality of an image datasets. The developed procedure for image pre-processing and enhancement consists of color balancing (Section 2), image denoising (Section 3), color-to-gray conversion (Section 4) and image content enrichment (Section 5). The pre-processing methodology could be really useful in the architectural, built heritage and archaeological fields, where automated 3D modeling procedures have become very common whereas skills in image acquisition and data processing are often missing. Our methodology could even be embedded in different types of processing scenarios, like completely automated web-based applications (e.g., Autodesk ReCap and Arc 3D Webservice) or offline desktop-based applications (e.g., Agisoft Photoscan, Photometrix iWitness, VisualSFM, and Pix4D Mapper).

The pipeline (Figure 4) is afterwards presented and evaluated using some datasets of architectural scenarios.

(a) *Without Pre-Processing*   (b) *With Pre-Processing*

Figure 3. 3D reconstruction of complex architectures without (**a**) and with (**b**) a suitable pre-processing methodology applied to improve the radiometric image quality and provide more complete and noise-free 3D point clouds.

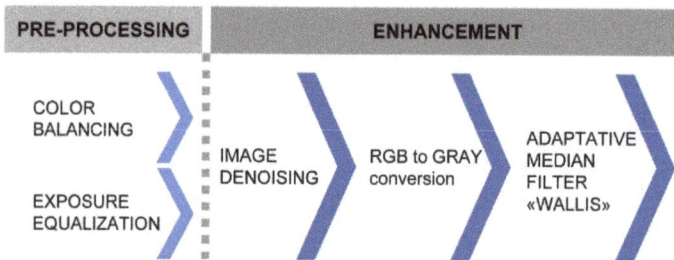

Figure 4. The proposed image pre-processing and enhancement pipeline.

Related Works

Image pre-processing is a set of methods used to increase the quality of images for successive processing purposes [9]. The aim is thus to enhance some image features important, e.g., for 3D reconstruction algorithms or to remove unwanted disturbs or degradations in the image. A pre-processing can be a simple histogram's stretching or a more complex approach like denoising or filtering [10,11]. Image pre-processing normally comprises enhancement (*i.e.*, the improvement of the image quality) and restoration (*i.e.*, the removal of degraded areas). The former is more subjective, whereas the latter is an objective process that involves the modeling of degradation (possibly from prior knowledge) and applying an inverse process to recover the original signal. Image pre-processing is a fundamental task for many successive tasks in applications like medical imaging [12], computer vision, underwater photogrammetry [13] or 3D modeling [14].

Maini and Aggarwal [15] provide an overview of concepts and algorithms commonly used for image enhancement. Stamos *et al.* [16] presents some metrics to estimate the amount of blur in image sequence, based on color saturation, local auto-correlation and gradient distribution. Feature tracking and camera poses recovery methods in blurry image sequences can be improved using edgelets [17] or blurring the previous frame in order to obtain a consistent tracking [18] or deblurring a current frame with a blur kernel [19]. Guidi *et al.*, [14] analyses how image pre-processing with polarizing filters and HDR imaging may improve indoor automated 3D reconstruction processes based on SfM methods. Verhoeven *et al.* [20] investigated the use of different grayscale conversion algorithms to decolorize color images as input for SfM software packages. Bellavia *et al.* [21] presented an online pre-processing strategy to detect and discard bad frames in video sequences. The method is based on the Double Window Adaptive Frame Selection (DWAFS) algorithm which works on a simple gradient statistic (gradient magnitude distribution). The percentile statistic of each frame is used to develop an adaptive decision strategy based on a dangling sample window according to the time series of the ongoing percentile values and the last best ones.

2. Color Balance and Exposure Equalization

Color balance is the global adjustment of the intensity of the (red, green, and blue) colors in order to render them correctly. Color balance and exposure equalization is a key step to ensure: (i) faithful color appearance of a digitized artifact; and (ii) consistency of the color-to-gray conversion. This latter one (see Section 4) is a fundamental step as all feature extraction and image matching algorithms works using only the luminance channel. A correct color balance allows minimizing the typical problem of incorrectly detected areas (e.g., different luminance value for the same color and/or isoluminant colors) that strongly appear in case of surfaces of the same color or colors with the same luminance value. Therefore, the color balance procedure aims to produce radiometrically-calibrated images ensuring the consistency of surface colors in all the images (*i.e.*, as much as possible similar RGB values for homologous pixels). Starting from captured RAW images, our workflow includes (Section 2.2): exposure compensation, optical correction, sharpen, and color balance.

2.1. Color Spaces

The use of an appropriate color space to work and render images on screen is fundamental.

sRGB, a standard RGB color space created cooperatively by HP and Microsoft in 1996, is certainly the best choice as output color space for textures and to display rendered 3D models for several reasons including:

- sRGB is the default color space for HTML, CSS, SMIL and other web standards;
- sRGB is consistent among different monitors or video-projectors; and
- sRGB is implemented in the OpenGL graphic libraries, used in many rendering software.

However, the sRGB color space is very narrow and may produce loss of information, mainly in the acquisition and processing phases. To avoid these problems, a broader rendered color space is used, such as the Adobe-RGB (1998), which represents an excellent compromise between the amount of colors that can be codified and the possibility of displaying them on the screen. The use of Adobe-RGB (1998) allows avoiding possible inaccuracies of the sRGB color space in shadows (~25% luminance) as well as highlights (~75% luminance). Adobe-RGB (1998) expands its advantages to areas of intense orange, yellow and magenta color. As the sRGB is a "*de facto*" standard for consumer cameras storing images in JPEG format, it is advisable to use the RAW format, which normally map to a rendered color space as the Adobe-RGB or sRGB color space.

2.2. Proposed Approach

Between the two general approaches (color characterization *vs.* spectral sensitivities based on color targets) [22] we adopted this last technique that uses a set of differently colored samples measured with a spectrophotometer.

The most precise characterization for any given camera requires recording its output for all possible *stimuli* and comparing it with separately measured values for the same *stimuli* [23]. However, storage of such a quantity of data is impractical, and, therefore, the response of the device is captured for only a limited set of *stimuli*—normally for the acquisition conditions. The responses to these representative *stimuli* can then be used to calibrate the device for input *stimuli* that were not measured, finding the transformation between measured CIExyz values and stored RGB values. To find this transformation, several techniques have been developed, including look-up tables [24].

The method for evaluating and expressing color accuracy ("color characterization") includes a physical reference chart acquired under standard conditions, a reference chart color space with the ideal data values for the chart, a way to relate or convert the device color space to the reference chart color space and, finally, a way to measure and show errors in the device's rendering of the reference chart. The target GretagMacbeth ColourChecker [25] is employed during the image acquisitions, considering the measurements of each patch as reported in Pascale [26].

A captured color image containing the GretagMacbeth ColourChecker is neutralized, balanced and properly exposed. Using in-house software, an ICC (International Color Consortium) profile—assigned together with the Adobe-RGB (1998) color space of the RAW image—is generated. Before creating ICC profiles, a standard gamma correction ($\gamma = 2.2$) is applied, converting all images to the camera's native linear color space, thus improving the quality of the profiles. A protocol is developed to use the same calibration for groups of images with the same features (*i.e.*, orientation, exposure and framed surfaces) thus to maintain consistency in the process and results.

The color accuracy is computed in terms of the mean camera chroma relative to the mean ideal chroma in the CIE color metric (ΔE^*_{00}) as defined in 2000 by CIE [27]:

$$\Delta E^*_{00} = \sqrt{\left(\frac{\Delta L'}{k_L S_L}\right)^2 + \left(\frac{\Delta C'}{k_C S_C}\right)^2 + \left(\frac{\Delta H'}{k_H S_H}\right)^2 + R_T \frac{\Delta C'}{k_C S_C} \frac{\Delta H'}{k_H S_H}} \tag{1}$$

This formula is a new version of the original one (1976) and is more suitable for our uses. It takes into consideration the problem of non-perceptual uniformity of the colors for which ΔE^*_{00} varies the weight of L^* depending on where the brightness range falls. Song and Luo [28] showed that the perceptible and acceptable color differences in complex images presented on a CRT (Cathode Ray Tube) monitor are approximately 2.2 and 4.5, respectively. In our case, the latter value was used as a strict reference for accuracy, defined from perception tests on the results obtained using this value.

Exposure error in f-stops was also evaluated on the plane of the target assumed as one the main object captured in the image. The ΔE^*_{00} and the exposure error calculations was computed using Imatest Studio software version 3.9.

From an operational point of view, the preservation of color fidelity throughout the image processing is ensured by:

- taking pictures in the most homogeneous operative conditions (aperture/exposure direction and intensity of light);
- including ColourChecker target inside the photographed scenes in order to correct the image radiometry;
- storing photos in RAW format; and
- using an appropriate color space from the beginning of the image processing.

An important and critical issue is the acquisition of the color target. In order to maintain uniform lighting in an external environment, for each image, we need to consider: (i) surfaces illuminated and oriented as the ColourChecker and that presents an angle of incidence with sunlight of approximately 20°–45° or (ii) image acquisitions performed with overcast sky. To minimize the light glare, that would give unexpected results in the calibration process, the ColourChecker is normally placed on a tripod with a dark background and orthogonal to the camera optical axis. Finally, we verified that a ColourChecker image width of 500 to 1500 pixels is sufficient for ΔE^*_{00} analysis, as also suggested in the Imatest user guide.

3. Image Denoising

Image noise is defined in the ISO 15739 standard as "unwanted variations in the response of an imaging system" [29]. It is formed when incoming light is converted from photons to an electrical signal and originates from the camera sensor, its sensitivity and the exposure time as well as by digital processing (or all these factors together). Noise can appear in different ways:

- *Fixed pattern noise* ("hot" and "cold" pixels): It is due to sensor defects or long time exposure, especially with high temperatures. Fixed pattern noise always appears in the same position.
- *Random noise*: It includes intensity and color fluctuations above and below the actual image intensity. They are always random at any exposure and more influenced by ISO speed.
- *Banding noise*: It is caused by unstable voltage power and is characterized by the straight band in frequency on the image. It is highly camera-dependent and more visible at high ISO speed and in dark image. Brightening the image or white balancing can increase the problem.
- *Luminance noise* (*i.e.*, a variation in brightness): It is composed of noisy bright pixels that give the image a grainy appearance. High-frequency noise is prevalent in the luminance channel, which can range from fine grain to more distinct speckle noise. This type of noise does not significantly affect the image quality and can be left untreated or only minimally treated if needed.
- *Chrominance noise* (*i.e.*, a variation in hue): It appears as clusters of colored pixels, usually green and magenta. It occurs when the luminance is low due to the inability of the sensor to differentiate color in low light levels. As a result, errors in the way color is recorded are visible and hence the appearance of color artifacts in the de-mosaicked image.

Starting from these considerations, the noise model can be approximated with two components:

(a)   A signal-independent Gaussian noise to compensate for the fixed pattern noise (FPN).

(b) A signal-dependent Poisson noise to compensate for the temporal (random) noise, called Shot Noise.

A denoise processing basically attempts to eliminate—or at least minimize—these two components.

Several denoising methods [30–32] deal directly with Poisson noise. Wavelet-based denoising methods [33,34] adapt the transform threshold to the local noise level of the Poisson process. Recent papers on the Anscombe transform by Makitalo and Foi [35] and Foi [36], argue that, when combined with suitable forward and inverse variance-stabilizing transformations (VST), algorithms designed for homoscedastic Gaussian noise work just as well as *ad-hoc* algorithms based on signal-dependent noise models. This explains why the noise is assumed to be uniform, white and Gaussian, having previously applied a VST to the noisy image to take into account the Poisson component.

An effective restoration of image signals will require methods that either model the signal a-priori (*i.e.*, Bayesian) or learn the underlying characteristics of the signal from the given data (*i.e.*, learning, non-parametric, or empirical Bayes' methods). Most recently, the latter approach has become very popular, mainly using patch-based methods that exploit both local and non-local redundancies and "self-similarities" in the images [24]. A patch-based algorithm denoises each pixel by using knowledge of (a) the patch surrounding it and (b) the probability density of all existing patches.

Typical noise reduction software reduces the visibility of noise by smoothing the image, while preserving its details. The classic methods estimate white homoscedastic noise only, but they can be adapted easily to estimate signal- and scale-dependent noise.

The main goals of image denoising algorithms are:

- perceptually flat regions should be as smooth as possible and noise should be completely removed from these regions;
- image boundaries should be well preserved and not blurred;
- texture detail should not be lost;
- the global contrast should be preserved (*i.e.*, the low-frequencies of denoised and input images should be equal); and
- no artifacts should appear in the denoised image.

All these goals are appropriate also for our cases were we need not to remove signal, nor distort blob shape and intensity areas to have efficient keypoint extraction and image matching processing.

Numerous methods were developed to meet these goals, but they all rely on the same basic method to eliminate noise: averaging. The concept of averaging is simple, but determining which pixels to average is not.

103

In summary:

- the noise model is different for each image;
- the noise is signal-dependent;
- the noise is scale-dependent; and
- the knowledge of each dependence is crucial to proper denoising of any given image which is not raw, and for which the camera model is available.

To meet this challenge, four denoising principles are normally considered:

- transform thresholding (sparsity of patches in a fixed basis);
- sparse coding (sparsity on a learned dictionary);
- pixel averaging and block averaging (image self-similarity); and
- Bayesian patch-based methods (Gaussian patch model).

Each principle implies a model for the ideal noiseless image. The current state-of-the-art denoising recipes are in fact a smart combination of all these ingredients.

3.1. Evaluated Methods

We investigated different denoise algorithms, some commercial, namely:

- *Imagenomic Noiseware* [37,38]: It uses hierarchical noise reduction algorithms, subdividing the image noise into two categories: luminance noise and color noise, furthermore divided into frequencies ranging from very low to high. The method includes detection of edges and processing at different spatial frequencies, using the YCbCr color space. Noiseware presents good quality results and it is easy to set-up.
- *Adobe Camera RAW* denoise [39]: Noise-reduction available in Camera Raw 6 with Process 2010 (in Section 6 simply called *Adobe*) uses a luminance noise-reduction technique based on a wavelet algorithm that seeks to determine extremely high-frequency noise and to separate it from high-frequency image texture. The method is capable of denoising large noisy areas of an image as well as to find and fix "outliers", *i.e.*, localized noisy areas. Unfortunately, the method is a global filter and it needs a skilled manual intervention for each image to set-up the right parameters.
- *Non-Local Bayesian* filter [40–42]: It is an improved patch-based variant of the Non Local-means (NL-means) algorithm, a relatively simple generalization of the Bilateral Filter. In the NL-Bayes algorithm, each patch is replaced by a weighted mean of the most similar patches present in a neighborhood. To each patch is associated a mean (which would be the result of NL-means), but also a covariance matrix estimating the variability of the patch group. This allows

computing an optimal (Bayesian minimal mean square error) estimate of each noisy patch in the group, by a simple matrix inversion. The implementation proceeds in two identical iterations, but the second iteration uses the denoised image of the first iteration to estimate better the mean and covariance of the patch models.

- *Noise Clinic* [43–46]: It is the conjunction of a noise estimation method and of a denoising method. Noise estimation is with an extension of [47] method to be able to estimate signal-dependent noise, followed by multiscale NL-Bayes denoising method. The multiscale denoising follow these principles: (a) signal dependent noise estimated at each scale; and (b) zoom down followed by Anscombe transform to whiten the noise at each scale; denoising performed at each scale, bottom-up (coarse to fine). Noise Clinic is implemented in DxO Optics Pro with the name of Prime (Probabilistic Raw IMage Enhancement), and it is useful for very noisy and high-ISO RAW images, or for photos taken with an old camera that could not shoot good-quality images at ISO higher than 1600 ISO.

- *Color Block Matching 3D* (*CBM3D*) filter [48]: A color variant of Block Matching 3D (BM3D) filter [49]. BM3D is a sliding-window denoising method extending the Discrete Cosine Transform (DCT) [25] and NL-means algorithms. BM3D, instead of adapting locally a basis or choosing from a large dictionary, uses a fixed basis. The main difference from DCT denoising is that a set of similar patches is used to form a 3D block, which is filtered by using a 3D transform, hence the name "collaborative filtering". The algorithm works in two stages: "basic estimate" of the image and the creation of the final image, and with four steps each stage: (a) finding the image patches similar to a given image patch and grouping them in a three-dimensional block: (b) 3D linear transform of the 3D block; (c) shrinkage of the transform spectrum coefficients; and (d) inverse three-dimensional transformation. This second step mimics the first step, with two differences. The first difference is that it compares the filtered patches instead of the original patches. The second difference is that the new 3D group is processed by an oracle Wiener filter, using coefficients from the denoised image obtained at the first step to approximate the true coefficients. The final aggregation step is identical to that of the first step. CBM3D extends the multi-stage approach of BM3D via the YoUoVo color system. CBM3D produces a basic estimate of the image, using the luminance data, and delivers the denoised image performing a second stage on each of the three color channels separately. This generalization of the BM3D is non-trivial because authors do not apply the grayscale BM3D independently on the three luminance-chrominance channels but they impose a grouping constraint on both chrominance. The grouping constraint means that the grouping is done only once, in the luminance (which typically has a higher SNR than the chrominance), and exactly the same grouping is reused for filtering both chrominance. The constraint on the chrominance increases the stability of the grouping with respect to noise. With this solution, the quality of denoised images is also excellent for moderate noise levels.

3.2. Proposed Approach

Following the experiment results, an in-house method (named *CBM3D-new*) was developed starting from the CBM3D approach [48]. For every image of a dataset, the method automatically select the necessary parameters based on the type of camera, ISO sensitivity and stored color profiles. In particular, the processing selection of the latter one is based on image features and camera capabilities: dealing with professional or prosumer setups, when source images are stored as RAW images or in non-RAW formats characterized by a wide color space such as the Adobe-RGB (1998), then opponent color space are chosen. When source images are stored in JPG format using a relatively narrower color space, such as sRGB—the most used in consumer cameras—then YCbCr color space is chosen.

The camera ISO is strictly related to the image noise. The sigma parameter, *i.e.*, the standard deviation of the noise, increases when the ISO increases, ranging from lower values ($\sigma = 1$ for images shot at less than 100 ISO) to higher ones ($\sigma = 10$ for images shot at more than 800 ISO). ISO sensitivity similarly influences other filtering parameters, such as the number of sliding step to process every image block (ranging from 3 to 6), the length of the side of the search neighborhood for full-search block-matching (ranging from 25 to 39) as well as the number of step forcing to switch to neighborhood full-search (ranging from 1 to 36).

4. Color-to-Gray

Most of the algorithms involved in the image-based 3D reconstruction pipeline (mainly feature extraction for tie points identification and dense image matching) are conceptually designed to work on grayscale images (*i.e.*, single-band images) instead of the RGB triple. This is basically done to highly reduce the computational complexity of the algorithms compared to the utilization of the three channels. Color to grayscale conversion (or decolorization) can be seen as a dimensionality reduction problem and it should not be underestimated, as there are many different properties that need to be preserved. Over the past decades, different color-to-gray algorithms have been developed to derive the best possible decolorized version of a color image [20]. All of them focus on the reproduction of color images with grayscale mediums, with the goal of: (i) a perceptual accuracy in terms of the fidelity of the converted image; and (ii) a preservation of the color contrast and image structure contained in the original color also in the final decolorized image. Nevertheless, these kinds of approaches are not designed to fulfill the needs of image matching algorithms where local contrast preservation is crucial during the matching process. This was also observed in Lowe [50] where the candidate key points with low contrast are rejected in order to decrease the ambiguity of the matching process.

Color-to-gray conversion methods can be classified according to their working space:

- *Color Space* (linear or non-linear): The CIE Y method is a widely used conversion, based on the CIE 1931 XYZ color space. It takes the XYZ representation of the image and uses Y as the grayscale value.
- *Image Space* (also called functional): Following Benedetti *et al.* [51], these methods can be divided in three groups.

    (a) *Trivial methods*: They are the most basic and simple ones, as they do not take into account the spectral power distribution (SPD) of the color channels. They lose a lot of image information as for every pixel they discard two of the three color values, or discard one value averaging the remaining ones, not taking into account any color properties. Despite this loss, they are commonly used for their simplicity and speed. A typical trivial method is the RGB Channel Filter that selects a channel between R, G or B and uses this channel as the grayscale value (this method is afterwards called *GREEN2GRAY*).

    (b) *Direct methods*: The conversion is a linear function of the pixel's color values. Typically, this class of functions takes into account the spectrum of different colors. The Naive Mean direct method takes the mean of the color channels. With respect to trivial methods, it takes information from every channel, though it does not consider the relative spectral power distribution (SPD) of the RGB channels. The most popular of these methods is *RGB2GRAY* that uses the NTSC CCIR (National Television System Committee—Consultative Committee on International Radio) 601 luma weights, with the formula:

$$Y' = 0.2989R' + 0.5870G' + 0.1140B' \tag{2}$$

Other weights can be used, according to the users and software (e.g., Adobe Photoshop uses these specific weights for the channels R, G, and B: 0.4, 0.4, 0.2).

    (c) *Chrominance direct methods*: They are based on more advanced algorithms, trying to mitigate the problem related to isoluminant colors. They assign different grayscale values to isoluminant colors, altering the luminance information and using the chrominance information. In order to increase or decrease the "correct" luminance to differentiate isoluminant colors, these methods exploit a result from studies on human color perception, known as the Helmholtz–Kohlrausch (H-K) effect [52]. The H-K effect states that the perceived lightness of a stimulus changes as a function of the chroma. This phenomenon is predicted by a chromatic lightness term that corrects the luminance based on the color's chromatic component and on starting color space. Chrominance direct methods can be performed either locally [53,54] or globally [55]. Local methods make pixels in the color image not processed in the same way and usually rely on the local chrominance edges for enhancement. Global methods strive to produce one mapping function for the whole image thus producing same luminance for the same RGB triplets and high-speed conversion.

4.1. Evaluated Methods

We investigated different color-to-gray methods, namely:

- *GREEN2GRAY*: It is a trivial method working in Image Space where the green channel is extracted from a RGB image and used to create the final grayscale image.
- *Matlab RGB2GRAY*: It is a direct method implemented in Matlab and based on the above mentioned weighted sum of the three separate channels.
- *Decolorize* [55]: The technique performs a global grayscale conversion by expressing the grayscale as a continuous, image-dependent, piecewise linear mapping of the primary RGB colors and their saturation. Their algorithm works in the YPQ color opponent space and aims to perform a contrast enhancement too. The color differences in this color space are projected onto the two predominant chromatic contrast axes and are then added to the luminance image. Unlike a principal component analysis, which optimizes the variability of observations, a predominant component analysis optimizes the differences between observations. The predominant chromatic axis aims to capture, with a single chromatic coordinate, the color contrast information that is lost in the luminance channel. The luminance channel Y is obtained with the NTSC CCIR 601 luma weights. The method is very sensitive to the issue of gamma compression with some risks of decrease of the quality of the results mainly in light areas or dark areas where many features will be lost because the saturation balancing interacts incorrectly with the outlier detection.
- *Realtime* [56–58]: This method is based on the consideration that in the human visual system the relationship to the adjacent context plays a vital role to order the different colors. Therefore, the method relaxes the color order constraint and seeks better preservation of color contrast and significant enhancement

of visual distinctiveness for edges. For color pairs without a clear order in brightness, a bimodal distribution (*i.e.*, a mixture of two Gaussians) is performed to automatically find suitable orders with respect to the visual context in optimization. This strategy enables automatically finding suitable grayscales and preserves significant color changes. Practically the method uses a global mapping scheme where all color pixels in the input are converted to grayscale using the same mapping function (a finite multivariate polynomial function). Therefore, two pixels with the same color will have the same grayscale. The technique is today implemented in OpenCV 3.0. In order to achieve real-time performance, a discrete searching optimization can be used.

- *Adobe Photoshop.*

To evaluate the performances of the aforementioned methods (Figure 5), we applied the pixel-by-pixel difference method applying an offset of 127 levels of brightness to better identify the differences. This technique is the most appropriate to evaluate a method's efficiency for machine readable process. The simple image subtraction can rapidly provide visual results rather than using CIELAB ΔE^*ab or other perceptually-based image comparison methods.

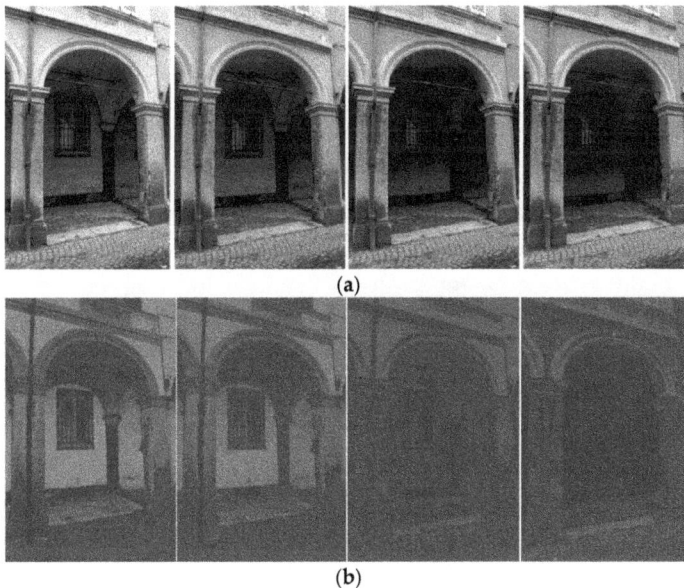

Figure 5. Results of color-to-gray conversion methods: produced gray scale images (**a**) and image difference (**b**) between (from left to right) *Adobe Photoshop* and *Decolourize, Adobe Photoshop* and *Realtime, Adobe Photoshop* and *Matlab RGB2GRAY*, and *Adobe Photoshop* and *BID*.

4.2. Proposed Approach

Based on the results achieved with the aforementioned methods, a new decolorization technique, named Bruteforce Isoluminants Decrease (*BID*), was developed. The aim of BID is to preserve the consistency between different images considering the following requirements.

- *Feature discriminability*: The decolorization method should preserve the image features discriminability in order to match them in as many images as possible.
- *Chrominance awareness*: The method should distinguish between isoluminant colors.
- *Global mapping*: While the algorithm can use spatial information to determine the mapping, the same color should be mapped to the same grayscale value for every pixel in the image.
- *Color consistency*: The same color should be mapped to the same grayscale value in every image of the dataset.
- *Grayscale preservation*: If a pixel in the color image is already achromatic, it should maintain the same gray level in the grayscale image.
- *Unsupervised algorithm*: It should not need user tuning to work properly, in particular for large datasets.

BID computes the statistical properties of the input dataset with the help of a representative collection of image patches. Differently from the *Multi-Image Decolourize* method [51], *BID* is a generalization of the *Matlab RGB2GRAY* algorithm, which simultaneously takes in input and analyses the whole set of images that need to be decolorized. *BID* has its foundation in the statistics of extreme-value distributions of the considered images and presents a more flexible strategy, adapting dynamically channel weights depending on specific input images, in order to find the most appropriate weights for a given color image. *BID* preserves as much as possible the amount of the conveyed information. The algorithm behind *BID* tries to maximize the number of peaks obtained in the converted image and to distribute as evenly as possible the amount of tones by evaluating the goodness of a fitting distribution. To calculate the best rectangular fitting, we assumed a 0 slope regression line. The general equation of the regression line is:

$$\beta = \bar{y} - m\bar{x} \tag{3}$$

where β is equivalent to the average of the histogram points. After calculating the average, the minimum error within all the calculated combinations of channel mixings if sought. The error is calculated as a least squares error:

$$S = \sum_{i=1}^{n} (y_i - \beta)^2 \tag{4}$$

where y_i are the actual points, while β is the best linear fitting of the histogram. *BID* cyclically varies the amount of red, green and blue and for each variation calculates the distribution of the resulting grayscale image and assesses the fitting quality with respect to a rectangular distribution. Then, *BID* chooses the mixing that maximizes the number of tones obtained in the converted image. Finally, similarly to Song *et al.* [59], *BID* uses a measurement criterion to evaluate the decolorization quality, *i.e.*, the newly defined dominant color hypothesis.

Figure 6 reports an example of *BID* results with respect to *Matlab RGB2GRAY* method. The main disadvantage of the developed method is the high computational pre-processing time due to the sampled patches on each image of the dataset.

Figure 6. A comparison between *Matlab RGB2GRAY* (**left**) and *BID* (**right**) that shows a smoother histogram.

5. Image Content Enhancement with Wallis Filtering

Image contents play a fundamental role in many processing and feature extraction methods. There are various enhancement algorithms to sharp and increase the image quality [9,60–62]. For image-based 3D reconstruction purposes, low-texture surface (such as plaster building facades) causes difficulties to feature detection methods (such as the Difference-of-Gaussian (DoG) function) and matching algorithms, leading to outliers and unsuccessful matching results. Among the proposed methods to enhance image contents, the Wallis filter [11] showed very successful performances in the photogrammetric community [63–67]. Jazayeri *et al.* [68] tested the Wallis filter for different parameters to evaluate its performances for interest point detection and description. Those results demonstrated that an optimum range of values exists and depending on the requirements of the user, but automatic value selection remains undetermined.

The filter is a digital image processing function that enhances the contrast levels and flattens the different exposure to achieve similar brightness of gray level values. The filter uses two parameters to control the enhancement's amount, the *contrast expansion factor A* and the *brightness forcing factor B*. The algorithm is adaptive and adjusts pixel brightness values in local areas only, contrary to a global contrast filter, which applies the same level of contrast throughout an entire image. The resulting enhanced image contains greater detail in both low and high-level contrast regions concurrently, ensuring that good local enhancement is achieved throughout the entire image. The Wallis filter requires the user to accurately set a target mean and standard deviation in order to locally adjust areas and match the user-specified target values. Firstly, the filter divides the input image into neighboring square blocks with a user-defined size ("window size") in order to calculate local statistics. Then, mean (M) and standard deviation (S) of the unfiltered image are calculated for each individual block based on the gray values of the pixels and the resulting value is assigned to the central cell of each block. The mean and standard deviation values of all other cells in the block are calculated from this central cell by bilinear interpolation. In this way, each individual pixel gets its own initial local mean and standard deviation based on surrounding pixel values. The user-defined mean and standard deviation values are then used to adjust the brightness and the contrast of the input cells. The resulting enhanced image is thus a weighted combination of the original and user-defined mean and standard deviation of the image. The implementation of the filter of Wallis, given the aforementioned factor A and B, can be summarized as follows:

- let S be the standard deviation for the input image;
- let M be the mean for the input image;
- for each (x,y) pixel in the image,
- calculate local mean m and standard deviation s using a NxN neighborhood; and finally
- calculate the enhanced output image as

$$(x, y) = S \times (\text{input}(x, y) - m)/(s + A) + M * B + m \times (1 - B) \qquad (5)$$

The quality of the Wallis filter procedure relies on two parameters: the *contrast expansion factor A* and the *brightness forcing factor B*. The main difficulty when using the Wallis filter is the correct selection of these parameters, in particular for large datasets, where a unique value of A or B could lead to unsuitable enhanced images. Although several authors reported parameters for successful projects, the filter is more an *"ad-hoc"* recipe than an easily deployable system for an automatic photogrammetric pipelines. To overcome this problem and following the achievement presented in [69], a Wallis parameters characterization study was carried out to automatically determine them. Three different datasets, each one composed of three images and involving the majority of possible surveying case studies were used (Figure 7):

1. a cross vault characterized by smooth bright-colored plaster;
2. a building facade and porticoes with smooth plaster; and
3. a Venetian floor, with asphalt and porphyry cubes with a background facade overexposed and an octagonal pillar in the foreground coated with smooth plaster.

Figure 7. Employed datasets (cross vault with plaster, building façade and Venetian floor, respectively) to study the automatic selection of the Wallis filter parameters.

For every dataset, the images were enhanced using different Wallis parameters and then matched to find homologues points using a calibrated version [7] of the SIFT operator available in Vedaldi's implementation [69]. This characterization procedure delivered the following considerations.

1. The number of extracted tie points is inversely proportional to the value of the parameter A, but the number of correct matches remains basically

stable when varying A, which can then be set at high values to speed up the computation (6–8).

2. Varying the user-specified standard deviation, the number of tie points and correct matches increases substantially linearly up to a value of 100 and then remains constant (Figure 8a).

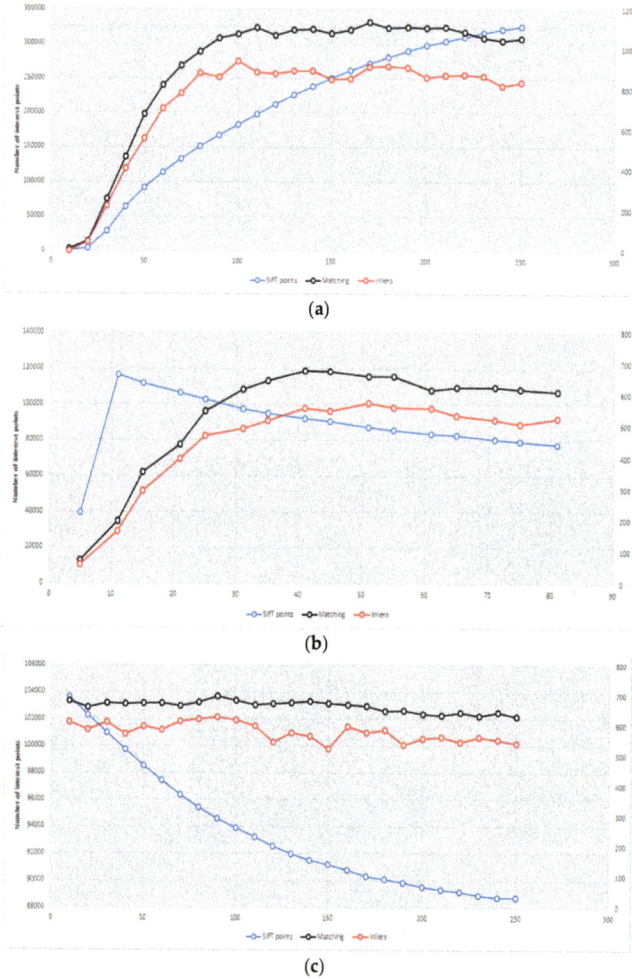

(a)

(b)

(c)

Figure 8. Results of the Wallis evaluation on the cross vault dataset. The charts report the number of extracted interest points with respect to the considered parameter values: (**a**) cross vault dataset—performance evaluation varying the standard deviation; (**b**) cross vault dataset—performance evaluation varying the window size; and (**c**) cross vault dataset—performance evaluation varying the mean.

114

3. Sensor resolution and window size are linearly related and the increasing of the window size beyond the optimal value does not involve any improvement in either the number of positive matches and in the number of extracted tie points (Figure 8b).
4. The mean presents optimal values between 100 and 150 with a decay afterwards (Figure 8c).

Starting from these observations, a new implementation of the Wallis filter was realized to select the optimal filter parameters and achieve the highest possible ratio of corrected matches with respect to the number of extracted tie points. In particular, the window size parameter is chosen according to the sensor resolution, and presents a linear variation starting from the experimental best values of 41 for a 14 MPixel sensor and 24 for a 10 MPixel sensor. According to our experimental trials, the standard deviation was forcefully set to 60 and the mean to 127. The Contrast Expansion Constant parameter (A) was set to 0.8 to increase the number of detected interest points located in homogeneous and texture-less areas and, alongside, to speed up the computation. The brightness forcing factor (B) according to the experimental results, was set to 1 if the image mean was lower than 150, linearly decreased otherwise, evaluating the entropy of the image.

Figure 9 shows the results of Wallis filtering: lower image contents are boosted, whereas a better histogram is achieved.

Figure 9. Example of the Wallis filtering procedure: note the much smoother and Gaussian-like histogram.

6. Assessment of the Proposed Methodology

The implemented pre-processing procedure was evaluated on various image networks featuring different imaging configurations, textureless areas and repeated pattern/features. The employed datasets try to verify the efficiency of different techniques in different situations (scale variation, camera rotation, affine transformations, *etc.*). The datasets contain convergent images, some orthogonal camera rolls and a variety of situations emblematic of failure cases, *i.e.*, 3D scenes (non-coplanar) with homogeneous regions, distinctive edge boundaries (e.g., buildings, windows, doors, cornices, arcades), repeated patterns (recurrent architectural elements, bricks, *etc.*), textureless surfaces and illumination changes. With respect to other evaluations where synthetic datasets, indoor scenarios, low resolution images, flat objects or simple two-view matching procedures are used and tested, such datasets are more varied with the aim of a complete and precise scene's 3D reconstruction.

All algorithms are tested and applied to raw images, *i.e.*, images as close as possible to the direct camera output retaining only the basic in-camera processing: black point subtraction, bad pixel removal, dark frame, bias subtraction and flat-field correction, green channel equilibrium correction, and Bayer interpolation. The datasets are processed with different image orientation software (Visual SFM, Eos Photomodeler and Agisoft Photoscan), trying to keep a uniform number of extracted key points and tie points. Then, dense point clouds are extracted with a unique tool (nFrames SURE). The performances of the pre-processing strategies are reported using:

(i) pairwise matching efficiency *i.e.*, number of correct inlier matches after the RANSAC (RANdom SAmple Consensus) phase normalized with all putative correspondences (Section 6.1);

(ii) the statistical output of the bundle adjustment (Sections 6.2 and 6.3);

(iii) the number and density of points in the dense point cloud (Sections 6.2 and 6.3); and

(iv) an accuracy evaluation of the dense matching results (Section 6.2).

6.1. Dataset 1

The first dataset (four images acquired with a Nikon D3100, sensor size 23.1 × 15.4 mm, 18 mm nominal focal length) shows part of Palazzo Albergati (Bologna, Italy) characterized by repeated brick walls, stone cornices and a flat facade. The camera was moved along the façade of the building, then tilted and rotated (Figure 10). This set of images is used to evaluate the denoise and color-to-gray techniques with respect to the tie points extraction procedure. The pairwise matching is assessed using three camera movements: (i) parallel with short baseline (a,b); (ii) rotation of *ca.* 90° (a–d); and (iii) tilt of *ca.* 45° (b,c).

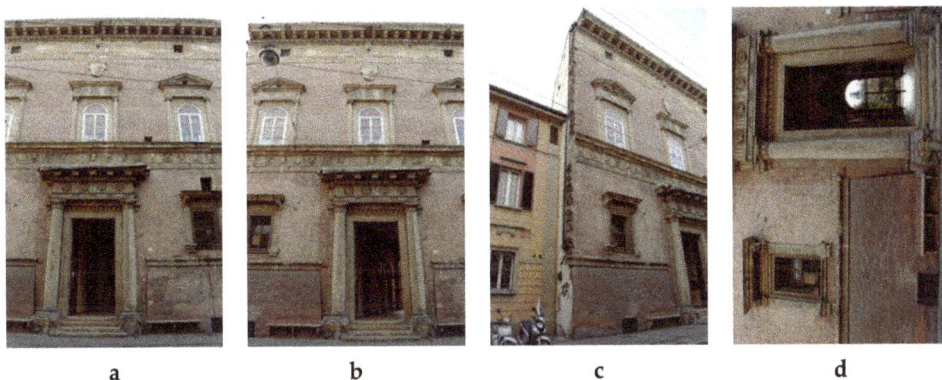

a   b   c   d

Figure 10. The images used to assess the pairwise matching efficiency of the tie point extraction procedure: parallel acquisitions with short baseline (**a,b**); tilt of *ca.* 45° (**c**) and rotation of *ca.* 90° (**d**).

Table 1 reports the computed pairwise matching efficiency after applying the various denoising methods and color-to-gray techniques. The developed methods (*CBM3D-new* and *BID*) demonstrate a better efficiency in the tie point extraction.

Table 1. Efficiency of each denoise and color-to-gray technique for the image pairs of Dataset 1. *Adobe* method is Camera Raw 6 with Process 2010 (see Section 3.1).

	Denoising				Color-to-Gray		
	Parallel 00-01	Rotate 90° 00-03	Tilt 45° 01-02		Parallel 00-01	Rotate 90° 00-03	Tilt 45° 01-02
Adobe	0.982	0.812	0.5	*Adobe*	0.992	0.821	0.630
CBM3D	**0.991**	0.837	0.561	*REALTIME*	0.992	0.827	0.618
NLBayes	0.978	0.801	**0.651**	*Decolourize*	0.992	**0.863**	0.626
Noiseclinic	0.984	0.769	0.473	*RGB2GRAY*	0.980	0.690	0.329
IMAGENOMIC	0.980	0.690	0.329	*GREEN2GRAY*	0.992	0.786	0.640
Nodenoise	0.975	0.679	0.335	*BID*	**0.993**	0.825	**0.676**

6.2. Dataset 2

The second dataset (35 image acquired with a Nikon D3100, sensor size 23.1 × 15.4 mm, 18 mm nominal focal length) concerns two spans of a three floors building (6 × 11 m) characterized by some arches, pillars, cross vaults and plastered walls with uniform texture. The camera was moved along the porticoes, with some closer shots of the columns (Figure 11). With this dataset we report how color balancing and denoising methodologies help improving the bundle adjustment and dense matching procedures. The accuracy evaluation of the dense matching results is done using a Terrestrial Laser Scanning (TLS) survey as reference (a Faro Focus3D was employed). Three regions (Figure 12A1–A3) are identified and compared with the photogrammetric dense clouds. The average image GSD (Ground Sample Distance) in the three regions of interest is *ca.* 2 mm but the dense matching was carried out using the second-level image pyramid, *i.e.*, at a quarter of the original image resolution. Therefore, in order to have a reference comparable to the dense matching results, the range data are subsampled to a grid of 5 × 5 mm.

Figure 11. Dataset 2 (35 images)—two arches of a portico.

6.2.1. Color Balance Results

The results of the orientation and dense matching steps are reported in Table 2. The color balancing procedure generally helps in increasing the number of oriented images, except with PS where the dataset is entirely oriented at every run. Furthermore, it helps in deriving denser point clouds.

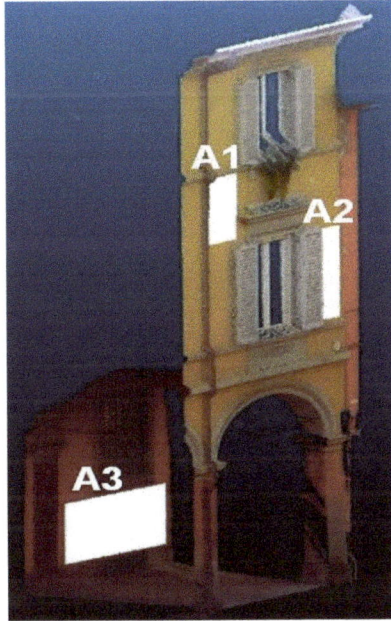

Figure 12. The three regions of interest (A1–A3) surveyed with TLS as ground truth.

Table 2. Bundle adjustment (BA) results and dense matching improvement on Dataset 2.

	Not Enhanced	*Color Balanced*
VisualSFM (VSFM)		
Numb. oriented images	31	33
BA quality (px)	0.48	0.48
EOS Photomodeler (PM)		
Numb. oriented images	31	33
BA reprojection error (px)	1.44	0.89
Agisoft Photoscan (PS)		
Numb. oriented images	35	35
BA reprojection error (px)	0.51	0.54
Dense Matching (SURE)		
# 3D points	1,259,795	1,626,267

6.2.2. Image Denoising Results

The denoising methods are coupled to *Matlab RGB2GRAY* and Wallis filtering before running the automated orientation and matching procedures. The achieved adjustment results, according to the different denoising methods, show that more images can be oriented (Table 3) and denser point clouds can be retrieved (Table 4).

Table 3. Bundle adjustment (BA) results for every denoising procedure on Dataset 2. *Adobe* method is Camera Raw 6 with Process 2010 (see Section 3.1). On such a small dataset, Photoscan seems to be quite robust even without any pre-processing method.

	No Denoise	Adobe	CBM3D-New	NL-Bayes	Noise Clinic	Image Nomic
VisualSFM (VSFM)						
Numb. oriented images	32	35	35	35	35	35
BA quality (px)	0.29	0.48	0.48	0.45	0.69	0.37
EOS PhotoModeler (PM)						
Numb. oriented images	33	33	33	33	33	31
BA reprojection error (px)	0.89	0.84	0.83	0.86	0.87	0.88
Agisoft Photoscan (PS)						
Numb. oriented images	35	35	35	35	35	35
BA reprojection error (px)	0.49	0.49	0.49	0.49	0.53	0.55

Table 4. Evaluation of the denoising procedures on the dense matching phase. The ground truth is given by a TLS survey resampled to 5×5 mm grid. The histograms of the cloud-to-cloud point distribution errors are also reported.

		No Denoise	Adobe	CBM3D-New	NL-Bayes	Noise Clinic	Image Nomic
Numb. 3D points		998,995	1,308,768	1,456,024	1,456,561	1,428,996	1,346,559
A1 (0.8 × 1.4 m): 36.294 pts	# 3D points	1227	27,508	16,257	31,386	23,419	8194
	Std Dev (mm)	N/A	12.26	13.61	11.73	12.07	17.22
A2 (0.5 × 1.9 m): 33.868 pts	# 3D points	495	25,327	17,966	29,632	23,541	7548
	Std Dev (mm)	N/A	11.18	11.73	11.90	11.04	19.00
A3 (2.6 × 1.3 m): 120.222 pts	# 3D points	32.014	176,478	179,553	183,166	184,835	160,120
	Std Dev (mm)	N/A	11.76	12.35	11.44	10.03	10.99

6.2.3. Color-to-Gray Results

The color-to-gray conversion (coupled with the Wallis filtering) shows how algorithms are differently affecting the BA procedure (Table 5) as well as the dense matching results (Table 6). It can be generally noticed that the proposed *BID* method allows retrieving a larger number of oriented images, a better re-projection errors and denser point clouds.

Table 5. Bundle adjustment (BA) results of the various color-to-gray procedures on Dataset 2.

	GREEN2GRAYAdobe	Realtime	Decolourize	RGB2GRAYBID	
VisualSFM (VSFM)					
Numb. oriented images	35	33	30	16	33	35
BA reprojection error (px)	0.55	0.42	0.37	0.38	0.35	0.58
EOS PhotoModeler (PM)					
Numb. oriented images	35	33	33	28	32	35
BA reprojection error (px)	0.87	0.89	0.87	0.89	0.88	0.86
Agisoft Photoscan (PS)					
Numb. oriented images	35	35	35	35	35	35
BA reprojection error (px)	0.52	0.53	0.54	0.54	0.52	0.51

6.3. Dataset 3

The last dataset (265 images acquired with a Nikon D3100, sensor size 23.1 × 15.4 mm, 18 mm nominal focal length) regards a three floors historical building (19 m height × 10 m width), characterized by arcades with four arches, columns, cross vaults and plastered walls with uniform texture. The camera was moved along the porticoes, with some closer shots of the columns and of the vaults (Figure 13). The dataset is used to show how the various steps of the proposed pre-processing pipeline positively influence the 3D reconstruction procedure. Table 7 reports the achieved image orientation results without and with pre-processing. The developed methods (*CBM3D-new* for the noise reduction and *BID* for the color-to-gray conversion) allow orienting larger numbers of images.

Table 6. Dense matching results of the various color-to-gray procedures on Dataset 2. Comparison with respect to the ground truth TLS data are also shown. The histograms represent the point distribution errors of the cloud-to-cloud alignments.

		GREEN2GRAY	Adobe	Realtime	Decolourize	RGB2GRAY	BID
Numb. 3D points		1,703,607	1,444,269	1,522,044	1,375,971	1,184,432	1,964,397
A1—36,294 pts	# 3D points	27,127	12,776	10,890	N/A	6863	11,360
	Std Dev (mm)	11.28	14.83	12.78	N/A	15.13	18.83
					N/A		
A2—33,868 pts	# 3D points	32,619	12,853	15,264	N/A	6520	9950
	Std Dev (mm)	10.76	13.82	10.77	N/A	12.39	22.29
					N/A		
A3—120,222 pts	# 3D points	162,563	153,281	163,246	N/A	152,923	157,837
	Std Dev (mm)	11.95	10.82	8.71	N/A	8.89	10.82
					N/A		

The dense matching procedure is then applied starting from the orientation results achieved in Photoscan results (Table 8). In this case, it is also shown how an appropriate pre-processing procedure allows deriving denser point clouds. The point density distribution for the different dense clouds (named Local Density Computation) was estimated. The density has been computed using a tool able to count, for each 3D point of the cloud, the number of neighbors N (inside a sphere of a radius R, fixed at 20 cm). The results of the Local Density Computation (shown as color-coded maps and histograms) show that the successive combination of the proposed pre-processing methods gradually achieve, beside an increasing amount of 3D points, a higher density and a more uniform distribution of points.

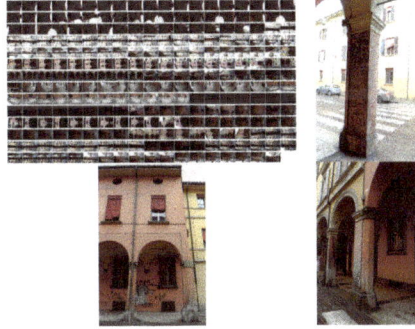

Figure 13. Dataset 3 (265 images)—a portico with arcades, columns, vaults and plastered walls.

Table 7. Result of the image orientation on Datasets 3 (265 images) without or with various pre-processing methods.

	No Pre-Processing	No Color Correction, Only Wallis	Color Correction, No Denoise, RGB2GRAY, Wallis	CBM3D-New, RGB2GRAY, Wallis	CBM3D-New, BID, Wallis
VisualSFM (VSFM)					
Numb. oriented images	90 (2nd run)	218 (4th run)	214 (4th run)	232 (4th run)	249 (1st run)
BA quality (px)	0.739	0.650	0.285	0.411	0.750
PhotoModeler Scanner (PM)					
Numb. oriented images	55	N/A	180	247	247
BA quality (px)	0.687	N/A	0.779	0.804	0.790
Agisoft Photoscan (PS)					
Numb. oriented images	262	264	265	265	265
BA quality (px)	0.695	0.696	0.689	0.687	0.679

Table 8. Results of dense matching on Dataset 3 without and with various pre-processing methods. The Local Density Computation (radius: 20 cm) is also reported. The color-code maps show the matched point distribution on the surveyed scene. The histogram reports the number of neighbors (X-axis) *versus* the number of matched points in the cloud (Y-axis).

	No Pre-Processing	No Color Correction, Only Wallis	Color Correction, No Denoise, RGB2GRAY, Wallis	CBM3D-New, RGB2GRAY, Wallis	CBM3D-New, BID, Wallis
Dense Matching (nFrames SURE)					
# 3D points	7,086,643	13,741,089	14,248,488	14,770,940	18,266,571
Maps of the number of neighbors					
Histograms of the number of neighbors					

123

7. Conclusions

The paper reported a pre-processing methodology to improve the results of the automated photogrammetric pipeline for 3D scene reconstruction. The developed pipeline consists of color balancing, image denoising, color-to-gray conversion and image content enrichment. Two new methods for image denoising (*CBM3D-new*) and grayscale reduction (*BID*) were also presented. The pipeline was evaluated using some datasets of architectural scenarios and advantages were reported. From the results in Section 6, it is clear that the pre-processing procedure, which requires very limited processing time, generally positively influences the performances of the orientation and dense matching algorithms. The evaluation shows how the combination of the various methods is indeed helping in achieving complete orientation results, sometimes better BA re-projection errors (although it is not a real measure of better quality) and, above all, denser and complete 3D dense point clouds.

As shown in the results, the best strategy implies applying a color enhancement, a denoise procedure based on the *CBM3D-new* method, the *BID* method for grayscale conversion and the Wallis filtering. This latter filtering seems to be fundamental also in the orientation procedure and not only when applying dense matching algorithms (as reported in the literature).

The developed pre-processing method is quite flexible and features the following characteristics.

- It is derived from many experiments and merging various state-of-the-art methods.
- The setting parameters can be fixed reading the image metadata (EXIF header).
- The denoising and grayscale conversion consider the entire dataset and they are not image-dependent.
- It is customized for improving the automated 3D reconstruction procedure.
- It is not based only on perceptual criteria typically used in image enhancement algorithms.
- It also gives advantages to the texture mapping phase.

The presented research substantially improved our 3D reconstruction pipeline and allows us to model large architectural scenarios for documentation, conservation and communication purposes (Figure 14).

Figure 14. Example of dense 3D point cloud of a portico obtained with the presented photogrammetric methodology based on an advanced pre-processing pipeline.

Author Contributions: All authors equally contributed to the procedure development, data processing and paper preparation.

Conflicts of Interest: The authors declare no conflict of interest.

References

1. Snavely, N.; Seitz, S.M.; Szeliski, R. Modeling the world from internet photo collections. *Int. J. Comput. Vis.* **2008**, *80*, 189–210.
2. Barazzetti, L.; Scaioni, M.; Remondino, F. Orientation and 3D modeling from markerless terrestrial images: Combining accuracy with automation. *Photogramm. Rec.* **2010**, *25*, 356–381.
3. Heinly, J.; Schönberger, J.L.; Dunn, E.; Frahm, J.M. Reconstructing the World* in Six Days*(As Captured by the Yahoo 100 Million Image Dataset). In Proceedings of the IEEE CVPR, Boston, MA, USA, 7–12 June 2015.
4. Pierrot-Deseilligny, M.; de Luca, L.; Remondino, F. Automated image-based procedures for accurate artifacts 3D modeling and orthoimage generation. *Geoinform. FCE CTU J.* **2011**, *6*, 291–299.
5. Remondino, F.; Del Pizzo, S.; Kersten, T.P.; Troisi, S. Low-cost and open-source solutions for automated image orientation—A critical overview. In Progress in Cultural Heritage Preservation, Proceedings of the 4th International Conference, EuroMed 2012, Lemessos, Cyprus, 29 October–3 November 2012; pp. 40–54.
6. Crandall, D.; Owens, A.; Snavely, N.; Huttenlocher, D. SfM with MRFs: Discrete-Continuous Optimization for Large-Scale Structure from Motion. *IEEE PAMI* **2013**, *35*, 2841–2853.

7.   Apollonio, F.I.; Ballabeni, A.; Gaiani, M.; Remondino, F. Evaluation of feature-based methods for automated network orientation. *Int. Arch. Photogramm. Remote Sens. Spat. Inf. Sci.* **2014**, *XL-5*, 47–54.

8.   Remondino, F.; Spera, M.G.; Nocerino, E.; Menna, F.; Nex, F. State of the art in high density image matching. *Photogramm. Rec.* **2014**, *29*, 144–166.

9.   Gonzalez, R.F.; Woods, R. *Digital Image Preprocessing*; Prentice Hall: Upper Saddle River, NJ, USA, 2007; p. 976.

10. Milanfar, P. A tour of modern image filtering: New insights and methods, both practical and theoretical. *Signal Process. Mag.* **2013**, *30*, 106–128.

11. Wallis, R. An approach to the space variant restoration and enhancement of images. In Proceedings of the Symposium on Current Mathematical Problems in Image Science, Monterey, CA, USA, 10–12 November 1976; pp. 329–340.

12. Degenhard, A.; Tanner, C.; Hayes, C.; Hawkes, D.J.; Leach, M.O. Pre-processed image reconstruction applied to breast and brain MR imaging. *Physiol. Meas.* **2001**, *22*, 589–604.

13. Mahiddine, A.; Seinturier, J.; Boi, D.P.J.; Drap, P.; Merad, D.; Long, L. Underwater image preprocessing for automated photogrammetry in high turbidity water: An application on the Arles-Rhone XIII roman wreck in the Rhodano River, France. In Proceedings of the IEEE VSMM Conference, Milan, Italy, 2–5 September 2012; pp. 189–194.

14. Guidi, G.; Gonizzi, S.; Micoli, L.L. Image pre-processing for optimizing automated photogrammetry performances. *ISPRS Int. Ann. Photogramm. Remote Sens. Spat. Inf. Sci.* **2014**, *II-5*, 145–152.

15. Maini, R.; Aggarwal, H. A comprehensive review of image enhancement techniques. *J. Comput.* **2010**, *2*, 8–13.

16. Stamos, I.; Liu, L.; Chen, C.; Wolberg, G.; Yu, G.; Zokai, S. Integrating automated range registration with multiview geometry for the photorealistic modeling of large-scale scenes. *Int. J. Comput. Vis.* **2008**, *78*, 237–260.

17. Klein, G.; Murray, D. Improving the agility of keyframe-based SLAM. In Proceedings of the 10th ECCV Conference, Marseille, France, 12–18 October 2008; pp. 802–815.

18. Mei, C.; Reid, I. Modeling and generating complex motion blur for real-time tracking. In Proceedings of the IEEE CVPR, Anchorage, AK, USA, 23–28 June 2008; pp. 1–8.

19. Lee, H.S.; Kwon, J.; Lee, K.M. Simultaneous localization, mapping and deblurring. In Proceedings of the IEEE ICCV Conference, Barcelona, Spain, 6–13 November 2011; pp. 1203–1210.

20. Verhoeven, G.; Karel, W.; Štuhec, S.; Doneus, M.; Trinks, I.; Pfeifer, N. Mind your gray tones—Examining the influence of decolourization methods on interest point extraction and matching for architectural image-based modelling. *Int. Arch. Photogramm. Remote Sens. Spat. Inf. Sci.* **2015**, *XL-5/W4*, 307–314.

21. Bellavia, F.; Fanfani, M.; Colombo, C. Fast Adaptive Frame Preprocessing for 3D Reconstruction. In Proceedings of the VISAPP Conference, Berlin, Germany, 11–14 March 2015; pp. 260–267.

22. Hong, G.; Luo, M.R.; Rhodes, P.A. A study of digital camera colourimetric characterization based on polynomial modeling. *Colour Res. Appl.* **2001**, *26*, 76–84.

23. Yaroslavsky, L.P. Local adaptive image restoration and enhancement with the use of DFT and DCT in a running window. *Proc. SPIE* **1996**, *2825*, 2–13.

24. Reinhard, E.; Arif Khan, E.; Oguz Akyüz, A.; Johnson, G. *Colour Imaging Fundamentals and Applications*; A K Peters: Wellesley, MA, USA, 2008.

25. McCamy, C.S.; Marcus, H.; Davidson, J.G. A colour rendition chart. *J. Appl. Photogr. Eng.* **1976**, *11*, 95–99.

26. Pascale, D. *RGB Coordinates of the Macbeth ColourChecker*; The BabelColour Company: Montreal, QC, Canada, 2006.

27. Sharma, G.; Wu, W.; Dalal, E.N. The CIEDE2000 Colour-difference formula: Implementation notes, supplementary test data and mathematical observations. *Colour Res. Appl.* **2005**, *30*, 21–30.

28. Song, T.; Luo, M.R. Testing color-difference formulae on complex images using a CRT monitor. In Proceedings of the IS & T and SID Eighth Color Imaging Conference, Scottsdale, AZ, USA, 7–10 November 2000; pp. 44–48.

29. ISO 15739: 2003 Photography—Electronic Still-Picture Imaging—Noise Measurements. 2013. Available online: http://www.iso.org/iso/home/store/catalogue_ics/catalogue_detail_ics.htm?csnumber=59420 (accessed on 15 January 2016).

30. Motwani, M.; Gadiya, M.; Motwani, R.; Harris, F. Survey of image denoising techniques. In Proceedings of Global Signal Processing Expo Conference (GSPx), Santa Clara, CA, USA, 2004; pp. 27–30.

31. Patil, J.; Jadhav, S. A comparative study of image denoising techniques. *Int. J. Innov. Res. Sci. Eng. Technol.* **2013**, *2*, 787–794.

32. Lebrun, M.; Colom, M.; Buades, A.; Morel, J.M. Secrets of image denoising cuisine. *Acta Numer.* **2012**, *21*, 475–576.

33. Nowak, R.; Baraniuk, R. Wavelet-domain filtering for photon imaging systems. *IEEE Trans. Image Process.* **1997**, *8*, 666–678.

34. Kolaczyk, E. Wavelet shrinkage estimation of certain Poisson intensity signals using corrected thresholds. *Stat. Sin.* **1999**, *9*, 119–135.

35. Makitalo, M.; Foi, A. Optimal inversion of the Anscombe transformation in low-count Poisson image denoising. *IEEE Trans. Image Process.* **2011**, *20*, 99–109.

36. Foi, A. Noise estimation and removal in MR imaging: The variance-stabilization approach. In Proceedings of the IEEE International Symposium on Biomedical Imaging: From Nano to Macro, Chicago, IL, USA, 30 March 2011–2 April 2011; pp. 1809–1814.

37. Imagenomic LLC. *Noiseware 5 Plug-In User's Guide*; Imagenomic LLC: Alexandria, VA, USA, 2012.

38. Petrosyan, A.; Ghazaryan, A. Method and System for Digital Image Enhancement. U.S. Patent 7751641 B2, 28 April 2005.

39. Seiz, G.; Baltsavias, E.P.; Grün, A. Cloud mapping from ground: Use of photogrammetric methods. *Photogram. Eng. Remote Sens.* **2002**, *68*, 941–951.

40. Buades, A.; Coll, B.; Morel, J.M. A review of image denoising algorithms, with a new one. *Multiscale Model. Simul.* **2005**, *4*, 490–530.

41. Kervramm, C.; Boulanger, J. Optimal spatial adaptation for patch-based denoising. *IEEE Trans. Image Process.* **2006**, *15*, 2866–2878.

42. Awate, S.P.; Whitaker, R.T. Unsupervised, information-theoretic, adaptive image filtering for image restoration. *IEEE Trans. Pattern Anal. Mach. Intell.* **2006**, *28*, 364–376.

43. Lebrun, M.; Colom, M.; Morel, J.M. The noise clinic: A blind image denoising algorithm. *Ipol J.* **2015**, *5*, 1–54.

44. Colom, M.; Buades, A.; Morel, J. Nonparametric noise estimation method for raw images. *J. Opt. Soc. Am. A* **2014**, *31*, 863–871.

45. Lebrun, M.; Colom, M.; Morel, J.M. The noise clinic: A universal blind denoising algorithm. In Proceedings of the IEEE International Conference on Image Processing, Paris, France, 27–30 October 2014; pp. 2674–2678.

46. Lebrun, M.; Colom, M.; Morel, J.M. Multiscale image blind denoising. *IEEE Trans. Image Process.* **2015**, *24*, 3149–3161.

47. Ponomarenko, N.N.; Lukin, V.V.; Zriakhov, M.S.; Kaarna, A.; Astola, J.T. An automatic approach to lossy compression of AVIRIS images. In Proceedings of the IEEE International Geoscience and Remote Sensing Symposium, Barcelona, Spain, 23–28 July 2007; pp. 472–475.

48. Dabov, K.; Foi, A.; Katkovnik, V.; Egiazarian, K. Colour image denoising via sparse 3D collaborative filtering with grouping constraint in luminance-chrominance space. In Proceedings of the IEEE International Conference on Image Processing, San Antonio, TX, USA, 16 September–19 October 2007; pp. 313–316.

49. Dabov, K.; Foi, A.; Katkovnik, V.; Egiazarian, K. Image denoising by sparse 3D transform-domain collaborative Filtering. *IEEE Trans. Image Process.* **2007**, *16*, 2080–2095.

50. Lowe, D. Distinctive image features from scale-invariant keypoints. *Int. J. Comput. Vis.* **2004**, *60*, 91–110.

51. Benedetti, L.; Corsini, M.; Cignoni, P.; Callieri, M.; Scopigno, R. Colour to gray conversions in the context of stereo matching algorithms: An analysis and comparison of current methods and an ad-hoc theoretically-motivated technique for image matching. *Mach. Vis. Appl.* **2012**, *23*, 327–348.

52. Shizume, T.; Ohashi, G.; Takamatsu, H.; Shimodaira, Y. Estimation of the Helmholtz-Kohlrausch effect for natural images. *J. Soc. Inf. Disp.* **2014**, *22*, 588–596.

53. Smith, K.; Landes, P.E.; Thollot, J.; Myszkowski, K. Apparent greyscale: A simple and fast conversion to perceptually accurate images and video. *Comput. Graph. Forum* **2008**, *27*, 193–200.

54. Kim, Y.; Jang, C.; Demouth, J.; Lee, S. Robust colour-to-gray via nonlinear global mapping. *ACM Trans. Graph.* **2009**, *28*.

55. Grundland, M.; Dodgson, N.A. Decolourize: Fast, contrast enhancing, colour to grayscale conversion. *Pattern Recognit.* **2007**, *40*, 2891–2896.

56. Lu, C.; Xu, L.; Jia, J. Contrast preserving decolourization. In Proceedings of the IEEE International Conference on ICCP, Seattle, WA, USA, 28–29 April 2012; pp. 1–7.

57. Lu, C.; Xu, L.; Jia, J. Real-time contrast preserving decolourization. In Proceedings of the SIGGRAPH Asia 2012 Technical Briefs, Singapore, 28 November–1 December 2012; ACM: New York, NY, USA, 2012.

58. Lu, C.; Xu, L.; Jia, J. Contrast preserving decolourization with perception-based quality metrics. *Int. J. Comput. Vis.* **2014**, *110*, 222–239.

59. Song, Y.; Bao, L.; Xu, X.; Yang, Q. Decolourization: Is rgb2gray() out? In Proceedings of the SIGGRAPH Asia 2013 Technical Briefs, Hong Kong, China, 19–22 November 2013; ACM: New York, NY, USA, 2013.

60. Ciocca, G.; Cusano, C.; Gasparini, F.; Schettini, R. Content-aware image enhancement. In Proceedings of the Artificial Intelligence and Human-Oriented Computing, Rome, Italy, 10–13 September 2007; pp. 686–697.

61. Kou, F.; Chen, W.; Li, Z.; Wen, C. Content adaptive image detail enhancement. *IEEE Signal Process. Lett.* **2015**, *22*, 211–215.

62. Sarkar, A.; Fairchild, M.D.; Caviedes, J.; Subedar, M. A comparative study of colour and contrast enhancement for still images and consumer video applications. In Proceedings of the 16th Colour Imaging Conference: Colour Science and Engineering Systems, Technologies and Applications, Portland, OR, USA, 10–15 November 2008; pp. 170–175.

63. Baltavias, E.P. Multiphoto Geometrically Constrained Matching. Ph.D. Thesis, Institute of Geodesy and Photogrammetry, ETH, Zurich, Switzerland, December 1991.

64. Baltavias, E.P.; Li, H.; Mason, S.; Stefanidis, A.; Sinning, M. Comparison of two digital photogrammetric systems with emphasis on DTM generation: Case study glacier measurement. *Int. Arch. Photogramm. Remote Sens. Spat. Inf. Sci.* **1996**, *31*, 104–109.

65. Ohdake, T.; Chikatsu, H. 3D modeling of high relief sculpture using image based integrated measurement system. *Int. Arch. Photogramm. Remote Sens. Spat. Inf. Sci.* **2005**, *36*, 6.

66. Remondino, F.; El-Hakim, S.; Gruen, A.; Zhang, L. Turning images into 3D models—Development and performance analysis of image matching for detailed surface reconstruction of heritage objects. *IEEE Signal Process. Mag.* **2008**, *25*, 55–65.

67. MacDonald, L.; Hindmarch, J.; Robson, S.; Terras, M. Modelling the appearance of heritage metallic surfaces. *Int. Arch. Photogramm. Remote Sens. Spat. Inf. Sci.* **2014**, *XL-5*, 371–377.

68. Jazayeri, I.; Fraser, C.S. Interest operators for feature-based matching in close range photogrammetry. *Photogramm. Rec.* **2010**, *25*, 24–41.

69. Vedaldi, A.; Fulkerson, B. VLFeat—An open and portable library of computer vision algorithms. In Proceedings of the 18th ACM International Conference on Multimedia, Firenze, Italy, 25–29 October 2010.

Data Product Specification Proposal for Architectural Heritage Documentation with Photogrammetric Techniques: A Case Study in Brazil

Mariana B. Campos, Antonio M. G. Tommaselli, Ivana Ivánová and Roland Billen

Abstract: Photogrammetric documentation can provide a sound database for the needs of architectural heritage preservation. However, the major part of photogrammetric documentation production is not used for subsequent architectural heritage projects, due to lack of knowledge of photogrammetric documentation accuracy. In addition, there are only a few studies with rigorous analysis of the requirements for photogrammetric documentation of architectural heritage. In particular, requirements focusing on the geometry of the models generated by fully digital photogrammetric processes are missing. Considering these needs, this paper presents a procedure for architectural heritage documentation with photogrammetric techniques based on a previous review of existing standards of architectural heritage documentation. The data product specification proposed was elaborated conforming to ISO 19131 recommendations. We present the procedure with two case studies in the context of Brazilian architectural heritage documentation. Quality analysis of the produced models were performed considering ISO 19157 elements, such as positional accuracy, logical consistency and completeness, meeting the requirements. Our results confirm that the proposed requirements for photogrammetric documentation are viable.

Reprinted from *Remote Sens.* Cite as: Campos, M.B.; Tommaselli, A.M.G.; Ivánová, I.; Billen, R. Data Product Specification Proposal for Architectural Heritage Documentation with Photogrammetric Techniques: A Case Study in Brazil. *Remote Sens.* **2015**, *7*, 13337–13359.

1. Introduction

Photogrammetric documentation of architectural heritage can be understood as a non-subjective data record of the historical, physical and temporal features of cultural monuments and buildings. This data record is a form of preservation that represents a permanent record of the state of architectural heritage at a specific time or period [1]. Photogrammetric documentation is composed of descriptive information and graphic representation of the architectural heritage structure developed with photogrammetric techniques. In our paper we call this data record 'a model'. Photogrammetry has become faster and more affordable with the advance of digital cameras, the development of digital photogrammetric platforms and automated solutions. Digital technology advantages reminded the Venice Charter principles [2], promoting new approaches in photogrammetric documentation (e.g., 3D reconstruction) and encouraging several projects and research in this domain [3].

Digital technology advantages in photogrammetric techniques do not always guarantee accurate models. As discussed by Nocerino *et al.* [4], some digital methods focus on fully automatic reconstruction and they are often not concerned with the accuracy and reliability of the generated model, resulting in heritage models mostly for visual applications, which causes a level of mistrust in the end users of photogrammetric documentation.

The main objective of photogrammetric documentation is to support architectural heritage preservation projects. However, due to lack of knowledge about reliability in the accuracy of digital photogrammetric models, the application of photogrammetric documentation for its main purpose is discouraged. All data and products derived from these data (e.g., photogrammetric documentation) are associated with a level of uncertainty. Description of data quality is necessary to help users understand the level of uncertainty associated with the product and evaluate whether the data product is fit for their use [5].

Selecting appropriate photogrammetric documentation to support architectural heritage preservation projects is not easily done, in particular by non-specialists, due to lack of understanding of standard terminologies and specifications in this domain. A Data Product Specification (DPS) for photogrammetric documentation can help with communication between data producers and users. Standardization promotes technological, economic and societal benefits, preventing information loss and providing knowledge transfer, quality improvement and effectiveness in data production [6].

Data product specification is a precise technical description of the data product in terms of the requirements that will enable the data product to be created, supplied to and used by another party [7]. The data product discussed in this paper is restricted to a set of points with three-dimensional coordinates, which enable architectural heritage surface modeling. Photogrammetric techniques provide other products, such as orthoimages and digital terrain model, which require different procedures. These products will be not discussed in this paper.

Some DPS for photogrammetric documentation are recognized by the heritage preservation community, including recommendations from the International Committee for Documentation of Cultural Heritage (CIPA) [8]. However, these specifications were elaborated before recent digital advances in photogrammetry and in most cases they do not include essential requirements, such as data capture, data quality and metadata. Hence, there is a need to update photogrammetric documentation requirements for architectural heritage.

The development of a DPS for photogrammetric documentation of architectural heritage is a challenge due to the architectural uniqueness of heritage structures, which makes standardization in the heritage documentation a complex task.

Motivated by the need for technical specification in photogrammetric documentation of architectural heritage, and the importance of communication between geomatics and heritage experts, in this paper we propose a procedure for architectural heritage documentation with photogrammetric techniques based on a review of existing architectural heritage documentation standards. The recommendations are focused on geometric aspects of the data product. We applied the suggested requirements in two case studies in the context of Brazilian architectural heritage documentation, which exemplifies the proposed DPS usage.

2. Review of Digital Documentation Techniques for Architectural Heritage

The importance of a sound database for architectural heritage preservation is well recognized at international level. Architectural heritage documentation can be used to provide a permanent record of monuments and buildings, ensure that the maintenance and conservation of the heritage is sensitive to changes in architectural heritage structure and acquire knowledge about heritage values [9]. Patias and Santana [10] define the documentation as a combination of a data report and a dossier of measured representations that can include a site plan, sections, elevations, three-dimensional models, among other documentation data. Digital heritage documentation is defined by Letellier [11] as a production and storage of computerized digital information, measured drawings, photogrammetric records, and other electronic data to form a cultural heritage record.

It is desirable that the documentation method be accurate, portable (due to the accessibility problem in architectural heritage locations), flexible (because of the variety of architectural heritage structures), low cost and with fast acquisition [12]. Digital technological advantages in survey and modeling help to achieve these objectives, with the new possibilities of digital procedures, product and storage.

Survey can be performed by direct or indirect measurements. Direct measurements (e.g., tape measure) demand contact with the structure, which, for preservation reasons, is not recommended for architectural heritage survey. Indirect measurement techniques are advantageous because no contact with the structure is required. Examples of such techniques include topographic surveying [13], photogrammetry [14], computer vision, such as Structure from Motion (SfM) [15], laser scanning [16], range imaging [17,18], reconstruction with shape from structured light [19,20] and multi-sensor integration [3]. These techniques have become faster and more affordable with technological advantages. Andrés and Pozuelo [21] presented an overview of the evolution of indirect techniques for architectural heritage documentation survey. In the same direction, other authors reviewed methods for 3D digitalization of architectural heritage [22–24].

Classic surveying techniques, such as topographic mapping, provide high accuracy measurements. However, these techniques can be lengthy and costly when a massive acquisition is demanded due to the high level of detail required [23,25]. In this case, classic surveying is combined with other indirect measurement techniques [26]. For instance, in complex architectural heritage modeling, a dense point cloud is needed to complete the coarse model produced by the topographic mapping technique. This combination is used in many architectural heritage documentation projects, as shown by Giuliano [27] who combined photogrammetry and classic survey to develop a model of the ruins of the mausoleum 'Torre del Ballerino'. Scherer and Lerma [28] presented a review of topographic equipment development, from conventional total stations to photogrammetric scanning stations.

Among the indirect measurement techniques mentioned above, the most widely used techniques for architectural heritage survey are photogrammetry and laser scanning, especially for mapping large and complex monuments and buildings, where there is hardly any alternative [8].

Photogrammetry was the first indirect measurement technique applied to architectural heritage documentation [29]. Photogrammetric technique has numerous advantages in architectural heritage documentation: it provides geometric and radiometric information, produces a suitable level of details across the whole façade-even with scale variations due to the different camera viewing angles-enables high accuracy models (e.g., up to millimeter level), identifies borders, has fast results, is low-cost and the photographs have documentation value [30]. However, loss of information caused by occlusions and image acquisition only during daylight could be some limitations of this technique, which could be circumvented, for instance, with additional images and artificial illumination. More details about photogrammetry advantages and limitations were discussed by Dallas [30].

Nowadays, terrestrial laser scanning systems are very popular for architectural heritage documentation. The main advantage of laser scanning is the fast collection of a large number of 3D coordinates of the cultural heritage structure. Nonetheless, the high density of points can be a disadvantage, due to a complexity of data processing.

The architectural heritage model developed by photogrammetry can be as accurate as the laser scanning models [31] and, compared to the model developed by the laser scanning technique, has lower costs [32]. Furthermore, photogrammetry provides object edges while laser scanning provides random point clouds, hindering intuitive interpretation. Boehler and Marbs [32] presented a complete comparison of photogrammetry and laser scanning, concluding that the techniques are complementary. As shown in several studies [33–36], the combination of these techniques brings positive results for architectural heritage documentation.

After the architectural heritage survey, the numerical model obtained should be converted to a geometric model. This process is known as reverse modeling [37]. Reverse modeling is a complex process that could be done using different modeling techniques, as discussed below.

For years, the objective in graphic representations was to reduce the three-dimensional surfaces to a two-dimensional representation, using projective geometry principles. Advances in computer graphic techniques created a new scenario with 3D possibilities for graphic representation of objects. With these new possibilities, architectural heritage modeling for documentation can be performed by several modeling methods [38], for example, surface-based methods or volumetric methods. The most common modeling techniques used to generate architectural heritage models are the Delaunay-based method [39], constructive solid geometry (CSG) [40], boundary representation (B-REP) [41] and voxel-based object reconstruction [42].

Choice of modeling method depends mainly on the complexity of the architectural heritage model and the required accuracy. The CSG method, for example, has an intuitive modeling process and is frequently applied for the representation of simple objects. On other hand, this method has a limited set of primitive operations that hinders the modeling of complex structures. The most frequently used method for complex architectural heritage is the B-REP, based on irregular mesh. Despite B-REP being computationally more complex than the CSG method, it enables more detailed representation of the dense point cloud. More details about computer graphic modeling methods were described by Watt [43]. A discussion of the principles for computer-based visualization application in heritage documentation was presented in the London charter, Section 2.1 [44].

3. Review of Existing Specifications

The purpose of the review of existing specifications for architectural heritage documentation is to identify normative references for data product specification for photogrammetric documentation of architectural heritage proposed in this paper. In the first instance, we reviewed specifications with international significance, accepted by the geomatics and heritage community. However, each country has its own heritage preservation policy and legislation, requiring an adaptation of international standards to the national scenario. Therefore, in a second instance we analyzed specifications for the documentation of architectural heritage with national significance.

In the 1980s, the International Committee of Architectural Photogrammetry realized the need for reflection about photogrammetric documentation quality and elaborated the Advice and Suggestions for the furtherance of Optimum Practice in Architectural Photogrammetry surveys (AS-OPAP) [8]. The main contribution of this specification was the recommendations for final quality control of the model. However, quality recommendations for photogrammetric processes, such as interior and exterior orientation, are not part of CIPA's recommendations. Quality control during data creation enables to achieve the desired final quality of the model.

A decade later, Waldhaeusl and Ogleby [45] presented the 3×3 rules for simple photogrammetric documentation of architecture, structured in three geometric rules (preparation of control information, multiple photographic all-around coverage and taking stereopairs for stereo-restitution), three photographic rules (keeping the inner geometry of the camera constant, selecting homogenous illumination and stable camera format) and three organizational rules (making proper sketches, writing proper protocols and making a final check). These guidelines were elaborated before recent advances in digital photogrammetry, especially for cameras devices, and updating them in line with rapid technological advancement is problematic. The same problem was identified in the requirements presented by Buchanan in Photographing Historic Buildings for the Record [46] that focus on analog image acquisition.

Accuracy Standards for Digital Geospatial Data (ASPRS) [47], the ISO TC 211 for geographic information [48], the International Heritage Documentation Standards (IHDS) [49] and the Recording, Documentation and Information Management for the Conservation of Heritage Places [11] are among the most recent international specifications to be applied to architectural heritage documentation. The last two specifications mentioned were supported by RecorDIM (Recording and Documentation Information Management). The IHDS emphasize the difficulty of international standardization for architectural heritage documentation requirements, due to the architectural uniqueness of the structures, which requires the use of various documentation techniques and shows the need for national specifications.

Historic American Building Survey (HABS) [50–52] presents a series of requirements for historical reports production and photographic survey for USA architectural heritage documentation. However, HABS specification does not provide requirements for digital modeling, hindering digital products analysis, such as, performing analysis of digital models considering analog requirements or the reduction of 3D to 2D models because of analog storage.

Standards and Guidelines for the Conservation of Historic Places in Canada (SGC) [53] present a set of recommendations for preservation, conservation and documentation of Canadian heritage. SGC does not include techniques for data surveying and this is a limitation for heritage documentation, since the quality of the data depends directly on the techniques used to survey the data.

Metric Survey Specification for Cultural Heritage (MSSCH) [54] contains recommendations for photogrammetric and laser scanning procedures and data quality analysis. However, it also has some problems with updating recommendations for digital cameras.

A guideline for photogrammetric survey focusing on architectural heritage applications can be found in "twelve tips for Metric Photography of Architectural and Archaeological Cultural Heritage" by GIFLE [55]. However, this advice is not intended for data processing and data quality analysis.

Finally, we note the Spanish recommendation of the Andalusian Institute: Technical Recommendations for Geometric Documentation of Heritage Entities [56] (*Recomendaciones técnicas para la documentacion geométrica de entidades patrimoniales*), which discusses the techniques for geometric documentation of architectural heritage, such as photogrammetry and laser scanning, and presents standards of data acquisition and data delivery for both techniques. However, requirements for data quality analysis are missing.

To date, Brazilian specifications for photogrammetric documentation of architectural heritage do not exist. However, there are some related specifications that were used as normative reference for the requirements for photogrammetric documentation of architectural heritage proposed in this paper. These references include specifications for graphical representation by the Brazilian Association of Technical Standards (ABNT), Manual for Cultural Heritage Preservation Project [57] (MCHPP) by the Brazilian Institute of Cultural Heritage (IPHAN) and Brazilian specifications developed for geographical information by the National Cartography Committee (CONCAR).

Besides the identification of international and national specifications related to photogrammetric documentation of architectural heritage, it is necessary to verify gaps in these specifications that affect their application for architectural heritage documentation production.

For this purpose, one international specification (AS-OPAP) and two national specifications (MSSCH and MCHPP) that have more requirements for heritage documentation with photogrammetric techniques than the other identified specifications were selected. The selected specifications were compared to ISO 19131:2007 Geographic information—Data product specification (ISO 19131) [7], which provides guidelines for the development of geographical data product specifications. The aim of this comparison was to analyze the completeness of the most relevant specifications with respect to the international standard for specification for a geographical data product. Table 1 shows the content suggested by ISO 19131, and the presence (x) or absence ( ) of the same content in the three selected specifications.

Among other ISO 19131 specification content elements, AS-OPAP, the CIPA's international specification contains sections on data quality control. AS-OPAP presents 44% of the content. However, it misses information about the reference system, data product delivery and metadata, which directly affect the use of the model. At the national level, the content of the MSSCH is closer to ISO 19131 than MCHPP. MSSCH presents 72% of the content required by ISO while MCHPP presents only 50%. It is relevant to evaluate which content is missing. In the MSSCH, Abbreviations, Spatial schema and Data maintenance are omitted, which affect the user less than the omission of Reference Systems, Data Quality, Metadata and Data capture that are missing in MCHPP. Drawing from the results of this analysis, we believe there is a need for a proper Brazilian specification for the photogrammetric documentation of architectural heritage. We present our proposal for specification for architectural heritage documentation with photogrammetric techniques in the following section.

Table 1. Completeness analysis of Advice and Suggestions for the furtherance of Optimum Practice in Architectural Photogrammetry surveys (AS-OPAP), Metric Survey Specification for Cultural Heritage (MSSCH) and Manual for Cultural Heritage Preservation Project (MCHPP) specifications.

Contents	Sub Contents	AS-OPAP	MSSCH	MCHPP
Overview	General information about the data	X	X	X
	Terms and definitions		X	X
	Abbreviations			
	Name and acronyms of the data product	X		X
Specification scope			X	X
Data product identification	Title	X	X	X
	Abstract	X	X	
	Topic category	X		X
	Geographical description		X	X
Data content	Spatial schema			
Reference Systems	Spatial		X	
	Temporal		X	
Data Quality		X	X	
Data product delivery			X	X
Metadata			X	
Data capture		X	X	
Data maintenance				
Portrayal		X	X	X

4. Data Product Specification

According to ISO 19131 a data product specification (DPS) can be defined as a description of a dataset, operational procedures and additional information that will provide information to users to create, supply and use this dataset [7].

ISO 19131 presents general recommendations for structure and content of a data product specification, with requirements based on technical coherence and relevance for geographic data product. These recommendations can be adapted for the development of DPS for photogrammetric documentation, providing photogrammetric documentation requirements in conformity with ISO standards. Consequently, datasets produced based on this specification will respect ISO standards as well. We present in this section the content and the structure of a DPS proposal for photogrammetric documentation of Brazilian architectural heritage, based on ISO 19131.

4.1. Data Product Identification and Specification Scope

Table 2 shows the identification information of the data product, such as, title, main theme, extent of the geographic area covered and the form of the spatial representation [7]. The specification scope is presented in Table 3, which is defined in terms of spatial or temporal extent, feature type, property value, spatial representation and product hierarchy.

Table 2. Data Product Identification.

Information	Description
Title	Technical Specification for Photogrammetric Documentation of Architectural Heritage
Alternative title	ET/DOC-FOPARQ
Topic category	Society (code 016) and structure (code 017) (as defined in ISO 19115 [58]).
Geographic description	Country code BR [59]; Data type code 003 [58].
Spatial representation title	Vector (code 001), text (code 003) and stereoscopic model (code 005). Theses codes are defined in ISO 19115 [58].

Table 3. Specification Scope.

Information	Description
Scope Identification	The Technical Specification for Photogrammetric Documentation of Architectural Heritage (ET/DOC-FOPARQ) describes requirements for documentation of Brazilian architectural heritage with photogrammetric techniques and digital technology
Hierarchical level code	015-Model. This code is defined in ISO 19115 [58].
Hierarchical level name	BCH/TCH-MB.
Scope description	This specification does not cover all Brazilian cultural heritage. ET/DOC-FOPARQ includes the tangible cultural heritage limited only for monuments and buildings.
Spatial extent	National level
Temporal extent	This technical specification depends on the temporal extension of the normative reference used to support this specification: ISO 19131 [7], MCHPP [56], ISO 19115 [58], ISO 19157 [60], NBR 6492 [61] and Geospatial Metadata Profile of Brazil [62]. Therefore, ET/DOC-FOPARQ recommendations are valid until the normative references are also valid.
Coverage	Brazilian territory

4.2. Data Content and Structure

The diagram in Figure 1 shows the content and structure of photogrammetric documentation of a Brazilian cultural heritage documentation model. In this case, the photogrammetric documentation is divided into two classes: Descriptive information about architectural heritage based on ICOMOS recommendations [9], and architectural heritage model generated by photogrammetry.

The architectural heritage model could be classified as class A or class B, according to photogrammetric documentation purpose. Class A is comprised of architectural heritage models which aim to support current and future projects that require metric models. However, not all applications need metric models (e.g., illustrative promotion of architectural heritage for the population and visual projects). The recommendation in the Section 4.3 to Section 4.5 depends on the architectural heritage application (Class_code). In these cases, where the model of the architectural heritage documentation is used only for visualization, the architectural heritage model can be classified as Class B. The motivation for class B is to value the projects that do not have metric purposes but are relevant for society as a preservation tool, enabling architectural heritage disclosure to the population.

4.3. Reference System

The architectural heritage models can be associated with a local or spatial reference system. Usage of a local reference system is suggested for Class A and Class B models. Initial errors from GNSS (Global Navigation Satellite System) positioning are thus avoided. Assuming that the architectural heritage model needs to be geo-referenced, performing the whole photogrammetric process in a local reference system is recommended and, at the end of the process, applying a transformation to the desired spatial reference system, considering the error propagation involved in this transformation.

4.4. Data Quality

ISO 19157 defines the principles for describing geographical data quality [60] with six data quality elements: positional accuracy, logical consistency, completeness, temporal quality, thematic accuracy and usability. Data quality information is essential for evaluation of the product's conformance to the product specification and its fitness for use. It was considered, in this research, that only positional accuracy (Class A), logical consistency (Class A and B) and completeness (Class A and B) are applicable for photogrammetric documentation of architectural heritage.

4.4.1. Positional Accuracy

Positional accuracy of architectural heritage model developed with photogrammetric techniques consists of analysis of two data quality sub-elements: absolute positional accuracy and relative positional accuracy (only Class A, not applicable for Class B). Absolute and relative accuracy give different insights about the positional accuracy of architectural heritage model, such as accuracy of coordinates and local positional consistency, respectively. For instance, the case of unacceptable absolute positional accuracy and acceptable relative positional accuracy may indicate a systematic error in the architectural heritage model, which was unnoticed earlier.

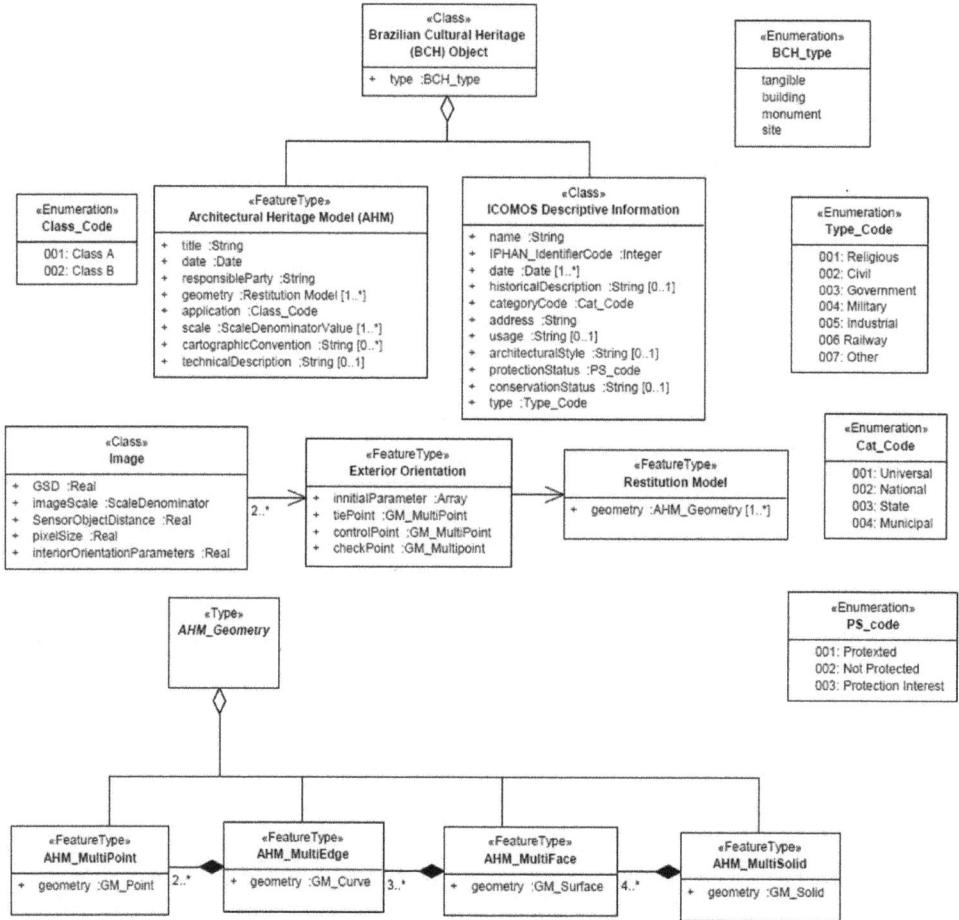

Figure 1. Brazilian cultural heritage documentation-content and structure.

Absolute positional accuracy can be evaluated considering how close the measured value is to the "true" value (reference value), in other words, the accuracy of the position of features within a spatial reference system [60]. Systematic and random errors in the photogrammetric measurement determine the magnitude of the absolute positional accuracy. The measure used for expressing the absolute positional accuracy is the Root Mean Square Error ($RMSE_{(a)}$), in which the errors are obtained from the differences between the estimated coordinates and independent surveyed coordinates. For an architectural heritage model, it is necessary to establish an error limit to determine whether the measured value is close enough to the value to be accepted as true. If the calculated error in each component (X, Y and Z) of the absolute positional accuracy is less than the error limit, we can accept the architectural heritage model in terms of absolute positional accuracy.

The error limit of the absolute positional accuracy (ε_a) is based on the error theory in photogrammetric process (measure errors, orientation errors and projection errors) and the graphic error, which represent 0.3 mm in the graphic representation scale (k) [8], as shown in Equation (1).

$$\varepsilon_a = 0.3 \, mm \times \left(\frac{1}{k}\right) \tag{1}$$

Relative positional accuracy is defined as the closeness of the relative positions of the features in a data set to their respective positions accepted as true [60]. In the same way as the absolute positional accuracy, it is necessary to establish an error limit to relative positional accuracy (ε_r). Considering the photogrammetric process, an acceptable error limit to relative accuracy is 0.2 mm in the graphic representation scale (k) (Equation (2)). The calculated error of the relative positional accuracy ($RMSE_{(b)}$) is obtained with the differences from the estimated distances between points on the model and the same distances surveyed independently. If the calculated error is less than the error limit, we can accept the architectural heritage model in terms of relative positional accuracy.

$$\varepsilon_r = 0.2 \, mm \times \left(\frac{1}{k}\right) \tag{2}$$

In summary, if $RMSE_{(a)} < \varepsilon_a$ and $RMSE_{(b)} < \varepsilon_b$, we can accept the architectural heritage model in terms of positional accuracy as acceptable for class A.

4.4.2. Logical Consistency

Logical consistency is defined as the degree of agreement of data with the dataset's structure, attributes and relationships, respecting defined logical rules [60]. A data set can be analyzed in logical consistency considering conceptual consistency, topological consistency, domain consistency and format consistency. The most important data quality element for an architectural heritage model (class A and B) developed with photogrammetric techniques is topological consistency (correctness of the topological feature in a data set). Topological consistency analysis can detect errors (e.g., overshoot, undershoot, overlap, gap and others) that could be interpreted incorrectly as positional errors in the model, since these quality principles are correlated. Figure 2 shows examples of topological errors. More details can be found in ISO 19157 [60].

According to ISO 2859-1 for sampling procedures for inspection by attributes [63], the samples that follow data set conformity in an acceptance quality limit (AQL) should be higher than 90%. Thus, it is suggested that 90% of the data set (architectural heritage model, class A or B) should be consistent.

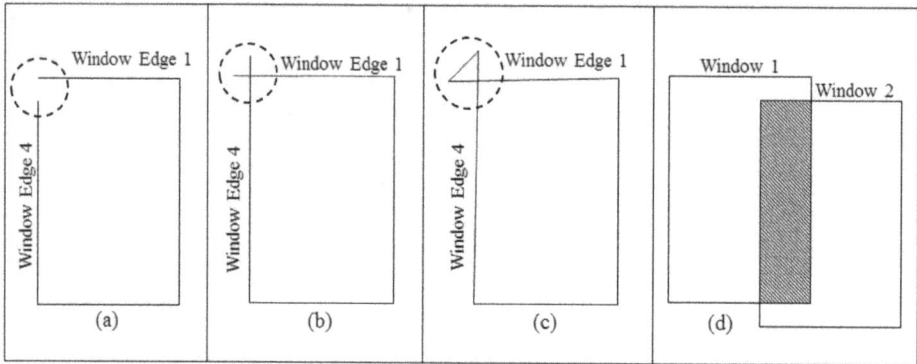

Figure 2. Topological errors examples: (**a**) undershoot, (**b**) overshoot, (**c**) self-intersection, (**d**) overlap.

4.4.3. Completeness

Completeness analysis consists of identifying the excess (commission) and the absence (omission) of data (features, attributes and relationships) in a dataset compared with its specification [60].

For architectural heritage models, exclude any excess information is suggested, in other words, the commission should be 0%. The acceptable quality level for omission is more permissive due to the limitations of photogrammetry technique (e.g., data absence caused by occlusions). For this reason, we suggest the value 5% of the total number of architectural heritage model features for omission to class A and B.

4.5. Data Capture

It is not easy to achieve the acceptance quality limit for positional accuracy, logical consistency and completeness, especially in architectural heritage survey. Therefore, some precautions in data acquisition and data processing are required. The following recommendations are guidelines based on the photogrammetric process applied to architectural heritage survey, that aim to help users achieving the acceptance quality limit for applicable quality elements in Class A or Class B. These guidelines are divided into data acquisition, interior and exterior orientations and feature restitution or modeling.

Data acquisition recommendations to class A: (1) Ground Sample Distance (GSD) [47] must be smaller than one third of the error limit for absolute positional accuracy (GSD $< \varepsilon_a$ /3). GSD depends on the sensor size, focal length, sensor-object distance and image scale; (2) Interior Orientation Parameters (IOP) must be stable in time gap between calibration and image acquisition. Focusing ring must be mechanically locked and autofocus features must be turned off. Zoom lens should be avoided even when it is locked to a fixed focal length. Cameras with automatic stabilization mechanism should also be avoided or this feature needs to be turned off; (3) Depending on the depth variation of the surveyed architectural heritage building, the depth of field can be increased using a suitable f/stop, while maintaining diffraction effects under the adopted circle of confusion (CoC); (4) A camera tripod must be used to prevent the motion blur effect; (5) Multiple images of the architectural heritage facade, preferably stereo pairs (normal case) complemented by oblique orientations, must be captured [64]; (6) Whenever it is possible, occlusion of features by natural and anthropic objects must be avoided; (7) Lossless image compression formats should be used, to prevent loss of information (e.g., RAW or TIFF formats); (8) It is recommended that photographs should be taken on a clear, cloudy day. This condition reduces the high contrast caused by shadows and radiometric difference between stereo pairs; (9) Control points and checkpoints for bundle triangulation and checkpoints for modeling quality control should be acquired with accuracy of one third of the error limit of the absolute positional accuracy; (10) Length and orientation of the distances for relative accuracy analysis should be decided considering the dimension and shape of the surveyed architectural heritage object. These distances should be large enough to identify possible deformations in the model.

Interior orientation recommendations to Class A: (11) Standard deviation of the estimated focal length should be less than 1 pixel; (12) Standard deviation of the estimated principal point coordinates should be less than 1 pixel; (13) Evaluation of the IOP's significance [65] by comparing the parameter magnitude with its standard deviation is recommended and verifying whether the effects of a particular parameter in the image limits are less than the image measurement error; (14) Automatic and semiautomatic methods for measurement of image points are suggested, for instance using coded targets [66]. These methods enable subpixel precision; (15) Whenever feasible, the use of 3D calibration field is recommended, especially when the architectural heritage has significant variations in depth [67].

Exterior orientation recommendations to Class A: (16) Exterior Orientation Parameters (EOP) should preferably be determined by indirect methods (image bundle triangulation). Especially for large representation scales, direct methods based on GNSS and inertial measurement unit (IMU) for EOP's determination are not yet compatible with the required accuracy for photogrammetric documentation of architectural heritage applications. Furthermore, in the case of digital cameras, some adaptations for direct methods must be done, such as determining the nodal point physically. (17) Image measurement of tie points, control points and checkpoints should be automated or semi-automated [68–70], whenever possible. (18) RMSE of the checkpoints after bundle adjustment should be less than two thirds of the error limit for absolute positional accuracy. (19) A trend test, for example t-student test, should be applied to assess bias in the estimated coordinates.

Modeling recommendations to class A: (20) The selected modeling technique should consider the project requirements. The technique and applied software are limited by level of detail, cost, accuracy, format and other requirements of the project. An example is the classic process in photogrammetry, the restitution of features. Considering the restitution, the use of the stereoscopic method is recommended instead of the monoscopic method. The stereoscopic method allows visualization of variations in depth, helping border identification.

Recommendations for class B are more flexible because this class of product is derived mainly for visualization. Data acquisition can be done following recommendations 1 to 8 for class A. For interior orientation recommendations 11 to 15 should be adopted, for exterior orientation recommendations 16 to 17, and recommendation 20 should be followed for modeling.

4.6. Data Product Delivery

This section presents recommendations for layout of an architectural heritage model and delivery format of photogrammetric documentation.

Layout should follow NBR 6492 [61] (recommendation for graphic representation of architectural project), NBR 10068/87 (layout dimension) and NBR 8403/84 (features of drawing lines) developed by the Brazilian Association of Technical Standards (ABNT).

The data product delivery follows the recommendations from the Brazilian institute of Cultural Heritage (IPHAN). These recommendations can be found in the Manual for Cultural Heritage Preservation Project, which include specific scales, paper format, layout content and delivery format (analog or digital).

4.7. Metadata

Metadata should follow the Geospatial Metadata Profile of Brazil (*Perfil de Metadados Geoespaciais do Brasil-Perfil MGB*) [62], which is the national adoption of the ISO 19115, the international standard for geospatial metadata [58].

5. Case Study: Presidente Prudente Railway Station—An Example of Class a Product

Section 5 presents a case study to prove the applicability of the data product specification for photogrammetric documentation of architectural heritage, focusing on data quality analysis for class A in 1:50 scale. The content and structure of this chapter follow recommendations defined in Section 4 (ET/DOC-FOPARQ).

5.1. Descriptive Information of Presidente Prudente Railway Station

Nowadays, many 20th century monuments and buildings form part of Brazilian Cultural Heritage, for example, the railway station in Presidente Prudente. This construction is part of a set of buildings with historical and architectural value to Presidente Prudente city, representing the beginning of the city's development. Because of the historical interest of Presidente Prudente Railway Station, photogrammetric documentation to record the state of this architectural heritage was required. Photogrammetric modeling of this cultural heritage monument is challenging, because Presidente Prudente Railway Station has façades with homogeneous texture and low level of details, which complicate the matching between features. Furthermore, this building has dominant horizontal shape, requiring a careful planning of the coverage to ensure suitable images geometry. Due to these difficulties in the photogrammetric process, Presidente Prudente Railway Station is an interesting example of the applicability of the data product specification proposal for photogrammetric architectural documentation.

As previously discussed, photogrammetric documentation requires descriptive information about architectural heritage and the architectural heritage model generated by photogrammetry. Table 4 presents descriptive information records of Presidente Prudente Railway Station.

5.2. Architectural Heritage Model Development (Class A)

5.2.1. Data Acquisition

A Nikon 3200 digital camera with tripod was used to acquire the case study images (see its specifications in Table 5). A set of 20 images was acquired over the 12 stations.

First, the position of camera stations were planned, considering suitable base distances approximately parallel to the heritage façade and ensuring 60% overlap between images. On average, the camera stations was 19 meters away from the façade, ensuring values of GSD ranging from 2 mm to 3 mm. GSD is less than one third of the error limit for absolute positional accuracy for a 1:50 scale (5 mm), as recommended in Section 4.5. Camera stations were then ground marked.

Next, a local reference system was realized. The position of the origin [0, 0, 0] was defined close to the left corner of the façade and the axis was north oriented, considering a calculated azimuth between the origin and one camera station with known coordinates. Then, topographic methods, such as polygonal and double-intersections, were used to determine the 3D coordinates of the camera stations, control points and checkpoints. The mean positional accuracy of the control and checkpoints in the façade was estimated with error analyses, resulting in accuracy values around 3 mm. As recommended in Section 4.5, the accuracy of control and checkpoints was less than one third of the error limit for absolute positional accuracy for a 1:50 scale (5 mm).

Table 4. Descriptive information records of Presidente Prudente Railway station.

Name Presidente Prudente Railway Station	**Identifier Code** Not applicable
Date Original building 1919; First reconstruction 1926 Second reconstruction and current state 1944	**Category code** Cat_code: 004

History
The railway station in Presidente Prudente is part of a set of buildings with historical and architectural value to Presidente Prudente city, symbolizing the beginning of the city's development.

Adress
St. JúlioTiezzi 220, Presidente Prudente, São Paulo, Brazil.
Geographic coordinates (22°7′23″ W; 51°22′56″ S)

Usage
Original use—Railway station
Current use—seat of a governmental institution

Architectural style
The building of the Presidente Prudente railway station has features of the 1940s in Brazil, represented specially by the geometric volumes arrangement. The building has *Art Decó* influence.

Protection status PS_code: 003	**Conservation status** Not applicable	**Typology** Typo_code: 006

Photos (Source: Presidente Prudente municipal collection)

(1944)	(1970)	(2014)
1944	1970	2014

Finally, distances to endpoints of 29 edges in the façade were collected in an independent survey (tape measured) for the analysis of the relative positional accuracy in the architectural heritage model. The collected edges were pre-selected considering the dimension and shape of the surveyed architectural heritage monument, the distribution in the façade and the image contrast (low or high), which interfere with the quality of the restitution process. These distances, which vary in horizontal and vertical directions, between 0.5 to 3 meters, with an estimated measured accuracy of 10 mm, enable evaluation of the relative positional accuracy between stereoscopic models.

Table 5. Technical specifications for the camera model.

Camera Model	Sensor Size	Nominal Focal Length	Image Dimension	Pixel size
Nikon 3200	CMOS APS-C (23.1 × 15.4)	28 mm	6016 × 4000 pixels (24 megapixels)	0.0038 mm

5.2.2. Camera Calibration, Orientation and Object Modeling

A 3D terrestrial calibration field with coded targets in ARUCO style [71] was used for the camera calibration process. As shown in Figure 3a, the targets were regularly distributed in the calibration field (floor and walls) and the coordinates of four corners for each target had previously been measured using topographic and photogrammetric methods, with 3 mm accuracy used as control points. The ARUCO target corners can be automatically located over the images [72]. In this case, a set of 28 images was taken from four camera stations, providing 3600 observations from 162 control points. The acquired images were horizontal and convergent, with changes in position and rotation, minimizing linear dependency between the interior and exterior orientation parameters.

The camera calibration was performed using the in-house-developed software, Calibration with Multi-Cameras (CMC), in which the IOP were determined by bundle adjustment with the Conrady-Brown lens distortion model [73]. Analysis of the IOP significance was performed and it was concluded that affinity parameters are not significant for this camera calibration case. Therefore, only the focal length (f), the principal point coordinates (x_0, y_0), the symmetric radial lens distortion coefficients (k_1, k_2, k_3) and the decentering lens distortion coefficients (p1, p2) were determined.

Table 6 presents the estimated interior orientation parameters and the corresponding standard deviations. The standard deviation of the focal length was determined with less than 1 pixel, as well as, the standard deviation of the principal point coordinates—as recommended in Section 4.5. This result was achieved due to subpixel target measurement techniques. A 3D terrestrial calibration field was used because the architectural heritage under study, Presidente Prudente Railway Station, has significant variations in depth.

Table 6. Estimated interior orientation parameters and standard deviations.

IOP	f (mm)	x_0 (mm)	y_0 (mm)	k_1 (mm^{-2})
Value	28.099	0.1038	−0.0254	-1.5398×10^{-4}
Standard deviation	0.0018	0.0002	0.0002	1.51×10^{-6}
IOP	k2 (mm^{-4})	k3 (mm^{-6})	p1 (mm^{-2})	p2 (mm^{-2})
Value	-1.7623×10^{-7}	-1.12×10^{-10}	-5.68×10^{-6}	-7.11×10^{-6}
Standard deviation	2.060×10^{-8}	8.5×10^{-11}	3.5×10^{-7}	4.4×10^{-7}

The 20 images of the façade were acquired immediately after camera calibration to avoid IOP changes. In the post-processing, these images were later resampled to correct lens distortion and then bundle adjustment was performed using the Leica Photogrammetry Suite (LPS).

(a)

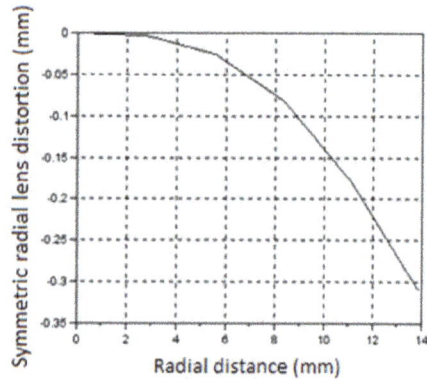

(b)

Figure 3. (a) Symmetric radial lens distortion effect and (b) Terrestrial calibration field.

Raw images were post processed and resampled in the in-house-developed software, known as P_retif. This resampling step was required because LPS software presented some unexpected results with high distortion images (Figure 3b).

In the bundle adjustment, the camera station coordinates (camera position) measured directly during topographic survey were used as initial parameters for the coordinates of the camera perspective center (X_0, Y_0, Z_0) with a constraint of 0.5 m for standard deviation. Tie points were generated automatically with image matching techniques, amounting to 232 points. A total of 11 control points with an accuracy of 3 mm, were manually measured in stereo model and transferred to neighbor images by least-squares matching.

The exterior orientation quality control was accomplished with 7 independent checkpoints with the same characteristics of the control points. Table 7 shows the resulting statistics: average, standard deviation and RMSE of the checkpoints' coordinate discrepancies. A t-student trend test for the checkpoints was performed. The trend analysis for 95% confidence level showed that the coordinate discrepancies have no trend.

Table 7. Statistics of the discrepancies in checkpoint after bundle adjustment.

Statistics	ΔXt (m)	ΔYt (m)	ΔZt (m)
Average	−0.0003	−0.00014	0.0011
Standard deviation	0.0012	0.0012	0.0030
RMSE	0.0012	0.0011	0.0030

The RMSE of the obtained discrepancies in checkpoints is less than two thirds of the error limit to absolute accuracy in all coordinates (<10 mm). Considering that the accuracy in the orientation estimation step is acceptable, it is possible to proceed to the modeling step.

Presidente Prudente Railway Station façades have a simplified architecture, with a low level of details, thus, the modeling method applied was the restitution of features, a classic method in photogrammetry. The restitution process was developed in a stereo environment (LPS PRO600 for MicroStation). Figure 4 presents the architectural heritage model of the Presidente Prudente Railway Station. The data quality assessment (absolute positional accuracy, relative positional accuracy, logical consistency and completeness) of this architectural heritage model is presented in Section 5.2.3.

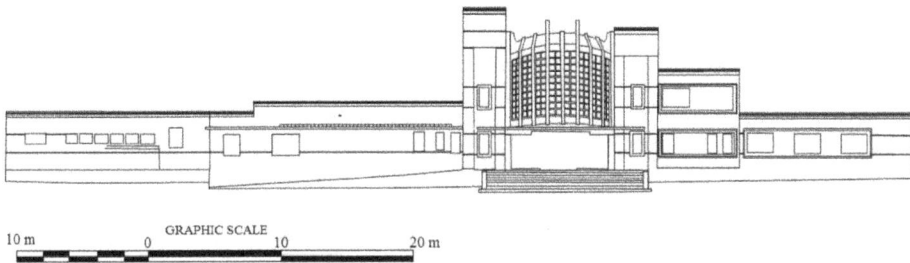

Figure 4. Presidente Prudente Railway Station model.

5.2.3. Data Quality Analysis

Absolute positional accuracy was analyzed considering seven checkpoints, well distributed over the model and unique to this process (not the same checkpoints used in the bundle adjustment). The checkpoints were also determined with 3 mm

149

accuracy, using topographic methods. Table 8 presents the statistics, average, standard deviation and RMSE of restitution checkpoint discrepancies. Figure 5 shows control points used in the bundle adjustment and checkpoints used in the final model accuracy assessment with the corresponding resultant of residues in X and Y coordinates.

Table 8. Average, standard deviation and RMSE of the discrepancies in the restitution checkpoints.

Statistics	ΔXr (m)	ΔYr (m)	ΔZr (m)
Average	0.0021	0.0004	0.0064
Standard deviation	0.0039	0.0011	0.0129
$RMSE_{(a)}$	0.0042	0.0011	0.0135

Notice that the RMSE of the checkpoints in X, Y and Z coordinates are less than the error limit for absolute positional accuracy for a 1:50 scale (15 mm). We conclude that the model of the Presidente Prudente Railway Station is adequate in absolute positional accuracy ($RMSE_{(a)} < \varepsilon_a$), thus the recommendations in Section 4.5 are applicable to achieve the proposed absolute positional accuracy. Furthermore, a t-student trend analysis for 95% confidence level showed that the coordinate discrepancies have no trend.

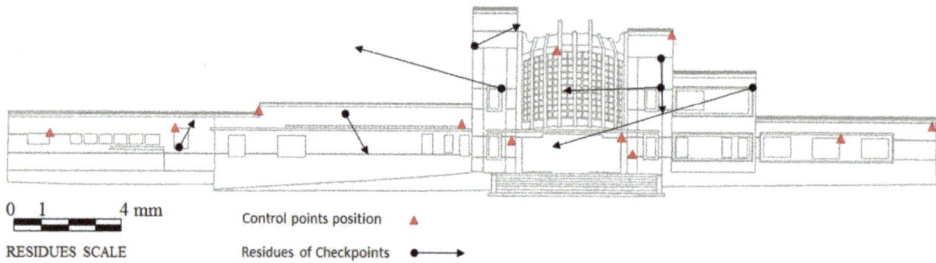

Figure 5. Checkpoints distribution and residues.

The distances between endpoints of distinguishable features were measured in the model of the Presidente Prudente Railway Station and compared to reference values to evaluate relative positional accuracy. An analysis of the distances showed that the acquired data follow a normal distribution, taking into account the Anderson-Darling normality test with 95% confidence level (P-value 0.05), and an obtained a p-value of 0.204. The error limit to relative accuracy for 1:50 scale is 10 mm. Table 9 shows that the calculated error to relative accuracy ($RMSE_{(b)}$) was acceptable.

150

Table 9. Relative positional accuracy analysis for the Presidente Prudente Railway Station model.

Number of Edges	Average (m)	Standard Deviation (m)	RMSE(b) (m)
29	0.0001	0.010	0.010

However, it was observed that the $RMSE_{(b)}$ is close to the threshold, leading us to believe the distances, collected for relative positional accuracy analysis during an independent survey (tape measurements), had errors with a magnitude of 10 mm. The acquisition of accurate reference values, such as distances and checkpoints, is a major concern in an architectural heritage documentation project, where millimeter accuracy is required, mainly because few survey techniques can reach this level of accuracy at affordable cost.

The $RMSE_{(a)} < \varepsilon_a$ and the $RMSE_{(b)} < \varepsilon_b$, thus we can accept the model of the Presidente Prudente Railway Station as Class A product in terms of positional accuracy. This case study shows the importance of quality analysis during the whole photogrammetric process to achieve the required positional accuracy. It can be also concluded that the proposed requirements for positional accuracy are feasible.

Nevertheless, to consider the model as Class A product, it is suggested that logical consistency and completeness should also be evaluated (Section 4.4). The logical consistency analysis was performed with the standards tools Quantum GIS software offers [74]. The topological errors were automatically identified by the software and corrected manually. The following errors were investigated: overlap, overshoot, undershoot and gap. A set of 976 features compose the model of the Presidente Prudente Railway Station. From the total of 976 valid features, there were 0 gaps, and 2 overlaps (0.2%). Sixty-four overshoots and undershoots (6.5%) were identified. The identified errors were eliminated and were not identified again in a new test. We therefore have reason to believe that there are no topological errors in the final model.

Completeness is related to the project's specification, thus the architectural heritage features that will be represented in the model need to be defined during the initial planning process. In this regard, a number of significant features were selected, including the number of windows, doors, stairs, building borders, plumbing, window details and others architectural details. Considering all the features specified, 3.52% were not represented, mainly because of occlusions and borders of low resolution. Commission has not been identified. The model of the Presidente Prudente Railway Station is therefore admissible for an acceptance quality limit of 5% of the total number of architectural heritage model features for omission and 0% of commission.

Positional accuracy, logical consistency and completeness were evaluated and the acceptance quality limits for Class A in a 1:50 scale, in each of these data quality principles were achieved. The model of Presidente Prudente Railway Station can therefore be classified as Class A product.

6. Case Study: Prudente de Morais Monument—An Example of Class B Product

The Prudente de Morais Monument, the original construction dating from 1944, is part of the architectural heritage complex that comprises the Presidente Prudente Railway Station. The main purpose of this case study is to exemplify models that fit into Class B products (non-metric models) in a scale 1:10. The architectural heritage models classified as Class B products are used principally for visualization, being applied, for example, in preliminary studies of architectural heritage preservation projects.

Classification as Class A or B aims to specify the use of the models. The difference between Class A and B is essentially in the architectural heritage model development, defined by the positional accuracy requirement. The descriptive information about architectural heritage follows the same structure for both classes and they will not be presented in this section. The focus of this section is data quality control for Class B.

The images were acquired with a calibrated low-cost Sony DSC-W520 camera (4.7 mm nominal focal length), from different viewpoints and with 70% overlap between images. The 3D model was processed in Autodesk 123D Catch software [75] (desktop version) which is based on the structure from motion technique, followed by mesh generation and rendering methods.

The application of this software was motivated by the increase in its use for models of non-metric purposes, specially developed by non-experts. Furthermore, this case study exemplifies the use of software with different levels of automation than the software used for the Presidente Prudente Railway Station modeling. Santagati, Inzerillo and Di Paola [76] presented a comparison between 3D models generated with terrestrial LIDAR and 3D models obtained with 123D Catch, concluding that on average, in most applications, positional accuracy has a magnitude of 1 to 2 cm. With this is mind, six distances from the statue were collected to verify the relative positional accuracy of the model. Table 10 presents the average, standard deviation and RMSE of the differences between the estimated distances between points on the model and independently surveyed distances. Figure 6 shows the Prudente de Morais model with the distances measured in the model (in black) and the corresponding reference value for these distances obtained in an independent survey (in red).

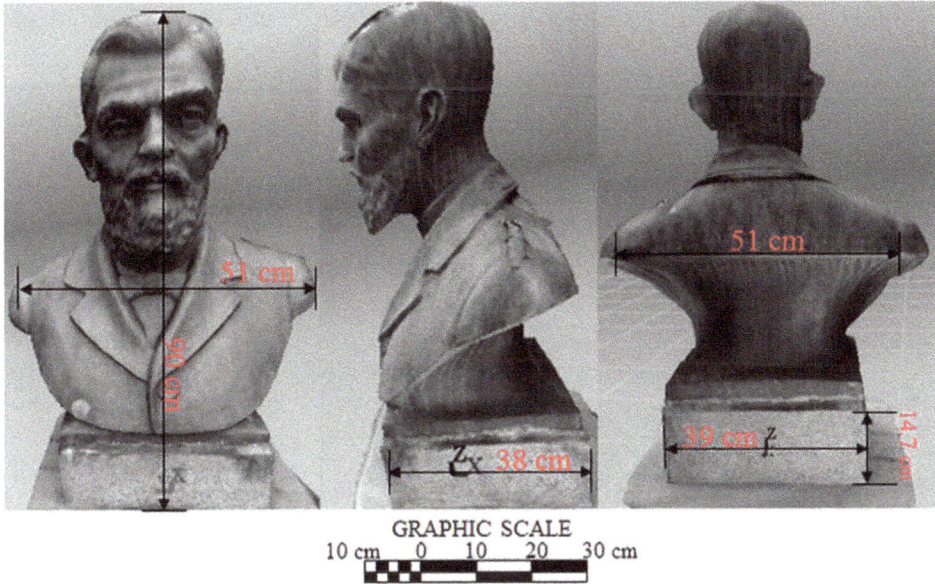

Figure 6. Prudente de Morais model.

Table 10. Results of relative positional accuracy analysis for the Prudente de Morais model.

Statistics	Average	Standard Deviation	RMSE
(cm)	0.13	2.00	1.83

RMSE of the differences between the estimated distances between points on the model and independently surveyed distances should be less than the error limit for relative positional accuracy for a 1:10 scale (2 mm), as recommended in Section 4.4.1. These results are not sufficient to support the photogrammetric documentation project which needs high positional accuracy (Class A), showing the importance of quality control in the photogrammetric process. However, there are several solutions for reconstruction of 3D models based on structure from motion technique that enables orientations and modeling control, such as PhotoModeler, PhotoScan, VisualSfM, ARC3D, among others [76,77], which could be applied to Class A and B.

In Class B, the recommendations are more flexible for positional accuracy analysis, since this proposed product category is intended mainly for visualization. Nonetheless, analysis of the logical consistency and completeness for class B are mandatory.

The logical consistency analysis assumes that, in this case, the modeling process consists of generating a triangular mesh from a point cloud. This point cloud was obtained from the calculation of the homologous points coordinates. In the mesh generation process, some parameters that ensure logical consistency of the surface are considered, avoiding mistakes such as overlap, overshoot and undershoot. Topological analysis using the Quantum GIS software [73] confirmed that these errors were not identified in a resulting model. Invalid geometries were not identified either. We therefore concluded that the model of the Prudente de Morais monument is consistent.

The data product specification includes representation of the bust, represented by 82,650 features in total. The major problem identified was a lack of definition in the model borders. A set of 200 features were missing (0.26%), within the limit of 5% for omission. Excess features have been identified and excluded, resulting to 0% of commission. The model of the Prudente de Morais monument can therefore be considered complete, according to the specification (ET/DOC-FOPARQ).

Logical consistency and completeness were evaluated and the acceptance quality limits for Class B in each of these data quality elements were achieved.

7. Conclusions

Architectural heritage should be passed to future generations in its historical and cultural authenticity. Photogrammetric documentation is a feasible technique for architectural heritage documentation and preservation. Nowadays, with the availability of affordable digital equipment, there are an increasing number of photogrammetric documentation initiatives. The development of a DPS for photogrammetric documentation significantly contributes to data product reliability and, consequently, to the preservation of heritage information. In this context, motivated by the need of specifications for the photogrammetric documentation of architectural heritage to approach geomatics and heritage experts and ensure photogrammetric documentation application, this paper presented a procedure of recommendations for photogrammetric documentation of architectural heritage, based on Brazilian case study experience. The proposed data product specification is a result of an analysis of existing specifications related to photogrammetric documentation focused on architectural heritage, as well as, the main problems in the photogrammetric documentation specifications that needed to be improved. We proposed recommendations for photogrammetric documentation of architectural heritage and tested these recommendations in case studies for photogrammetric products of Class A and B.

In the case study of the railway station in Presidente Prudente, which is an example of Class A product, we analyzed absolute and relative positional accuracy, topological consistency, commission and omission. Acceptance quality limits recommended in Class A for these elements, considering 1:50 scale, are 15 mm, 10 mm, 10%, 0% and 5%, respectively. RMSE of the checkpoints in X, Y and Z coordinates are within the error limit for absolute positional accuracy (4.2 mm, 1.1 mm and 13.5 mm, respectively) and RMSE of the differences from the estimated distances between points on the model and the same distances surveyed independently is compatible with the error limit for relative positional accuracy (10 mm). Topological errors were identified and eliminated, thus, there are no topological errors in the final model (0%). Commission has not been found (0%) and the omission (3.52%) is within the acceptance quality limit (5%).

In the case study of the Prudente de Morais statue, which is an example of a Class B product, we evaluated topological consistency, commission and omission, considering 10%, 0% and 5% as an acceptance quality limit, respectively. Topological errors were not identified in the resulting model (0%). Excess of features have been identified and excluded, resulting in 0% of commission. Considering all the features specified, 0.26% were not represented, within the limit for omission. The results of our case studies confirm that the proposed requirements are viable.

Our research contributes to the development of standards for photogrammetric documentation and applied photogrammetric method in context of architectural heritage. The advantages of the procedure presented are the application of an international standard for data product specification adapted for digital photogrammetry and the classification of the photogrammetric documentation model in classes, which assists product reliability and application.

Future work includes update of DPS for photogrammetric documentation of other types of tangible cultural heritage, which we did not discuss in this paper. These types of objects, such as archaeological sites, require other techniques and consequently other recommendations. Requirements (for all types of objects of the architectural cultural heritage) about other data quality aspects (e.g., thematic accuracy) will be evaluated for photogrammetric documentation application. Moreover, requirements for the radiometric quality of acquired images and of a resulting model, and procedures for the representation of texture of surfaces in the digital model of the architectural heritage, should be analyzed.

Acknowledgments: This work was jointly funded by Fundação de Amparo à Pesquisa do Estado de São Paulo (FAPESP—Grant: 2013/15940-9) and Conselho Nacional de Desenvolvimento Científico e Tecnológico (CNPq—Grant: 130505/2013).

Author Contributions: This research was performed by Mariana B. Campos supervised, with significant contributions, by Antonio M.G. Tommaselli and Ivana Ivánová. These authors contributed extensively and equally to prepare this paper. This research was complemented with a research internship in the University of Liège, supervised by Roland Billen, who discussed and commented on the drafts.

Conflicts of Interest: The authors declare no conflict of interest.

References

1. Yilmaz, H.M.; Yakar, M.; Gulec, S.A.; Dulgerler, O.N. Importance of digital close-range photogrammetry in documentation of cultural heritage. *J. Cult. Herit.* **2007**, *8*, 428–433.
2. International Council on Monuments and Sites (ICOMOS). International Charter for the Conservation and Restauration of Monuments and Sites (Venice Charter). In Proceedings of the Second International Congress of Architects and Technicians of Historic Monuments, Venice, Italy, 25–31 May 1964; p. 4.
3. Remondino, F. Heritage Recording and 3D Modeling with Photogrammetry and 3D scanning. *Remote Sens.* **2011**, *3*, 1104–1138.
4. Nocerino, E.; Menna, F.; Remondino, F. Accuracy of typical photogrammetric network in cultural heritage 3D modeling. *Int. Arch. Photogramm. Remote Sens. Spat. Inf. Sci.* **2014**, *1*, 465–472.
5. Juran, J.M.; Godfrey, A.B.; Hoogstoel, R.E.; Schilling, E.G. *Juran's Quality Handbook*, 5th ed.; McGraw-Hill Professional: New York, NY, USA, 1998.
6. International Organization for Standardization (ISO). *Economic Benefits of Standard*, 1st ed.; ISO: Geneve, Switzerland, 2004; p. 65.
7. International Organization for Standardization (ISO). *ISO 19131:2006 Data Production Specification*; ISO: Geneva, Switzerland, 2006; p. 47.
8. International Committee of Architectural Photogrammetry (CIPA). *Advice and Suggestions for the Furtherance of Optimum Practice in Architectural Photogrammetry Survey*, 1st ed.; UNESCO: Paris, France, 1981; p. 100.
9. International Council on Monuments and Sites (ICOMOS). Principles for the Recording of Monuments, Group of Buildings and Sites. In Proceedings of the 11th ICOMOS General Assembly, Sofia, Bulgaria, 5–9 October 1996.
10. Patias, P.; Santana, M. Introduction to heritage documentation. In *Cipa Heritage Documentation: Best Practices and Applications*, 1st ed.; Stylianidis, E., Patias, P., Santana Quintero, M., Eds.; CIPA (Ziti Publications): Thessaloniki, Greece, 2011; Volume 38, pp. 9–13.
11. Letellier, R.; Schimid, W.; LeBlanc, F. *Recording, Documentation and Information Management for the Conservation of Heritage Places: Guiding Principles*, 1st ed.; The Getty Conservation Institute: Los Angeles, CA, USA, 2007; p. 174.
12. Remondino, F.; Rizzi, A. Reality-based 3D documentation of natural and cultural heritage sites—Techniques, problems, and examples. *Appl. Geomat.* **2010**, *2*, 85–100.
13. Anderson, J.M.; Mikhail, E.M. *Surveying Theory and Practice*, 7th ed.; McGraw-Hill Science/Engineering/Math: New York, NY, USA, 1998; p. 1167.
14. Thompson, M.M.; Gruner, H. *Manual of Photogrammetry*, 4th ed.; ASPRS: Falls Church, VA, USA, 1980; p. 1150.
15. Spetsakis, M.; Aloimonos, J. A multi-frame approach to visual motion perception. *Int. J. Comput. Vis.* **1991**, *6*, 245–255.
16. Petrie, G.; Toth, C.K. Terrestrial Laser Scanners. In *Topographic Laser Ranging and Scanning*, 1st ed.; Shan, J., Toth, C.K., Eds.; Taylor & Francis Group: New York, NY, USA, 2008; pp. 87–126.

17. Lange, R.; Seitz, P. Solid-state time-of-flight range camera. *IEEE J. Quantum. Electron.* **2001**, *37*, 390–397.

18. Perenzoni, M.; Stoppa, D. Figure of metric for indirect Time-of-flight 3D cameras: Definition and experimental evaluation. *Remote Sens.* **2011**, *3*, 2461–2472.

19. Valkenburg, R.J.; McIvor, A.M. Accurate 3D measurement using light system. *Image Vis. Comput.* **1998**, *16*, 99–110.

20. Reiss, M.L.L.; Tommaselli, A.M.G. A low-cost 3D reconstruction system using a single-shot projection of a pattern matrix. *Photogramm. Rec.* **2011**, *26*, 91–110.

21. Andrés, M.A.N.; Pozuelo, F.B. Evolution of the architectural and heritage representation. *Landsc. Urban Plan.* **2009**, *91*, 105–112.

22. Pieraccini, M.; Guidi, G.; Atzenni, C. 3D digitizing of cultural heritage. *J. Cult. Herit.* **2001**, *2*, 63–70.

23. Pavlidis, G.; Koutsoudis, A.; Arnaoutoglou, F.; Tsioukas, V.; Chamzas, C. Methods for 3D digitization of cultural heritage. *J. Cult. Herit.* **2007**, *8*, 93–98.

24. Sansoni, G.; Trebeschi, M.; Docchio, F. State-of-the-art and application of 3D Imaging Sensors in Industry, Cultural Heritage, Medicine and Criminal investigation. *Sensors* **2009**, *9*, 568–601.

25. Livieratos, E. Empiric, topographic or photogrammetric recording? Answers to properly phrased questions. In Proceedings of the Terrestrial Photogrammetry and Geographic Information Systems for the documentation of the National Cultural Heritage, Thessaloniki, Greece, 1992.

26. Grussenmeyer, P.; Landes, T.; Voegtle, T.; Ringle, K. Comparison methods of Terrestrial Laser Scanning, Photogrammetry and Tacheometry data for recording of cultural heritage buildings. *Int. Arch. Photogramm. Remote Sens. Spat. Inf. Sci.* **2008**, *37*, 213–218.

27. Giuliano, M.G. Cultural Heritage: An example of graphic documentation with automated photogrammetric systems. *Int. Arch. Photogramm. Remote Sens. Spat. Inf. Sci.* **2014**, *40*, 251–255.

28. Scherer, M.; Lerma, L. From the conventional total station to the prospective image assisted photogrammetric scanning total station: Comprehensive review. *J. Survey. Eng.* **2009**, *135*, 173–178.

29. Albertz, J. Albrecht Meydenbauer: Pioneer of photogrammetric documentation of the cultural heritage. *Int. Arch. Photogramm. Remote Sens. Spat. Inf. Sci.* **2001**, *18*, 19–25.

30. Dallas, R.W.A. Architectural and archaeological photogrammetry. In *Close Range Photogrammetry and Machine Vision*, 1st ed.; Athinson, K.B., Ed.; Whittles Publishing: Caithness, Scotland, 1996; pp. 283–303.

31. Fassi, F.; Fregonese, L.; Achermann, S.; De Troia, V. Comparison between laser scanning and automated 3D modelling techniques to reconstruct complex and extensive cultural heritage areas. *Int. Arch. Photogramm. Remote Sens. Spat. Inf. Sci.* **2013**, *40*, 73–80.

32. Boehler, W.; Marbs, A. 3D scanning and photogrammetry for heritage recording: A comparison. In Proceedings of the 12th International Conference on Geoinformatics, Geospatial Information Research: Bridging the Pacific and Atlantic, Gävle, Sweden, 7–9 June 2004; pp. 291–298.

33. Yastikli, N. Documentation of cultural heritage using digital photogrammetry and laser scanning. *J. Cult. Herit.* **2007**, *8*, 423–427.

34. Lerma, L.J.; Navarro, S.; Cabrelles, M.; Villaverde, V. Terrestrial laser scanning and close range photogrammetry for 3D archaeological documentation: The Upper Palaeolothic Cave of Parpalló as a case study. *J. Archaeol. Sci.* **2010**, *37*, 499–507.

35. Mateus, L.; Ferreira, V.; Barbosa, M. TLS and digital photogrammetry as tools for conservation assessment. In Proceedings of the 3rd International Conference on Heritage and Sustainable Development, Porto, Portugal, 19–22 June 2012.

36. Xu, Z.; Wu, L.; Shen, Y.; Li, F.; Wang, Q.; Wang, R. Tridimensional reconstruction applied to cultural heritage with the use of camera-equipped UAV and terrestrial laser scanner. *Remote Sens.* **2014**, *6*, 10413–10434.

37. De Luca, L.; Veron, P.; Florenzano, M. Reverse engineering of architectural buildings based on a hybrid modeling approach. *Comput. Graph.* **2006**, *30*, 160–176.

38. Gomes, L.; Bellon, O.R.P.; Silva, L. 3D Reconstruction methods for digital preservation of cultural heritage: A survey. *Pattern Recognit. Lett.* **2014**, *50*, 3–14.

39. Edelsbrunner, H. Shape reconstruction with Delaunay complex. In *LATIN'98: Theoretical Informatics: 3rd Latin American Symposium Campinas, Brazil, April 20–24, 1998 Proceedings*; Springer: Berlin, Germany, 1998; pp. 119–132.

40. Vilbrandt, C.; Pasko, G. Cultural heritage preservation using constructive shape modeling. *Comput. Graph. Forum* **2004**, *23*, 25–41.

41. Saygi, G.; Agugiaro, G.; Hamamcioglu-Turan, M.; Remondino, F. Evaluation of GIS and BIM roles for the information management of historical buildings. *ISPRS Ann. Photogramm. Remote Sens. Spat. Inf. Sci.* **2013**, *2*, 283–288.

42. Müller, K.; Smolic, A.; Kaspar, B.; Merkle, P.; Rein, T.; Eisert, P.; Wiegand, T. 3D reconstruction of natural scenes with view-adaptive multi-texturing. In Proceedings of the 2nd International Symposium on IEEE 3D Data Processing, Visualization and Transmission, Thessaloniki, Greece, 6–9 September 2004; pp. 116–123.

43. Watt, A. *3D Computer Graphics*, 3rd ed.; Pearson Education Limited: Harlow, UK, 2000; p. 569.

44. Denard, H. *London Charter for the Computer-Based Visualisation of Cultural Heritage*, 1st ed.; King's College London: London, UK, 2009; p. 13.

45. Waldhaeusl, P.; Ogleby, C. 3×3-Rules for Simple Photogrammetric documentation of Architecture. *Int. Arch. Photogramm. Remote Sens. Spat. Inf. Sci.* **1994**, *30*, 426–429.

46. Buchanan, C.D. *Photographing Historic Buildings for the Record*, 1st ed.; HMSO: London, UK, 1983; pp. 154–196.

47. American Society for Photogrammetry and Remote Sensing. *ASPRS Positional Accuracy Standards for Digital Geospatial Data*; ASPRS: Bethesda, MD, USA, 2014; Volume 81, pp. A1–A26.

48. International Organization for Standardization (ISO). *ISO/TC 211 Geographic information/Geomatics*; ISO: Geneva, Switzerland, 1994.

49. Recording and Documentation Information Management (RecorDIM). *Draft Report on: International Heritage Documentation Standards*, 1st ed.; RecorDIM: London, UK, 2007; p. 25.

50. National Park Service. *HABS/ HAER/ HALS Photography Guidelines*, 2nd ed.; NPS: Washington, DC, USA, 2015; p. 306.

51. National Park Service. *Recording Historic Structures and Sites with HABS Measured Drawings*, 3rd ed.; NPS: Washington, DC, USA, 2008.

52. National Park Service. *Recording Historic Structure*, 2nd ed.; John Willey & Sons: Washington, DC, USA, 2004.

53. Park Canada. *Standards and Guidelines for the Conservation of Historic Places in Canada*, 2nd ed.; Park Canada: Ottawa, ON, Canada, 2010.

54. Bryan, P.G.; Blake, B.; Bedford, J. *Metric Survey Specification for Cultural Heritage*, 2nd ed.; English Heritage: London, UK, 2009; p. 111.

55. GIFLE (Grupo de Investigación en Fotogrametría y Láser Escáner). *12 Tips for Metric Photography of Architectural and Archeological Cultural Heritage to Extract 2D/ 3D/ 4D Measurements*, 1st ed.; GIFLE: Valencia, Spain, 2014.

56. Instituto Andaluz del Patrimonio Histórico. *Recomendaciones Técnicas Para la Documentación Geométrica de Entidade Patrimoniales*, 1st ed.; Consejería de cultura: Andaluz, Spain, 2011; p. 23.

57. Gomide, J.H.; da Silva, P.R.; Braga, S.M.N. *Manual de Elaboração de Projetos de Preservação do Patrimônio Cultural: Caderno 1*, 1st ed.; Instituto do programa monumenta: Brasília, Brazil, 2005; p. 76.

58. International Organization for Standardization (ISO). *ISO 19115:2003 Geographic Information—Metadata*; ISO: Geneva, Switzerland, 2006; p. 140.

59. International Organization for Standardization (ISO). *ISO 3166-1:1997 Country Code*; ISO: Geneva, Switzerland, 1997.

60. International Organization for Standardization (ISO). *ISO 19157:2011 Geographic Information—Data quality*; ISO: Geneva, Switzerland, 2011; p. 160.

61. Associação Brasileira de Normas Técnicas (ABNT). *NBR 6492-Representação de Projetos Arquitetura*, 1st ed.; ABNT: Rio de Janeiro, Brazil, 1994.

62. Comissão Nacional de Cartografia (CONCAR). *Perfil de Metadados Geoespacial do Brasil (Perfil MGB)*, 1st ed.; CONCAR: Rio de Janeiro, Brazil, 2009; p. 194.

63. International Organization for Standardization (ISO). *ISO 2859-1:1999 Sampling Procedures for Inspection by Attributes-Part 1*; ISO: Geneva, Switzerland, 1999; p. 87.

64. Luhmann, T.; Robson, S.; Kyle, S.; Harley, I. *Close Range Photogrammetry: Principle, Techniques and Applications*, 1st ed.; John Wiley & Sons: Hoboken, NJ, USA, 2006; p. 510.

65. Chandler, J.H.; Fryer, J.G.; Jack, A. Metric capabilities of low-cost digital cameras for close range surface measurements. *Photogramm. Rec.* **2005**, *20*, 12–26.

66. Li, Z.; Liu, M. Research on decoding method of coded target in close-range photogrammetry. *J. Comput. Syst.* **2010**, *6*, 2699–2705.

67. Fraser, C. Automatic camera calibration in close-range photogrammetry. *Photogramm. Eng. Remote Sens.* **2013**, *79*, 381–388.

68. Hahn, M. Automatic control point measurement. In Proceedings of the Photogrammetric Week'97, Heidelberg, Germany, 1997; pp. 115–126.

69. Barazzetti, L.; Scaioni, M.; Remondino, F. Orientation and 3D modeling from markerless terrestrial images: Combining accuracy with automation. *Photogramm. Rec.* **2010**, *25*, 356–381.

70. Tommaselli, A.M.G.; Bervegliere, A. Automatic Orientation of Multi-Scale Terrestrial Images for 3D Reconstruction. *Remote Sens.* **2014**, *6*, 3020–3040.

71. Muñoz-Salinas, R. ARUCO: A minimal library for Augmented Reality applications based on OpenCv. 2012. Available online: http://www.uco.es/investiga/grupos/ava/node/26 (accessed on 20 June 2013).

72. Tommaselli, A.M.G.; Marcato, J., Jr.; Morais, M.V.A.; Silva, S.L.A.; Artero, A.O. Calibration of panoramic cameras with coded targets and 3D calibration field. *Int. Arch. Photogramm. Remote Sens. Spat. Inf. Sci.* **2014**, *40*, 137–142.

73. Brown, D.C. Close-Range Camera Calibration. *Photogramm. Eng.* **1971**, *37*, 855–866.

74. QGIS Project. QGIS User Guide 2.0. 2014. Available online: http://www.qgis.org/en/docs/index.html (Accessed on 28 August 2015).

75. Autodesk 123D Catch. Available online: http://www.123dapp.com/catch (accessed on 28 August 2015).

76. Santagati, C.; Inzerillo, L.; Di Paola, F. Image-based modeling techniques for architectural heritage 3D digitalization: Limits and potentialities. *Int. Arch. Photogramm. Remote Sens. Spat. Inf. Sci.* **2013**, *40*, 555–560.

77. Roncella, R.; Re, C.; Forlani, G. Performance evaluation of a structure and motion strategy in architectural and cultural heritage. *Int. Arch. Photogramm. Remote Sens. Spat. Inf. Sci.* **2011**, *38*.

Scanning Photogrammetry for Measuring Large Targets in Close Range

Shan Huang, Zuxun Zhang, Tao Ke, Min Tang and Xuan Xu

Abstract: In close-range photogrammetry, images are difficult to acquire and organize primarily because of the limited field of view (FOV) of digital cameras when long focal lenses are used to measure large targets. To overcome this problem, we apply a scanning photography method that acquires images by rotating the camera in both horizontal and vertical directions at one station. This approach not only enlarges the FOV of each station but also ensures that all stations are distributed in order without coverage gap. We also conduct a modified triangulation according to the traits of the data overlapping among images from the same station to avoid matching all images with one another. This algorithm synthesizes the images acquired from the same station into synthetic images, which are then used to generate a free network. Consequently, we solve the exterior orientation elements of each original camera image in the free network and perform image matching among original images to obtain tie points. Finally, all original images are combined in self-calibration bundle adjustment with control points. The feasibility and precision of the proposed method are validated by testing it on two fields using 300 and 600 mm lenses. The results confirm that even with a small amount of control points, the developed scanning photogrammetry can steadily achieve millimeter scale accuracy at distances ranging from 40 m to 250 m.

Reprinted from *Remote Sens.* Cite as: Huang, S.; Zhang, Z.; Ke, T.; Tang, M.; Xu, X. Scanning Photogrammetry for Measuring Large Targets in Close Range. *Remote Sens.* **2015**, *7*, 10042–10075.

1. Introduction

The development of electric sensors and image processing techniques has contributed to the rapid growth of the application of photogrammetry. Since 2000, photogrammetric mapping using digital photographic systems has become popular because of the application of digital cameras [1–6]. Given their changeable focus lens, various photography methods, and low cost, digital single-lens reflex cameras (DSLRs) are widely applied in close-range measurement tasks [7–13]. The high resolution of DSLRs is helpful for high-precision measurements. These advantages suggest the need to develop photogrammetry systems that can help inexperienced users accomplish close-range measurements well [11,12,14]. The size of a single digital image sensor (e.g., charge-coupled device (CCD) detector) is limited. As such, the field of view (FOV) of a digital image captured by this sensor is restricted compared with that of film image with the same focal length. Therefore, when digital cameras with long focal lens are used for measurements, the limited intersection angle decreases the measurement precision if a normal photography method is employed. Oblique photography can ensure intersection angles and maintain high measurement accuracy [8,15–18]. However, when the intersection angle increases, the parallax of correspondences changes in a large range and may become discontinuous as the occlusion occurs. Large intersection angles make image matching difficult, whereas small ones result in low intersection precision. Therefore, in many measurement methods, multi-view photography is applied to solve this contradiction [19–21]. Nonetheless, this technique introduces new problems, such as the difficult acquisition of images in order, particularly in large measurement ranges. Likewise, organizing images for data processing is complicated. There is a lot of work to be done to solve these issues.

In aerial photogrammetry, several approaches for expanding the format of cameras have been proposed (e.g., UltraCam and digital mapping cameras) to facilitate the continuous acquisition of nadir images. These aerial cameras consist of a specific number of fixed CCD detectors and employ a special software to make the frames completely stitched into a whole image. The stitched images are then used in traditional triangulation method. However, the portability of these aerial cameras is crucial to fulfilling close-range measurements tasks. Therefore, multi-view photography by one camera is a widely adopted alternative technique to "normal" photography to ensure intersection angles and maintain high measurement accuracy [8,15–18]. In this case, the approach of matching all images to one another and certain improvements of the method are proposed to determine the relationship among images for automatic image organization. To ensure measurement accuracy, numerous ground control points are employed as in V-STARS [22] and ACION [23]. Nevertheless, these methods augment the complexity of data processing and increase the workloads of field operation. Moreover, oblique photography is generally implemented manually to control oblique angles, which usually result in photographic gaps and weak intersection angles.

Cameras with novel design have also been proposed to address the problem of the limited format of single frames. Scanning photography is the photography approach in which the camera is controlled and rotated around one point to acquire images in multiple rows and columns to enlarge the FOV of one station. The images in numerous rows and columns are referred to as an image matrix. The core idea of scanning photogrammetry is that the view of stations can alternatively be enlarged given that the view of cameras is difficult to do so. This method is easier to exploit than to expand the format by multiple CCD detectors. In this approach, camera parameters, including focal length and pixel size of each projection, are kept constant as well [24,25]. A3 Edge, a contemporary aerial mapping camera developed and introduced by VisionMap, is an example of commercially available cameras for aerial photogrammetry. This camera takes images when it is rotated in a cross-track direction to enlarge FOVs [24]. In close-range measurements, a surveying robot carrying an image sensor and a theodolite or total station are typically applied, allowing the immediate acquisition of the position and posture of images [26]. In view of the limited image format used in surveying robots, theodolite scanning photogrammetry is proposed by controlling the robot to rotate the camera and acquire images in multiple rows and columns to enlarge the FOV of one station [27]. However, the use of surveying robots for measurements is expensive. Zhang *et al.* [16,28,29] proposed panning and multi-baseline photogrammetry to obtain rotating images during measurements. This photography approach specifies the overlaps among images, making it easy to organize images in order. Nevertheless, this technique is operated by hand, and only experienced photographers can accomplish the work well. L. Barazzetti *et al.* [30–32] generated gnomonic projective images from several pinhole images acquired by rotation platform with long focal lens to reconstruct a 3D model. This method provides a general network around the object using central perspective images with normal focal lens and reconstructs "sharp" details through gnomonic projections. Spherical panoramas have also been employed to extract 3D information [25,33–35]. G. Fangi [33] captured panoramas with the use of both normal and long focal lenses. The normal focal lens spherical panorama was used to establish the station coordinates, whereas the long one was employed to ensure accuracy.

In this study, we improve the solution of scanning photography. This method allows stations to be distributed in order as that in aerial photogrammetry and correspondingly allows standardized data to be acquired in close range. A photo scanner consisting of a non-metric camera, a rotation platform, and a controller is developed for convenient data acquisition without coverage gap. Once the operator sets the required parameters and specifies the ground coverage of the station, this machine can automatically control the camera to rotate in horizontal and vertical directions and obtain images as designed overlaps.

Our scanning photogrammetry method is expected to extract 3D information from long focal images immediately and robustly without the use of normal focal images to provide the general network. Therefore, we propose a modified triangulation according to the traits of data obtained by scanning photography. This developed approach can directly achieve high-precision measurement results regardless of the adopted focal lens (*i.e.*, long or normal). Likewise, this approach is similar to processing images acquired with multiple CCD detectors. The main rationale underlying this technique is that the overlap between the ground coverages of adjacent image matrices is known and larger than 60% based on station distribution, whereas the overlap between stereo original images from different stations is unknown and may be less than 50%. Thus, the relative orientation should be executed between the images synthesized from image matrices than among the original images. We then generate a free network of synthetic images and solve the exterior orientation elements of original images from elements of the former and their relative rotation angles recorded by the scanner. Consequently, we determine the overlap relationship among the original images. This method avoids identifying the overlap relationship among images by matching all of them to one another, which is a time-consuming approach. Finally, all of the original images are employed for final bundle adjustment because the multi-directional images obtained by scanning photography could affect the accuracy and robustness of the method. Considering the unknown interior elements and instable distortion in using DSLRs, we apply self-calibration bundle adjustment with control points to solve the interior elements of the camera; this approach has been adopted by most researchers [13,20,36,37].

The proposed method is applied in two test fields to validate the feasibility and precision of scanning photogrammetry. The experiments prove that scanning photogrammetry organizes image data well when large scenes are measured using long focal lens and achieves millimeter accuracy. Several comparative experiments are also performed to analyze the factors that may affect measurement accuracy.

The rest of this paper is organized as follows. Following the Introduction, Section 2 explains the theories of scanning photogrammetry. Section 3 evaluates the performance of the scanning photogrammetry system with the use of real datasets obtained with 300 and 600 mm lenses. Section 4 discusses the differences between our scanning photogrammetry technique and previous ones, and analyzes the reasons for the efficient measurement results. Finally, Section 5 presents the conclusions of the study and cites recommendations for future research.

2. Proposed Scheme

The proposed scheme comprises four parts, namely, Section 2.1 scanning photography, Section 2.2 station distribution, Section 2.3 scanning platform (photo scanner), and Section 2.4 data processing.

2.1. Scanning Photography

As previously mentioned, the size of a single image sensor is limited, thereby restricting the FOV of DSLRs when long focal lenses are used to yield comparatively high ground resolution images. Therefore, scanning photography was developed. This photography approach obtains sequences of images by rotating the camera in both horizontal and vertical directions to enlarge the FOV of one station. This technique is similar to VisionMap A3 camera, which acquires the flight sequences of frames in a cross-track direction to provide wide angular coverage of the ground [24]. When used in close-range photogrammetry, the presented scanning photography further improves traditional photography by exposing camera frames in both horizontal and vertical directions because the height of the camera is generally difficult to change in a wide range when high targets are photographed. This particular mechanism is similar to the concepts of "cross-track" and "along-track" in aerial photography. Each image captured by the camera is called an original image, and all original images taken at one station form an image matrix as displayed in Figure 1.

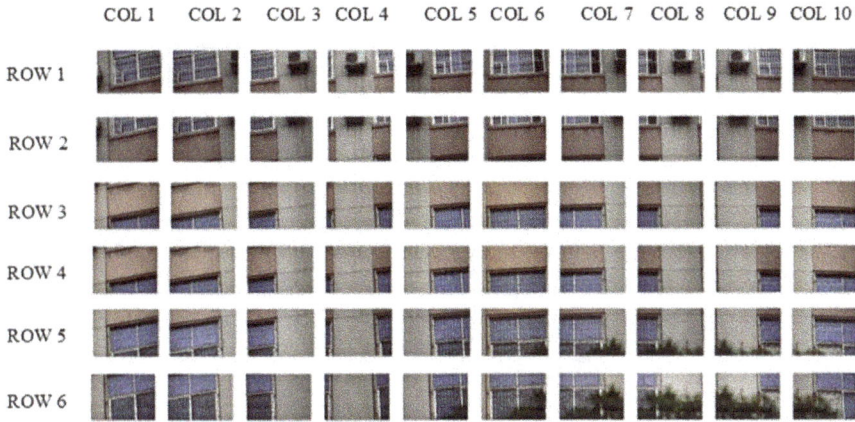

Figure 1. Illustration of image matrix at one station (The matrix contains 6 rows and 10 columns).

In scanning photography, rotation angles are calculated based on the principle that the relative rotation angles between two adjacent images in both directions are equal. However, in our experiments, this approach is inadequate because the images at the corners of image matrices may be covered by invalid areas as a result of the central perspective when photographing. Figure 2 illustrates how this scenario occurs. In this figure, points D, E, and S are in a horizontal plane, whereas points A, B, C, D and E are in another plane that is perpendicular to the horizontal one. Point B is in line AC, and $AD \perp DE$, $BE \perp DE$, $SE \perp DE$, and $SE \perp BE$. Angles $\angle BSE$ and $\angle ASD$ are defined as β_1 and β_2, respectively. Therefore, $\beta_1 > \beta_2$. Points A, B, C, D, and E represent the center points of images A, B, C, D, and E, respectively. Image E is the reference image in both horizontal and vertical directions and is named the "normal" image. β_1 denotes the vertical angle rotating from images E to B, and β_2 depicts the vertical angle rotating from images D to A. Correspondingly, when photos are taken at the same height, the relative vertical rotation angle between the adjacent image rows far away from the "normal" image is smaller than that between the image rows near the "normal" image. Invalid coverages appear when the relative rotation angle in the vertical direction remains constant. This condition also applies to the rotation angle in the horizontal direction.

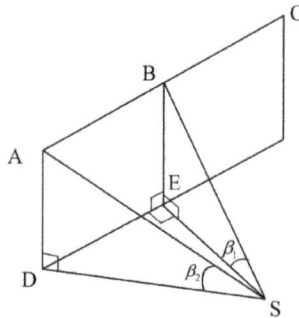

Figure 2. Rotation angles influenced as perspective projection.

In view of the abovementioned problem, we propose an improved means for calculating rotation angles. Figure 3 presents how rotation angles are calculated using this method. S is the perspective center. Horizontal rotation angle ranges from α_1 to α_2, and vertical rotation angle varies from β_1 to β_2. f is the focal length, and region $W \times H$ is the ground coverage of the station projected to the focal plane of the "normal" image. The rotation angle of each original image is determined based on the designed overlaps in both directions. In Figure 3, the junctions of grids represent the image centers, and the rays from the perspective center to the image center indicate the principal rays of each image. The mechanism for calculating rotation angles is cited below. By following these procedures, we can calculate the rotation angles of images as in Equation (1).

(1) Obtain the range of rotation angles of a station. As mentioned above, horizontal rotation angle ranges from α_1 to α_2, and vertical rotation angle varies between β_1 and β_2 .

(2) Region $W \times H$ can be calculated according to the angle range.

(3) Calculate the temporary horizontal and vertical distances between centers of the adjacent images in the same row and column in image matrix as ΔW_{temp} and ΔH_{temp}, respectively. Accordingly, determine the row and column number of image matrix as N_{row}, N_{col}, respectively.

(4) Recalculate the horizontal and vertical distances as ΔW and ΔH, respectively, to divide the region equally into rows N_{row} and columns N_{col}.

(5) Determine the location of each image center on focal plane of the "normal" images, and calculate the rotation angles of each image in both directions.

In our experiment, the absolute value of the horizontal and vertical rotation angles is assumed to be less than $60°$. Figure 4 shows the horizontal and vertical rotation angles of original images in one image matrix; where the black solid dots denote the centers of images. In particular, the figure demonstrates that the relative rotation angle between adjacent images is decreased when the rotation angle increases from the "zero position." The improved results of scanning photography are depicted in Figure 5. To ensure successful measurement, the overlap between any two adjacent images in one image matrix should not be less than 30% in both directions.

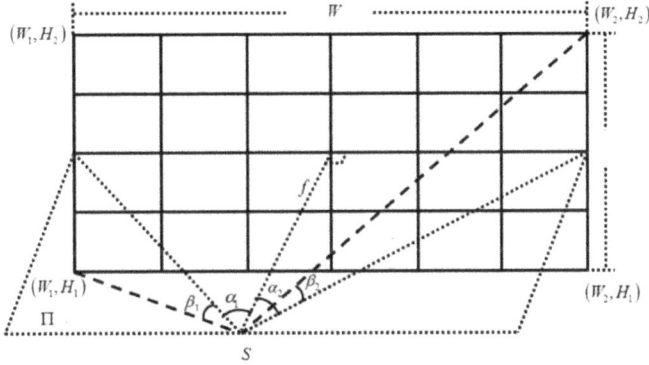

Figure 3. Calculation of rotation angles.

To limit image distortion, the horizontal rotation angle α should be in the range of $[-45°, 45°]$, and the vertical rotation angle β should be within $[-30°, 30°]$.

$$
\begin{aligned}
\Delta W_{temp} &= (1 - p_x) \bullet f \bullet 2\tan(\phi_H/2) \\
\Delta H_{temp} &= (1 - p_y) \bullet f \bullet 2\tan(\phi_V/2) \\
W_1 &= f \bullet \tan\alpha_1 \qquad\qquad W_2 = f \bullet \tan\alpha_2 \qquad\quad W = W_2 - W_1 \\
H_1 &= \sqrt{f^2 + W_1^2} \bullet \tan\beta_1 \qquad H_2 = H_1 = \sqrt{f^2 + W_2^2} \bullet \tan\beta_2 \quad H = H_2 - H_1 \\
N_{col} &= W/\Delta W_{temp} \qquad\qquad N_{row} = H/\Delta H_{temp} \\
\Delta W &= W/N_{col} \qquad\qquad\quad \Delta H = H/N_{row} \\
\alpha_{rc} &= \tan^{-1}\left((W_1 + c \bullet \Delta W)/f\right) \\
\beta_{rc} &= \tan^{-1}\left((H_1 + r \bullet \Delta H)/\sqrt{(W_1 + c \bullet \Delta W)^2 + f^2}\right)
\end{aligned}
\tag{1}
$$

where f is the focal length; p_x and p_y are the known horizontal and vertical overlaps between images in the image matrix, respectively; ϕ_H and ϕ_V are the horizontal and vertical views of the camera, respectively; α_1 and β_1 are the horizontal and vertical rotation angles of the image at the bottom left corner of the image matrix, respectively; α_2 and β_2 are the rotation angles of the image at the top right corner; W and H are the width and height of the projected region on the "normal" image, respectively; ΔW_{temp} denotes the temporary horizontal distance between centers of the adjacent images in the same row in image matrix; ΔH_{temp} is the temporary vertical distance between centers of the adjacent images in the same column; N_{row} and N_{col} are the row and column numbers of the image matrix, respectively; ΔW

represents the distance between adjacent image centers in the same row; ΔH is the final distance between adjacent image centers in the same column; and α_{rc} and β_{rc} depict the horizontal and vertical rotation angles of the image at row r and column c in the image matrix, respectively.

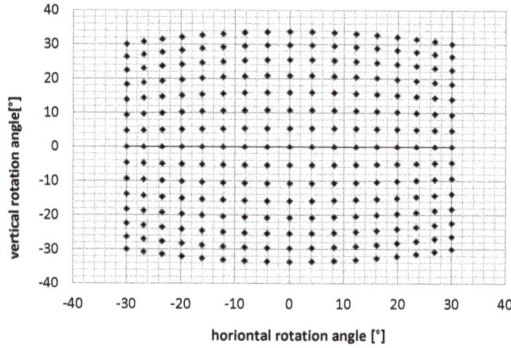

Figure 4. Distribution of horizontal and vertical rotation angles; the focal length is 100 mm, and the camera format is 36 mm × 24 mm. The rotation angles of the image at the bottom left corner and top right corner of the image matrix are $(-30°, -30°)$, and $(30°, 30°)$, respectively. The set overlaps in horizontal and vertical directions are 80% and 60%, respectively.

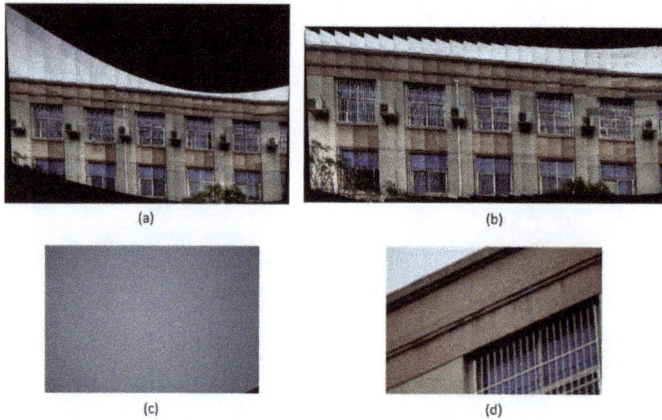

Figure 5. Synthetic images from image matrices with rotation angles calculated in different approaches; (**a**) is the synthetic image from the image matrix acquired using the same approach in determining the relative rotation angle between adjacent images; (**b**) is the synthetic image from the image matrix acquired with the improved method introduced in this paper; (**c**) is the image at the top left corner of the image matrix presented in Figure 5a; (**d**) is the image at the top left corner of the image matrix presented in Figure 5b.

168

2.2. Station Distribution

To facilitate data acquisition and processing, our station distribution is set similar to the flight strip in aerial photogrammetry. The top view of this distribution is displayed in Figure 6, where the horizontal line denotes the measured target, the rectangles represent the stations, and the corresponding two solid lines stretching from the station to the object are the FOV of the station. S denotes the length of the target, and D is the photographic distance (referred to as photo distance hereafter for brevity). For easy manipulation, the baseline denoted by B between adjacent stations is designed the same, which is the same as aerial photogrammetry. Then, the operator will easily find out the location of the next station during data acquisition. To ensure the maximum utilization of images, the columns and rows of each station are varied because the FOV of each station is different. θ is the least FOV of stations determined by operators. N represents the total number of stations, and M is the given number of least stations from which any interest point on the target would be photographed. Figure 6a shows the station distribution when the target has a large width. In such a case, the distance from the first station to the last is the same as the width of the target. In our scanning photography mechanism, the FOV of stations ranges from θ to 2θ according to the location of the station. Similarly, the largest intersection angle of the measured points ranges from θ to 2θ. We limit θ between $20°$ and $45°$ to maintain a good balance between measurement and matching precisions. Photo distance D is determined according to the selected camera focus, required measurement precision, and scene environment. However, certain scenarios exist (e.g., Figure 6b) in which the target is extremely narrow such that the FOV of a station is greatly limited in the given photo distance. Figure 6b depicts the solution for this condition, ensuring the intersection angle at measuring points. The distance from the first station to the last is larger than the width of the target. As such, the intersection angle at the measuring points is larger than θ, and the view angle of stations can either be smaller or larger than θ. To maintain measurement precision, every point on targets should be photographed from at least three stations. In other words, M should be selected, but it should not be less than 3. Then, the overlap between two adjacent stations is set above 67% to enable stereo photogrammetric mapping. It is known that large overlap will result in small baseline. However, as in Figure 6, our scanning photogrammetry is a kind of photogrammetry method using multi-view images. Image correspondences of each tie point for bundle adjustment are from multiple stations, which can form considerable intersection angle. Even if the baseline of the adjacent stations is small, it will not affect the measurement precision. Nevertheless, a small baseline will lead to consuming more time for data obtainment and processing. Therefore, we advise the operators to determine the parameter after balancing the data acquisition time and variation of the target in the depth direction. These parameters are computed as in Equation (2).

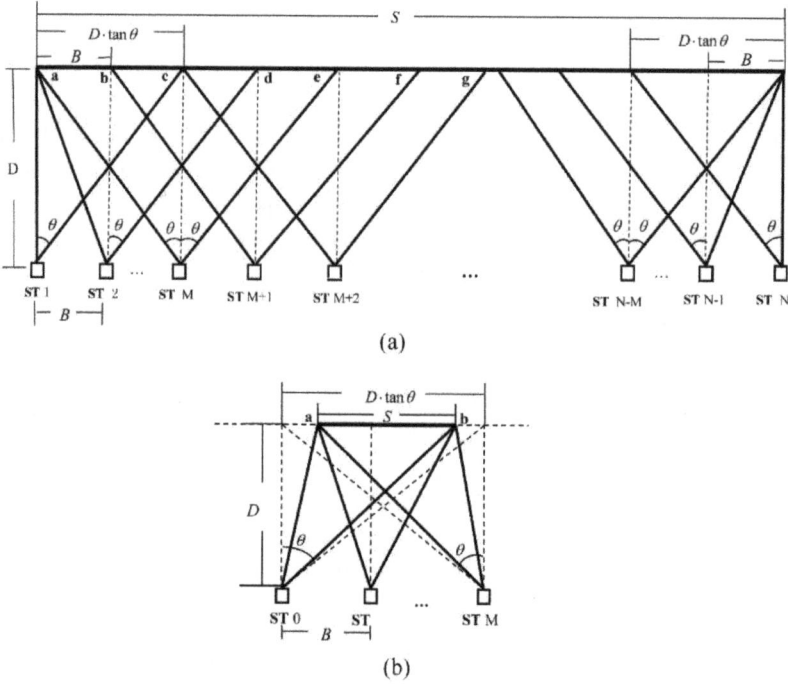

Figure 6. Station distribution: (**a**) station distribution of wide target measuring; (**b**) station distribution of narrow target measuring.

$$
\begin{aligned}
&B_{temp} = D \bullet \tan\theta / (M-1) \\
&N = \lfloor S/B_{temp} + 0.5 \rfloor + 1 \\
&\begin{cases}
\quad if \ N \geqslant M, \ \ B = S/(N-1) \\
if \ N < M, \ \ B = D \bullet \tan\theta \bullet (M-1), \ \ N = M
\end{cases}
\end{aligned} \qquad (2)
$$

where B_{temp} and B are the temporary and final baselines, respectively; D is the photo distance θ; θ is the given least FOV of a station; S is the length of the target; M is the given number of least stations from which any interest point on the target would be acquired; and N is the total number of stations.

Stations should be distributed in adherence to the following rules:

(1) The first station should be aligned to the left edge of the target shown in Figure 6a. However, when faced with the situation as in Figure 6b, the first station should be located at the distance of $(D \bullet \tan\theta - S)/2$, away from the left edge of the target.

(2)   The photo distances of every station are approximately equal when the variation of the target in depth direction is not large. Otherwise, the stations should be adjusted to the variation of the target in depth direction to maintain equal photo distances.

(3)   The length of the baselines should be the same.

After the location of the first station is identified, the location of the remaining stations can be determined according to the abovementioned principles. As demonstrated in Figure 6a, the horizontal coverage of the first, second, and third stations ranges from points a to c, a to d, and a to e, respectively. Contrarily, the fourth station covers points b to f. The remaining stations follow the same concept. In Figure 6b, points a and b denote the beginning and end of the target, respectively, and the horizontal coverage of each station begins from point a to b.

Since the height of the different parts of the target is variable, the vertical view of each station is based on the height of the photography region. Given that a large pitch angle leads to severe image distortion, the vertical rotation angle of each station should be limited in the range of $-30°$ to $30°$. In other words, when the target is extremely high, the height of the instrument or the photo distance must be increased to maintain the vertical view within range.

2.3. Photo Scanner

To guarantee image quality and measurement efficiency, we develop a photo scanner for automatic data acquisition. This scanner shown in Figure 7 consists of a non-metric camera, a rotation platform, a controller (e.g., PC or tablet PC with Win7/8), and a lithium battery. The camera is mounted on the rotation platform composed of a motor unit and transmission mechanism; this platform controls the camera scanning across horizontal and vertical directions. Figure 7 indicates that the rotation center of the platform is not the perspective center of the camera. However, when the scanner is used for large engineering measurements, the offset values are relatively tiny compared with photo distance. A module for determining rotation angles is also integrated into the software to transmit data and control platform rotation. The scanner automatically obtains the image once the camera parameters (*i.e.*, focus and image format), data storage path, horizontal and vertical overlaps between images in each image matrix, and ground coverage of stations are inputted into the software by the operators. For convenient measurement, the scanner is designed to be mounted on a tripod for the total station, and the battery unit for the rotation platform can be hung on the tripod. Thus, the developed photo scanner can be carried and operated by only one person when conducting measurements, providing great convenience.

The mechanism for applying the developed scanner is as follows:

(1)   The instrument is installed and placed at the proper position.

(2)   The photographic plane is paralleled to the average plane of the needed photographing region, and the rotation platform is leveled with the bubble on the instrument base. This position of the camera is defined as the "zero posture" in this station and is considered the origin of rotation angles in both horizontal and vertical directions.

(3)   The FOV of this station is specified, and the required parameters are inputted into the controller.

(4)   The number of rows and columns of the image matrix as well as the rotation angles in horizontal and vertical directions for each image are computed, and the signal is sent to manage the rotation and exposure of the camera for automatic image acquisition.

Data can be stored by transferring images into the controller (e.g., PC or tablet PC) and storing them in a compact flash card. In the second method, the images from the camera are not required to be transmitted to the PC or tablet PC. As such, the photo interval is shorter. During photographing, the controller enables a real-time quick view of the captured images. At the same time, the rotation angles of each captured image are stored in the controller as a log file, which can be used as auxiliary information for data processing.

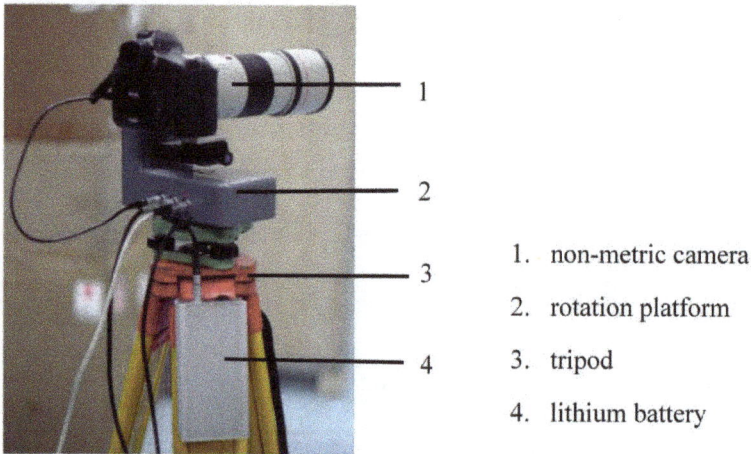

1. non-metric camera
2. rotation platform
3. tripod
4. lithium battery

Figure 7. Parts of the photo scanner

2.4. Data Processing

In our proposed method, data are automatically processed once the user specifies the data directory and required outputs. To achieve high accuracy and stable solution, all original images are used for bundle adjustment because redundant observations are helpful in obtaining a robust solution, and information loss is unavoidable during image synthesis. Considering that all original images are needed for bundle adjustment, we employ modified triangulation to improve the efficiency of data organization and processing. As previously noted, the core idea of modified triangulation is that although the overlaps between original images from different stations remain unknown, we can estimate the overlaps between coverages of image matrices according to the station distribution. Thus, we use the images synthesized from the image matrix to generate a free network. We then compute the initial exterior elements of original images in this free network from the elements of synthetic images and rotation angles of original images. After the matching relationships among the original images from neighboring image matrices are identified according to the matching results of the two corresponding synthetic images, image matching is executed among original images to determine tie points. Self-calibration bundle adjustment with control points is then performed. In establishing the initial exterior parameters of original images, we assume that the perspective centers of original images and their corresponding synthetic images are the same even though they are actually different. The accurate exterior parameters of original images that vary from those of the corresponding synthetic images can be obtained after bundle adjustment. Figure 8 exhibits the flow of data processing. The processes of taking original images with long focal lens into bundle adjustment are determined to be similar to that using the VisionMap A3 system. Thus, this procedure is valid and can be used in both aerial and close-range photogrammetry.

2.4.1. Synthetic Image

In practical measurements in scanning photography, the unknown overlaps between stereo pairs of original images from different stations are not always large enough for relative orientation. This condition therefore leads to an unstable relative orientation among stereo pairs. Considering that the overlaps between station image matrices are large enough for relative orientation, we apply a simplified method of synthesizing image matrices for free network generation. In view of information loss when original images are synthesized from image matrices, synthetic images are not used for final bundle adjustment but are only used for computing the initial exterior elements of original images. Although the camera is not rotated around the perspective center, the offset values from its center to the center of rotation are relatively tiny compared with the photo distance when performing large engineering measurements using the developed photo scanner. Therefore, we simplify the model by ignoring the offset values and projecting the image matrix to the focal plane of the "normal" image to synthesize the image matrix. Figure 9 illustrates the image synthesis model. For simplicity, only one row of the image matrix is graphed. The rotation angles of the images can be presented as (H_i, V_i) $(i = 1, 2, \ldots, m)$. H_i and V_i are the horizontal and vertical angles, respectively, and m is the number of images in the image matrix. The method used to generate a synthetic image is to re-project all original images in the image matrix to the equivalent focal length plane with $H = 0$ and $V = 0$, as shown in Figure 9. Equation (3) is employed to obtain synthetic images given that the scanner is designed to rotate first in the horizontal direction and then in the vertical direction.

$$\begin{bmatrix} x' \\ y' \\ -f \end{bmatrix} = R_i^{HV} \begin{bmatrix} x_i \\ y_i \\ -f \end{bmatrix}$$

$$R_i^{HV} = \begin{bmatrix} \cos H_i & 0 & -\sin H_i \\ 0 & 1 & 0 \\ \sin H_i & 0 & \cos H_i \end{bmatrix} \bullet \begin{bmatrix} 1 & 0 & 0 \\ 0 & \cos V_i & -\sin V_i \\ 0 & \sin V_i & \cos V_i \end{bmatrix}$$

(3)

where x and y denote the image coordinates in image i, x' and y' are the coordinates in the synthetic image, and R_i^{HV} is the rotation matrix from the synthetic image to frame i.

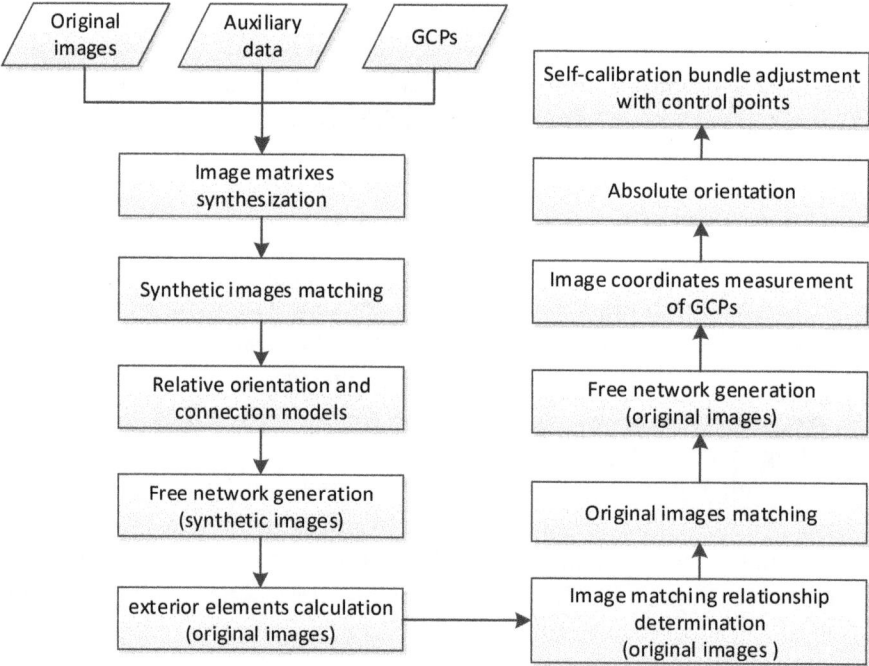

Figure 8. Flow chart of data processing

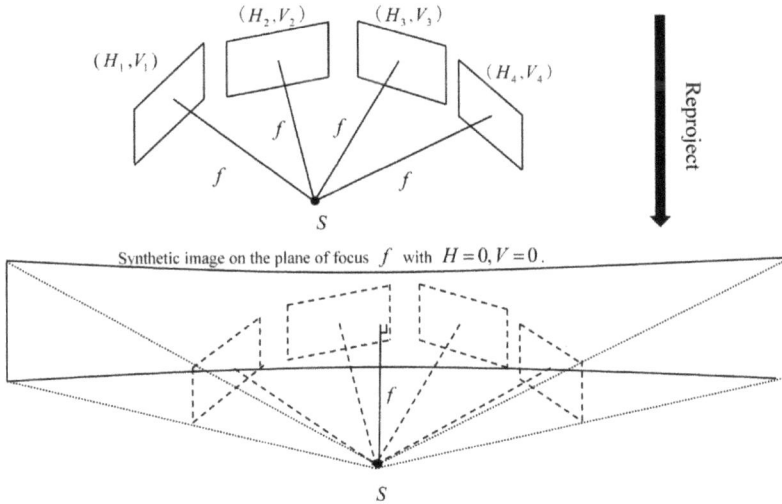

Figure 9. Model of image synthesis

The synthesis method explained above is suitable only when the view of the image matrix is not significantly large ($\leq 120°$).

2.4.2. Image Matching and Error Detection

In general, the parallax may change in a wide range and be volatile because of occlusions in close-range image pairs. When the relative rotation angle is larger than 15°, which commonly occurs in oblique photography, traditional image matching is unsuitable. Alternatively, scale-invariant feature transform (SIFT) match can be steadily applied in many cases when it involves an angle that is less than 50° [38]. Given that the SIFT features presented by 128-dimensional vectors are used for image matching, a long computational time is needed when a large number of feature points are involved. Thus, graphics processing unit (GPU) acceleration is employed. Considering that the memory of GPUs is limited, we perform a block-matching algorithm. In this algorithm, we simply use each block from one image to match all of the blocks from another image. The block matching results are accepted only when the number of correspondences is larger than the threshold.

Mismatching inevitably occurs in image matching methods. Thus, we should pay attention to error detection. The outliers in SIFT point matches are generally discarded with random sample consensus (RANSAC) for estimating model parameters. In our scanning photography, two kinds of error detection can be used after image matching, and the method to be applied is selected based on whether the corresponding points are from the same station. For matching images from the same station, the object coordinates of perspective centers are approximately equal.

175

Therefore, the homography model [39] in Equation (4) is used as the estimation model. The geometric distortions between images are severe when the matching images are acquired from different stations, suggesting that transformation models (e.g., affine and projective transforms) are no longer suitable for detecting matching errors [40]. After performing several experiments, we adopt the quadric polynomial model in Equation (5) to estimate geometric distortions. This model is the superior choice in terms of computational time and measurement precision. The threshold of residual for RANSAC should be relaxed because the interior elements of original images are inaccurate, and the estimation model is not rigid.

$$
\begin{bmatrix} x\prime \\ y\prime \\ 1 \end{bmatrix} \begin{bmatrix} h_1 & h_2 & h_3 \\ h_4 & h_5 & h_6 \\ h_7 & h_8 & h_9 \end{bmatrix} \bullet \begin{bmatrix} x \\ y \\ 1 \end{bmatrix}
\tag{4}
$$

where (x, y) and (x', y') are the image coordinates on the left and right images in an image pair, respectively, $h_1, h_2, \ldots, h_9$ are elements of the homography matrix, and $h_9 = 1$.

$$
\begin{aligned}
x' &= a_1 x^2 + a_2 xy + a_3 y^2 + a_4 x + a_5 y + a_6 \\
y' &= b_1 x^2 + b_2 xy + b_3 y^2 + b_4 x + b_5 y + b_6
\end{aligned}
\tag{5}
$$

where $a_1, a_2, \ldots, a_6$, and $b_1, b_2, \ldots, b_6$ are quadric polynomial coefficients.

2.4.3. Modified Aerial Triangulation

The overlaps between synthetic images are known based on station distribution. Therefore, these images, instead of the original ones, are more suitable for interpreting epipolar geometry. The reason is that the overlaps among original images are unknown before the pre-processing of image matching. Thus, we use synthetic images in the triangulation of relative orientation and model connection to generate a free network. In close-range photogrammetry, the initial elements of relative orientation are crucial to achieving good solution in triangulation when intersection angles are large. In this event, a stable algorithm of relative orientation is required. Compared with other direct methods (e.g., 6-, 7-, and 8-point methods), the 5-point method performs best in most cases [41]. For this reason, we apply the 5-point relative orientation algorithm proposed by Stewenius [42] from the perspective of algebraic geometry. This algorithm employs a Gröbner basis to easily explain the solution [42]. To improve precision, an iterative scheme for refining relative orientation is adopted after the calculation of initial parameters. Each successive model is then connected with one another similar to the case in aerial triangulation. Correspondingly, the free network of synthetic images is established.

The free network provides the exterior orientation elements of synthetic images. However, the exterior orientation elements of each original image are the required parameters for final bundle adjustment. Given that the changes are small among the space coordinates of each original image perspective center in one image matrix, the initial space coordinates of the original image are considered similar to those of their corresponding synthetic image. Accordingly, we can solve the azimuth elements of original images by decomposing the rotation matrix calculated with the azimuth elements of synthetic images and rotation angles of original images as depicted in Equation (6).

$$R_{ji} = R_j \bullet R_{ji}^{HV} \tag{6}$$

where R_{ji} is the rotation matrix of image i in station j, R_j is the rotation matrix of synthetic image j in the free network, and R_{ji}^{HV} is the rotation matrix of relative rotation angles from the synthetic image of station j to image i.

Before bundle adjustment with all original images, image matching should be executed among original images to obtain the corresponding points as tie points for bundle adjustment. Stereo matching is performed between the original images from different stations as well as between the original images from the same station. We find six images, namely, two from the same station and four from the next station, which overlap with each original image to proceed with matching. For the images from the same station, we chose the two images at the right of and below the processing image in the image matrix. Meanwhile, for the images from the next station, we chose four images with the highest overlapping rate with the image for stereo matching. The overlapping rate is calculated according to the matching results of the two corresponding synthetic images. Then, we tie all matching results together and then use multi-image forward intersection to compute the space coordinates of the tie points in the free network. As a result, the free network of original images is established. Absolute orientation is performed to establish the relationship between the space coordinate system of free network and the object space coordinate system before bundle block adjustment. We then compute the exterior elements of original images and space coordinates of tie points in the object space coordinate system.

In general, traditional bundle adjustment is the most mature approach because the collinearity condition for all tie points (TPs) and ground control points (GCPs) are satisfied simultaneously [43]. Considering the instability of intrinsic parameters of non-metric cameras, we utilize the on-line self-calibration bundle adjustment with control points to obtain the final result. In consideration of the camera intrinsic parameters, radial distortion, and tangential distortion, the mathematical model of self-calibration bundle adjustment based on the collinearity equation is established as in Equation (7) [44].

$$x - x_0 - \Delta x = -f \frac{a_1\left(X - X_{S_i}\right) + b_1\left(Y - Y_{S_i}\right) + c_1\left(Z - Z_{S_i}\right)}{a_3\left(X - X_{S_i}\right) + b_3\left(Y - Y_{S_i}\right) + c_3\left(Z - Z_{S_i}\right)}$$

$$y - y_0 - \Delta y = -f \frac{a_2\left(X - X_{S_i}\right) + b_2\left(Y - Y_{S_i}\right) + c_2\left(Z - Z_{S_i}\right)}{a_3\left(X - X_{S_i}\right) + b_3\left(Y - Y_{S_i}\right) + c_3\left(Z - Z_{S_i}\right)}$$

$$(7)$$

where

$$\Delta x = (x - x_0)\left(k_1 r^2 + k_2 r^4\right) + p_1\left[r^2 + 2\left(x - x_0\right)^2\right] + 2p_2\left(x - x_0\right)\left(y - y_0\right)$$

$$\Delta y = (y - y_0)\left(k_1 r^2 + k_2 r^4\right) + p_2\left[r^2 + 2\left(y - y_0\right)^2\right] + 2p_1\left(x - x_0\right)\left(y - y_0\right) \quad (8)$$

$$r^2 = (x - x_0)^2 + (y - y_0)^2$$

where (x, y) is the image coordinate of a point in image i; $[X, Y, Z]$ is the object coordinate of the point; $\left[X_{S_i}, Y_{S_i}, Y_{S_i}\right]$ is the object coordinate of perspective center of image i; $a_1, a_2, a_3, b_1, b_2, b_3, c_1, c_2$, and c_3 are elements of the rotation matrix of image i; f is the focal length; (x_0, y_0) denotes the coordinate of the principal point of image i; $(\Delta x, \Delta y)$ denotes the correction of the lens distortion; k_1 and k_2 are radial distortion coefficients; and p_1 and p_2 are tangential distortion coefficients.

During bundle adjustment, we can establish several correspondence pairs for each tie point, and the corresponding points can be either from the same station or from different stations. The pairs from the different stations are characterized by large intersection angles that are beneficial to the accuracy of depth direction, whereas the pairs from the same station are characterized by very small intersection angles that may lower accuracy. Thus, we consider the observations of image coordinates of tie points as correlative rather than independent, and the correlation coefficient is inversely proportional to the intersection angle. The weight of observations is based on intersection angles as considered in bundle adjustment and the weight of the ground points is set to 10 times the tie points. Iterative least squares adjustment is exploited for solving the unknown parameters of original images. Finally, accurate solutions are yielded for all images.

3. Experimental Results

3.1. Experimental Data

Scanning photogrammetry has been proposed mainly for acquiring data when large targets are measured and for obtaining accurate measurements of large objects at any focal length. Here, we emphasize in our experiments the application of cameras with long focal lens. Our method has been proven effective in experiments involving measuring the flag platform of Wuhan University and the high slope of Three Gorges Project permanent lock, as shown in Figure 10. The Cannon 5D Mark II camera (format size: 5616 × 3744, pixel size: 6.4 μm) is used for the tests. In the experiments, reflectors for the total station are used as control points. Object surfaces will cause reflectors to be attached unevenly because the surfaces may be rough. We use plexiglass boards, which are thicker, as the baseplates of reflectors to ensure that the reflectors attached to the object surfaces are even. Figure 10a shows a quick view of the flag platform. The height, width, and depth of this area are 10, 35, and 12 m. As shown in the figure, there are stairways leading to the flag platform from the playground, resulting in several layers in the measurement area, which brings difficulty in data processing because of the discrete relief displacement. Control points are measured by Sokkia SET1130R3 total station with 3 mm accuracy. Figure 10b shows a quick view of the high slope of Three Gorges Project permanent lock. The height, width, and depth of this region is 70, 150, and 70 m. Owing to the special environment in this field, the best position for measurement is the opposite side of the lock. The photo distance can reach up to 250 m. The Leica TCRA1201 R300 total station is used to measure control points with an accuracy of 2.9 mm.

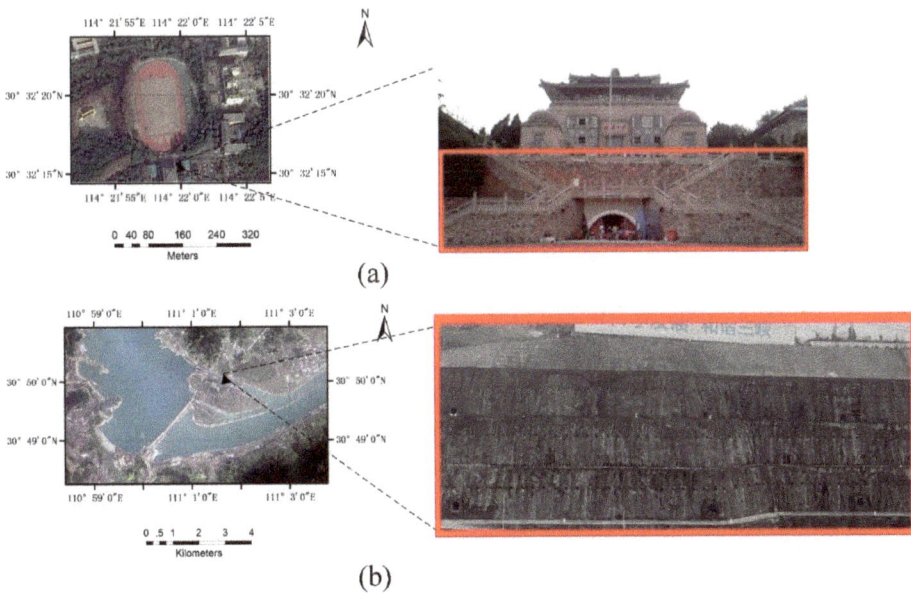

Figure 10. Quick view of test fields. ((a) flag platform of Wuhan University.; (b) high slope of Three Gorges Project permanent lock. The regions in the red rectangles show the measuring ranges.)

3.2. Results of Synthetic Images

Figure 11 illustrates the synthetic images of the two test fields. Figure 11a shows the synthetic image of one station for the field test of the flag platform with 300 mm focus lens, and the photo distance is 40 m. Figure 11b shows the synthetic image of one station for test field of the high slope of Three Gorges Project permanent lock with 300 mm focus lens at a distance of 250 m. The method applied to yield synthetic images is described in Section 2.4.1. As shown, the invalid area in the synthetic images is slightly due to the use of the improved method. The performance is not that good, because no post-processes, such as unifying color or finding best seam lines, are executed. However, these factors do not have much influence on the matching of synthetic images.

3.3. Results of Image Matching (Synthetic Images and Original Images)

Three kinds of image matching exist in this process: one for synthetic images, one for original images from different stations, and one for original images from the same station. For synthetic images, every two overlapping images is a stereo pair. However, for original images, only two overlapping images from different stations can form a stereo pair. Both stereo pairs and non-stereo pairs are needed to be matched with the SIFT matching method. After SIFT matching is performed, error detection methods are used to refine the matching results. Depending on whether the image pair is a stereo pair or a non-stereo pair, RANSAC with estimation by quadric polynomial or homography model is used for detecting errors.

Figures 12 and 13 show the results of matching and mismatching points in synthetic images from two stations. The red crosses denote the detected correct corresponding points, whereas the blue crosses denote the detected errors. The detection model is a quadric polynomial model. The matching results of the original images are shown in Figures 14–17. Image pairs in Figures 14 and 16 are stereo pairs and are processed by quadric polynomial models. Meanwhile, image pairs in Figures 15 and 17 are non-stereo pairs and are processed by homography models. The results verify that these error detection methods are valid. Given that an amount of feature points are extracted from the images showed in Figures 14 and 15, we perform block-matching algorithm. Thus, matching errors clustering exists because of the mechanism of our block-matching algorithm. However, our error detection method can eliminate outliers efficiently.

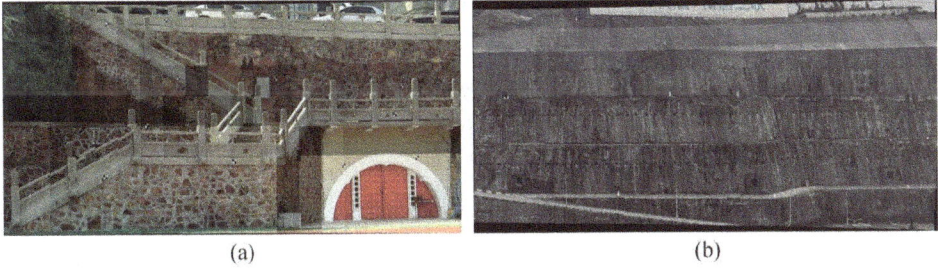

(a)   (b)

Figure 11. Synthetic images: (**a**) synthetic image of one station for test field showed in Figure 10a; (**b**) synthetic image of one station for test field showed in Figure 10b.

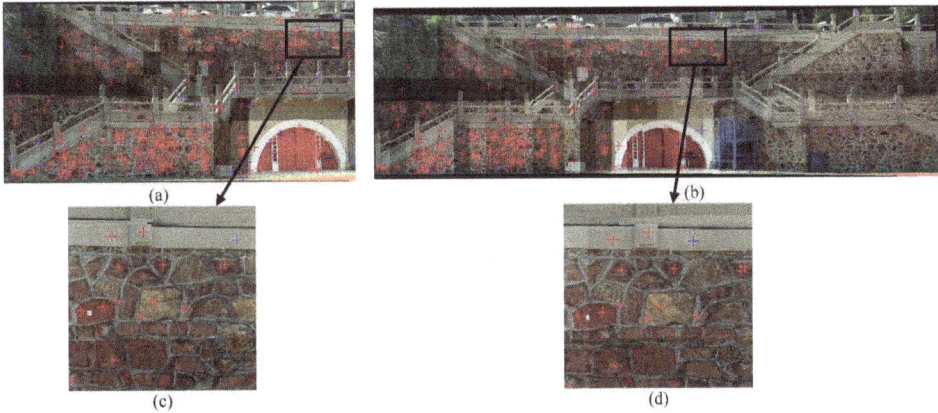

Figure 12. Synthetic images matching. (**a**,**b**) shows the matching results of synthetic images from the first and second station for test field in Figure 10a; (**c**,**d**) illustrates part of the results.

3.4. Results of Point Clouds

The free network of original images is generated after the correspondences of original images are tied together. The generated point cloud is a mess, because exterior parameters of images are not computed accurately and interior parameters of the camera have not been calibrated at the moment. After self-calibration bundle adjustment with control points, the point cloud will become regular and present the sketchy model of the target. Figures 18 and 19 show the difference of the point cloud of two test fields before and after self-calibration bundle adjustment with control points.

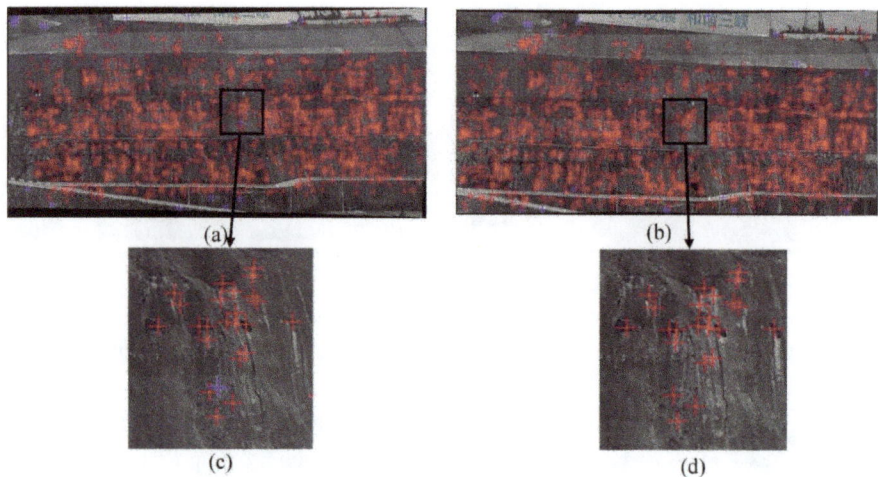

Figure 13. Synthetic images matching. (**a,b**) shows the matching results of synthetic images from the first and second station for test field in Figure 10b; (**c,d**) illustrates part of the results.

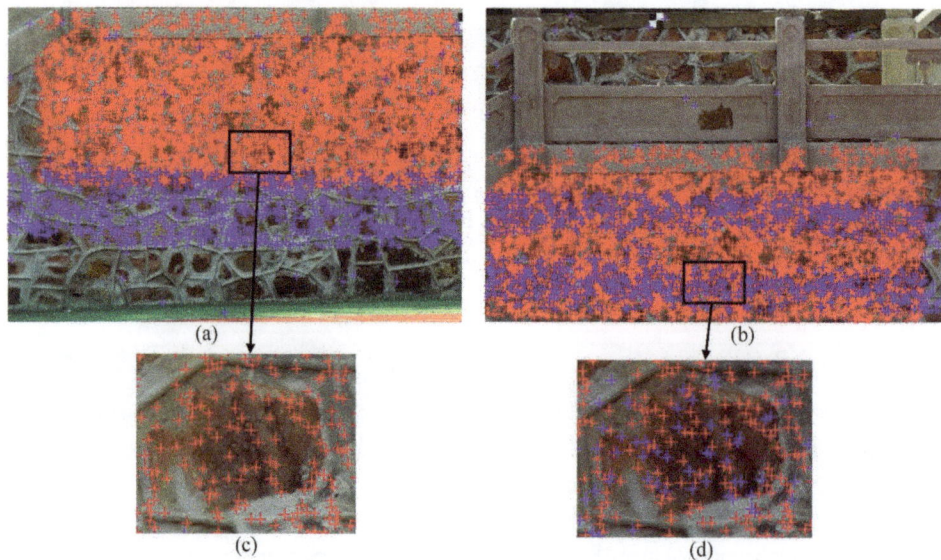

Figure 14. Original images matching. (**a,b**) shows the matching results of images from the adjacent stations as measuring the first field showed in Figure 10a; (**c,d**) illustrates part of the results.

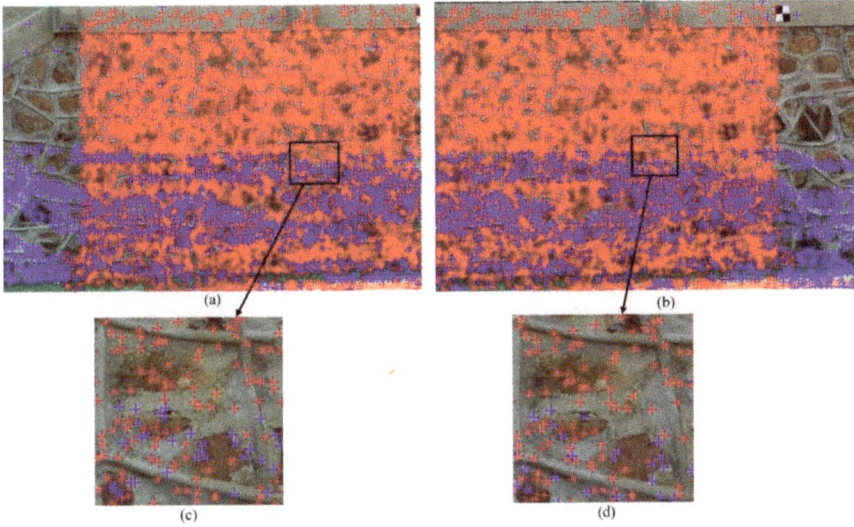

Figure 15. Original images matching. (**a,b**) shows the matching results of adjacent images in the same row of one image matrix as measuring the first field showed in Figure 10a; (**c,d**) illustrates part of the results.

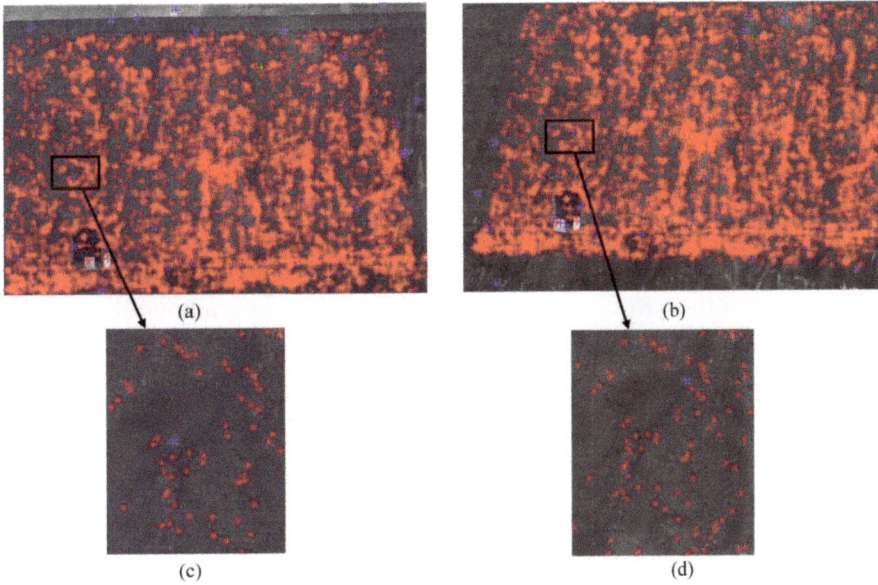

Figure 16. Results of stereo images matching. (**a,b**) shows the matching results of images from the adjacent stations as measuring the second field showed in Figure 10b; (**c,d**) illustrates part of the results.

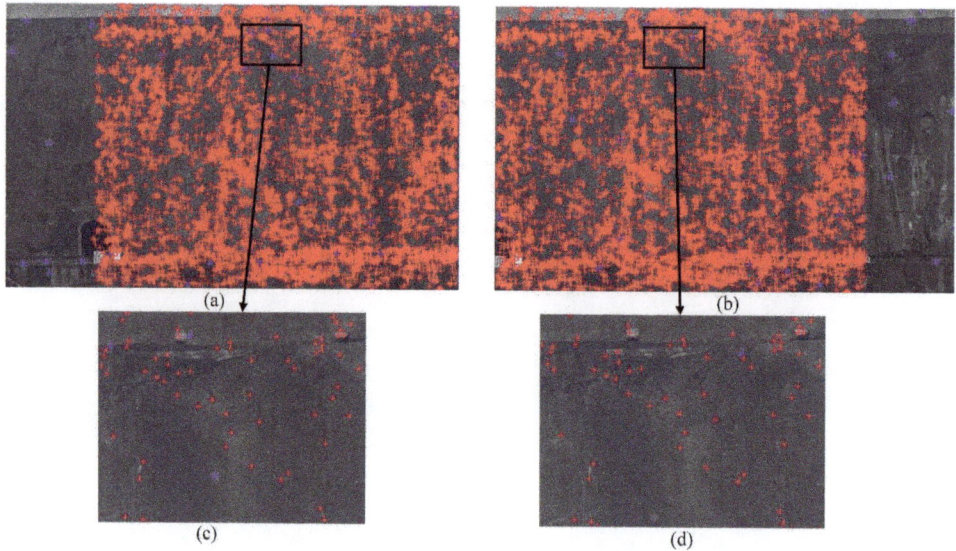

Figure 17. Results of stereo images matching. (**a,b**) shows the matching results of adjacent images in the same row of one image matrix as measuring the second field showed in Figure 10b; (**c,d**) illustrates part of the results.

3.5. Coordinate Residuals of Triangulation

We conduct several groups of comparative experiments to evaluate the effects of different factors, such as photo distance, overlapping rate between original images in image matrices, given number of least stations (M) from which any interest point on the target are acquired, set intersection angle(θ), and focal length, on the measurement accuracy in scanning photogrammetry. In the first test field of the flag platform, we analyze the effects of the four factors by using a camera with a 300 mm lens. In the second test field, the high slope at permanent lock of Three Gorges Project, two lenses with 300 and 600 mm lenses are used in evaluating the effects of focal length.

3.5.1. Comparison of Different Photo Distances

To evaluate the measurement accuracy from different photo distances when using scanning photogrammetry, the images of the first test field are obtained from 40, 80, and 150 m, and the ground resolutions are 0.9, 1.7, and 3.2 mm, respectively. Experiments are correspondingly named as cases I, II, and III. The horizontal and vertical overlaps between original images in image matrices in the three cases are set the same, 80% and 60%, respectively. We adjust the baseline of the stations to ensure that the station numbers of the three experiments are equal; that is, the intersection angles of the three experiments are different.

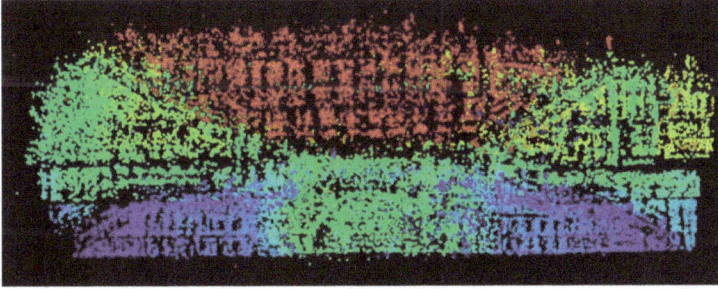

(a)

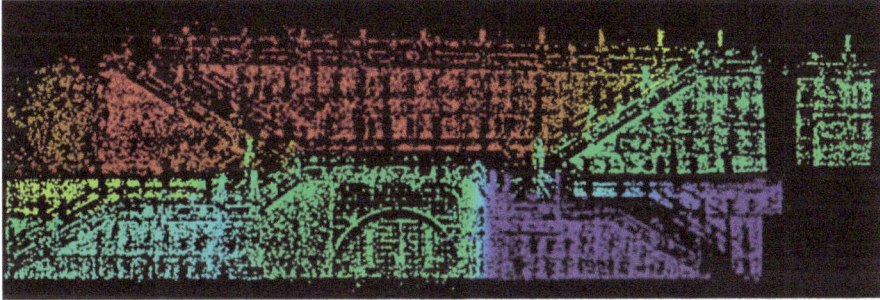

(b)

Figure 18. Point clouds. (**a,b**) shows the point clouds before and after self-calibration bundle adjustment with control points in the experiment of the first field showed in Figure 10a.

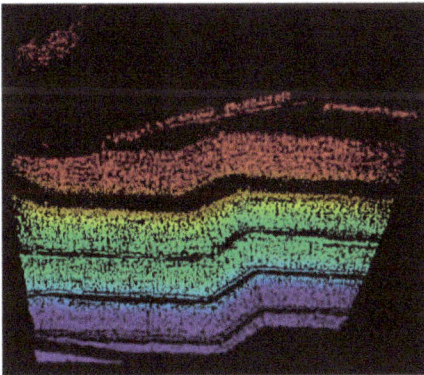

(a)

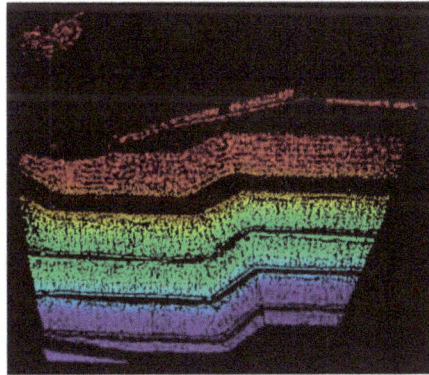

(b)

Figure 19. Point clouds. (**a,b**) shows the point clouds before and after bundle adjustment in the experiment of the second field showed in Figure 10b.

Table 1 displays the details of this experiment, and Figure 20 shows the error vectors. To compare the influence of photo distances in scanning photogrammetry, we choose 12 points as control points from 26 points distributed evenly in the test area, and the rest are check points. Owing to the angle existing between the average plane of the measured target and the XY plane in the chosen object coordinates, the distribution of control and check points on the right side of the picture is denser than that on the left side of pictures. Moreover, as shown in Table 1, the mean intersection angle of tie points is smaller than the set least intersection angle. The reason is that many of the tie points only appear in two adjacent stations.

Table 1. Parameters and coordinate residuals of experiments for first test field at different photo distances.

Cases		I	II	II	III	III
Photo distance (m)		40	80	80	150	150
Focal length (mm)				300		
Ground resolution (mm)		0.9	1.7	1.7	3.2	3.2
Total station number N				5		
Give least intersection angle θ (°)		26	35	35	35	35
Baseline (m)		8.8	14	14	26.3	26.3
Image amount		858	171	171	41	41
RMSE of image point residuals (pixel)		1/2	1/2	1/2	1/2	1/2
Mean intersection angle of tie points (°)		23.7	18.5	18.5	18.1	18.1
Maximum intersection angle of tie points (°)		43.2	37.6	36.4	35.8	35.6
Minimum intersection angle of tie points (°)		4.2	4.4	4.8	6.3	6.5
Number of control points		12	12	10	12	8
	X	1.2	1.8	1.6	1.5	1.5
	Y	1.0	1.3	0.7	1.2	0.4
Accuracy (mm)	Z	1.1	2.2	2.7	2.5	2.9
	XY	1.5	2.2	1.7	1.9	1.5
	XYZ	1.9	3.1	3.2	3.1	3.3
Number of check points		14	14	16	14	18
	X	2.0	2.1	2.4	2.5	2.5
	Y	0.8	1.8	1.9	1.6	1.7
Accuracy (mm)	Z	1.0	2.5	2.5	2.8	2.7
	XY	2.1	2.8	3.0	2.9	3.0
	XYZ	2.3	3.7	3.9	4.0	4.1

As shown by the results in Table 1, scanning photogrammetry can achieve high accuracy even when measurement is performed from different photo distances. If ground sampling distance increases (which occurs if the photo distance also increases), then the pixel becomes larger when projected on the object, and less detail can be captured. As ground sampling distance increases, fine details become visible, which improves measurement accuracy. In this experiment, the accuracy of the control points measured by total station and the ground sampling distance are in the same magnitude, and the accuracy of the control points measured by total station is even larger than that of the ground sampling distance, thus causing the measurement accuracy to not be proportional to the photo distance. At the same time, the accuracy measured by the total station in the X-direction is the lowest. Therefore, in case I, the

186

final measurement accuracy in the X-direction is lower than that in other directions because of its dependence on the accuracy measured by the total station. To further analyze the necessary amount of control points, we decrease the number of control points to 10 and 8 in cases II and III, respectively. The measurement results in Table 1 reveal that necessary control points decrease when the photo distance increases, which is related to the relief displacement of the target on images.

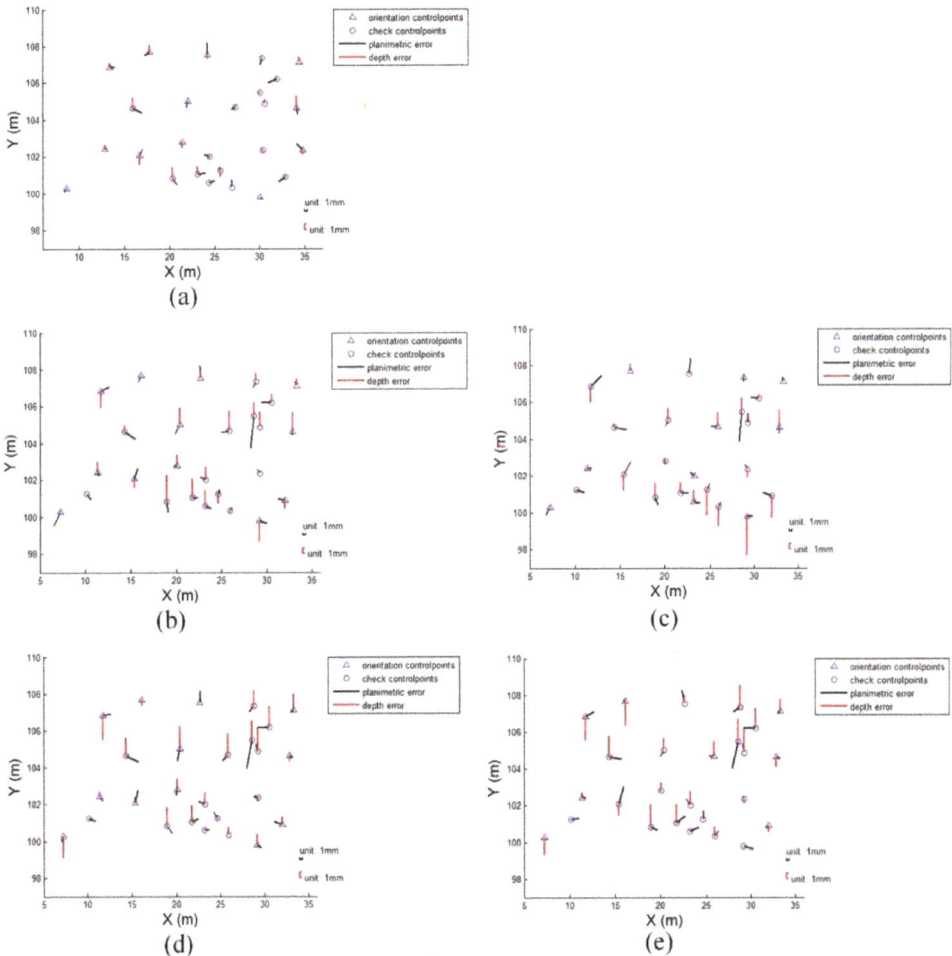

Figure 20. Error vectors of control points and check points. (**a**) is the error vectors of points measured at a distance of 40 m; (**b**) and (**c**) denote error vectors of points measured at a photo distance of 80 m. (**b**) is the result with 12 control points, and (**c**) is 10 control points; (**d**) and (**e**) show error vectors at a photo distance of 150 m, and they are the results with 12 control points and eight control points, respectively.

3.5.2. Comparison of Different Overlaps between Images in Image Matrices

In data acquisition, the number of original images that should be taken at one station is determined by the overlaps between images in image matrices. However, more images will relate to additional time spent on data acquisition and processing. Therefore, we prefer to minimize overlaps without affecting measurement accuracy. To analyze the influence of the horizontal overlapping rate in the image matrix on measurement accuracy, data with different horizontal overlaps (80%, 60%, and 30%) are obtained in the flag platform at a photo distance of 40 m. The detailed parameters and coordinate residuals of control and check points are shown in Table 2. Figure 21 displays the error vectors. The experiment results reveal that horizontal overlaps in the image matrix affect measurement accuracy; that is, accuracy is lowered with the reduction of horizontal overlap. However, the influence is insignificant. We can deduce that the vertical overlap follows a similar way. Therefore, we can conclude that when the required precision is met, the horizontal and vertical overlaps can be decreased to improve the efficiency of data acquisition and processing. Meanwhile, experience dictates that the overlapping rate cannot be less than 30% in both directions to guarantee successive data processing.

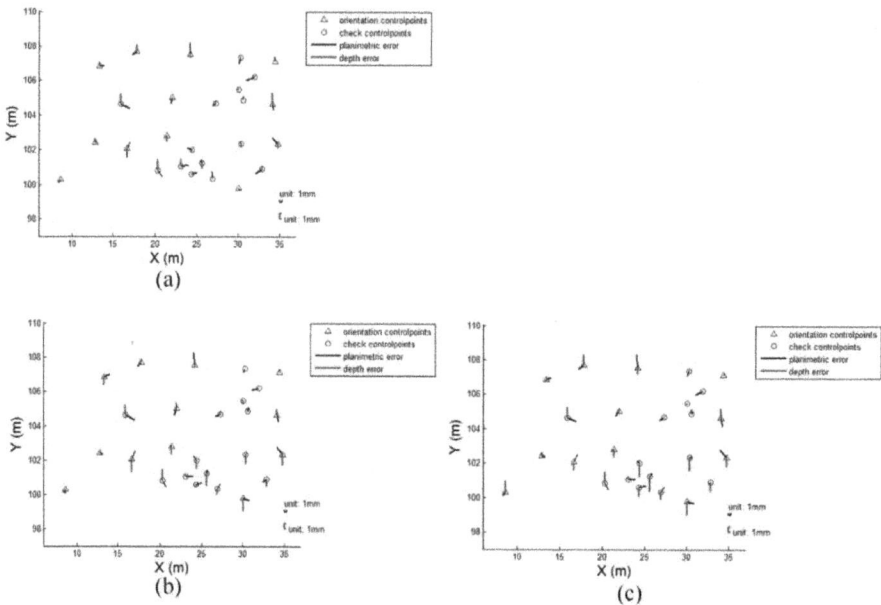

Figure 21. Error vectors of control points and check points. (**a**), (**b**), and (**c**) denote error vectors of points with horizontal overlap of 80%, 60%, and 30%, respectively. The coordinate residuals of check points in the experiments with image horizontal overlap 80%, 60%, and 30%, respectively.

Table 2. Parameters and coordinate residuals of experiments for first test field at different horizontal overlaps of image matrix.

Cases		I	II	III
Photo distance (m)		40	40	40
overlap in horizontal direction		80%	60%	30%
overlap in vertical direction		60%	60%	60%
Focal length (mm)			300	
Ground resolution (mm)			0.9	
Total station number N			5	
Give least intersection angle θ (°)			26	
Baseline (m)			8.8	
Image amount		858	444	232
RMSE of image point residuals (pixel)		1/2	1/2	1/2
Mean intersection angle of tie points (°)		23.7	23.4	22.5
Maximum intersection angle of tie points (°)		43.2	43.2	43.0
Minimum intersection angle of tie points (°)		4.2	5.6	2.4
Number of control points		12	12	12
	X	1.2	1.5	1.6
	Y	1.0	1.1	1.2
Accuracy (mm)	Z	1.1	1.5	1.6
	XY	1.5	1.9	2.0
	XYZ	1.9	2.4	2.6
Number of check points		14	14	14
	X	2.0	2.0	1.9
	Y	0.8	0.7	0.7
Accuracy (mm)	Z	1.0	1.4	1.8
	XY	2.1	2.1	2.1
	XYZ	**2.3**	**2.5**	**2.7**

3.5.3. Comparison of M

According to the station distribution in our scanning photogrammetry, the baseline between stations decreases as the value of the given number of least stations from which any interest point on the target would be photographed (M) increases. A shorter baseline is known to be conducive to image matching; however, the number of images in the project will be increased, which would require more time for data acquisition and processing. To analyze the influence of M, we conduct two experiments that measure the flag platform from three and five stations each at a photo distance of 80 m. The parameters and coordinate residuals of control points and check points are listed in Table 3. Figure 22 shows the error vectors of these points. In this experiment, the relief is small from this photo distance, suggesting that M rarely influences measurement accuracy. However, if the relief is large, M should be increased. We recommend that the value of M should be no less than 3.

Table 3. Parameters and coordinate residuals of experiments for first test field at different values of M.

Cases		I	II
Photo distance (m)		80	80
overlap in horizontal direction		80%	80%
overlap in vertical direction		60%	60%
Focal length (mm)		300	
Ground resolution (mm)		1.7	
Total station number N		3	5
Give least intersection angle θ (°)		35	35
Baseline(m)		28	14
Image amount		90	171
RMSE of image point residuals(pixel)		1/2	1/2
Mean intersection angle of tie points(°)		25.2	18.5
Maximum intersection angle of tie points(°)		36.3	36.4
Minimum intersection angle of tie points(°)		3.4	4.8
Number of control points		10	10
	X	1.6	1.6
	Y	0.8	0.7
Accuracy (mm)	Z	2.6	2.7
	XY	1.8	1.7
	XYZ	3.2	3.2
Number of check points		16	16
	X	2.4	2.4
	Y	1.7	1.9
Accuracy (mm)	Z	2.5	2.5
	XY	3.0	3.0
	XYZ	3.9	3.9

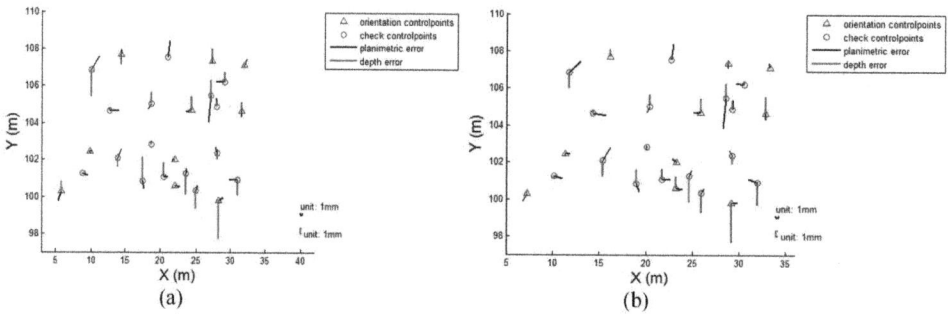

Figure 22. Error vectors of control points and check points. (a) and (b) denote error vectors of points in the experiments of obtaining images from three and five stations Coordinates residuals of check points in the experiments with M as 3 and 5, respectively.

3.5.4. Comparison of Set Intersection Angle

In this part, experiments with different intersection angles are conducted. The intersection angle is set to 26° and 35° in the two tests at a photo distance of 40 m. According to the abovementioned rules of station distribution, the amounts of total stations are 3 and 5, respectively. Table 4 lists the detailed parameters of the experiments and the coordinate residuals. Figure 23 shows the error vectors. The results unexpectedly demonstrate that a larger intersection angle corresponds to a lower measurement accuracy. We speculate that the main cause of this situation is increase in the photo distortion when the intersection angle is enlarged. Therefore, a large intersection angle is not always better than a smaller intersection angle and should be determined according to the relief displacement of the measuring target.

Table 4. Parameters and coordinate residuals of experiments for first test field at different intersection angles.

Cases		I	II
Photo distance (m)		40	40
overlap in horizontal direction		80%	80%
overlap in vertical direction		60%	60%
Focal length (mm)		300	
Ground resolution (mm)		0.9	
Total station number N		5	3
Give least intersection angle θ (°)		26	35
Baseline (m)		8.8	17.6
Image amount		858	738
RMSE of image point residuals (pixel)		1/2	1/2
Mean intersection angle of tie points (°)		23.7	29.9
Maximum intersection angle of tie points (°)		43.2	44.8
Minimum intersection angle of tie points (°)		4.2	6.1
Number of control points		12	12
	X	1.1	1.3
	Y	1.0	1.1
Accuracy (mm)	Z	1.2	1.4
	XY	1.4	1.7
	XYZ	1.9	2.2
Number of check points		14	16
	X	2.0	1.8
	Y	0.8	0.8
Accuracy (mm)	Z	1.0	1.9
	XY	2.1	2.0
	XYZ	2.3	2.7

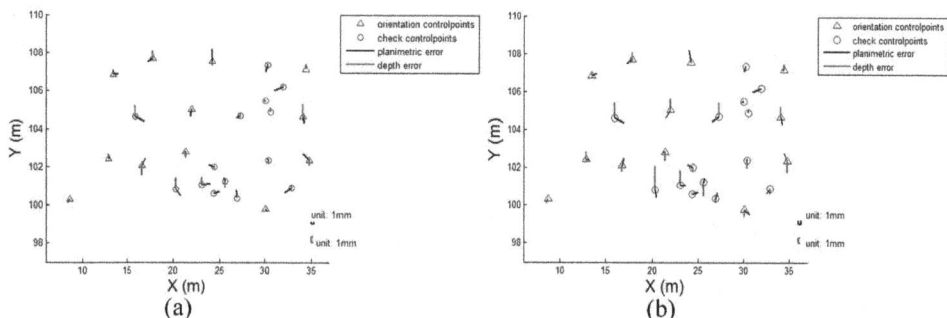

Figure 23. Error vectors of control points and check points. (**a**) and (**b**) demonstrate error vectors of points in the experiments with designed intersection angle as 26°and 35°.

3.5.5. Comparison of Different Focuses

Ground sampling distance can be improved by increasing the focal length of the camera, which can be beneficial to measurement accuracy. In this part, two lenses with focus lenses of 300 mm and 600 mm are used for measuring the high slope in the second field (Figure 10b). The tests are named Cases I and II, respectively. Table 5 shows the detailed parameters of the experiments and the coordinate residuals, and Figure 24 illustrates the error vectors. The scene of the field is complex and thus can only be surveyed at a long distance of 250 m. A total of 14 control points are selected for orientation. Table 5 shows that our scanning photogrammetry can achieve high precision when a large target from a long distance is measured. This efficient measurement in Case II reveals that our method is valid and can apply telephoto lens to achieve high precision with a few control points. However, the residuals in case II is expected to improve much more than that in case I because the ground sampling distance in case II is improved to half of that in case I. However, the accuracy was not improved substantially because the accuracy was limited by the accuracy of the control points measured by the total station. Another reason is the systematic errors that cannot been corrected completely because the distortion in images obtained with telephoto lenses is complicated. Therefore, finding ways to solve this problem is the main concern of our future works.

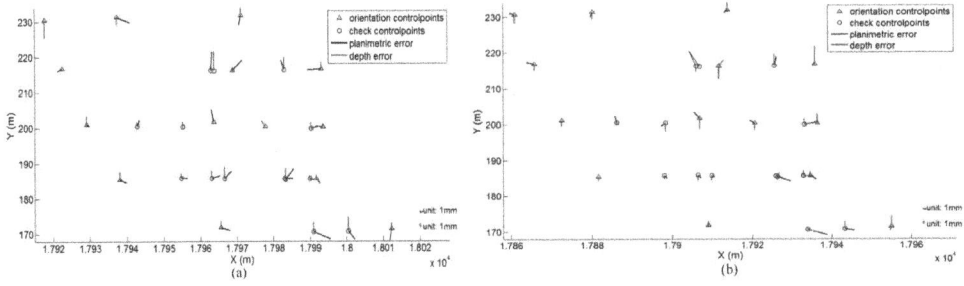

Figure 24. Error vectors of control points and check points. (**a**) and (**b**) demonstrate error vectors of points in the experiments with 300 and 600 mm focal lens.

Table 5. Parameters and coordinate residuals of experiments for second test field at different focuses.

Cases		I	II
Photo distance (m)		250	250
overlap in horizontal direction		80%	60%
overlap in vertical direction		60%	40%
Focal length (mm)		300	600
Ground resolution (mm)		5.3	2.7
Total station number N		3	3
Give least intersection angle θ (°)		30	30
Baselines (m)		55	55
Image amount		601	982
RMS of image point residuals (pixel)		2/5	1/2
Mean intersection angle of tie points(°)		22.2	23.4
Maximum intersection angle of tie points(°)		33.1	29.4
Minimum intersection angle of tie points(°)		6.1	6.0
Number of control points		14	14
Accuracy(mm)	X	3.3	1.7
	Y	3.3	1.6
	Z	3.9	4.4
	XY	4.6	2.3
	XYZ	6.0	5.0
Number of check points		14	14
Accuracy (mm)	X	3.2	2.7
	Y	4.0	3.2
	Z	4.4	3.9
	XY	5.1	4.2
	XYZ	6.8	5.7

4. Discussion

This paper uses scanning photography to solve the problem of data acquisition and organization caused by the limited FOV of digital cameras when measuring large targets with long focal lens. When cameras with long focal lens are used to measure large targets, the ground coverage is small, resulting in the need to take lots of images. If all these images are taken manually, then only experienced photographers can accomplish the work well. Otherwise, coverage gaps will occur, especially when applying multi-view photography to obtain images to maintain measurement precision. Scanning photography combined with the isometric station distribution proposed in this paper is designed to obtain images regularly, as in data acquisition in aerial photogrammetry. Also, it is a kind of photogrammetry method using multi-view images. After operators input the required parameters, the photo scanner automatically obtains images, thus making it easy to acquire images without coverage gap and to organize images for data processing. Moreover, the photo scanner can record the rotation angle of each image when obtaining images, which can be used in data processing as auxiliary information. A modified triangulation is conducted according to the traits of the data acquired by scanning photography that the overlaps among images from the same station are known, improving the efficiency and stability of data processing. The modified triangulation can be considered as a kind of global structure from motion (SFM) with auxiliary information.

Successfully used in aerial and close-range photogrammetry, scanning photography is the approach of controlling the camera to rotate around one point to acquire images to enlarge the field-view-of one station. Vision Map A3 system is an example of scanning photography in aerial photogrammetry. The camera used in this system consists of dual CCD detectors with two 300 mm lenses. As the flight moves when obtaining images, the camera can only be controlled to rotate in a cross-track direction, which provides a field-of-view of 104°. In the along-track direction, the field-of-view is enlarged by using dual CCDs. Traditional triangulation with GPS information is utilized for data processing. After that, super large frames synthesized from all pairs of original images of one sweep are generated for stereo photogrammetric mapping [24]. In close-range, the platform for obtaining images is static and always controls only one camera rotating in both horizontal and vertical directions (the same as along-crack and cross-crack direction) to enlarge the field of view. While, several systems take images both with normal and long focal lenses in measurements; normal focal lens images are used to provide a general network around the object, whereas the long focal lens images are employed to ensure accuracy and reconstruct the 'sharp' details [30–33]. And, the relative angles between adjacent images from one station are always the same in these systems.

The differences of our scanning photogrammetry and the previous ones are listed as follows:

(1)   Difference in scanning photography

To avoid the situation when images at the corners of the image matrix are covered by invalid areas (as in Figure 5), which results from central perspective when photographing, the relative rotation angle between adjacent images decreases as the rotation angle increases from the "zero position", instead of staying the same in the previous scanning photography. Considering our scanning photogrammetry is used for measuring targets outside, which always allow only one side of the object towards the camera, we recommend that the horizontal rotation angle should be in the range of $[-45°, 45°]$, and that the vertical rotation angle should be within $[-30°, 30°]$ to limit image distortion. So that, our simplified synthesis method of projecting the image matrix on a plane is enough to actually deliver the application.

(2)   Difference in rotation platform

For easy data acquisition, the photo scanner is designed to obtain images automatically. The weight of the long focal lens, particularly telephoto, may be heavier than that of the camera body; hence, the center of rotation and the center of camera are designed in different locations to enable the photo scanner to be successfully used with telephoto, as shown in Figure 7. However, the offset values of the two centers are small compared with the photo distance when large engineering measurements are measured using this photo scanner. Thus, we ignore the offset values when image matrices are synthesized.

(3)   Difference in data processing

According to the mechanism of our scanning photography and station distribution, the overlap between ground coverages of adjacent image matrices is known and large enough for relative orientation, whereas the overlap between original stereo images from adjacent stations are unknown and may be insufficient for relative orientation. Therefore, we modify the traditional triangulation by utilizing synthetic images to calculate the initial parameters for bundle adjustment. The offset in X, Y, Z coordinates from the center of the camera to the center of rotation is very small compared to the photo distance in most cases; here, we ignore the difference when image matrices are synthesized. Then, a simplified approach is employed for generating synthetic images, which do not need the surface model of the target. Although model errors exist in the synthetic image, they do not have much influence on the matching of synthetic images, especially when the photo distance is large. After synthetic images are generated, they are used to generate a free network, and the initial exterior orientation parameters of the original images are calculated accordingly. Original images are employed when bundle adjustment is performed because perspective centers of each image from the same station are actually different and multi-directional images will contribute to accuracy and robustness. The accurate exterior parameters of the original images different from parameters of the corresponding synthetic images can be obtained after bundle adjustment.

Two reasons that our scanning photogrammetry can effectively process images with long focal length.

(1) The modified triangulation ensures the initial parameters are stable, which benefits for the convergence of bundle adjustment.

(2) In consideration of the unstable intrinsic parameters of the non-metric cameras with long focal lens, especially telephoto, the on-line self-calibration bundler adjustment with control points is employed in our scanning photogrammetry. We hold that the intrinsic parameters are constant during the short time of data acquisition. Further, we fixed the focus of lens with a tape to avoid small changes caused by vibration when the photo scanner moves. During bundler adjustment, the weights of observations based on intersection angles are considered to ensure its convergence.

However, the narrow field of view of long focal length can impact adversely on the linear dependencies between the interior and exterior orientation parameters [45]. Although bundle adjustment convergences in the experiments, the results do not meet expectations. The precision can be improved with proper calibration model. In future works, we will research this aspect.

5. Conclusions

In this paper, scanning photogrammetry was employed to improve the efficiency of large object measurement in close range. The limited format of digital cameras prevents normal photography when large targets are measured with long focal lens, which leads to low efficiency because of the difficult data acquisition and organization when using oblique photography as a substitution. This method was derived to acquire images in order as in traditional aerial photogrammetry, which was easily conducted and could avoid coverage gaps. Therefore, we improved the scanning photography, provided a design of station distribution, and developed the corresponding data acquisition instrument to ensure image data were obtained in an orderly way. Then, considering the traits of data acquired in our method, a modified triangulation is performed by utilizing synthetic images from image matrices to generate the free network to avoid matching all original images with one another firstly; then, yielding a free network from original images according to the free network from synthetic images; finally, executing self-calibration bundle adjustment with control points to calculate the final results. This modified triangulation method highly improved the efficiency and stability of data processing when using long focal lens for measurements. Also, the results of experiments confirmed that our method could achieve high measurement precision with 300 and 600 mm lenses, when using a small number of control points.

However, the following problems remain to be solved in future works.

(1)   To apply our method to more kinds of targets, such as tunnels, we consider projecting the image matrix on a cylinder or a sphere in future research.

(2)   To enhance measurement accuracy, we will continue to work on dealing with the image distortion using telephoto lens.

(3)   In the next step, we will work towards calibration of the rotation platform to give more precise initial parameters for bundle adjustment.

Acknowledgments: This work was supported in part by the National Natural Science Foundation of China under Grants 41201482, the Research Funds of the Key Laboratory of Mapping from Space, National Administration of Surveying, Mapping and Geoinformation with project number K201405, the Fundamental Research Funds for the Central Universities with project number 2014213020201, and the Geographic National Condition Monitoring Engineering Research Center of Sichuan Province with project number GC201514. The authors would like to thank the anonymous reviewers, members of the editorial team, He Huang and Jianan He from Wuhan University, China, for their comments and contributions, which have significantly improved this paper.

Author Contributions: Shan Huang designed and implemented the algorithm and performed the experiments, she also wrote the paper; Zuxun Zhang directed the algorithm and experiment design and proposed to do the scanning photogrammetry; Tao Ke directed the algorithm and experiment design; Min Tang and Xuan Xu designed and made the photo-scanner.

Conflicts of Interest: The authors declare no conflict of interest.

References and Notes

1.   Kim, K.Y.; Kim, C.Y.; Lee, S.D.; Seo, Y.S.; Lee, C.I. Measurement of tunnel 3-D displacement using digital photogrammetry. *J. Korean Soc. Eng. Geol.* **2007**, *17*, 567–576.

2.   Baldi, P.; Fabris, M.; Marsella, M.; Monticelli, R. Monitoring the morphological evolution of the Sciara del Fuoco during the 2002–2003 Stromboli eruption using multi-temporal photogrammetry. *ISPRS J. Photogramm.* **2005**, *59*, 199–211.

3.   Nakai, T.; Ryu, M.; Miyauchi, H.; Miura, S.; Ohnishi, Y.; Nishiyama, S. *Underground Space Use: Analysis of the Past and Lessons for the Future*; Taylor & Francis: Istanbul, Turkey, 2005; pp. 1203–1209.

4.   Previtali, M.; Barazzetti, L.; Scaioni, M.; Tian, Y. An automatic multi-image procedure for accurate 3D object reconstruction. In Proceedings of the 2011 IEEE 4th International Congress on Image and Signal Processing (CISP), Shanghai, China, 15–17 October 2011; Volume 3, pp. 1400–1404.

5. Balletti, C.; Guerra, F.; Tsioukas, V.; Vernier, P. Calibration of Action Cameras for Photogrammetric Purposes. *Sensors* **2014**, *14*, 17471–17490.

6. Flener, C.; Vaaja, M.; Jaakkola, A.; Krooks, A.; Kaartinen, H.; Kukko, A.; Kasvi, E.; Hyyppä, H.; Hyyppä, J.; Alho, P. Seamless Mapping of river channels at high resolution using mobile LiDAR and UAV-Photography. *Remote Sens.* **2013**, *5*, 6382–6407.

7. Chandler, J. Effective application of automated digital photogrammetry for geomorphological research. *Earth Surf. Proc. Land* **1999**, *24*, 51–63.

8. Fraser, C.S.; Shortis, M.R.; Ganci, G. Multi-sensor system self-calibration. *Proc. SPIE* **1995**, *2598*.

9. Peipe, J.; Schneider, C.T. High-resolution still video camera for industrial photogrammetry. *Photogramm. Rec.* **1995**, *15*, 135–139.

10. Feng, W. The application of non-metric camera in close-range photogrammetry. *Railway Investig. Surv.* **1982**, *4*, 43–54.

11. Rieke-Zapp, D.H. A Digital Medium-Format Camera for Metric Applications. *Photogramm. Rec.* **2010**, *25*, 283–298.

12. Yakar, M. Using close range photogrammetry to measure the position of inaccessible geological features. *Exp. Tech.* **2011**, *35*, 54–59.

13. Yilmaz, H.M. Close range photogrammetry in volume computing. *Exp. Tech.* **2010**, *34*, 48–54.

14. Luhmann, T. Close range photogrammetry for industrial applications. *ISPRS J. Photogramm.* **2010**, *65*, 558–569.

15. Fraser, C.S.; Cronk, S. A hybrid measurement approach for close-range photogrammetry. *ISPRS J. Photogramm.* **2009**, *64*, 328–333.

16. Ke, T.; Zhang, Z.X.; Zhang, J.Q. Panning and multi-baseline digital close-range photogrammetry. *Proc. SPIE* **2007**, *6788*.

17. Ordonez, C.; Martinez, J.; Arias, P.; Armesto, J. Measuring building facades with a low-cost close-range photogrammetry system. *Automat. Constr.* **2010**, *19*, 742–749.

18. Dall'Asta, E.; Thoeni, K.; Santise, M.; Forlani, G.; Giacomini, A.; Roncella, R. Network Design and Quality Checks in Automatic Orientation of Close-Range Photogrammetric Blocks. *Sensors* **2015**, *15*, 7985–8008.

19. Fraser, C.S.; Edmundson, K.L. Design and implementation of a computational processing system for off-line digital close-range photogrammetry. *ISPRS J. Photogramm.* **2000**, *55*, 94–104.

20. Fraser, C.S.; Woods, A.; Brizzi, D. Hyper redundancy for accuracy enhancement in automated close range photogrammetry. *Photogramm. Rec.* **2005**, *20*, 205–217.

21. Jiang, R.N.; Jauregui, D.V. Development of a digital close-range photogrammetric bridge deflection measurement system. *Measurement* **2010**, *43*, 1431–1438.

22. Geodetic Systems, Inc. Available online: http://www.geodetic.com/v-stars.aspx (accessed on 27 June 2015).

23. AICON 3D Systems. Available online: http://aicon3d.com/start.html (accessed on 27 June 2015).

24. Pechatnikov, M.; Shor, E.; Raizman, Y. VisionMap A3 - super wide angle mapping system basic principles and workflow. In Proceeding of the 21th ISPRS Congress, Beijing, China, 3–11 July 2008; pp. 1735–1740.

25. Fangi, G.; Nardinocchi, C. Photogrammetric Processing of Spherical Panoramas. *Photogram. Rec.* **2013**, *28*, 293–311.

26. Takasu, M.; Sato, T.; Kojima, S.; Hamada, K. Development of Surveying Robot. In Proceedings of the 13th International Association for Automation and Robotics in Construction, Tokyo, Japan, 11–13 June 1996; pp. 709–716.

27. Huang, Y.D. 3-D measuring systems based on theodolite-CCD cameras. *Int. Arch. Photogramm. Remote Sens. Spat. Inf. Sci.* **1993**, *29*, 541.

28. Ke, T.; Zhang, Z.; Zhang, J. Panning and multi-baseline digital close-range photogrammetry. *Geomat. Inf. Sci. Wuhan Univ.* **2009**, *34*, 44–47.

29. Ke, T. Panning and Multi-baseline Digital Close-range Photogrammetry. Ph.D. Thesis, Wuhan University, Wuhan, China, 2008.

30. Barazzetti, L.; Previtali, M.; Scaioni, M. 3D modeling from gnomonic projections. *ISPRS Ann. Photogramm. Remote Sens. Spat. Inf. Sci.* **2012**, *1*, 19–24.

31. Barazzetti, L.; Previtali, M.; Scaioni, M. Simultaneous registration of gnomonic projections and central perspectives. *Photogramm. Rec.* **2014**, *29*, 278–296.

32. Barazzetti, L.; Previtali, M.; Scaioni, M. Stitching and processing gnomonic projections for close-range photogrammetry. *Photogramm. Eng. Remote Sens.* **2013**, *79*, 573–582.

33. Fangi, G. Multiscale Multiresolution Spherical Photogrammetry with Long Focal Lenses for Architectural Surveys. *Int. Arch. Photogramm. Remote Sens. Spat. Inf. Sci.* **2010**, *38*, 1–6.

34. Amiri Parian, J.; Gruen, A. Sensor modeling, self-calibration and accuracy testing of panoramic cameras and laser scanners. *ISPRS J. Photogramm.* **2010**, *65*, 60–76.

35. Schneider, D.; Hans, M. A geometric model for linear-array-based terrestrial panoramic cameras. *Photogram. Rec.* **2006**, *21*, 198–210.

36. Mikhail, E.M.; Bethel, J.S.; McGlone, J.C. *Introduction to Modern Photogrammetry*; Wiley: New York, NY, USA, 2001; p. 479.

37. Maas, H.G.; Hampel, U. Photogrammetric techniques in civil engineering material testing and structure monitoring. *Photogramm. Eng. Remote Sens.* **2006**, *72*, 39–45.

38. Lowe, D.G. Distinctive Image Features from Scale-Invariant Keypoints. *Int. J. Comput. Vis.* **2004**, *60*, 91–110.

39. Hartley, R.; Zisserman, A. *Multiple View Geometry in Computer Vision*, 2nd ed.; Cambridge University Press: New York, NY, USA, 2003; pp. 87–131.

40. Ma, J.L.; Chan, J.C.; Canters, F. Fully Automatic Subpixel Image Registration of Multiangle CHRIS/Proba Data. *IEEE Trans. Geosci. Remote Sens.* **2010**, *48*, 2829–2839.

41. Nistér, D. An efficient solution to the five-point relative pose problem. *IEEE Trans. Pattern Anal. Mach. Intell.* **2004**, *26*, 756–770.

42. Stewenius, H.; Engels, C.; Nister, D. Recent developments on direct relative orientation. *ISPRS J. Photogramm.* **2006**, *60*, 284–294.

43. Tee-Ann, T.; Liang-Chien, C.; Chien-Liang, L.; Yi-Chung, T.; Wan-Yu, W. DEM-Aided Block Adjustment for Satellite Images With Weak Convergence Geometry. *IEEE Trans. Geosci. Remote Sens.* **2010**, *48*, 1907–1918.
44. Slama, C.C.; Theurer, C.; Hendrikson, S.W. *Manual of Photogrammetry*, 4th ed.; American Society of Photogrammetry: Falls Church, VA, USA, 1980; p. 1059.
45. Stamatopoulos, C.; Fraser, C.S. Calibration of long focal length cameras in close range photogrammetry. *Photogramm. Rec.* **2011**, *26*, 339–360.

Optimized 3D Street Scene Reconstruction from Driving Recorder Images

Yongjun Zhang, Qian Li, Hongshu Lu, Xinyi Liu, Xu Huang, Chao Song, Shan Huang and Jingyi Huang

Abstract: The paper presents an automatic region detection based method to reconstruct street scenes from driving recorder images. The driving recorder in this paper is a dashboard camera that collects images while the motor vehicle is moving. An enormous number of moving vehicles are included in the collected data because the typical recorders are often mounted in the front of moving vehicles and face the forward direction, which can make matching points on vehicles and guardrails unreliable. Believing that utilizing these image data can reduce street scene reconstruction and updating costs because of their low price, wide use, and extensive shooting coverage, we therefore proposed a new method, which is called the *Mask automatic detecting method*, to improve the structure results from the motion reconstruction. Note that we define vehicle and guardrail regions as "mask" in this paper since the features on them should be masked out to avoid poor matches. After removing the feature points in our new method, the camera poses and sparse 3D points that are reconstructed with the remaining matches. Our contrast experiments with the typical pipeline of structure from motion (SfM) reconstruction methods, such as Photosynth and VisualSFM, demonstrated that the Mask decreased the root-mean-square error (RMSE) of the pairwise matching results, which led to more accurate recovering results from the camera-relative poses. Removing features from the Mask also increased the accuracy of point clouds by nearly 30%–40% and corrected the problems of the typical methods on repeatedly reconstructing several buildings when there was only one target building.

Reprinted from *Remote Sens*. Cite as: Zhang, Y.; Li, Q.; Lu, H.; Liu, X.; Huang, X.; Song, C.; Huang, S.; Huang, J. Optimized 3D Street Scene Reconstruction from Driving Recorder Images. *Remote Sens*. **2015**, *7*, 9091–9119.

1. Introduction

Due to the increasing popularity of using reconstruction technologies, more 3D supports are needed. Researchers have proposed many methods to generate 3D models. Building models from aerial images is a traditional method to reconstruct a 3D city. For example, Habib proposed a building reconstruction method from aerial mapping by utilizing a low-cost digital camera [1]. Digital map, Light Detection and Ranging (LIDAR) data, and video aerial image sequences have been used to build models combined [2]. These methods can reconstruct the model of large-area cities at a high efficiency; however, the models reconstructed from aerial data always have lacked detailed information, which constrains their further applications. In order to reconstruct city models with rich details, the terrestrial data based reconstruction also has been explored [3–5]; and street scenes have been reconstructed with imagery taken from different view angles [3,4]. These images were captured by a moving vehicle that carried a GPS/INS navigation system. Mobile LIDAR was used to reconstruct buildings with progressively refined point clouds by incrementally updating the data [5]. Even though mobile LIDAR can acquire 3D points quickly, it clearly has some limitations. For example, the density of the point clouds can be easily affected by the driving speeds, number of scanners, multiple returns, range to target, *etc.* The advantages and disadvantages of mobile LIDAR and its abundant applications in city reconstructions have been summarized [6]. The integrated GPS/Inertial Navigation Systems (INS) navigation system and mobile LIDAR play an important role in most classical city-scale reconstruction methods. However, in urban areas, the accuracy of GPS/GNSS is sometimes limited by the presence of large buildings. Although this limitation can be minimized by using Wi-Fi or telephone connections, we cannot neglect the necessity of INS in the above methods that has made city-scale reconstruction very expensive.

The urban area in China has grown rapidly in recent years, which has brought a large number of tasks for road surveying. However, the development of 3D street reconstruction is limited by the lack of mobile mapping equipment carrying stable GPS/INS systems or mobile LIDAR in China. In order to address this issue, it is crucial to be able to reconstruct sparse 3D street scenes without the assistance of GPS/INS systems. The Structure from Motion (SfM) technique [7] was recently used to reconstruct buildings from unstructured and unordered data sets without GPS/INS information [8,9], and the results should be scaled and georeferenced into object space coordinate systems. For example, the photo tourism system [10,11] is one of the typical SfM methods which can recover 3D point clouds, camera-relative positions, and orientations from either personal photo collections or Internet photos that do not rely on a GPS/INS system or any other equipment to provide location, orientation, or geometry. The image data used by the above typical SfM method characteristically have less repetition and obvious objects in the foreground, which allows processing by the typical SfM method have no additional steps.

With the heavier traffic density nowadays, especially more buses, taxis, and private vehicles are equipped with driving recorders to avoid traffic accident disputes. A driving recorder is a dashboard-mounted camera, which can collect images while a vehicle is operating. Figure 1 shows a typical driving recorder and recorded image.

(a)   (b)

Figure 1. Driving recorder and recorded image. (**a**) Photo of one type of driving recorder obtained from the Internet. (**b**) Test data recorded by the driving recorder in this paper.

More and more uses of driving recorders allow them to replace mobile mapping equipment to collect images, reconstruct, and update street scene point clouds at a lower cost yet in a shorter update time. Images of street views are typically captured by driving recorders mounted in the front of a moving vehicle, facing the forward direction along the street. Large quantities of vehicles are captured in the video images. However, due to the relative motion among the vehicles and the repeating patterns of guardrails, without the assistance of GPS/INS information, the matching pairs of images of vehicles and guardrails may be outliers. These outliers often take a dominant position, which cannot be removed by the epipolar constraint method effectively, thereby causing the typical SfM process to fail. Hence, this paper mainly focuses on detecting vehicles and guardrail regions, and then removing the feature points on them to reduce the number and negative effects of the outliers present in the driving recorder data. After removal, the remaining points can be used to reconstruct the street scene.

In order to reduce the cost of reconstructing point clouds, the SfM method is proposed in this paper to reconstruct the street scene based on driving recorder images without GPS/INS information. However, we reconstruct the results only in the relative coordinate system, rather than georeferencing it to the absolute coordinates. This paper focuses on removing the feature points on the vehicle and guardrail regions, which can improve the performance of the recovered camera-tracks and the accuracy of the reconstructed sparse 3D points. Vehicle and guardrail region automatic detection methods are proposed in Sections 2.1–2.4. The features removing and reconstruction method is described in Section 2.5; and the improved reconstruction effects are shown in Section 3 from the following three aspects: Section 3.2 addresses the precision of pairwise orientation; Section 3.3 shows the camera poses recovering results, and the sparse 3D point clouds reconstructing results are introduced in Section 3.4. The results and implications of this research are discussed in Section 4; and the limitations of the proposed method and future research directions are described in Section 5.

2. Methodology

The paper proposed guardrail and vehicle region detection methods, and then masked feature points on guardrail and vehicle regions to improve the reconstruction result. We propose to "mask" out the vehicle and guardrail regions before reconstruction because guardrails have repeating patterns and vehicles move between frames, which subsequently always produce outliers on the image of the guardrail and vehicle regions. In this paper, the images of the vehicle and guardrail regions are collectively called the Mask. The pipeline of 3D reconstruction that utilizes driving recorder data is illustrated in Figure 2. We can first detect the SIFT [12] feature points and the Mask in each image, and then we remove the features on the Mask and match the remaining feature points between the pairs of images. Based on the epipolar constraint [7], we will remove the outliers to further refine the results and finally conduct an incremental SfM procedure [7] to recover the camera parameters and sparse points.

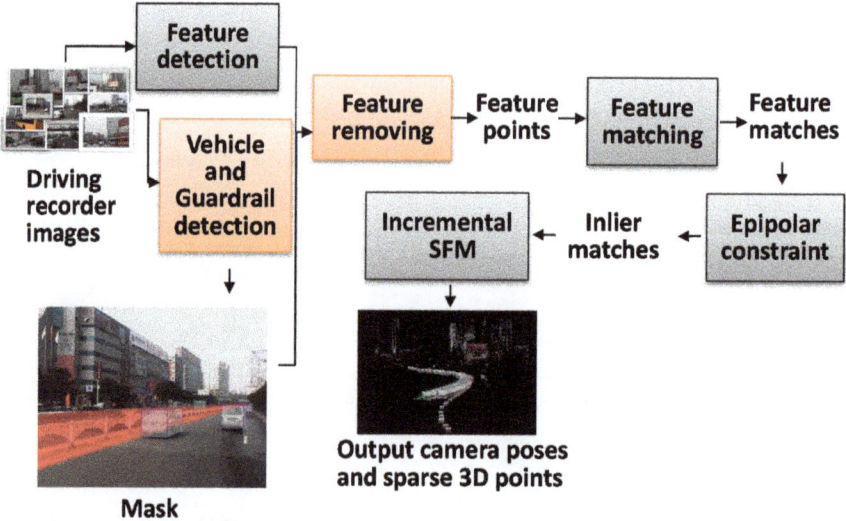

Figure 2. The pipeline of 3D reconstruction from driving recorder data. The grey frames show the typical SfM process. The two orange frames are the main improvement steps proposed in this paper.

It is challenging to detect the Mask with object detection methods due to the following difficulties:

1. Both the cameras and the objects are in motion, which changes the relative pose of the objects. Moreover, the appearance of vehicles varies significantly (e.g., color, size, and difference between back/front appearances).
2. The environment of the scene (e.g., illumination and background) often changes, and events such as occlusions are common.
3. Guardrails are strip distributions on images, which make the detection of whole regions difficult.

Haar-like features [13,14] based on Adaboost classifiers [15] were used to address the above challenges. With the help of classifiers, the front/back surfaces of vehicle and some parts of guardrails are automatic detected within a few seconds. The classifiers also could robust against the changing of light condition and environment.

In order to diminish the adverse impact of outliers on reconstruction, the Mask requires detection as entirely as possible. Therefore, based on the typical vehicle front/back surface detection method in Section 2.1, the design of the vehicle side surfaces detection method and the blocked-vehicle detection method are described in Sections 2.2 and Section 2.4, respectively. The blocked-vehicle is a vehicle moving in the opposite direction partially overlapped by the guardrail. The guardrail region detection method is introduced in Section 2.3, which is based on the Haar-like classifiers and the position of the vanishing point. Finally, the Mask and the reconstruction process are introduced in Section 2.5.

2.1. Vehicle Front/Back Surfaces Detection

As the system of vehicle back surface detection [16] by Haar-like feature-based Adaboost classifier is described in details, we only summarize its main steps here. Classifiers based on Haar-like features can detect objects with a similar appearance. There is a big difference between the front and back surfaces of vehicles and buses; therefore, four types of classifiers were trained to detect the front and back surfaces of vehicles and buses, respectively.

The classifier was trained with sample data. After the initial training, the trained classifier was used to independently detect vehicles. There are two types of samples, positive and negative. A positive sample is a rectangular region cut from an image that contains the target object, and a negative sample is a rectangular region without the target object. Figure 3 shows the relation of the four classifier types and their trained samples. Each classifier is trained with 1000–2000 positive samples and at least 8000 negative samples. All the samples were manually compiled; and we separated the images containing vehicles as positive samples and the remaining images were used as negative samples. Although diverse samples can produce better classifier performance, a small amount of duplications are acceptable. Therefore, the positive samples of the same vehicle cut from different images are effective samples, and the samples from the same images with a slightly adjusted position are allowable as well. Two samples can even be totally duplicated, which will have a minimal adverse effect on the performance of the classifier when the number of samples is large enough. After inputting the samples into the OpenCV 2.4.9 [17] training procedure, the classifier can be trained with the default parameters automatically. A cascade classifier is composed of many weak classifiers. Each classifier is trained by adding features until the overall samples are correctly classified. This process can be iterated up to construct a cascade of classification rules that can achieve the desired classification ratios [16]. Adaboost classifier is more likely to overfit on small and noisy training data. Too many iterative training processes may cause the overfitting problem, too. Therefore, we need to control the maximum number of iteration in the training processes. In OpenCV training procedure, there are some constraints designed to avoid the overfitting problem. For example, the numStages parameter limits the stage number of classifier, and the maxWeakCount parameter helps to limit the count of trees. These parameters could prevent classifiers from the overfitting. Besides these parameter-constraints, we can also use more training data to minimize the possibility of overfitting.

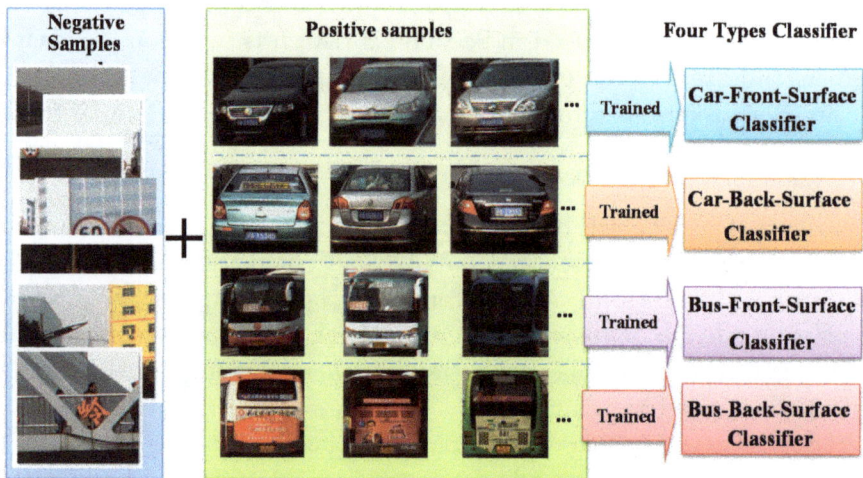

Figure 3. Example of samples and classifiers.

A strong cascade classifier consists of a series of weak classifiers in the order of sparse to strict. A sparse classifier has few constraints and low classification accuracy but a high computational speed; while a strict classifier has many constraints and high classification accuracy but a low computational speed. When an image area is input into a strong cascade classifier, it is first detected by the initial sparse classifier. Only a positive result from the previous classifier triggers the evaluation of the next classifier. Negative results (e.g., background regions of the image) therefore are quickly discarded so the classifier can spend more computational time on more promising object-like regions [13]. Most image areas without the target object can be easily identified and eliminated at the very beginning of the process with minimal effort. Therefore, a cascade classifier is able to enhance computational efficiency [18].

2.2. Vehicle Side-Surface Detection

The side-surfaces of vehicles cannot be detected by feature-based classifiers since a vehicle's appearance changes with the angle of view. Poor matching points on these regions inevitably have adverse effects on the reconstruction, especially the side-surfaces of large vehicles that are close to the survey vehicle.

The side-surface region can be determined if the interior orientation parameters, the rough size of the vehicles, and the position of the front/back surfaces of vehicles on the images are known. However, most driving recorders do not contain accurate calibration parameters so we deduced the equations described in this section to compute the rough position of the vehicle side-surface region based on the position of the front/back-surfaces and the vanishing point in the image, the approximate height H of the recorder, the rough value of focal length f, and the pitch angle of recorder θ. The vanishing point used in this section was located using the [19,20] method and the position of the vehicle front/back-surface was detected with the method described in Section 2.1. The vanishing point is considered a point in the picture plane that is the intersection of a set of parallel lines in space on the picture plane. Although the vehicle side-surface detection method proposed in this section can only locate the approximate position of the vehicle side-surface, it is adequate for masking out the features on vehicles to improve the reconstruction results. The length of M"N" is the key step in the vehicle side-surface detection method. The process of computing the length of M"N" is described below:

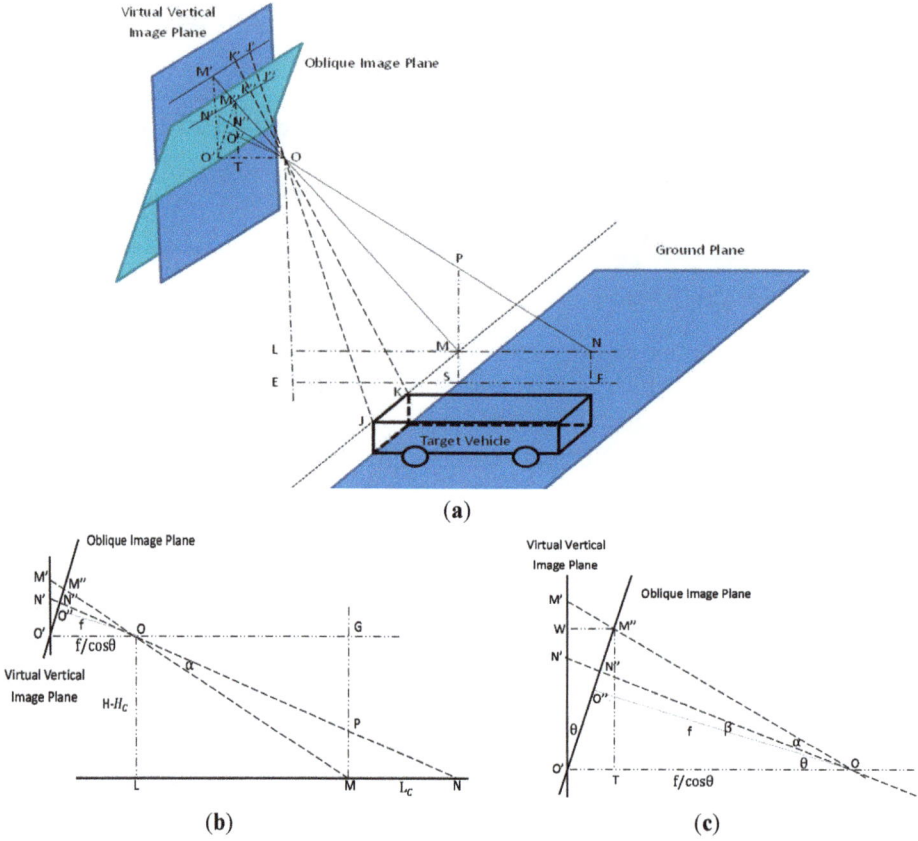

Figure 4. Photographic model of driving recorder. (**a**) Integrated photographic model of driving recorder. (**b**) Side view of model. (**c**) Partial enlargement of side view model. The oblique image plane is the driving recorder image plane. Point O is the projective center, and O″ is the principal point on the driving recorder image plane. The focal length f is OO″. Point O′ is the principal point on the virtual vertical image plane. Line OE is perpendicular to the ground. Point E is the intersection point of the ground and line OE. The plane M′O′OOEF can be drawn perpendicular to both the image plane and the ground. OM′ is perpendicular to M′J′, and OM is perpendicular to MJ. Line LN is perpendicular to OE. MP is a vertical line for the ground, and P is the intersection point of line MP and line ON. Line M″T is perpendicular to O′O. The angle between the oblique plane and the vertical plane is θ. Angles MON and O″ON″ are α and β, respectively.

In Figure 4a, we suppose that the real length, height, and width of the vehicle are L_C, H_C and W_C, respectively. The width of the target on image $K″J″$ is W_P. H is the height of projective center O to ground OE. Therefore, it can be seen that the length of target MN is L_C, the length of LE is H_C and OL is $H - H_C$. Figure 4a shows that

208

triangle M″TO is similar to OLM and triangle $K″J″$ O is similar to KJO. We therefore can deduce the following equations from the triangle similarity theorem:

$$\frac{K″J″}{KJ} = \frac{OM″}{OM} = \frac{M″T}{OL} \tag{1}$$

So the length of M″T can be described with Equation (2):

$$M″T = \frac{K″J″ \cdot OL}{KJ} = \frac{W_P \cdot (H - H_C)}{W_C} \tag{2}$$

It can be seen from $\triangle W$ OM′ in Figure 4c that, angle W O′M′ is θ, and the length of WM″ is equal to O′T in rectangle WM″ TO′. Then, Equation (3) can be established with the length of M″T in Equation (2):

$$WM″ = M″T \cdot \tan\theta = \frac{W_P \cdot (H - H_C) \cdot \tan\theta}{W_C} \tag{3}$$

In Figure 4c, the length of M″T is equal to W O′ in rectangle WM″ TO′, so the length of M′O′ is equal to M″T add M′W. Then, Equation (4) can be established based on $\triangle$ M′ WM″ $\frown \triangle$ M′ O′O:

$$\frac{WM″}{O′O} = \frac{M′W}{M′O′} = \frac{M′W}{M″T + M′W} \tag{4}$$

Equation (5) is transformed from Equation (4), and the length of M′W is expressed below:

$$M′W = \frac{WM″ \cdot M″T}{O′O - WM″} = \frac{W_P^2 \cdot (H - H_C)^2 \cdot \sin\theta}{W_C^2 \cdot f - W_P \cdot W_C \cdot (H - H_C) \cdot \sin\theta} \tag{5}$$

Equation (6) is established from rectangle WM″ TO′ in Figure 4c.

$$O′ M′ = O′ W + M′ W = M″ T + M′ W =$$
$$\frac{W_P \cdot (H - H_c)}{W_c} + \frac{W_P^2 \cdot (H - H_C)^2 \cdot \sin\theta}{W_C^2 \cdot f - W_P \cdot W_C \cdot (H - H_C) \cdot \sin\theta} \tag{6}$$

In Figure 4a, since OM″ is the height of triangle OK″J″, we can infer that:

$$\because \triangle K′J′O \frown \triangle K″J″O, \triangle OM″T \frown \triangle OM′O′ \tag{7}$$

$$\therefore \frac{K′J′}{K″J″} = \frac{OM′}{OM″} = \frac{O′M′}{M″T} \tag{8}$$

209

We know that, $K''J''$ is W_P, therefore with the calculations of $M''T$ (Equation (2)) and $O'M'$ (Equation (6)), the length of $K'J'$ can be established from the transformation of Equation (8):

$$K'J' = \frac{K''J'' \cdot O'M'}{M''T} = W_P + \frac{W_P^2 \cdot (H - H_C) \cdot sin\theta}{W_C \cdot f - W_P \cdot (H - H_C) \cdot sin\theta} \tag{9}$$

KJ and $K'J'$ are parallel; therefore, we can infer that triangle K' $M'O$ and triangle KMO are similar triangles from Figure 4a. OM' is the height of triangle K' $M'O$ and OM is the height of triangle KMO. Meanwhile, OO' and LN are parallel lines so triangle $M'OO'$ is similar to triangle OML. Therefore, based on the triangle similarity theorem, Equation (10) can be established:

$$\frac{K'J'}{KJ} = \frac{OM'}{OM} = \frac{OO'}{LM} \tag{10}$$

The length of OO' is $f/cos\theta$ and KJ is W_C so LM can be calculated based on Equations (9) and (10):

$$LM = \frac{OO' \cdot KJ}{K'J'} = \frac{W_C \cdot f - W_P \cdot (H - H_C) \cdot sin\theta}{W_P \cdot cos\theta} \tag{11}$$

In Figure 4b, Equation (12) can be established since triangle PMN is similar to OLM:

$$\frac{MP}{OL} = \frac{MN}{LN} = \frac{MN}{LM + MN} \tag{12}$$

OL and MN are $H - H_C$ and L_C, respectively. Then, MP can be described with Equations (11) and (12)

$$MP = \frac{OL \cdot MN}{LM + MN} = \frac{L_C \cdot W_P \cdot (H - H_C) \cdot cos\theta}{W_C \cdot f - W_P \cdot (H - H_C) \cdot sin\theta + L_C \cdot W_P \cdot cos\theta} \tag{13}$$

In order to compute the length of $M''N''$, we suppose that:

$$\angle M'ON' = \angle MON = \alpha, \angle N''OO'' = \beta, \angle M'O'M'' = \angle O'OO'' = \theta \tag{14}$$

Based on cosine theorem, Equation (15) can be established:

$$cos\alpha = \frac{OP^2 + OM^2 - MP^2}{2 \cdot OP \cdot OM} \tag{15}$$

In Figure 4b, OG is equal to LM, and OL has the same length as GM in rectangle OGML. The length of OL is $H - H_C$. Based on the Pythagoras theorem, Equations (16) and (17) were deduced from $\triangle$OGP and $\triangle$OLM.

$$OP^2 = OG^2 + GP^2 = LM^2 + (OL - MP)^2 = LM^2 + (H - H_C - MP)^2 \tag{16}$$

$$OM^2 = OL^2 + LM^2 = (H - H_C)^2 + LM^2 \tag{17}$$

Taking Equations (16) and (17) into Equation (15), angle α can be described as follow:

$$\alpha = \arccos\left(\frac{LM^2 + (H - H_C)^2 - (H - H_C) \cdot MP}{\sqrt{LM^2 + (H - H_C - MP)^2} \cdot \sqrt{(H - H_C)^2 + LM^2}}\right) \tag{18}$$

In Figure 4b,c, $\angle M'OO' = \angle OML = \alpha + \beta + \theta$ so in triangle OML:

$$\tan(\alpha + \beta + \theta) = \frac{OL}{LM} \tag{19}$$

Equation (20) is the transformation of Equation (19), with OL= $H - H_C$:

$$\beta = \arctan\left(\frac{H - H_C}{LM}\right) - \alpha - \theta \tag{20}$$

Since OO'' is perpendicular to $O'M'$, the following equations can be established based on thesine theorem in Figure 4c.

$$\tan(\alpha + \beta) = \frac{O''M''}{OO''}, \tan\beta = \frac{O''N''}{OO''} \tag{21}$$

Based on Equation (21), since OO'' is f, the following equation can be transformed:

$$M''N'' = O''M'' - O''N'' = f \cdot [\tan(\alpha + \beta) - \tan\beta] \tag{22}$$

Finally, the length of M''N'' can be calculated by taking Equations (11), (13), (18), and (20) into (22).

We have supposed that L_C is the length of the vehicle. In Figure 5, M''N'' on line l is the projection length of L_C, which can be computed by the Equation (22). With the known positions of the vehicle front/back-surfaces, the vanishing point on the image, the length of M''N'', and the rough regions of the vehicle side-surfaces can be located with the following step.

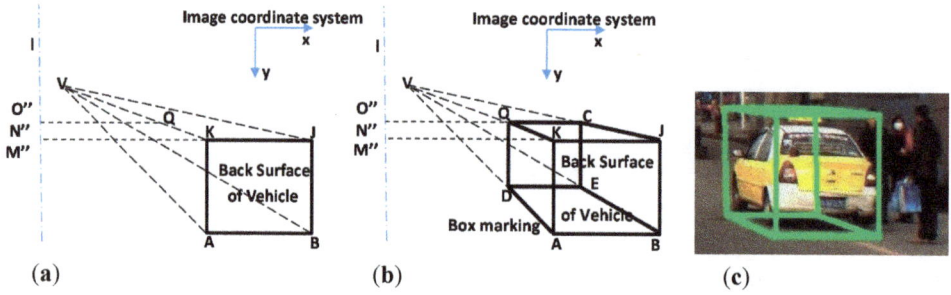

Figure 5. (a) and (b) depictions of the box marking drawing method. (c) Example of box marking in an image. The principal point O" is the center point of the image, and the black rectangle KJAB is the vehicle back surface in the image plane, which are detected by the classifier described in Section 2.1. Point V is the vanishing point in the image. Line l is the perpendicular bisector of the image passing through principal point O". Line KM" is parallel to the x axis of the image and M" is the intersection point on l. Line N"Q intersects lines VK and VJ at points Q and C, respectively. N"Q is parallel with M"K. Line QD intersects line VA at point D, and line DE intersects line VB at point E. Line QC and DE are parallel to the x axis and QD is parallel to the y axis of the image.

With the computed length of M"N", the position of point N" is known, then point C, D, and E can be located with the rules described in Figure 5. Thereafter, the black bolded-line region QCJBAD on Figure 5b can be determined. Based on the shape of the black bolded-line region, we defined it as the "box marking". The region surrounded by the box marking will contain the front, back, and side-surfaces of the vehicle generally. Therefore, according to the description below, the box markings of vehicle side-surfaces can be fixed by M"N".

Detecting vehicle side-surfaces by the box marking method has the advantage of having a fast speed and reliable results, but it relies on parameters H, f, and θ. These parameters can only be estimated crudely on driving recorders. Therefore, the above method can only locate the approximate position of the vehicle side surface. However, it is adequate for the method to reach the goal of eliminating features on the Mask.

2.3. Guardrail Detection

The guardrail is an isolation strip mounted on the center line of the road to separate vehicles running in opposite directions. It also can avoid pedestrian arbitrarily crossing the road. The photos of guardrail are shown in Figures 6d and 7c. There are two reasons for detecting and removing the guardrail regions. The main reason is that, due to the repeating patterns of guardrails, they always contribute to poor matches. Furthermore, the views of vehicles moving on the other side of the guardrail are always blocked by the guardrails. This shielding makes vehicles undetectable by the vehicle classifiers. Therefore, in order to detect the blocked-vehicle regions, it was necessary to locate the guardrail regions. The blocked-vehicle regions detection method, which is described in Section 2.4, is based on the guardrail detection method described below.

The guardrail regions are detected based on a specially-designed guardrail-classifier. Except for changes in the training parameters, the guardrail-classifier training process is similar to the vehicle training method, which is described in Section 2.1. In order to detect an entire region of guardrails, a special guardrail-classifier was trained based on OpenCV Object Detection Lib [17] with a nearly 0% missing object rate. The price of a low missing rate, however, inevitably is an increase in the false detection rate, which means that the classifier could detect thousands of results that included not only the guardrails but also some background. In the training process, two parameters, the Stages-Number and the desired Min-Hit-Rate of each stage, were decreased. One parameter, the MinNeighbor [17] (a parameter specifying how many neighbors each candidate rectangle should have to retain it), was set to 0 during the detecting process.

The special-designed guardrail classifier detection results are shown in Figure 6a as blue rectangles, and many of them are not guardrails. This is a side effect of guaranteeing a low missing object rate, but uses of the statistical analysis method can ensure that these false detections cannot influence the confirmation of the actual guardrail regions in further steps. The vanishing point was fixed by the [19,20] method, wherein a vanishing point is considered a point in the picture plane that is the intersection of a set of parallel lines in space on the picture plane. The lines are drawn from the vanishing point to each centre line of the rectangular regions at an interval of 2°. This drawing approach is shown in Figure 6b, and the drawing results are shown in Figure 6c with red lines. Since the heights of the guardrails were fixed and the models in the driving recorder were changed within a certain range, the intersection angles between the guardrail top and bottom edges on the image often changed from 10° to 15° as a general rule. An example of the intersection angles is shown in Figure 6d.

Based on the red lines drawn results, we made a triangle region which included an angle of 15° and a fixed vertex (the vanishing point). The threshold of 15° was the maximum potential intersection angle between the top and bottom edges of the guardrail. Then, we shifted the triangle region between 0° and 360° like Figure 7a. During the shift, the number of red lines included in every triangle region was counted. Then, we considered the triangle region that had the largest line numbers as the guardrail region. Figure 7b shows the position of the triangle region that had the largest line numbers, and Figure 7c shows the final detection results of the guardrail.

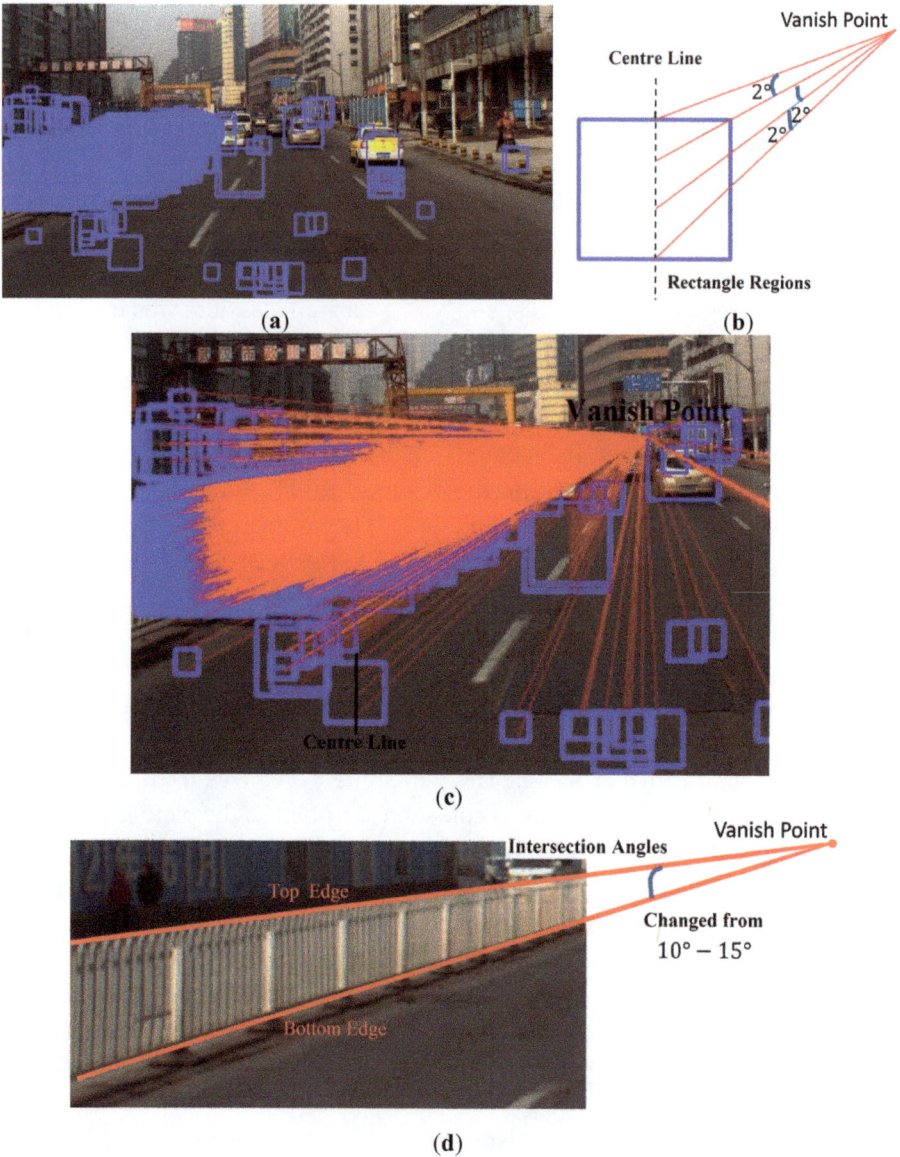

Figure 6. Guardrails detection process. (**a**) Detection results of a specially-designed guardrail-classifier which could detect thousands of results, including not only correct guardrails but many wrong detection regions as well. (**b**) Example of how to draw the red lines from the vanishing point to the detection regions. (**c**) Results of red lines drawn from the vanishing point to each centre line of the rectangle regions at an interval of 2°. An example of a rectangle region's centre line is shown in the bottom left corner of the (**c**); and (**d**) is an example of an intersection angle between the top and bottom edges of the guardrail.

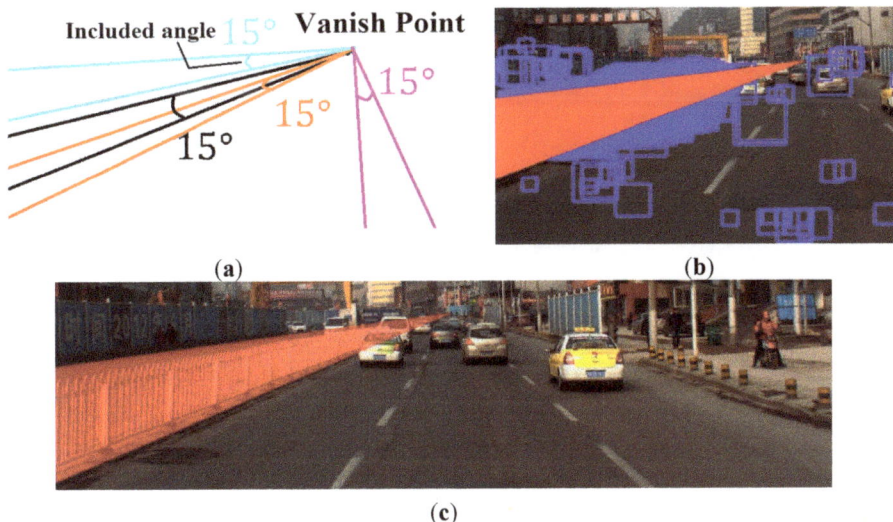

Figure 7. Guardrail location method. (**a**) Example of four triangle regions which included the angle of 15° and the fixed vertex (the vanishing point). (**b**) Triangle region that had the largest line numbers. (**c**) Final detection results of guardrail location method.

2.4. Blocked Vehicle Regions Detection

Sometimes, the vehicles that are moving in opposite direction can be blocked by guardrails, which results in the vehicle image overlapping with the guardrail partially. These occlusions make the vehicles undetectable by the front surface classifiers that were trained as described in Section 2.1. In this case, in order to detect blocked-vehicle regions, we increased the threshold of the intersection angle to broaden the guardrail region in order for the blocked vehicles to be included. In Figure 8, the blue box markings show the vehicle detection results based on the methods described in Sections 2.1 and 2.2. Two vehicles are missing from the detection, which are indicated by the yellow arrows. The red triangle region is the broadened guardrail region with a 20° intersection angle. The missing detections are included by the broadened guardrail regions, which are shown in Figure 8.

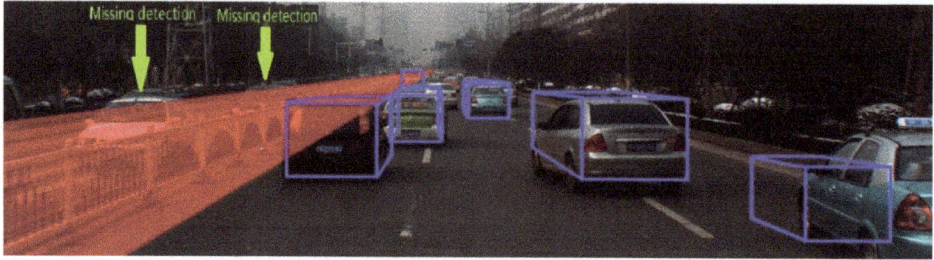

Figure 8. Blocked vehicles detection method (guardrail region broadening method). Two vehicles running in opposite directions are missed detection by the vehicle classifier, which are indicated by the yellow arrows. These missed detection vehicle regions are included in the broadened guardrail regions, which are shown as the red region.

2.5. Mask and Structure from Motion

In the SIFT matching algorithm, detecting the feature points on the images is the first step, and the correspondences are then matched between the features. The coordinates of the features were compared with the location of the Mask regions in the image, and then the features located in the Mask were removed from the feature point sets. An example of our removing results is shown in Figure 9. The Mask was obtained by merging the regions detected in Sections 2.1–2.4. After removing the SIFT feature points on the Mask, the remaining features were matched. Then, the QDEGSAC [21] algorithm was used to robustly estimate a fundamental matrix for each pair and the outliers were eliminated with a threshold of two pixels by the epipolar constraint [7,22]. The QDEGSAC algorithm is a robust model with a selection procedure that accounts for different types of camera motion and scene degeneracies. QDEGSAC is as robust as RANSAC [23] (the most common technique to deal with outliers in matches), even for (quasi-)degenerate data [21].

In a typical SfM reconstruction method, pairwise images are matched with the SIFT algorithm without any added process. Then, the inlier matches are determined by the epipolar constraint algorithm (similar to the QDEGSAC algorithm), and a sparse point cloud is reconstructed with the inliers by the SfM algorithm. However, in our method, during the pairwise matching process, the SIFT feature points on the vehicle and guardrail regions are masked out before matching. The remaining features are then matched by the SIFT algorithm. After the outliers were eliminated by the epipolar constraint algorithm, the SfM reconstruction process proceeds with the remaining matches.

(a)   (b)

Figure 9. SIFT feature points removing results. (**a**) Original SIFT feature points set on image. (**b**) Mask results, which show the masked out features on the vehicle and guardrail regions.

Both in the typical method and our proposed method, the QDEGSAC algorithm was used as an epipolar constraint algorithm to select the inliers, and the SfM process was conducted in VisualSFM [24,25]. Visual SFM is a GUI application of the incremental SfM system. It runs very fast by exploiting multi-core acceleration. The features mask out process in our method is the only difference from the typical method.

3. Experiment

3.1. Test Data and Platform

A driving recorder is a camera mounted on the dashboard of a vehicle that can record images when the vehicle is moving. The SfM reconstruction method can accept various image sizes from different types of recorders. We used 311 images taken by five recorders on roundabout as testing data to demonstrate the improved results with our method. We chose images taken on roundabout at large intervals to increase the complexity of the testing data. The Storm Media Player was used to extract images from videos. We manually extracted images with the intervals which are described in Table 1. The characteristics of the testing data are described below:

1. The testing images were taken by five recorders mounted on four vehicles, and the largest time interval between the two image sequences was nearly three years.
2. A total of 125 images were extracted from videos recorded by driving recorders 1, 2, and 3.

3. 186 images were recorded by recorders 4 and 5, which were mounted on the same vehicle with identical exposure intervals.
4. Roundabout was crowded during the recording time so the survey vehicles changed their lanes and speeds when necessary to move with the traffic.
5. The rest details of recorders and images are shown in Table 1.

Table 1. The characteristics of recorders and images.

Recorder NO	Sensor Type	Focus Style	Image Size	Image Extraction Intervals	Recording Date
1, 2, 3	Video	Zoom Lens	1920 × 1080	About 1 s	12/23/2014
4, 5	Camera	Fixed Focus	800 × 600	0.5 s	1/23/2012

We separated 311 images into sequences 1, 2, 3, 4, and 5 according to the recorder that recorded them. The results are shown in Table 2.

Table 2. The composition of three sets.

Set Number	Recorder Number	Image Number	Attribute
1	4, 5	186	Stereo images taken by two cameras mounted on the same vehicle with identical exposure intervals.
2	1, 2, 3, 4	218	The longest time interval between the two image sequences was nearly three years, and the images were two different sizes.
3	1, 2, 3, 4, 5	311	Three monocular and two stereo image sequences. The longest time interval between the two image sequences was nearly three years, and the images were two different sizes.

We conducted all the following experiments on a PC with an Intel Core i7-3770 3.4 GHz CPU (8cores), 4 GB RAM, and an AMD Radeon HD 7000 series GPU. The detection algorithm was implemented in a Visual C++ platform with the OpenCV 2.4.9 libraries. Training each vehicle classifier took nearly 75 h, and eight hours was required for training the guardrail classifier. Although training the classifier was a time-consuming process, the trained classifier could be used to detect vehicles at a fast speed after one-off training. The detection speed was affected by the number of targets. When running on the described PC, the average detecting time was 0.15 s for each classifier on a 1600 × 1200 pixel-sized image. In the following section, we compare the performance between the typical SfM reconstruction method and our method from three aspects: the precision of pairwise orientation, the recovered camera tracks, and the reconstructed point clouds, which are described in Sections 3.2–3.4, respectively.

3.2. Precision of Pairwise Orientation

In the SfM system, the accuracy of the reconstructed point clouds is determined by the quality of the correspondences. Hence, in this section, we evaluate and compare the matching results of the typical method and our method by the root-mean-square error (RMSE).

Based on the epipolar constraint, correspondences p_i, p_i' should be located on the corresponding epipolar line l_i, l_i' respectively (The epipolar line can be computed with the algorithm proposed by D. Nister [26]). However p_i' may deviate from epipolar line l_i' due to orientation errors. Thereafter, the RMSE is able to evaluate the accuracy of the pairwise orientation with following equation.

$$RMSE = \sqrt{\frac{d_1^2 + d_2^2 + \cdots d_n^2}{n}} \tag{23}$$

d_i in Equation (23) is the distance between point p_i' and the epipolar line l_i'. n is the number of matches. The pixel size of CCD was 0.0044 mm; therefore millimeter was used as the unit of RMSE.

The difference between the typical method and our method is that our method masked out the feature points on vehicles and guardrails before proceeding with matching. Then, in both the typical method and our method, the correspondences with d_i greater than the threshold (two pixels) were eliminated by the QDEGSAC algorithm [21] before inputting into Equation (23).

In order to demonstrate the improvement and robustness of our method, 666 image pairs of diverse street scenes were chosen randomly using the above image set. The RMSE of the pairwise orientation results by the typical method and our method are shown in Figure 10.

In Figure 10, the RMSEs in our method were less than in the typical method in general. The abnormity in the image pairs (the RMSEs in our method were larger than the typical method) for which we offer the following analysis. We found that the abnormal pairs were usually shot at long-range distances (more than 200 m) with little overlap, leading ultimately to a decrease in the number of accurate matches. A large proportion of the outliers led to an orientation failure, which produced abnormal RMSE results. In general, however, it can be concluded from Figure 10 that the Mask effectively improved the matching accuracy.

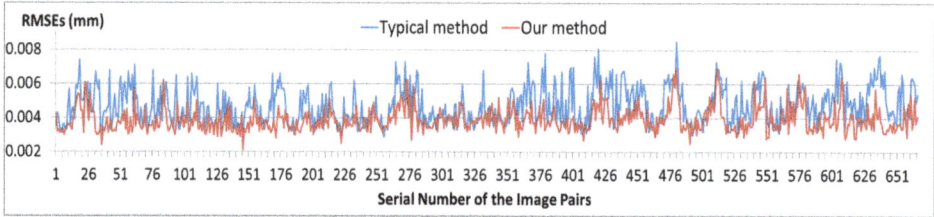

RMSEs (mm)

— Typical method — Our method

0.008
0.006
0.004
0.002

1 26 51 76 101 126 151 176 201 226 251 276 301 326 351 376 401 426 451 476 501 526 551 576 601 626 651

Serial Number of the Image Pairs

Figure 10. RMSEs of each image. The X-axis represents the serial number of the image pairs and the Y-axis represents the RMSEs, which are shown as millimeters. The blue and red lines show the RMSEs of the typical method and our method, respectively. The correspondences in our method were matched after removing the SIFT features on the Mask, and then the outliers were eliminated by the epipolar constraint (QDEGSAC) method. In the typical method, the correspondences were filtered only by the epipolar constraint (QDEGSAC) method.

3.3. Camera Poses Recovering Results

Figure 11 is an explanation of the reconstructed camera-pose-triangle in the following figures. The colored triangles represent the position of the recovered image/camera. Figures 12–14 shows the camera pose reconstruction results of three sets. The details and compositions of each set were described in Table 2. The difference between the typical method and our method is that the feature points on the Mask are removed before matching in our method. Since motor vehicles can only run in a smooth track, we were able to distinguish an unordered track as false reconstruction results easily.

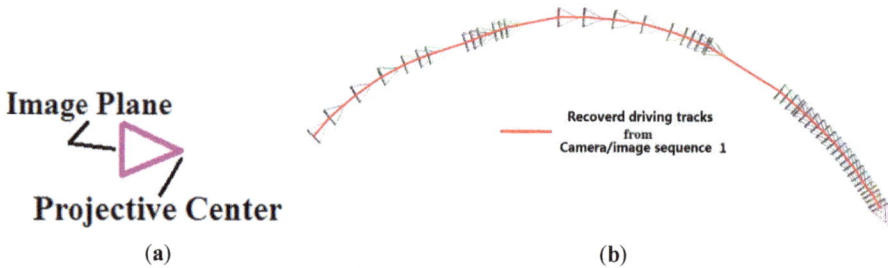

Image Plane

Projective Center

(a)

Recoverd driving tracks
from
Camera/image sequence 1

(b)

Figure 11. Explanation of the reconstructed camera-pose-triangle and driving tracks. (a) Colored triangle represents the position of the recovered image and the camera projective center. The size of the triangle is followed by the size of the image data. (b) Red line represents the recovered vehicle driving tracks that carried recorder 1. The colored triangles are the reconstructed results that represent the position of the images taken by recorder 1.

(a)

(b)

(c)

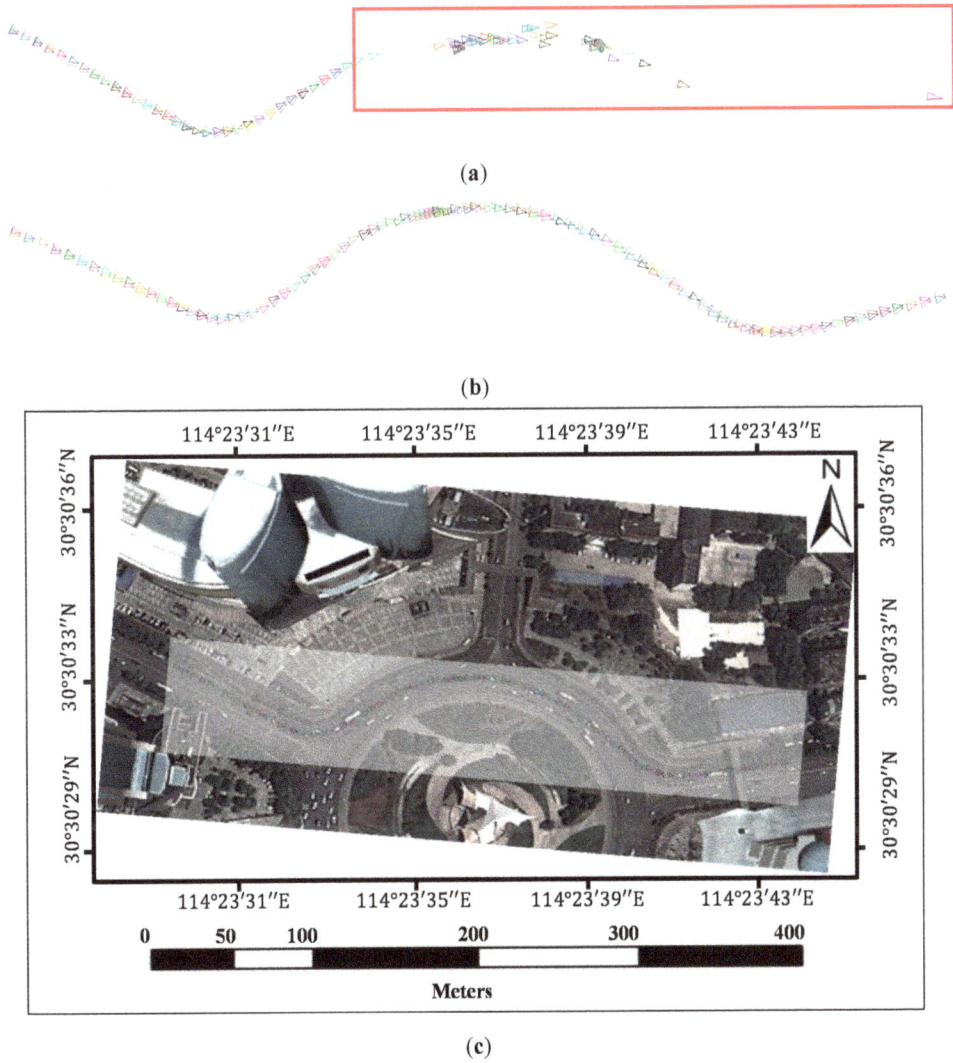

Figure 12. The recovered image positions of Set 1. These images were taken by recorders 4 and 5, which had the same exposure interval and were mounted on one vehicle. (**a**) and (**b**) are the recovered results from the same data with different methods. (a) depicts the reconstruction by the typical SfM method. The recovered images in the red rectangle of (a) are unordered obviously. (b) depicts the reconstruction by our method (features on vehicles and guardrails were masked out before matching and reconstruction). (**c**) is not a georeferenced result. We manually scaled the results of (b) and put it on the Google satellite map to help readers visualize the rough locations of the image sequences on roundabout.

221

(a)

(b)

(c)

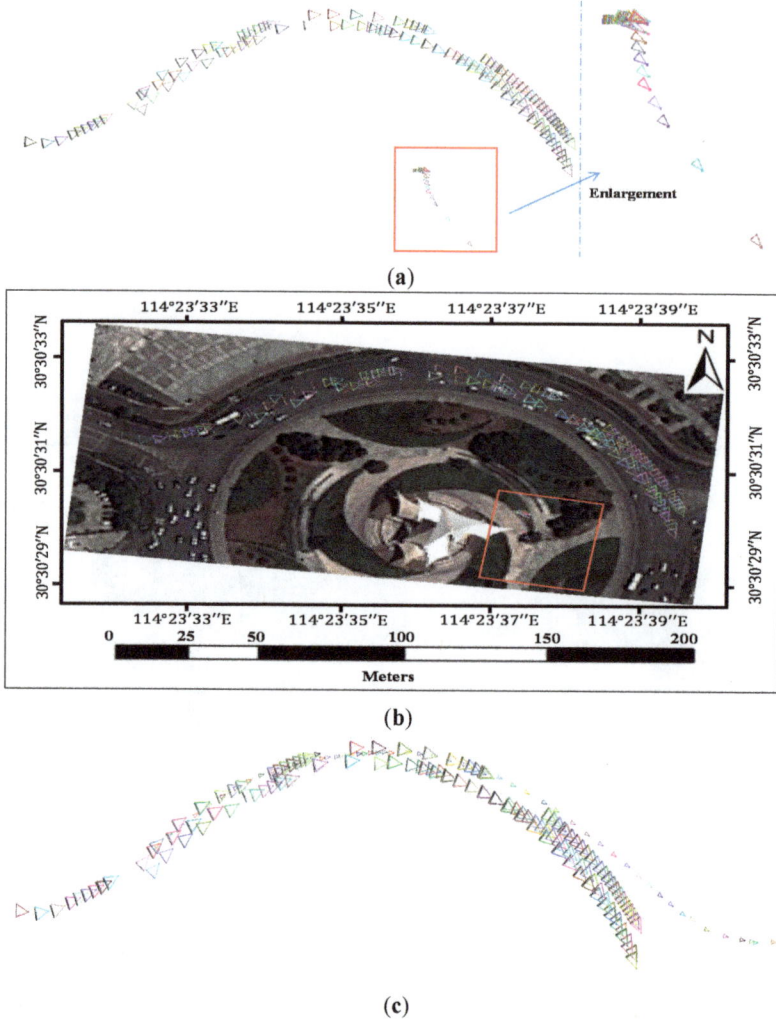

Figure 13. The recovered image positions of Set 2. These images were taken by recorders 1, 2, 3, and 4 mounted on their respective vehicles. (**a**) Reconstruction by the typical SfM method. The recovered disordered images in the red rectangles of (a) were recorded by recorder 4. (**b**) is not a georeferenced result. We manually scaled the results of (a) and put it on the Google satellite map. Based on the enlargement in (a) and the visualized rough location in (b), it can be seen that they were reconstructed in the wrong place. (**c**) Reconstruction by our method (features on vehicles and guardrails were masked out before matching and reconstruction). The recovered triangles of recorder 4 are smaller than the others because the sizes of the images taken by recorder 4 were smaller than those of the other recorders, which is reflected in (c) by the different reconstructed sizes of the triangles. (a) and (c) are the recovered results from the same data using different methods.

222

(a)

(b)

(c)

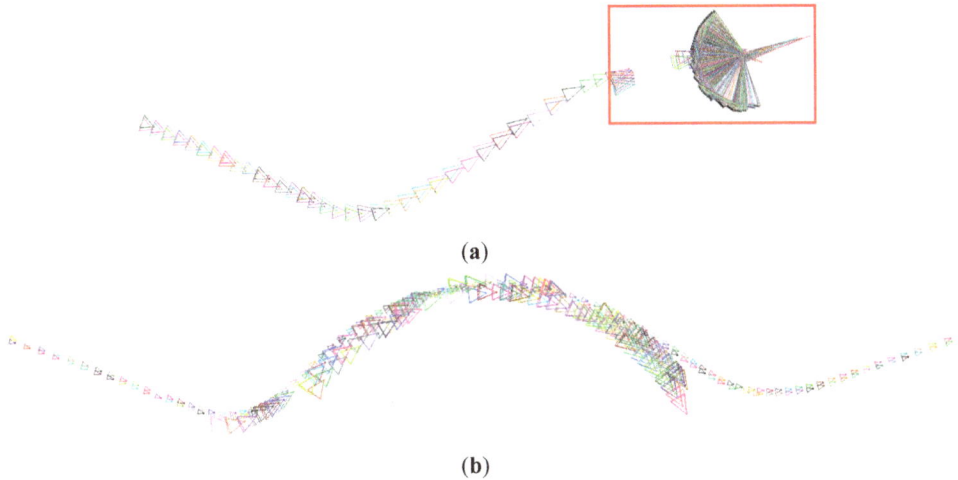

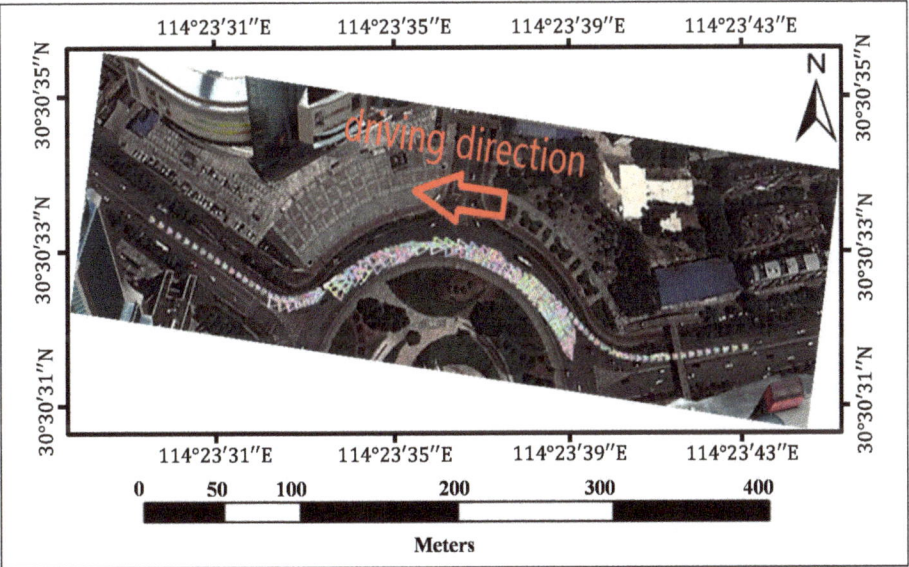

Figure 14. The recovered image positions of Set 3. These images were taken by recorders 1–5. (a) and (b) are the recovered results from the same data with different methods; (a) was reconstructed by the typical SfM method and (b) was reconstructed by our method (features on vehicles and guardrails were masked out before matching and reconstruction). The images in red rectangles in (a) were recovered in chaos. (c) is not a georeferenced result. We manually scaled the recovery results of our method and put it on the Google satellite map to help readers visualize the rough locations of the image sequences on roundabout.

223

The contrast experiment results show that the recovery performance of our method was better than the typical SfM method in each set. In contrast, the typical method was unable to recover an entire track of cameras in each set while the camera poses were recovered smoothly with our method.

We can infer from the above results that the typical method sometimes returns unreliable recovery results, especially for multi-sensors' data.

3.4. Sparse 3D Point Clouds Reconstruction Results

Sparse 3D point clouds can be reconstructed by the SfM algorithm with VisualSFM and Photosynth [27]. Photosynth is a powerful set of tools designed by Microsoft's Live Labs. It builds on a structure-from-motion system for unordered image collections, which is based on the Photo Tourism [10,11] research conducted by the University of Washington and Microsoft Research [27]. The structure from motion module in Photo Tourism comes from Bundler [10,11], which is one of the most developed SfM systems. As a useful tool, Bundler has been widely used in many point clouds reconstruction researches. This is the main reason why we chose the Photosynth as the contrast experiment tool. Furthermore, the high-level automation and widely using of Photosynth can also explain our choice. The reason why we chose VisualSFM is that VisualSFM is a powerful SfM tool; it has a flexible interface and stable performance. It is also frequently used in 3D reconstruction researches.

From the data all combined in Table 2, Set 3 is sufficient enough to cover the results from Set 1 and 2. Besides, the data collected by Recorder 1–5 in different image sizes had been lasting for as long as three years. Thus, Set 3 is able to contain various data from different cameras, which means that it is more representative than using Set 1 and 2 to evaluate the performance of reconstruction methods. Therefore, the following experiments were used VisualSFM and Photosynth based on the 311 images in Set 3. Figure 15 shows the model of main target buildings we aimed to reconstruct. Figures 16 and 17 show the side and vertical views of the results of the three methods, respectively.

The results in Figures 16 and 17 indicate that even the developed 3D reconstruction tool Photosynth is not capable of dealing with driving recorder data directly. However, the camera tracks and sparse point clouds were reconstructed successfully using the mask out correspondences as inputs to run SfM.

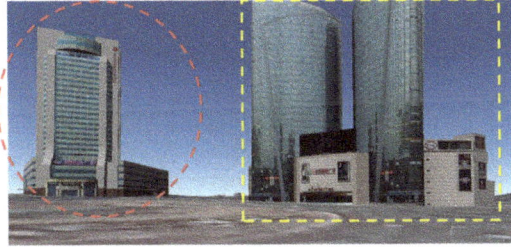

(a)

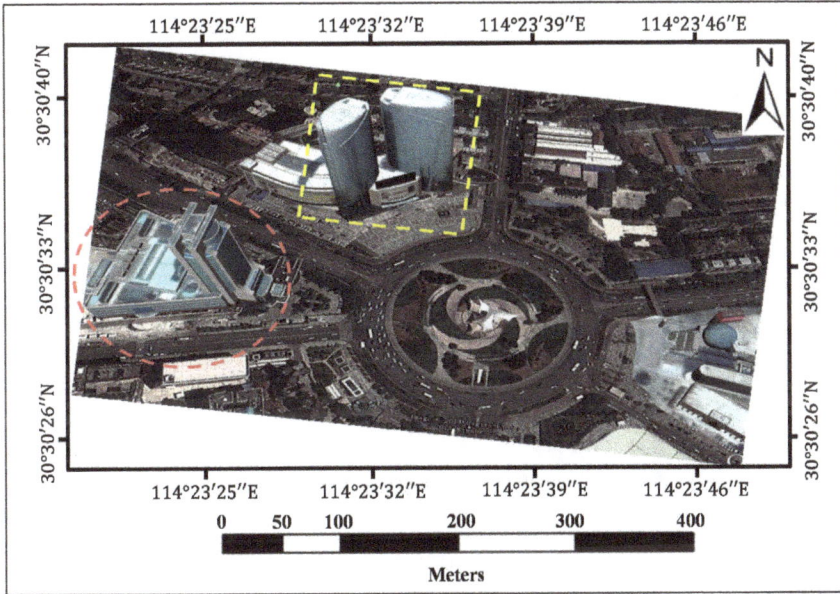

Figure 15. Main targets in the sparse point clouds reconstruction process. The two building models in (**a**) and (**b**) with red and yellow marks are the main reconstruction targets. (a) and (b) are the side and oblique bird's-eye view of two buildings from Google Earth, respectively.

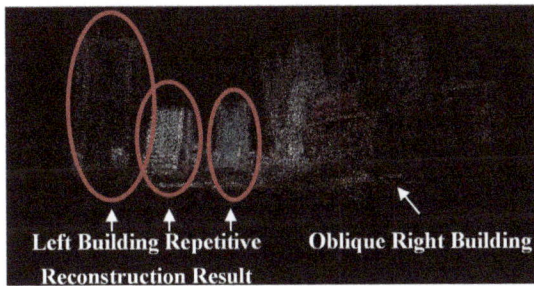

(a)

Figure 16. *Cont.*

225

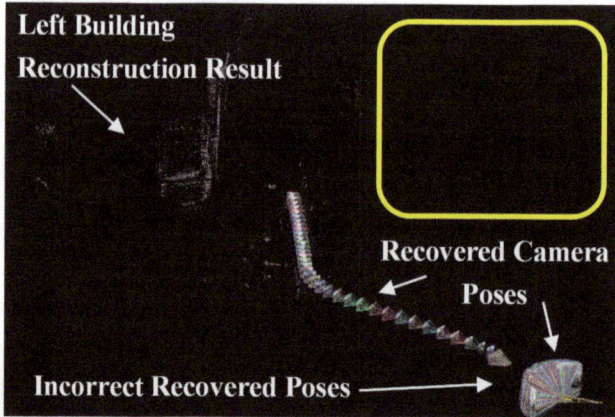

Left Building
Reconstruction Result

Recovered Camera
Poses

Incorrect Recovered Poses

(b)

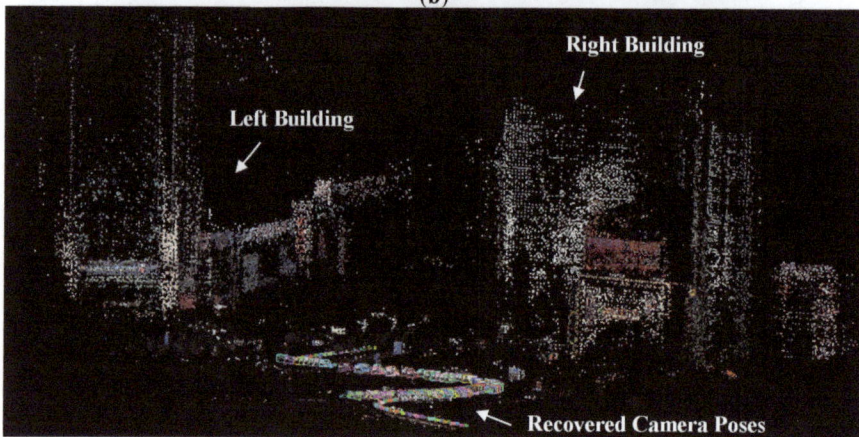

Right Building

Left Building

Recovered Camera Poses

(c)

Figure 16. Side view of main target reconstruction results with sparse point clouds. Each result was reconstructed with the same data of 311 images in Set 3. (**a**) Sparse point clouds reconstructed by Photosynth without any added processing. The building on the left marked in red was repetitively reconstructed. (**b**) Sparse point clouds reconstructed by VisualSFM with the typical method. The building on the right could not be reconstructed and should be positioned inside the yellow box. (**c**) Sparse point clouds reconstructed by VisualSFM with our method. The details of the differences between the typical method and our method are described in Section 2.5 but can be summarized by saying that our method removed the features on the Mask and matched the remaining feature points before reconstruction.

226

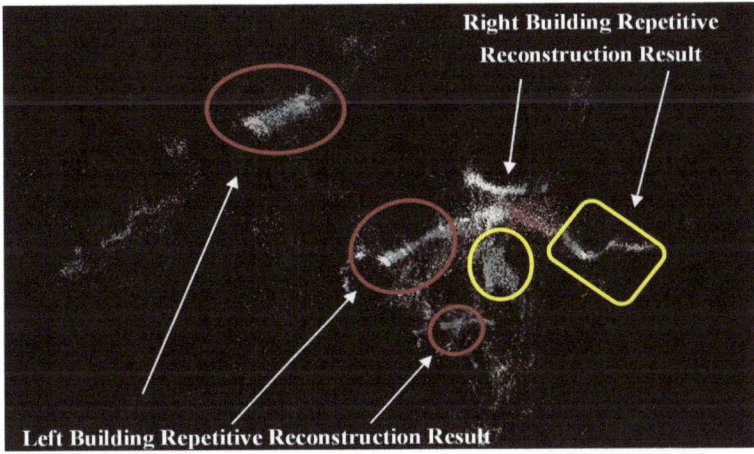

(a)

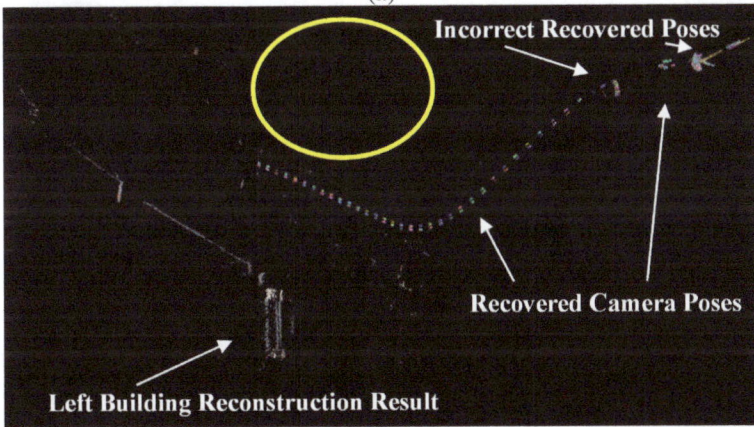

(b)

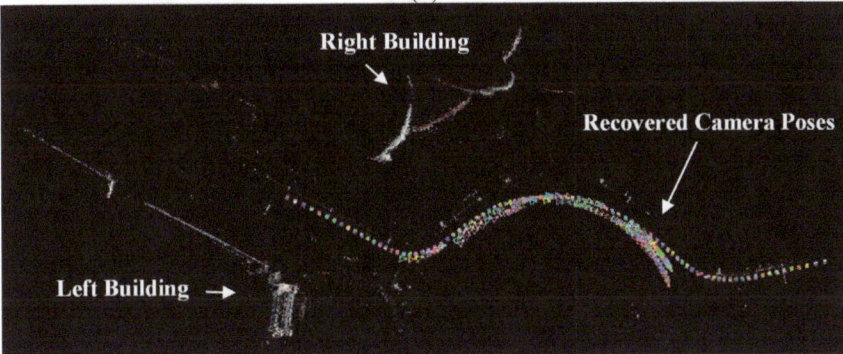

(c)

Figure 17. *Cont.*

227

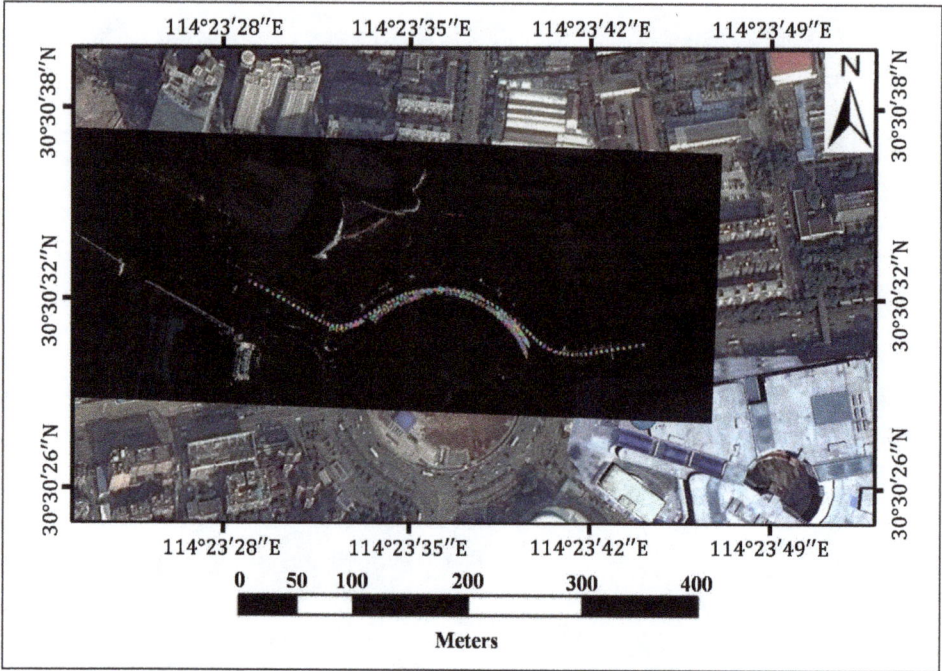

(d)

Figure 17. Vertical view of main target reconstruction results with sparse point clouds. Each result was reconstructed by same data of 311 images in Set 3. (**a**) Sparse point clouds reconstructed by Photosynth without any added processing. The result is chaos. Expecting the repetition we experienced as shown in Figure 16, it can be clearly seen that not only was the left building repeatedly reconstructed, but the right building was as well. The repetitive reconstructions of the buildings are marked in red for the left building and the right building is in yellow. (**b**) Sparse point clouds reconstructed by VisualSFM with the typical method. The right building was missed which should be reconstructed inside the yellow mark. (**c**) Sparse point clouds reconstructed by VisualSFM with our method. The details between the typical method and our method are described in Section 2.5, which can be summarized by saying that we removed the features on the Mask and matched the remaining feature points before reconstruction. (**d**) shows a more intuitive result. It is not a georeferenced result. We manually scaled the sparse point clouds of our method and put it on the Google satellite map, which can help readers visualize the high level of overlapping between the point clouds and the map, the rough relative positions of the two buildings, and the position of recovered images in roundabout.

228

Since the images recorded by a driving recorder have no GPS information, there is no other data available that can provide the absolute coordinates of the reconstructed point cloud. So in order to quantify the sparse point clouds, the plane fitting method is proposed below.

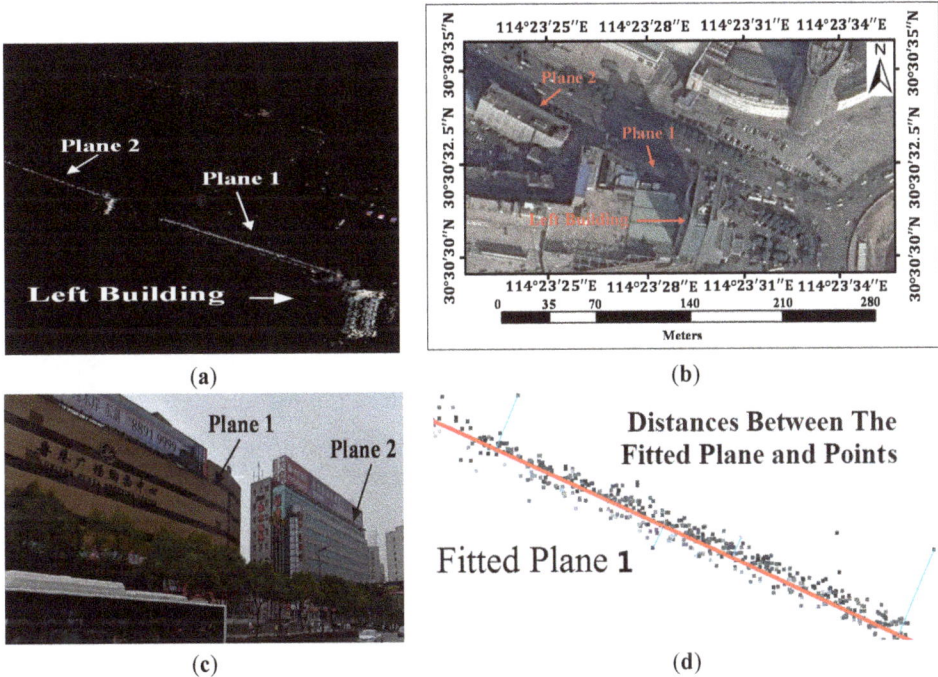

Figure 18. The vertical view of the two planes. (**a**) shows the sparse point clouds reconstructed by VisualSFM with our method. The Plane 1 and 2 are target planes we fitted. (**b**) shows the position of the target wall-planes in Google Map. (**c**) shows the Plane 1 and 2 in street view. (**d**) Example of plane fitted result in vertical view. The red line respects the vertical view of the plane fitted by wall points, and the blue lines are examples of the distances between the plane and points.

As we know, the points on the same wall also should lie on the same plane in the point cloud. Based on that principle, the distance between the fitted plane and the reconstructed wall point can be used to confirm whether the reconstruction performed well. Planes 1 and 2 in Figure 18 are the planes we aimed to fit. We manually chose the reconstructed points belonging to the above walls to fit the Plane 1 and 2. The fitting method is based on the Equation (24), the Plane Equation.

According to the Least Square method, the equation of each plane was computed with the coordinates of points.

$$Ax + By + Cz + D = 0 \qquad (24)$$

The x, y and z are coordinates of points; they are in relative coordinate system. In addition, the A, B, C and D are plane parameters that should be calculated by the Least Square method.

Based on the computed plane, the distances between the fitted plane and each wall point were then calculated. The maximum distance, minimum distance and RMSE of distance in the typical method and our method are shown in Table 3 below. Since there is no GPS information or other data available that can provide the geographical reference, the results are compared in the relative coordinates.

Table 3. The fitting results between the typical method and the proposed method. The results are in relative coordinates system.

Plane NO.	Typical Method			Proposed Method		
	RMSE	Maximum	Minimum	RMSE	Maximum	Minimum
1	0.0047	0.0170	4.420×10^{-6}	0.0031	0.0142	1.887×10^{-6}
2	0.0171	0.0994	2.133×10^{-5}	0.0095	0.0705	1.220×10^{-5}

Table 3 shows the RMSE decreased by 30–40 percentages in proposed method that indicated that the accuracies of reconstructed planes are improved in our method. Based on the above results, it is clearly seen that masking out the features from vehicles and guardrails can improve the reconstruction results. We also found our method to be robust enough for driving recorder data sets composed of different-sized images having nearly three years recovered intervals.

4. Discussion

This paper focused on sparse point cloud reconstruction with the features removed on the Mask. The Mask is the region where features should be eliminated before matching in order to avoid generating outliers. The Mask was first detected from the unstructured and uncontrolled driving recorder data automatically. Then, the feature points on Mask were eliminated before feature matching. Finally a SfM procedure with the remaining correspondences was performed.

The advantage of using driving recorder data is that a driving recorder can acquire city-scale street scenes less expensively. This low-cost data in larger quantities can support reconstructing and updating sparse street point clouds in shorter update periods of time. The improved reconstruction results are shown in Section 3 from three aspects: the precision of pairwise orientation, the recovered camera tracks, and the reconstructed point clouds. In the contrast experiment presented in Section 3.4, the right building was missing in the results of the typical SfM process [24,25] (shown in Figures 16b and 17b), which was caused by the disordered camera poses' recovered results shown in Section 3.3, Figure 14a. We found that these disordered images generally were crowded with a large number of moving vehicles, which led to poor matches. It was also proven that the commercial software Photosynth [27], which is based on the Photo Tourism [10,11] technology, was not able to provide good performance with the driving recorder data. The reconstructed points indicate obvious chaos, and the buildings were repeatedly reconstructed incorrectly. The disordered point clouds were shown in Figures 16a and 17a in Section 3.4. The overall quality of the point clouds is reflected in the comparison between the results shown in Figures 16 and 17. Furthermore, there were no other data to provide the absolute coordinates of the reconstructed point clouds so we qualified the point clouds with the plane fitting results in Table 3 and the overlap between the point clouds and Google satellite map in Figure 17d. In Figure 17d, the points are fitted with the map generally, which reflects the high level of overlapping between the point clouds and the map. The plane fitting results in Table 3 indicate that the reconstruction accuracy of the proposed method is higher than the typical method based on the decreased average distance between the fitted plane and the points.

The features mask method is based on one important factor: the SIFT algorithm can generate a large number of features that densely cover the image over the full range of scales and locations [12]. Approximately 1000–2000 correspondences can be matched with the SIFT algorithm on one pair of driving recorder data which cover more than 70% of the overlap. Generalized from ample experiments, the matching points on the Mask took a 21% proportion of the total matching points. Therefore, although the bad points on the Mask were removed, the remaining correspondences were adequate to recover the camera poses and point clouds of the street scene.

Although redundant data theoretically can produce better reconstruction results, a large number of data would adversely affect efficiency due to the fact that full pairwise matching takes $O\left(n^2\right)$ time for n input images, which is why the Mask method is used to improve the reconstruction results instead of simply increasing the data sets. Detecting and removing the Mask regions with our methods proposed in this paper only takes a few seconds in each image. The Mask results show that it can improve the quality of the matches, which may lead to a higher level of efficiency than the redundancy method.

The proposed reconstruction method is effective when the solution is scaled up. It also relies on the scalability of the SfM method. One of the most difficult problems of SfM is that it is time-intensive. There have been many relative research efforts aimed at shortening the reconstruction time, such as the vocabulary tree [28] and the Building Rome on a cloudless day [8]. The SfM method has been used to reconstruct an entire city with Internet photos. The paper aims to shorten the reconstruction time by the parallelism and throughput [8]. Our method can be a supplement to these methods, which means that, with the help of vocabulary tree and parallel computing, it can reconstruct an entire city with ample driving recorder images. Then, this method may replace mobile mapping technologies in some applications like updating 3D street data that have been georeferenced or reconstructing city-scale point clouds in relative coordinate systems.

The proposed method can reconstruct street scenes robustly with different-sized images, taken on different roads or with different lighting conditions; however, the images taken at night always contribute less to the reconstruction process. Obviously, the matches decrease in night images since the difference in building textures between day and night images are influenced by the city lighting. Similarly, lower quality will lead to fewer correspondences, which may not cause fatal mistakes to reconstruction but will increase the time consumed. Therefore if we ignore the efficiency and the data set is big enough, there is no strict requirement for the quality of image data.

The proposed method has one limitation. Since the driving recorder is always mounted to record traffic rather than buildings, the taller sections of nearby building cannot be recorded, thereby generating sparse reconstructed points for them. In general, three main innovations are presented in this paper:

1. We proposed a street scene reconstruction method from driving recorder data. This new method makes full use of the massive amount of data produced by driving recorders with shorter update time, which can reduce the costs of recovering 3D sparse point clouds compared to mobile mapping equipment carrying stable GPS/INS systems. In order to improve the recovery accuracy, we analyzed and summarized the distribution regularities of the outliers from the SIFT matching results through ample experiments.
2. Our work differs from the typical SfM approaches, in that, we eliminate the feature points on the Mask before matching is undertaken. We also proved through experiments that the relative orientation results and reconstruction results improved after removing the feature points on the Mask.
3. We designed guardrail and vehicle side region detecting methods based on the characteristics of the driving recorder data. The detection methods are based on the trained Haar-like-feature cascade classifiers, the position of the vanishing point, and some camera parameters.

5. Conclusions

This paper proposed a method to reconstruct street scenes with data from driving recorders, which are widely used in private and public vehicles. This low-cost method will be beneficial to reducing the cost and shortening the update time required for street scene reconstruction. However, using the unprocessed driving recorder data was found to contribute to the failure of reconstruction due to the large number of inevitable outliers on moving vehicles and guardrails with repeating patterns.

Based on our analysis from numerous SIFT matching results, we then proposed a method for removing the features on vehicle and guardrail regions, which is called the Mask in this paper. In order to remove the feature points on the Mask, an automatic detecting method was designed. As shown in Section 3, the proposed method improved the results in three areas: the precision of the pairwise orientation, the recovery performance of the camera poses, and the reconstruction results of the point clouds.

Our work differs from typical SfM approaches in that we remove the features on the Mask in order to improve the accuracy of the street scene reconstruction results from driving recorder data. The proposed method can be improved in the following areas, which will be the subjects of future research.

1. Reconstructing robust side surfaces in the vehicle detection method without camera parameters.
2. Extracting the most appropriate images from driving recorder videos.
3. Reducing the number of images in the time-consuming matching step with a reasonable strategy.
4. Increasing the density of reconstructed point clouds.
5. Detecting the blocked vehicles with more accuracy in a region.

Acknowledgments: This work was supported in part by the National Natural Science Foundation of China with project number 41322010, the Fundamental Research Funds for the Central Universities with project number 2014213020201, the Geographic National Condition Monitoring Engineering Research Center of Sichuan Province with project number GC201514 and the Research Funds of the Key Laboratory of Mapping from Space, National Administration of Surveying, Mapping and Geoinformation with project number K201405. We are very grateful also for the comments and contributions of the anonymous reviewers and members of the editorial team.

Author Contributions: Yongjun Zhang, Qian Li and Hongshu Lu designed the experiment and procedure. Xinyi Liu and Xu Huang performed the analyses. Chao Song, Shan Huang and Jingyi Huang performed the data and samples collection and preparation. All authors took part in manuscript preparation.

Conflicts of Interest: The authors declare no conflict of interest.

References

1. Habib, A.; Pullivelli, A.; Mitishita, E.; Ghanma, M.; Kim, E. Stability analysis of low-cost digital cameras for aerial mapping using different georeferencing techniques. *Photogramm. Rec.* **2006**, *21*, 29–43.

2. Zhang, Y.; Zhang, Z.; Zhang, J.; Wu, J. 3D building modelling with digital map, LiDAR data and video image sequences. *Photogramm. Rec.* **2005**, *20*, 285–302.

3. Xiao, J.; Fang, T.; Zhao, P.; Lhuillier, M.; Quan, L. Image-based street-side city modeling. *Acm Trans. Graph.* **2009**, *28*.

4. Cornelis, N.; Leibe, B.; Cornelis, K.; van Gool, L. 3D urban scene modeling integrating recognition and reconstruction. *Int. J. Comput. Vis.* **2008**, *78*, 121–141.

5. Aijazi, A.; Checchin, P.; Trassoudaine, L. Automatic removal of imperfections and change detection for accurate 3D urban cartography by classification and incremental updating. *Remote Sens.* **2013**, *5*, 3701–3728.

6. Williams, K.; Olsen, M.; Roe, G.; Glennie, C. Synthesis of transportation applications of mobile LiDAR. *Remote Sens.* **2013**, *5*, 4652–4692.

7. Hartley, R.; Zisserman, A. *Multiple View Geometry in Computer Vision*, 2nd ed.; Cambridge University Press: New York, NY, USA, 2003.

8. Frahm, J.; Lazebnik, S.; Fite-Georgel, P.; Gallup, D.; Johnson, T.; Raguram, W.C.; Jen, Y.; Dunn, E.; Clipp, B. Building Rome on a cloudless day. In Proceeding of the 2010 Europe Conference on Computer Vision (ECCV), Crete, Greece, 5–11 September 2010; Volume 6314, pp. 368–381.

9. Raguram, R.; Wu, C.; Frahm, J.; Lazebnik, S. Modeling and recognition of landmark image collections using iconic scene graphs. *Int. J. Comput. Vis.* **2011**, *95*, 213–239.

10. Snavely, N.; Seitz, S.M.; Szeliski, R. Photo tourism: Exploring photo collections in 3D. *ACM Trans. Graph.* **2006**, *25*.

11. Snavely, N.; Seitz, S.M.; Szeliski, R. Modeling the world from internet photo collections. *Int. J. Comput. Vis.* **2008**, *80*, 189–210.

12. Lowe, D.G. Distinctive image features from scale-invariant keypoints. *Int. J. Comput. Vis.* **2004**, *60*, 91–110.

13. Viola, P.; Jones, M. Rapid object detection using a boosted cascade of simple features. In Proceedings of the 2001 IEEE Computer Society Conference on Computer Vision and Pattern Recognition, Kauai, HI, USA, 8–14 December 2001; Volume 1, pp. 511–518.

14. Lienhart, R.; Maydt, J. An extended set of haar-like features for rapid object detection. In Proceedings of the 2002 International Conference on Image Processing, New York, NY, USA, 22–25 September 2002; pp. 900–903.

15. Schapire, R.E.; Singer, Y. Improved boosting algorithms using confidence-rated predictions. *Mach. Learn.* **1999**, *37*, 297–336.

16. Ponsa, D.; Lopez, A.; Lumbreras, F.; Serrat, J.; Graf, T. 3D vehicle sensor based on monocular vision. In Proceedings of the 2005 IEEE Intelligent Transportation Systems, Vienna, Austria, 13–16 September 2005; pp. 1096–1101.

17. OpencvTeam OpenCV 2.4.9.0 Documentation. Available online: http://docs.opencv.org/modules/objdetect/doc/cascade_classification.html (accessed on 18 October 2014).

18. Jiang, J.L.; Loe, K.F. S-AdaBoost and pattern detection in complex environment. In Proceedings of the 2003 IEEE Computer Society Conference on Computer Vision and Pattern Recognition (CVPR'03), Los Alamitos, CA, USA, 18–20 June 2003; pp. 413–418.

19. Kutulakos, K.N.; Sinha, S.N.; Steedly, D.; Szeliski, R. A multi-stage linear approach to structure from motion. In *Trends and Topics in Computer Vision*; Kutulakos, K.N., Ed.; Springer: Berlin, Germany, 2012; pp. 267–281.

20. Moghadam, P.; Starzyk, J.A.; Wijesoma, W.S. Fast vanishing-point detection in unstructured environments. *IEEE Trans. Image Process.* **2012**, *21*, 425–430.

21. Frahm, J.M.; Pollefeys, M. RANSAC for (Quasi-) degenerate data (QDEGSAC). In Proceedings of the 2006 IEEE Computer Society Conference on Computer Vision and Pattern Recognition (CVPR'06), New York, NY, USA, 17–22 June 2006; Volume 1, pp. 453–460.

22. Torr, P.; Zisserman, A. Robust computation and parametrization of multiple view relations. In Proceedings of the Sixth International Conference on Computer Vision, Bombay, India, 4–7 January 1998; pp. 727–732.

23. Fischler, M.A.; Bolles, R.C. Random sample consensus: A paradigm for model fitting with applications to image analysis and automated cartography. *Commun. ACM* **1981**, *24*, 381–395.

24. Wu, C. VisualSFM: A Visual Structure from Motion System. Available online: http://ccwu.me/vsfm/ (accessed on 8 October 2014).

25. Wu, C. Towards linear-time incremental structure from motion. In Proceedings of the 2013 International Conference on 3D Vision IEEE, Seattle, WA, USA, 29 June–1 July 2013; pp. 127–134.

26. Nister, D. An efficient solution to the five-point relative pose problem. *IEEE Trans. Pattern Anal. Mach. Intell.* **2004**, *26*, 756–777.

27. Microsoft Corporation. Photosynth. Available online: https://photosynth.net/Background.aspx (accessed on 8 October 2014).

28. Nister, D.; Stewenius, H. Scalable recognition with a vocabulary tree. In Proceedings of the 2006 IEEE Computer Society Conference on Computer Vision and Pattern Recognition (CVPR'06), New York, NY, USA, 17–22 June 2006; pp. 2161–2168.

Chapter 3:
Multi-Sensor Fusion

Multispectral Radiometric Analysis of Façades to Detect Pathologies from Active and Passive Remote Sensing

Susana Del Pozo, Jesús Herrero-Pascual, Beatriz Felipe-García,
David Hernández-López, Pablo Rodríguez-Gonzálvez and
Diego González-Aguilera

Abstract: This paper presents a radiometric study to recognize pathologies in façades of historical buildings by using two different remote sensing technologies covering part of the visible and very near infrared spectrum (530–905 nm). Building materials deteriorate over the years due to different extrinsic and intrinsic agents, so assessing these affections in a non-invasive way is crucial to help preserve them since in many cases they are valuable and some have been declared monuments of cultural interest. For the investigation, passive and active remote acquisition systems were applied operating at different wavelengths. A 6-band Mini-MCA multispectral camera (530–801 nm) and a FARO Focus3D terrestrial laser scanner (905 nm) were used with the dual purpose of detecting different materials and damages on building façades as well as determining which acquisition system and spectral range is more suitable for this kind of studies. The laser scan points were used as base to create orthoimages, the input of the two different classification processes performed. The set of all orthoimages from both sensors was classified under supervision. Furthermore, orthoimages from each individual sensor were automatically classified to compare results from each sensor with the reference supervised classification. Higher overall accuracy with the FARO Focus3D, 74.39%, was obtained with respect to the Mini MCA6, 66.04%. Finally, after applying the radiometric calibration, a minimum improvement of 24% in the image classification results was obtained in terms of overall accuracy.

Reprinted from *Remote Sens.* Cite as: Del Pozo, S.; Herrero-Pascual, J.; Felipe-García, B.; Hernández-López, D.; Rodríguez-Gonzálvez, P.; González-Aguilera, D. Multispectral Radiometric Analysis of Façades to Detect Pathologies from Active and Passive Remote Sensing. *Remote Sens.* **2016**, *8*, 80.

1. Introduction

Historical buildings and monuments are valuable constructions for the area where they are placed. The degradation of their construction materials is caused mainly by environmental factors such as pollution and meteorological conditions. Specifically, the presence of water plays an important role in stone deterioration processes [1]. It accelerates the weathering processes contributing to dissolution and frost/thaw cycles among others [2] allowing the formation of black crust on the rock surface resulting in mechanical and chemical degradations of stones. For that reason the use of non-contact and non-destructive technologies to study stone damages is important for the preservation of buildings and for the choice of the best technique for restoration [3,4].

Terrestrial laser scanners and multispectral digital cameras are two different technologies that are suitable for these studies. They are non-destructive and non-invasive sensors that allow researchers to acquire massive geometric and radiometric information across the building with high accuracy and in a short acquisition time. The geometrical information provided by laser scanner technology has been successfully applied in a large number of fields such as archaeology [5], civil engineering [6], geology [7] and geomorphological analysis [8]. On the other hand, radiometric information, provided by the laser intensity data and the multispectral digital cameras, is used less frequently. Even so, its high potential for classification tasks and recognition of different materials has been demonstrated [9]. Nowadays, in the literature, one can find works related to this issue ranging from methodologies of radiometric calibration [10] to corrections of intensity values [9,11] including applications of the intensity data [12]. Spectral classification methods are based on the properties of the reflected radiation from each surface and the fact that each specific material has wavelength dependent reflection characteristics. There are many classification methods, which vary in complexity. These methods include hard and soft classifiers, parametric and non-parametric methods and supervised and unsupervised techniques [13]. There are several works related to the application of these techniques to the identification of damage on building surfaces [14–18].

The main objective of this paper is the classification and mapping of pathologies and materials of a historical building façade from reflectance values at different wavelengths by combining intensity calibrated data from a FARO Focus3D laser scanner and calibrated images from a 6-band Mini-MCA multispectral camera. Additional goals were evaluating the degree of automation in the pathology detection process of façades. To achieve these objectives, the paper is divided into the following sections: Section 2 gives the details and specifications of the equipment employed and thoroughly describes the methods employed in the workflow methodology. Section 3 shows the classification maps and accuracy results for both unsupervised and supervised classifications, closing with Section 4 which summarizes the main conclusions and findings drawn from the study.

2. Material and Methods

The methodology developed to reach the objectives of the paper consists of three main stages: the data acquisition, the pre-processing and the processing of data as is outlined in Figure 1. For the data acquisition, two sensors with different operating principles were implemented: a passive multispectral camera and an active terrestrial laser scanner. The pre-processing step involved data filtering and several corrections applied to the spectral information to finally obtain data in reflectance values. During the last step and taking advantage of the metrics from the scan points, reflectance orthoimages were generated for both the multispectral images and the laser intensity. These orthoimages were the input for two different classifications processes: a clustering classification with data from each sensor and a supervised classification with the set of all data from both sensors.

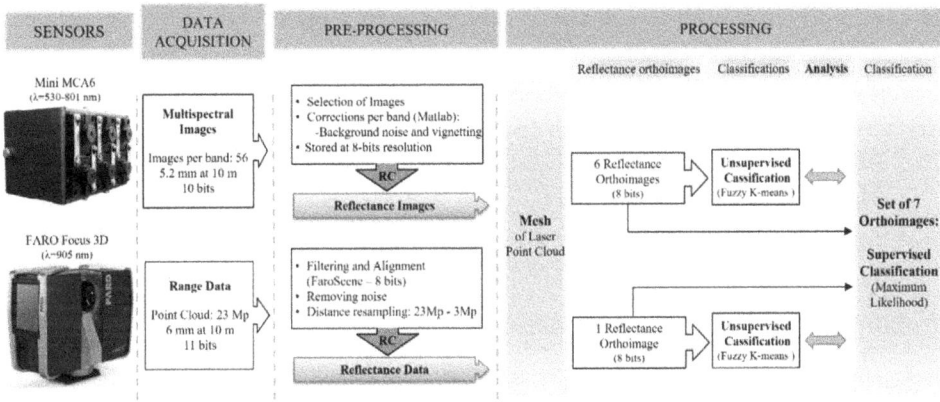

Figure 1. The workflow of the methodology presented. Acronyms: RC = Radiometric calibration and Mp = Millions of points.

2.1. Equipment

For the documentation of the façade, the following equipment was used: two radiometrically calibrated sensors with different characteristics and data acquisition principles, a passive multispectral camera and an active terrestrial laser scanner. Figure 1 shows the main characteristics of them and the different stages of the workflow followed in this research.

For the multispectral data acquisition, a calibrated lightweight Multiple Camera Array (MCA-Tetracam) was employed. This low-cost sensor allows versatility in data acquisition; however it requires the radiometric and geometric corrections to ensure the quality of the results [19]. It includes a total of 6 individual sensors with filters for the visible and near infrared spectrum data acquisition. More specifically, the individual bands of 530, 672, 700, 742, 778 and 801 nm were used. The longest wavelength was chosen taking into account that the multispectral sensor is not externally cooled. In spite of its 1280 × 1024 pixels of image resolution, the camera has a radiometric resolution of 10 bits. The focal length of 9.6 mm and the pixel size of 5.2 µm yield a façade sample distance (FSD) of 5.4 mm for a distance of 10 m, which should be taken into account for the pathology detection performance in small elements. The main limitation of this camera is the field of view (38° × 31°), so several captures were needed to keep the FSD.

The FARO Focus3D is a phase shift continuous wave terrestrial laser scanner (TLS) operating at a wavelength of 905 nm. It is not common to use this kind of sensor to perform radiometric studies but it guarantees a comprehensive data acquisition whose results are not influenced by changes in light. This device measures distances in a range of 0.60–120 m with a point measurement rate of 976,000 points per second. It has an accuracy of 0.015° in normal lighting and reflectivity conditions and a beam divergence of 0.19 mrad, equivalent to 19 mm per 100 m range. The field of view covers 320° vertically and 360° horizontally with a 0.009° of angular resolution and the returning intensity is recorded at 11 bits. This laser scanner includes, in addition, a double compensator in the horizontal and vertical axis that can be used as constraint for the scan alignment.

Additionally, a high resolution spectroradiometer (ASD FieldSpec3) (Figure 2) was used as a remote detector of radiant intensity from the visible to the shortwave infrared ranges (350 to 2500 nm with a maximum spectral resolution of 3 nm and ±1 nm wavelength accuracy) to validate the spectral results of the study [20]. Equipped with optical fiber cables, it measured reflectances from the different materials and covers of the façade with a 25° field of view. Measures were made by positioning the spectroradiometer gun (Figure 2a) as orthogonal as possible and at a distance of approximately 10 cm from the sample, trying to cover a relatively homogeneous area of the material.

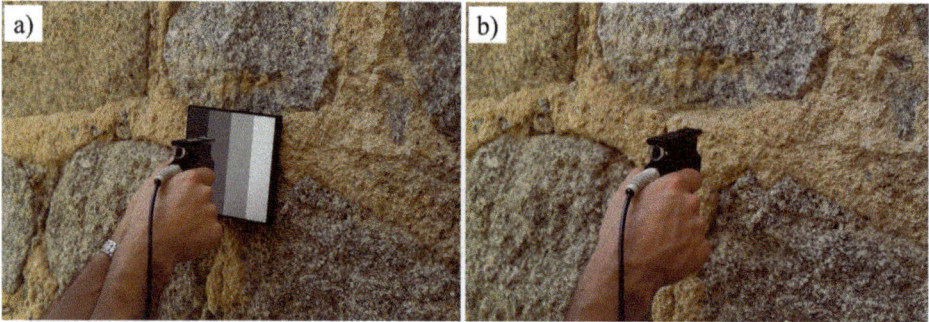

Figure 2. ASD FieldSpec3 spectroradiometer collecting spectral radiation reflected from (**a**) the Spectralon target and (**b**) mortar between contiguous stones of the examined façade.

2.2. Data Acquisition

Since each material has a unique reflectance behavior depending of the wavelength, the presence of pathologies on façades, such as moisture, moss or efflorescence, is likely to be successfully detected by analyzing the reflected visible and very near infrared radiation from the façades in reflectance values instead of digital levels (output digital format of the device). That is why these two sensors were radiometrically pre-calibrated and used to obtain orthoimages with surface reflectance values instead of digital levels. Since reflectance, for the specific case of a passive sensor, is a function of the solar incident radiance, a standard calibrated reflection target (Spectralon, Labsphere) was required and placed on the façade (Figure 2a), thus it appeared in every multispectral image to be able to calculate the solar irradiance (E) of each capture moment.

Illumination is a crucial parameter for data acquisition with passive sensor, particularly when several shot positions are required to cover the object of study. For that reason and to ensure the greatest resolution, taking fewer photos as far as possible was prioritized in this study. A total of 56 captures were collected with a FSD of 5.4 mm for the worst case so that the standard calibrated reflection target appeared in all of them.

On the other hand, the laser scanner data acquisition was designed so that the effect of the laser beam incidence angle [21,22] was minimized. Intensity data at 11-bit resolution was collected at an average distance of 10 m through three scans with scan area restrictions. Thus, 7 m of façade were covered for each of the scans assuming a maximum incidence angle error of 5.6% regarding the maximum oblique angle of incidence (19.29°). In addition, scanning positions were selected according to the different technical specifications of the scanner for an spatial resolution of 6 mm at the working distance. The laser network was adapted and filtered due to the presence of obstacles that hinder a single station data acquisition.

2.3. Pre-Processing

Before the reflectance orthoimage generation some corrections to raw data were applied to avoid error propagation in the radiometric calibration process. In this section, these radiometric corrections and the final radiometric calibration were described as well as the orthoimages generation process. Finally, the orthoimages were classified to obtain maps of different pathologies and building materials.

2.3.1. Multispectral Images Corrections

Low-cost sensors, such as the Mini MCA6, are more likely to be affected by different noise sources so that the actual value of radiation collected by them is altered (Equation (1)) [23]. Specifically, the Mini MCA6 was affected by two different sources errors: a background noise and a vignetting effect [20]. Both errors were studied under precise laboratory controlled conditions for each wavelength band.

The background noise is a systematic error caused by the sensor electronics of the camera. It was analyzed in a completely dark room in the absence of light determining the noise per band depending on the exposure time. For this study, the maximum background error was for the 801-nm band and involved a 1.07% increment of the actual digital level value. Regarding the vignetting effect [24], the radial attenuation of the brightness was studied taking images of a white pattern with uniform lighting conditions. Digital levels of each multispectral image were corrected for these two effects through a script developed in Matlab to improve the data quality before the radiometric calibration.

$$DL_{raw} = DL_{radiance} + (DL_{bn} + DL_v) \tag{1}$$

where DL_{raw} are the digital levels of the raw images, $DL_{radiance}$ are the digital levels from the radiance component, DL_{bn} are the digital levels from background noise and DL_v are the digital levels from the vignetting component.

2.3.2. Filtering and Alignment of the Point Clouds

The raw laser scanner data were filtered and segmented in order to remove those points that were not part of the object of study (adjacent building, artificial elements, trees, *etc.*). The individual point cloud alignment was done by a solid rigid transformation by the use of external artificial targets (spheres). The spheres were stationed in tripods at the plumb-line plane surveyed by the global navigation satellite system (GNSS). The laser local coordinate system could be transformed to a global coordinate system (UTM30N in ETRS89), allowing the geo-referencing of the subsequent classification for a global analysis and interpretation. This proposed workflow allowed a final relative precision of the coordinates of the artificial targets of 0.01 m and an absolute error of 0.03 m after post-processing. As a result, a unique point cloud in a local coordinate system with 11 mm precision (due to the error propagation of inherent error sources of laser scanner [25] and the error associated to the definition of the coordinate system) was generated.

2.3.3. Radiometric Calibrations

To perform the radiometric calibration of both sensors, auxiliary equipment such as lambertian surfaces with known spectral behavior (Spectralon) and/or a spectroradiometer are needed to solve the calibration. Thus, after the calibration process images values, in the case of the camera, and points' intensities, in the case of the laser scanner, correspond to the radiation emitted by the surface expressed in radiance or reflectance. The Mini MCA6 multispectral camera was calibrated in a previous field campaign [20] through *in situ* spectroradiometer measurements of artificial surfaces, with known and unknown reflectance behavior (Spectralon and polyvinyl chloride vinyl sheets respectively). Regarding the radiometric calibration of the TLS, it was carried out in laboratory by using a Spectralon and in absence of light.

The multispectral camera was calibrated by the radiance-based vicarious method [26–28], being the transformation equation from raw images into images with reflectances values Equation (2):

$$\rho_{MCA} = \frac{c0_\lambda + c1_\lambda \cdot DL/Fv_\lambda}{E_\lambda} \cdot \pi \tag{2}$$

where $c0_\lambda$ and $c1_\lambda$, offset and gain, are the calibration coefficients of each camera band, Fv_λ the shutter opening time factor and E_λ the solar irradiance at the ground level. Table 1 summarizes the multispectral camera calibration coefficients and the R^2 determination coefficient achieved per band.

Table 1. Calibration coefficients of the Mini MCA6 per band.

Bands	$c0_\lambda$	$c1_\lambda$	R^2
530 nm	0.000264	0.057718	0.9816
672 nm	−0.000795	0.050005	0.9823
700 nm	−0.000861	0.041353	0.9820
742 nm	−0.001205	0.074335	0.9843
778 nm	−0.001510	0.047292	0.9846
801 nm	−0.000834	0.047656	0.9827

In order to obtain reflectance values directly from laser data, a reflectance-based radiometric calibration [28] consisting of analyzing the distance-behavior of the intensity data (Figure 3) was performed (Equation (3)).

$$\rho_{FARO} = e^{a \cdot d} \cdot b \cdot d^2 \cdot e^{c1_F \cdot DL_F} \tag{3}$$

where a and b were the empirical coefficients related to the signal attenuation and internal TLS conversion from the received power to the final digital levels, d the distance between the laser scanner and the object, $c1_F$ the gain of the TLS and DL_F the raw intensity data in digital levels (11 bits). Please note that the empirical coefficients were obtained by a laboratory study, since the TLS internal electronics and intermediate signal processing is not disclosed.

In this case, a laboratory experiment from 5 to 36 m at one-meter intervals provided enough information to study the FARO Focus3D internal behavior (Figure 3). It was conducted in low-light conditions at a controlled temperature of 20 °C to model and simulate the system behavior. By positioning a Spectralon at each distance increment, intensity data were acquired at a quarter of the maximum resolution of the laser scanner (6 mm). The calibrated surface (Figure 2a) consists of four panels of 12%, 25%, 50% and 99% reflectance and it was assembled on a stable tripod to ensure its verticality. The raw intensity data from each reflectance panel were obtained by averaging the intensity values of the points belonging to each panel. The mean intensity value was plotted per panel and distance resulting in the Figure 4.

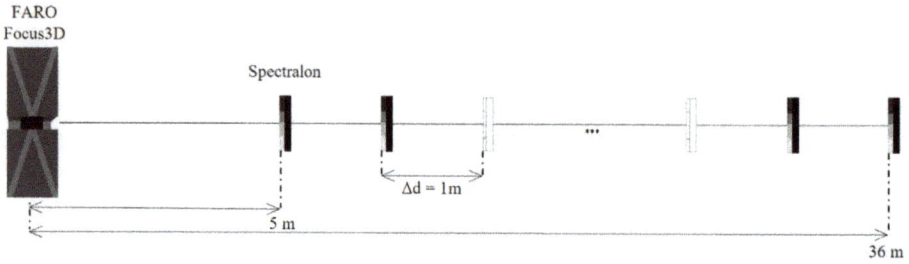

Figure 3. Sketch of the test performed to analyze the internal radiometric behavior of the FARO Focus3D.

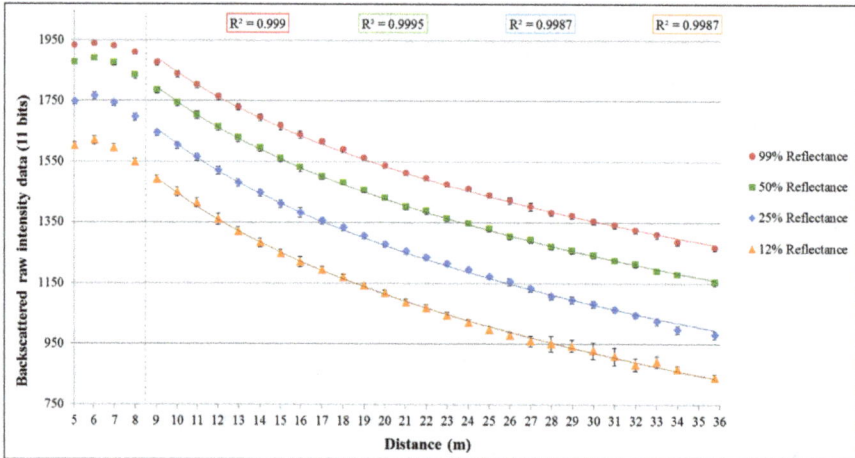

Figure 4. FARO Focus3D backscattered intensity behavior for the measurements of the four Spectralon panels at 1 m distance increments related to the signal attenuation (Equation (3)).

Figure 4 shows the signal attenuation of the FARO Focus 3D with distance as well as the logarithmic model that the measurements follow for distances up to 9 m. This particular behavior was noted in previous research works with similar sensors [29] and it is explained by the lidar equation [30]. By knowing the calibrated reflectance values of each Spectralon panel for 905 nm, the wavelength of the laser scanner, field measurements could be related with these reflectance values at each studied distance. Being 0.992, 0.560, 0.287 and 0.139 the normalized (0–1) reflectance values for the panel of 99%, 50%, 25% and 12% of reflectance respectively. Figure 5 shows how these values relate at a 10 m distance, and follow an exponential relationship which is shown in Equation (3). This distance was chosen as a threshold since for lower distances the calibration model changes due to the internal measurement system, involving alternative mathematical models.

As Figure 4 shows, the greater the distance the greater the intensity errors in the measurements. This behavior is related to the decrease of the received power due to the distance attenuation and signal scattering. Since the effective range of the employed TLS is higher than the studied distance, this error only appears significantly in the lower reflectance surface (12% panel).

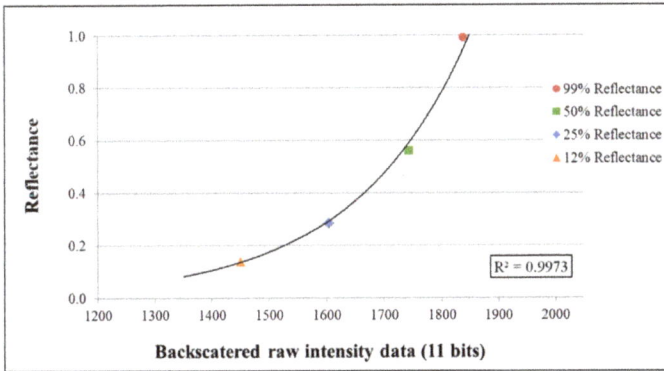

Figure 5. Relationship between TLS raw intensity data and reflectance for each spectralon panel at 10 m distance.

Based on the empirical study of the laser response, the attenuation of the signal with the distance (Figure 4) and the logarithmic behavior of the sensor [29], the relationship between digital levels and reflectances was finally approximated according to the Equation (4).

$$\rho_{FARO} = e^{0.214 \cdot d} \cdot 3.907 \cdot 10^{-7} d^2 \cdot e^{0.005415 \cdot DL_F} \tag{4}$$

This empirical equation can be applied only to objects at a distance over than 8 m since as can be shown in Figure 4, the FARO Focus3D has a completely different behavior for shorter distances.

2.4. Processing

In this subsection reflectance products are joined to achieve the orthoimages at each wavelength and they are finally classified to obtain maps of different building materials and pathologies.

2.4.1. Orthoimages Generation

Once the final point cloud was filtered, aligned and calibrated, a triangulation was applied to create the digital façade model (DFM). This step was required in order to generate continue 2D products (in the form of true orthoimages) and carry out the pathology detection by the classification process. For the DSM generation the incremental triangulation Delaunay algorithm was applied [31]. The output was refined to avoid artifact, meshing gaps, and other errors [32].

Orthoimages are highly demanded products that offer many benefits: metric accuracy and radiometric information useful to analyze different information quantitatively and qualitatively.

For the orthoimage generation, it was necessary to know the external orientation of the images with respect to the coordinate system of the laser point cloud model. For that purpose an average of 20 corresponding points between the point cloud and images were manually established. The image projection was characterized by a rigid transformation (rotation and translation) together with the internal camera parameters.

Orthoimages were generated based on the anchor point method [33]. This method consists of applying an affine transformation to each one of the planes formed by the optimized triangular mesh, which was obtained from the point cloud determined by the laser. Through the collinearity condition [34], the pixel coordinates of the vertices of the mesh were calculated, and the mathematical model of the affine transformation directly relates the pixel coordinates of the registered image and of the orthoimage.

2.4.2. Orthoimages Classifications

In order to categorize the orthoimages in different informational classes a previous automatic unsupervised classification and a posterior supervised classification were performed. The unsupervised classification was based on the Fuzzy K-means clustering algorithm where each observation can concurrently belong to multiple clusters [35]. For a set of n multidimensional pixels, the automatic management in l clusters iteratively minimizes the Equation (5) [36]:

$$J_m = \sum_{i=1}^{n} \sum_{l=1}^{\lambda} u^m{}_{i,l} \| x_i - c_l \|^2 ; \ 1 \leqslant m < \infty \qquad (5)$$

where m represents any real number greater than 1, x_i the i-th of d-dimensional measured data, u_{il} the degree of membership of x_i in the cluster l, c_l the d-dimensional center of the cluster and $\| ** \|$= Euclidean norm expressing the similarity between any measured data and the center.

Fuzzy partitioning is carried out through an iterative optimization of the objective function shown above, with the update of membership and the cluster centers by Equation (6).

$$u_{il} = \cfrac{1}{\sum\limits_{k=1}^{c} \left[\cfrac{\| x_i - c_l \|}{\| x_i - c_k \|} \right]^{\frac{2}{m-1}}} ; \quad c_l = \cfrac{\sum\limits_{i=1}^{n} u_{il}^{m} \cdot x_i}{\sum\limits_{i=1}^{n} u_{il}^{m}} \qquad (6)$$

This iteration will stop when $\max_{il} \left\{ \left| u_{il}^{(k+1)} - u_{il}^{(k)} \right| \right\} < \varepsilon$, where ε is the stop criterion between 0 and 1 and k represents the iteration steps.

After this classification, a first approach of the spectral classes and different construction materials was obtained. With a subsequently supervised classification and applying the expert knowledge of some classes, the final results improved. Furthermore, this supervised classification will serve as reference to discuss which sensor is the ideal one for detecting materials and pathologies in façades.

In this case, a maximum likelihood (ML) classification algorithm [37] was applied. The ML classifier quantitatively evaluates both the variance and covariance of the category spectral response patterns when classifying an unknown pixel. The resulting bell-shaped surfaces are called probability functions, and if the prior distributions of this function are not known, then it is possible to assume that all classes are equally probable. As a consequence, we can drop the probability in the computation of the discriminant function $F(g)$ (Equation (7)), and there is one such function for each spectral category [38].

$$F(g) = -\ln |\Sigma_p| - (g - \mu_p)^T \Sigma_p^{-1}(g - \mu_p) \qquad (7)$$

where p is the p-th cluster, Σ_p is the variance-covariance matrix and μ_p represents the class mean vector and g the observed pixel.

3. Experimental Results

The study area was the Shrine of San Segundo declared World Cultural Heritage in 1923 [39] (Figure 6). This Romanesque shrine is located in the west of the city of Ávila (Spain) and was built in the 12th century with unaltered grey granite plinths and walls with the alternation of granite blocks with different alteration degrees. The unaltered granite is mainly present in the blocks of low areas because of its high compressive strength and resistance to water absorption.

The field work was carried out on 27 July 2012 around the southern façade of the church (Figure 6), the most interesting façade from a historical point of view because it preserves the Romanesque main front. The five archivolts and capitals are decorated with plant and animal motifs. A total of 3 stations for the case of laser scanner were performed to cover the façade at a distance of 10 m (see Figure 6 right). The resolution of the data capture of the FARO Focus3D was a quarter of the full resolution provided by the manufacturer, 6 mm at 10 m. Moreover, the façade was photographed at the same distance with the Mini MCA6 multispectral camera with a FSD of 5.4 mm. A selection of 9 multispectral images of the 56 (7 per station) were used for the orthoimages generation. This selection was related with the most suitable images regarding the area of study and the optimal sharpness and quality of the set of images. The total volume of information generated amounted to 10.7 GB, where the great part was due to the meshes and orthoimages generation projects. Figure 7 shows the set of the 7 final orthoimages with a 6 mm FSD.

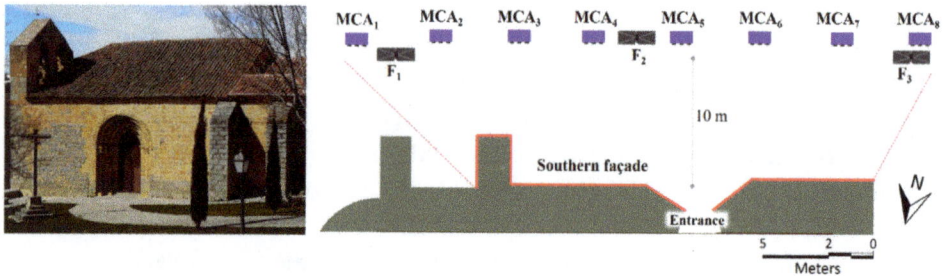

Figure 6. South façade of the Church of San Segundo in Ávila (Spain) **(left)** and a sketch of the acquisition setup with the different sensor's stations (MCA6-multispectral camera, FARO Focus3D) **(right)**.

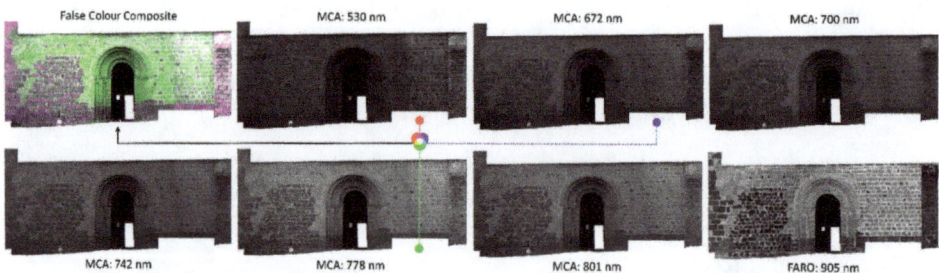

Figure 7. Set of 7 orthoimages of the façade in reflectance values from the two analyzed sensors (MCA6 multispectral camera and FARO Focus3D) and a false colorcolor composite orthoimage.

3.1. Reflectance Orthoimages

In order to compare the discrimination capability of both technologies to distinguish building materials and pathologies a first unsupervised classification of the orthoimages belonging to each sensor was performed (Figures 8 and 9). A final supervised classification with the complete set of 7 orthoimages was carried out. For each informational class manually representative areas distributed throughout the façade (between 5 and 10 polygons per class) were selected. This last classification serves as a reference with which to compare each individual unsupervised classification. The steps followed by the workflow are shown in Figure 1.

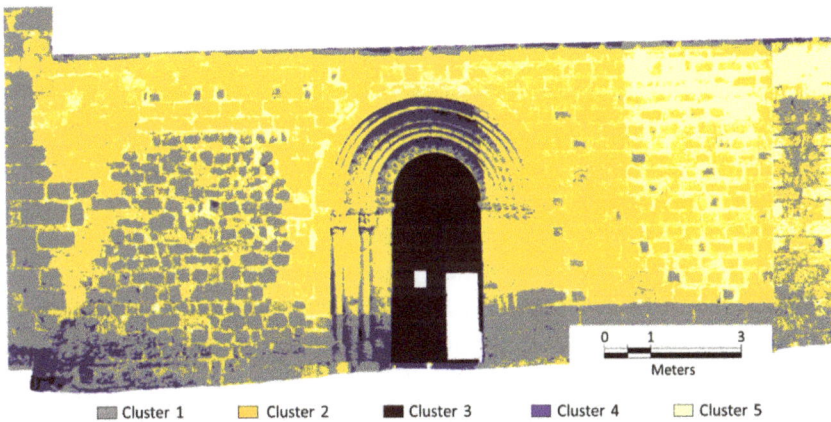

Figure 8. Mini MCA6 map for the 5-clusters unsupervised classification.

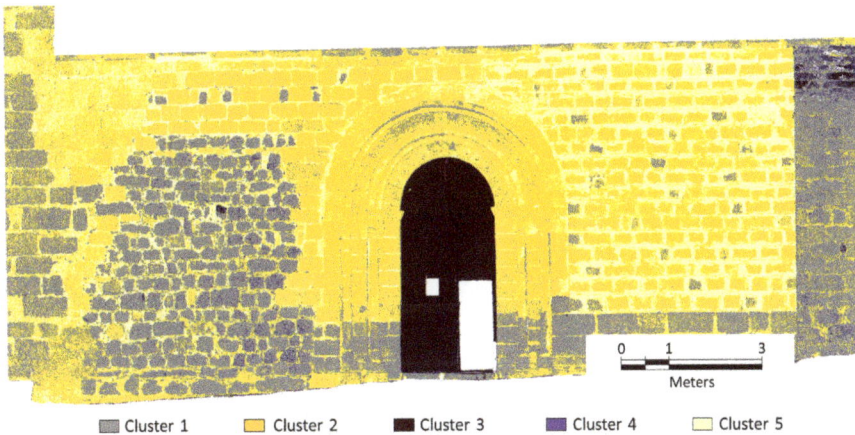

Figure 9. FARO Focus3D map for the 5-clusters unsupervised classification.

251

3.2. Orthoimages Classifications

Ten predefined clusters were used in each case for the unsupervised classification algorithm. In all of them, the resulting map showed the existence of affected areas. Post-analysis reduced the number of clusters. The number of clusters decreased from 10 (initial clusters) to 5 thematic classes with real meaning: (1) unaltered granite; (2) altered granite; (3) wood (door of the church); (4) areas with moisture evidences (caused by capillarity or filtration water) and (5) mortar between blocks.

It is noteworthy that results from Mini MCA6 are not fully satisfactory due to large variability in lighting conditions during the data acquisition. As mentioned at the beginning of Section 3, the fieldwork took place on 27 July 2012, with a 6-h total acquisition time. Although radiometric calibration reduces the effects of the lighting variability between different data acquisition time, passive sensors are really sensitive to shady areas. These areas could be seen in Figure 7, specifically in the orthoimages from the Mini MCA6, and also in the classification results of the entrance area in Figure 8 (blue color). However, this is not the case for the active sensor, FARO Focus3D, where the continuity of materials and pathologies is a remarkable aspect.

Comparing the results with a visual inspection, results correspond quite well to reality for both types of existing granites (unaltered and altered) and wood by three well differentiated clusters in all classification maps (Figures 8 and 9). Regarding pathologies detection, it was not possible to draw final conclusions with these first unsupervised classifications. However, this process served to perform a better defined supervised classification.

With the aim of having a reference with which to compare both unsupervised classification maps, a supervised classification of the full set of 7 orthoimages in reflectance values was performed (Figure 10) taking into account the two existing variants of granite, their pathologies derived primarily from moisture and the other informational classes.

The best overall accuracy for the Fuzzy K-means unsupervised classifications was 74.39%, achieved for the FARO Focus3D map in contrast with the 66.04% accuracy for the Mini MCA6 map. This indicates that the best correlation between the number of pixels correctly classified and the total number of pixels occurred for this near infrared active sensor.

Table 2 contrasts the results of the supervised classification (based on training areas) with the unsupervised classification for each sensor. The table shows the sum of pixels belonging to each class for each of the classifications performed. The count is expressed as a percentage of the total number of classified pixels (1,154,932 without taking the background class into account).

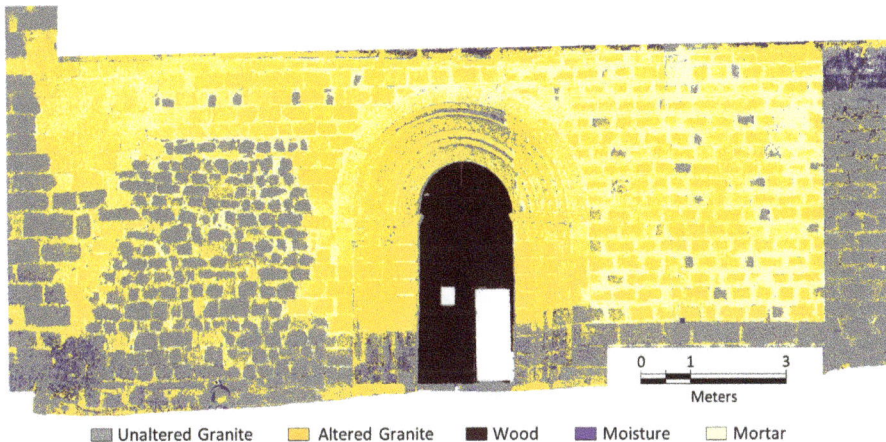

Figure 10. Multisensory map for the 5 informational classes supervised classification.

Table 2. Pixels computation belonging to each thematic class.

Class	Reference Map	Multispectral Map	Laser Map
Unaltered granite	30.04%	28.53%	27.33%
Altered granite	42.60%	48.06%	47.27%
Wood	5.35%	5.67%	5.82%
Moisture	1.88%	4.74%	1.31%
Mortar	20.13%	13.00%	18.27%

In a quantitative analysis for the estimation of the two types of granite and wood, results of both sensors are quite similar and really close to the reference map while intensity data from laser scanner are the closest to the reference map results for the estimation of moisture and mortar. Results show higher pixels classified as moisture in the case of multispectral map (2.86% higher with respect to the reference map) and few pixels classified as mortar (7.13% lower than the reference map) due mainly to the altered granite count (whose spectral response has the greatest similarity). Results from the laser sensor are quite similar, greater amount of altered granite by reducing the unaltered granite and mortar detected classes. Note that the best results for moisture detection are achieved with the FARO Focus3D, since humidity has a major interference with this wavelength [40]. Since the pathological classes (moisture and altered granite) are better recognized by the laser scanner and it is the most comprehensive sensor with results closer to the reference, it can be concluded that the active sensor has proven to be the best option to study and detect pathologies and different construction materials for studies with high variability in light conditions where passive sensors are greatly affected.

To evaluate the separability between classes the transformed divergence indicator [41], ranging from 0 to 2, was used as the most widely used quantitative estimator for this purpose [42]. Table 3 shows the separability between the final 5 classes.

Table 3. Transformed divergence for the supervised classification.

	Unaltered Granite	Altered Granite	Wood	Moisture
Altered granite	1.87	-	-	-
Wood	2.00	2.00	-	-
Moisture	1.99	1.99	2.00	-
Mortar	1.98	1.42	2.00	2.00

In general, a high separability was achieved for all 5 classes, highlighting the good separability between the spectral signatures of the two granite types. The worst results were for the mortar and the altered granite classes. This fact is explained by two reasons: on the one hand, the façade sample distance (FSD) of the orthoimages (11 mm in the worst case) was not enough to detect façade areas with smaller thickness of mortar; and on the other, altered granite class presented the closest spectral behavior regarding mortar. With respect to the moisture of the façade, it appeared in lower areas of the shrine (capillarity rising damp) and in the buttress, acting as a filter system for the water from the roof (filtration moisture). These areas are built with unaltered granite blocks since lower areas need to support the loads of the whole building (also in buttress). The radiometric misunderstanding between moisture and unaltered granite did not occur in the case of altered granite since the latter is part of the center of the façade, a low humidity area.

3.3. Accuracy Assessment

In order to assess the accuracy of the unsupervised classifications, the supervised classification approach based on maximum likelihood algorithm served as reference. Five classes and the seven bands available were considered in the classification process. Accuracy results for the case of the Mini MCA6 multispectral camera and the FARO Focus3D laser scanner were 66.04% and 74.39% respectively as mentioned above, and according to the Cohen's Kappa coefficient [43] the level of agreement was 0.50 and 0.621 respectively (excluding the null class).

Furthermore, as mentioned in Section 2.1, an ASD FieldSpec3 spectroradiometer was used to measure several samples of granite for a parallel study. Those measures, in this study, have been used as reference and as a complement to the above analysis to compare the spectral signatures of these construction materials with the discrete reflectance results obtained from the Mini MCA6 and the FARO Focus3D (Figure 11).

The spectral signatures and deviations of the two types of granite present in the façade are plotted for the wavelength range covered by both sensors (530–905 nm).

In Figure 11, the graph continuous lines show at any wavelength the mean value of the reflectances of unaltered and altered granite samples distributed along the façade and measured with the spectroradiometer (a total of 6 and 7 samples of granite, respectively). On the other hand, the colored areas represent the standard deviation of that spectroradiometer measurements. Regarding the discrete values of reflectance achieved with the sensors (discrete points) they result from the mean reflectance value of the "unaltered granite" and "altered granite" classes for each sensor's wavelength of the supervised classification map. The "mortar" class was not finally evaluated due to its variability in thickness along the façade and due to the fact that the FSD achieved was in many areas greater than its thickness.

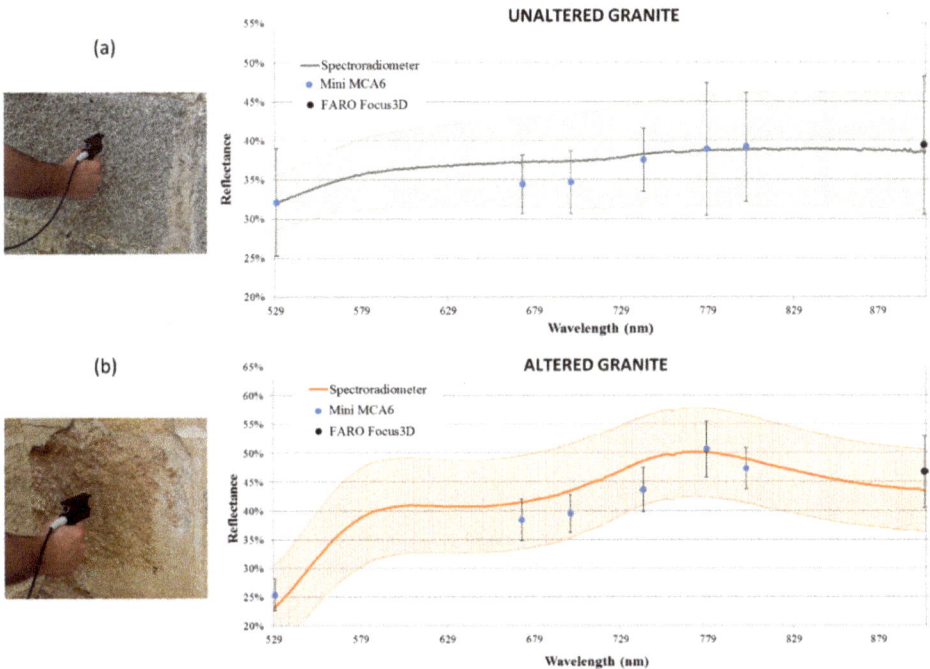

Figure 11. Spectral signatures of the two different types of granites, (a) unaltered and (b) altered, measured with the ASD spectroradiometer for the wavelength interval covered by the sensors used (Mini MCA6 and FARO Focus3D) where points are obtained from the orthoimages in reflectance values.

It should be mentioned that a great fit of the reflectance values from both sensors (discrete points) was achieved for both granite real spectral behaviors (spectroradiometer measurements) with admissible standard deviations associated (lower than those associated with spectroradiometer measurements). For both evaluated materials, the mean error was 0.007 (in the range 0–1), being the maximum 0.049 (in the range 0–1), which is better than the expected error for this vicarious calibration technique (around 5%).

The confusion matrices for the assessment of both sensors are shown in Tables 4 and 5 where the main diagonal indicates the percentage of pixels that have been correctly classify and the off-diagonal values represent misclassification. The producer and user accuracies as well as the overall accuravy are given. Regarding the moisture class, a significant performance improvement of the classifier is observed for this class for the operating wavelength of the FARO Focus3D. In the case of the mortar class, the Mini MCA6 do not bring good results mainly due to the errors produced during the 6-band registration process. Finally, we mention that in the case of the unaltered and altered granites, little variations were observed between both sensors.

Table 4. Confusion matrix of the Mini MCA6 unsupervised classification.

	Moisture	Mortar	Altered Granite	Unaltered Granite	Wood	User Accuracy
Moisture	40.14%	25.40%	0.54%	33.92%	0.01%	40.14%
Mortar	0.06%	39.11%	59.53%	1.31%	0.00%	39.11%
Altered granite	3.79%	7.69%	73.86%	14.61%	0.05%	73.86%
Unaltered granite	6.56%	5.30%	16.39%	70.75%	1.01%	70.75%
Wood	0.00%	0.00%	0.00%	0.00%	100.00%	100.00%
Producer accuracy	21.10%	57.71%	66.00%	74.28%	94.26%	
				Overall accuracy:		**66.04%**

Table 5. Confusion matrix of the FARO Focus3D unsupervised classification.

	Moisture	Mortar	Altered Granite	Unaltered granite	Wood	User Accuracy
Moisture	52.46%	1.04%	23.17%	19.81%	3.52%	52.46%
Mortar	0.00%	60.46%	37.36%	2.18%	0.00%	60.46%
Altered granite	0.05%	12.81%	78.29%	8.64%	0.20%	78.29%
Unaltered granite	0.29%	1.49%	21.70%	75.47%	1.04%	75.47%
Wood	0.48%	0.00%	0.00%	0.00%	99.52%	99.52%
Producer accuracy	89.64%	68.32%	68.68%	83.62%	91.81%	
				Overall accuracy:		**74.39%**

To conclude, it should be highlight that the improvement in both the overall accuracy and the Kappa coefficient is significant in the case of working with radiometrically calibrated sensors as opposed to the use uncalibrated ones [17].

The results for the Mini MCA6 have experienced a 24% improvement in terms of overall accuracy and 23% regarding the Kappa coefficient. Furthermore, the improvement from the use of the calibrated FARO Focus3D was of 29% and 35% regarding the overall accuracy and Kappa coefficient, respectively. Results worsen in the case of the Mini MCA6 due to two factors; the first is that the camera is a passive sensor, so it is sensitive to changes in light conditions and shadow areas during data acquisition. The second error factor is caused by the slave image registration process, as has been mentioned above, due to the errors in the determination of baselines, angular misalignments and the internal parameters of the camera. Any error in those parameters is propagated into the final multispectral orthoimage, being worsened for higher spatial resolutions, where the geometric pixel footprint in the object may differ depending on the wavelength.

4. Conclusions

The work presented in this paper shows a comparison of the classification results from the use of different radiometrically calibrated sensors to detect pathologies in materials of historical buildings façades. By combining the use of two different data acquisition techniques (active and passive), two sensors were examined: a multispectral camera and a 3D laser scanner. The results show the different radiometric responses of the ashlars of a church with different damages levels (mainly moisture). The classification algorithms used for the classification processes were the Fuzzy K-means and the maximum likelihood classification algorithms.

A complete description of the workflow followed is outlined describing the data acquisition, pre-processing (including sensors radiometric calibrations), orthoimages generation and the application of two classification algorithms to assess the final results. Our results show that the most comprehensive sensor for which the best results were obtained is the FARO Focus3D. This is possibly due to the advantage of working in an active way with no need of external radiation. As a result, classification maps were not affected by different lighting conditions during data acquisition. Furthermore, geometric models of the study object can be derived thanks to its data capture. With these models, physical pathologies (such as fissures, desquamations, *etc.*) could be analyzed and both these damages and chemical pathologies could be quantified. However, for the challenge of the registration of 6 wavelength bands, the results from the Mini MCA6 were quite good. Considering all those issues and with the experience of working with these sensors in previous studies, it is concluded that the radiometric calibration of the sensors is crucial since it contributes to improving the accuracy of the outcomes (a 35% Kappa coefficient improvement in the case of the FARO Focus3D). Thus, a sensor combination with laser scanning as a primary choice is the best solution for pathology detection and quantification. By adding the intensity information to visible or multispectral information, results of classification improve in a quantitative and a qualitative way.

In future work, the use of a hyperspectral camera or another laser scanner operating in the shortwave infrared as a complement of the sensors proposed will improve the pathologies detection and the overall accuracy results since the spectral resolution of the study would be increased. In addition, and for non-carved historical buildings, the roughness of the façade would be calculated from the scan points in order to have additional data of the materials so it can help in the discrimination process. Finally, and regarding the data acquisition of passive sensors, constant favorable climatic conditions will be planned so that the accuracy of its classification results may be significantly improved.

Acknowledgments: This work has been partially supported by the Spanish Ministry of Economy and Competitiveness through the ENERBIUS project "Integrated system for the energy optimization of buildings" (ENE2013-48015-C3-3-R).

Author Contributions: All authors conceived, designed and performed the experimental campaign. Susana del Pozo and Pablo Rodriguez-Gonzálvez implemented the methodology and analyzed the results. Susana del Pozo wrote the manuscript and all authors read and approved the final version.

Conflicts of Interest: The authors declare no conflict of interest.

References

1. Marszalek, M. Deterioration of stone in some monuments exposed to air pollution: A Cracow case study. In *Air Pollution and Cultural heritage*; Taylor and Francis: London, UK, 2004; pp. 151–154.

2. Corvo, F.; Reyes, J.; Valdes, C.; Villaseñor, F.; Cuesta, O.; Aguilar, D.; Quintana, P. Influence of air pollution and humidity on limestone materials degradation in historical buildings located in cities under tropical coastal climates. *Water Air Soil Pollut.* **2010**, *205*, 359–375.

3. Fort, R.; de Azcona, M.L.; Mingarro, F. Assessment of protective treatment based on their chromatic evolution: Limestone and granite in the Royal Palace of Madrid, Spain. In *Protection and Conservation of the Cultural Heritage in the Mediterranean Cities*; Galan, E., Zezza, F., Eds.; CRC Press/Balkema: Sevilla, Spain, 2002; pp. 437–441.

4. Weritz, F.; Kruschwitz, S.; Maierhofer, C.; Wendrich, A. Assessment of moisture and salt contents in brick masonry with microwave transmission, spectral-induced polarization, and laser-induced breakdown spectroscopy. *Int. J. Archit. Herit.* **2009**, *3*, 126–144.

5. Lambers, K.; Eisenbeiss, H.; Sauerbier, M.; Kupferschmidt, D.; Gaisecker, T.; Sotoodeh, S.; Hanusch, T. Combining photogrammetry and laser scanning for the recording and modelling of the late intermediate period site of pinchango alto, Palpa, Peru. *J. Archaeol. Sci.* **2007**, *34*, 1702–1712.

6. González-Aguilera, D.; Gómez-Lahoz, J.; Sánchez, J. A new approach for structural monitoring of large dams with a three-dimensional laser scanner. *Sensors* **2008**, *8*, 5866–5883.

7. Buckley, S.J.; Howell, J.; Enge, H.; Kurz, T. Terrestrial laser scanning in geology: Data acquisition, processing and accuracy considerations. *J. Geol. Soc.* **2008**, *165*, 625–638.

8. Armesto, J.; Ordóñez, C.; Alejano, L.; Arias, P. Terrestrial laser scanning used to determine the geometry of a granite boulder for stability analysis purposes. *Geomorphology* **2009**, *106*, 271–277.

9. Höfle, B.; Pfeifer, N. Correction of laser scanning intensity data: Data and model-driven approaches. *ISPRS J. Photogramm. Remote Sens.* **2007**, *62*, 415–433.

10. Kaasalainen, S.; Kukko, A.; Lindroos, T.; Litkey, P.; Kaartinen, H.; Hyyppä, J.; Ahokas, E. Brightness measurements and calibration with airborne and terrestrial laser scanners. *IEEE Trans. Geosci. Remote Sens.* **2008**, *46*, 528–534.

11. Franceschi, M.; Teza, G.; Preto, N.; Pesci, A.; Galgaro, A.; Girardi, S. Discrimination between marls and limestones using intensity data from terrestrial laser scanner. *ISPRS J. Photogramm. Remote Sens.* **2009**, *64*, 522–528.

12. Lichti, D.D. Spectral filtering and classification of terrestrial laser scanner point clouds. *Photogramm. Record* **2005**, *20*, 218–240.

13. Mather, P.; Tso, B. *Classification Methods for Remotely Sensed Data*, 2nd ed.; CRC Press: Boca Raton, FL, USA, 2009.

14. Lerma, J.L. Multiband *versus* multispectral supervised classification of architectural images. *Photogramm. Record* **2001**, *17*, 89–101.

15. Lerma, J.L. Automatic plotting of architectural facades with multispectral images. *J. Surv. Eng.* **2005**, *131*, 73–77.

16. Ruiz, L.; Lerma, J.; Gimeno, J. Application of computer vision techniques to support in the restoration of historical buildings. *ISPRS Int. Arch. Photogramm. Remote Sens. Spatl. Inf. Sci.* **2002**, *34*, 227–230.

17. Hemmleb, M.; Weritz, F.; Schiemenz, A.; Grote, A.; Maierhofer, C. Multi-spectral data acquisition and processing techniques for damage detection on building surfaces. In Proceedings of the ISPRS Commission V Symposium "Image Engineering and Vision Metrology", Dresden, Germany, 25–27 September 2006.

18. Del Pozo, S.; Herrero-Pascual, J.; Felipe-García, B.; Hernández-López, D.; Rodríguez-Gonzálvez, P.; González-Aguilera, D. Multi-sensor radiometric study to detect pathologies in historical buildings. *ISPRS Int. Arch. Photogramm. Remote Sens. Spat. Inf. Sci.* **2015**, *XL-5/W4*, 193–200.

19. López, D.H.; García, B.F.; Piqueras, J.G.; Alcázar, G.V. An approach to the radiometric aerotriangulation of photogrammetric images. *ISPRS J. Photogramm. Remote Sens.* **2011**, *66*, 883–893.

20. Del Pozo, S.; Rodríguez-Gonzálvez, P.; Hernández-López, D.; Felipe-García, B. Vicarious radiometric calibration of a multispectral camera on board an unmanned aerial system. *Remote Sens.* **2014**, *6*, 1918–1937.

21. Kaasalainen, S.; Ahokas, E.; Hyyppä, J.; Suomalainen, J. Study of surface brightness from backscattered laser intensity: Calibration of laser data. *IEEE Geosci. Remote Sens. Lett.* **2005**, *2*, 255–259.

22. Soudarissanane, S.; Lindenbergh, R.; Menenti, M.; Teunissen, P. Scanning geometry: Influencing factor on the quality of terrestrial laser scanning points. *ISPRS J. Photogramm. Remote Sens.* **2011**, *66*, 389–399.

23. Al-amri, S.S.; Kalyankar, N.V.; Khamitkar, S.D. A comparative study of removal noise from remote sensing image. *Int. J. Comput. Sci. Issue* **2010**, *7*, 32–36.

24. Zheng, Y.; Lin, S.; Kambhamettu, C.; Yu, J.; Kang, S.B. Single-image vignetting correction. *IEEE Trans. Pattern Anal. Mach. Intell.* **2009**, *31*, 2243–2256.

25. Reshetyuk, Y. *Self-Calibration and Direct Georeferencing in Terrestrial Laser Scanning. Doctoral Thesis in Infrastructure, Geodesy*; Royal Institute of Technology (KTH): Stockholm, Suecia, 2009.

26. Honkavaara, E.; Arbiol, R.; Markelin, L.; Martinez, L.; Cramer, M.; Bovet, S.; Chandelier, L.; Ilves, R.; Klonus, S.; Marshal, P. Digital airborne photogrammetry—A new tool for quantitative remote sensing? A state-of-the-art review on radiometric aspects of digital photogrammetric images. *Remote Sens.* **2009**, *1*, 577–605.

27. Biggar, S.; Slater, P.; Gellman, D. Uncertainties in the in-flight calibration of sensors with reference to measured ground sites in the 0.4–1.1 µm range. *Remote Sens. Environ.* **1994**, *48*, 245–252.

28. Dinguirard, M.; Slater, P.N. Calibration of space-multispectral imaging sensors: A review. *Remote Sens. Environ.* **1999**, *68*, 194–205.

29. Kaasalainen, S.; Krooks, A.; Kukko, A.; Kaartinen, H. Radiometric calibration of terrestrial laser scanners with external reference targets. *Remote Sens.* **2009**, *1*, 144–158.

30. Jelalian, A.V. *Laser Radar Systems*; Artech House: New York, NY, USA, 1992.

31. Bourke, P. An algorithm for interpolating irregularly-spaced data with applications in terrain modelling. In Proceedingsof the Pan Pacific Computer Conference, Beijing, China, 1 January 1989.

32. Attene, M. A lightweight approach to repairing digitized polygon meshes. *Vis. Comput.* **2010**, *26*, 1393–1406.

33. Kraus, K. *Photogrammetry: Geometry from Images and Laser Scans*, 2nd ed.; Walter de Gruyter: Berlin, Germany, 2007.

34. Albertz, J.; Kreiling, W. *Photogrammetrisches Taschenbuch*; Herbert Wichmann Verlag: Berlin, Germany, 1989.

35. Bezdek, J.C. *Pattern Recognition with Fuzzy Objective Function Algorithms*; Plenum Press: New York, NY, USA, 1981.

36. Kannan, S.; Sathya, A.; Ramathilagam, S.; Pandiyarajan, R. New robust fuzzy C-Means based gaussian function in classifying brain tissue regions. In *Contemporary Computing*; Ranka, S., Aluru, S., Buyya, R., Chung, Y.-C., Dua, C., Grama, A., Gupta, S.K.S., Kumar, R., Phoha, V.V., Eds.; Springer: Berlin, Germany, 2009; pp. 158–169.

37. Richards, J.A. *Remote Sensing Digital Image Analysis*; Springer: Berlin, Germany, 1999; Volume 3.

38. Lillesand, T.; Kiefer, R.W.; Chipman, J. *Remote Sensing and Image Interpretation*, 7th ed.; John Wiley & Sons: Hoboken, NJ, USA, 2015.

39. García, F.A.F. *La Invención de la Iglesia de san Segundo. Cofrades y Frailes Abulenses en los Siglos xvi y xvii*; Institución Gran Duque de Alba: Ávila, Spain, 2006.

40. Rantanen, J.; Antikainen, O.; Mannermaa, J.-P.; Yliruusi, J. Use of the near-infrared reflectance method for measurement of moisture content during granulation. *Pharm. Dev. Technol.* **2000**, *5*, 209–217.

41. Davis, S.M.; Landgrebe, D.A.; Phillips, T.L.; Swain, P.H.; Hoffer, R.M.; Lindenlaub, J.C.; Silva, L.F. *Remote Sensing: The Quantitative Approach*; McGraw-Hill International Book Co.: New York, NY, USA, 1978.

42. Tolpekin, V.; Stein, A. Quantification of the effects of land-cover-class spectral separability on the accuracy of markov-random-field-based superresolution mapping. *IEEE Trans. Geosci. Remote Sens.* **2009**, *47*, 3283–3297.

43. Cohen, J. Weighted kappa: Nominal scale agreement provision for scaled disagreement or partial credit. *Psychol. Bull.* **1968**, *70*, 213–220.

Large Scale Automatic Analysis and Classification of Roof Surfaces for the Installation of Solar Panels Using a Multi-Sensor Aerial Platform

Luis López-Fernández, Susana Lagüela, Inmaculada Picón and
Diego González-Aguilera

Abstract: A low-cost multi-sensor aerial platform, aerial trike, equipped with visible and thermographic sensors is used for the acquisition of all the data needed for the automatic analysis and classification of roof surfaces regarding their suitability to harbor solar panels. The geometry of a georeferenced 3D point cloud generated from visible images using photogrammetric and computer vision algorithms, and the temperatures measured on thermographic images are decisive to evaluate the areas, tilts, orientations and the existence of obstacles to locate the optimal zones inside each roof surface for the installation of solar panels. This information is complemented with the estimation of the solar irradiation received by each surface. This way, large areas may be efficiently analyzed obtaining as final result the optimal locations for the placement of solar panels as well as the information necessary (location, orientation, tilt, area and solar irradiation) to estimate the productivity of a solar panel from its technical characteristics.

Reprinted from *Remote Sens.* Cite as: López-Fernández, L.; Lagüela, S.; Picón, I.; González-Aguilera, D. Large Scale Automatic Analysis and Classification of Roof Surfaces for the Installation of Solar Panels Using a Multi-Sensor Aerial Platform. *Remote Sens.* **2015**, *7*, 11226–11246.

1. Introduction

Several techniques have been applied so far for the calculation of the solar incidence on roofs trying to find the most suitable roof surfaces for the installation of solar panels with optimal performance without assessing the possible obstacles that can prevent their installation. In most cases the analysis is done from Geographical Information Systems (GIS) fed with low-resolution raster or vector cadastral data sources [1–4], GML (Geography Markup Language) systems with simplified LOD (Level Of Detail) [5] or topologically consistent 3D city models obtained through the extrusion of building footprints to a stimated average building height [6]. These techniques, that generally use simplified models and approximations of the positions and orientations of the roofs, could imply errors in the calculation of the solar incidence on the surfaces of the roofs and thus cause large variations in their productivity [7]. In order to avoid these limitations, we developed a methodology able to identify the optimal location for the installation of solar panels [8] and the estimation of their solar irradiation from a precise dataset obtained with a low cost aerial platform equipped with RGB and thermographic cameras. The data processing methodology is scalable to any other aerial data sources like LiDAR (Light Detection and Ranging), acquired from piloted aerial vehicles or UAV's (Unmanned Aerial Vehicles). The use of geo-referenced 3D dense point clouds of roofs will allow the performance of accurate analysis of areas, orientations and tilts of the roofs. Furthermore, if these data are complemented with qualitative thermal information, which provides direct knowledge about the relative temperature difference between roofs and consequently about the influence of the solar radiation on them, the removal of surfaces that could present ideal geometric conditions to install solar panels will be possible. The thermal values recorded by the thermographic camera cannot be considered as the rigorous temperature values of the object due to the difficulty to perform an accurate emissivity correction on the measurement. However, these values provide useful information for the detection of differences in solar incidence over the object surface. For instance, those surfaces containing elements that could prevent the installation of solar panels such as skylights or chimneys will be automatically detected.

This work proposes and tests a methodology useful for the performance of the automatic classification of roofs regarding their suitability to harbor solar panels, the detection of the optimal location inside these surfaces and the estimation of the solar irradiation received. The methodology consists on the processing of visible and thermographic images acquired from a low-cost aerial trike equipped with a multi-sensor platform (MUSAS, which stands for MUltiSpectral Airborne Sensors) towards the generation of 3D point clouds of roofs.

The paper has been structured as follows: after this introduction, Section 2 includes a detailed explanation of the materials and methods used for data acquisition and processing towards the automatic segmentation and classification of the roof surfaces for the installation of solar panels. Section 3 is devoted to be an explanation of the methodology presented through its application to an urban area selected as case study; finally, Section 4 establishes the most relevant conclusions of the proposed approach. The procedure is presented through its application to a real case study, based on the classification of the roofs in a neighborhood in Ávila, located near the center of Spain (coordinates 40°38'27.6"N, 4°41'27.8"W). The location of the case study determines the restrictions applied for the installation of solar panels following the Spanish Regulation for Construction [9].

2. Materials and Methods

2.1. Equipment

2.1.1. Aerial Trike

The aerial trike is a driven low-cost aerial system equipped with a MUSAS platform able to accommodate different types of remote sensors, from imaging sensors (RGB and thermographic cameras) to navigation systems (GNSS/IMU). The use of these kind of aerial vehicles shall be regulated by the local legislation on air navigation as long as it does not conflict with the general air navigation law of the DGAC (General Guidance of Civil Aviation). The aerial trike is an experimental "Tandem Trike AIRGES" (Table 1), with a weight capacity up to 220 kg and flying altitudes up to 300 m, which is the limit established by the current Spanish legislation for this type of aerial vehicles [10]. The main advantages of this platform over the usual driven aerial vehicles are its low cost, ease of use and the possibility of flying below 300 m, consequently obtaining better GSD (Ground Sample Distance) with the imaging sensors. This increase in the spatial resolution of the RGB images regarding usual aerial photogrammetric flights allows the generation of dense point clouds, as well as improves the use of thermographic images, given that the thermographic sensor presents very limited resolution implying a high GSD that, for altitudes over 300 m could prevent the differentiation of characteristic elements of roofs. Regarding UAV platforms, the aerial trike is able to transport greater weights, which implies the possibility of using more and higher quality sensors with longer autonomy than copter-type platforms. Thus, this aerial platform is used to obtain good quality georeferenced images from a zenithal point of view, allowing for a better spatial perception given the absence of obstacles between the camera and the object.

Sensors are supported by a specific gyro-stabilized platform (MUSAS) (Figure 1) allowing for a full coverage of the study area with an appropriate GSD. This device includes two servomotors arranged on the x and y axes to maintain the vertical position of the camera along the flight path with precision. The servomotors are controlled by an Arduino board, which incorporates an IMU with 6 degrees of freedom: 3 accelerometers with a range of ± 3.6 G m/s^2, a double-shaft gyroscope (for pitch and roll control) and an additional gyroscope for yaw control (both gyroscopes have a measurement range of $\pm 300°$/s). The navigation devices allow the geolocation of each datum for the successive generation of geo-referenced point clouds with real and thermographic texture.

Last, the software developed for the control of the device was based on Quad1_mini V 20 software, with DCM (Direction Cosine Matrix) as the management algorithm for the IMU [11].

Table 1. Technical specifications of the manned aerial platform, aerial trike.

Parameter	Value
Empty weight	110 kg
Maximum load	220 kg
Autonomy	3.5 h
Maximum speed	60 km/h
Motor	Rotax 503
Tandem paraglide	MACPARA Pasha 4
Emergency system	Ballistic parachutes GRS 350
Gimbal	Stabilized with 2 degrees of freedom (MUSAS)
Minimum sink rate	1.10
Maximum glide rate	8.60
Plant surface	42.23 m^2
Projected area	37.80 m^2
Wingspan	15.03 m
Plant elongation rate	5.35
Central string	3.51 m
Boxes	54 boxes
Zoom factor	100%

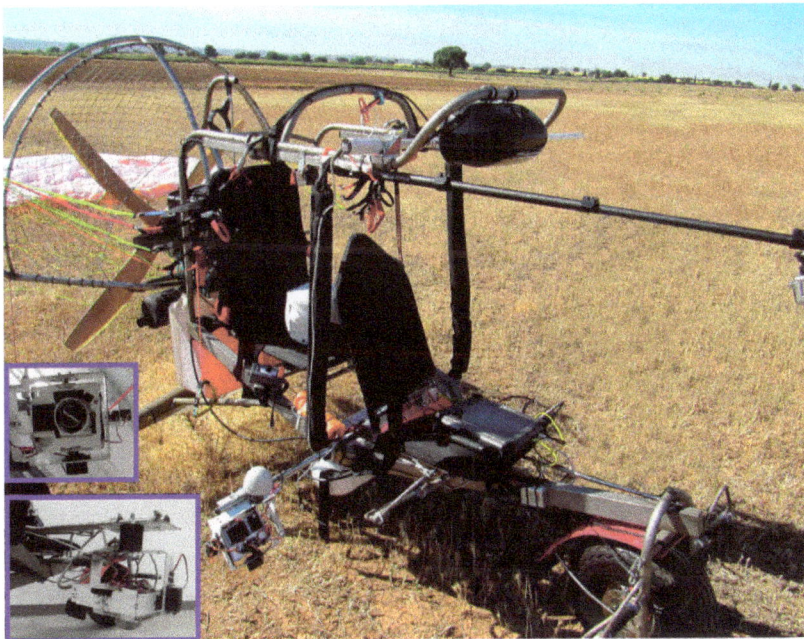

Figure 1. Aerial trike and MUSAS platform used in this study; bottom-left corner: detail of the gyro-stabilized platform for the installation of sensors.

2.1.2. RGB Camera

RGB cameras are used for the acquisition of images towards the reconstruction of 3D point clouds as well as to provide visual information of the state of the roofs. The visible camera selected for this work is a full frame reflex camera Canon 5D mkII. This camera has a CMOS sensor which size is 36 × 24 mm with a resolution of 21.1 megapixel and equipped with a 50 mm focal length lens. The size of the image captured with this sensor is 21 MP with a pixel size of 6.4 µm.

The camera is calibrated prior acquisition in order to allow the correction of the distortion and perspective effects from the data collected and the 3D reconstruction in the photogrammetric process. The calibration of cameras working in the visible band of the electromagnetic spectrum (Table 2) is performed through the acquisition of multiple convergent images of a geometric pattern (known as calibration grid) with different orientations of the camera. The adjustment of the rays ruling the position of the camera and the image in each acquisition allows the determination of the inner orientation parameters of the camera (focal length, format size, principal point, radial lens distortion and decentering lens distortion). The camera calibration was processed in the commercial software ImageMaster which performs the automatic detection of the targets in each image and computes and adjusts the orientation of each image, resulting in the computation of the internal calibration parameters of the camera. Since the flight was performed at medium speed and low altitude, there should not be any change in the "camera-lens" system caused by sudden movements of the platform or major changes in the atmosphere. For this reason, the camera calibrations performed in the laboratory have been considered as fixed for both cameras.

Table 2. Interior orientation parameters of visible camera Canon 5D mkII, result of its geometric calibration.

Parameter	–	Value
Focal length (mm)	Value	50.1
Format size (mm)	Value	34.819 × 23.213
Principal point	X value	−0.21
displacement (mm)	Y value	−0.11
Radial lens distortion	K_1 value (mm^{-2})	6.035546×10^{-5}
	K_2 value (mm^{-4})	-1.266639×10^{-8}
Decentering lens distortion	P_1 value (mm^{-1})	1.585075×10^{-5}
	P_2 value (mm^{-1})	6.541129×10^{-5}
Point marking residuals	Overall RMSE (pixels)	0.244

2.1.3. Thermographic Camera

The thermographic camera selected for this study is the FLIR SC655. This device has been specially developed for scientific applications. It allows the capture and recording of thermal variations (Figure 2) in real time, enabling the measurement of heat dissipation or leakage. Its sensor is an Uncooled Focal Plane Array (UFPA) 0.3 MP, capturing radiation with 7.5 to 13.0 μm wavelengths and measuring temperatures in a range from −20 °C to 60 °C. The Instant Field of View (IFOV) of the camera is 0.69 mrad, and its Field of View (FOV) is 25° (Horizontal) and 18.8° (Vertical) with the current lens of 25 mm focal length.

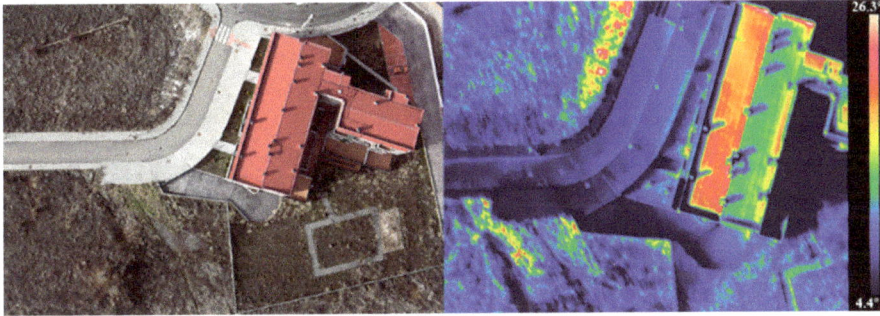

Figure 2. Comparison between the products generated by the different used imaging sensors. (**Left**) RGB image. (**Right**) Thermographic image with the thermal values represented using a color map.

Thermographic cameras capture radiation in the infrared range of the spectrum, in contrast to photographic cameras that work in the visible range. For this reason, the geometric calibration (Table 3) of the camera is performed using a specially designed calibration field, presented in [12], which is based on the capability of the thermographic cameras for the detection of objects being at different temperatures even if they are at the same temperature but with different emissivity values. This calibration field consists on a wooden plate with black background (high emissivity) on which foil targets are placed (low emissivity). In this case, the calibration was processed in the commercial photogrammetric station Photomodeler Pro5©, which not only computes the calibration parameters of the cameras in a procedure analogous to the one previously explained for ImageMaster, but it also provides the standard deviation of the calibration parameters as a value to test their quality.

Table 3. Interior orientation parameters of thermographic camera FLIR SC655, result of its geometric calibration.

Parameter	–	Value	Std. Deviation
Focal length (mm)	Value	25.063	0.022
Format size (mm)	Value	10.874 × 8.160	0.002
Principal point	X value	−0.174	0.022
displacement (mm)	Y value	0.024	0.026
Radial lens distortion	K_1 value (mm^{-2})	5.281×10^{-5}	2.4×10^{-5}
	K_2 value (mm^{-4})	7.798×10^{-7}	6.1×10^{-7}
Decentering	P_1 value (mm^{-1})	1.023×10^{-4}	1.2×10^{-5}
lens distortion	P_2 value (mm^{-1})	-3.401×10^{-5}	1.3×10^{-5}
Point marking residuals (pixels)	Overall RMSE	0.173	–

2.2. Methodology

2.2.1. Flight Planning and Data Acquisition

Proper flight planning is important to optimize available resources, ensuring a high quality of the images and minimizing capture time. The spatial information required for the flight planning can be obtained free of charge from the National Center of Geographic information in Spain (CNIG), from its National Aerial Orthoimage Plan (PNOA, 2009) with a GSD of 0.25 m and a Digital Terrain Model (DTM) with a 5 m grid resolution. The flight planning was carried out based on the classical aerial photogrammetric principles [13] but adapted to the new algorithms and structure from motion (SfM) strategies [14], ensuring image acquisition with forward and side overlaps of 70% and 30%, respectively. Given the format difference between the thermographic and RGB sensors, time between shots will be considered independently in order to ensure these overlaps.

The gyro-stabilized platform ensures the theoretical geometric setup of a photogrammetric aerial image capture in each shot, which stablishes that the optical axis of the camera should be zenithal. The theoretical definition of scale in digital aerial photogrammetry is related to the geometric resolution of the pixel size projected over the ground (GSD). This parameter can be calculated by considering the relationship between flight altitude over the ground, the GSD, the focal length of the sensor and the pixel size (Equation (1)). Considering that the objective of this study is not to identify small entities, and in order to allow the procedure to be scalable to sparse 3D point clouds from other measurement systems, we fixed the target GSD for the lower resolution sensor between 10 and 15 cm.

$$\frac{f}{H} = \frac{pixel\ size}{GSD} \tag{1}$$

where f is the focal length of the sensor; H is the flight altitude over the ground and GSD is the Ground Sample Distance.

Due to the lower resolution of the thermographic images, the flight planning is performed according to the characteristics of this sensor (Figure 3), considering the planning completely valid also for the RGB sensor.

In order to locate the solar incidence deficiencies over the objects under study in the higher production time zone of the solar panels, the survey should be performed when the Sun is located at the highest point of the solar path.

2.2.2. 3D Point Cloud Reconstruction

The image-based modelling technique based on the combination of photogrammetry and computer vision algorithms allows the reconstruction of dense 3D point clouds. The absolute orientation (position and attitude) of each image is known because the position of the imaging sensors is registered with respect to the GNSS/IMU navigation devices of the aerial trike, and data acquisition is synchronized with the navigation. For this reason, using these absolute orientation as initial approximations, only an orientation refinement was required for the precise geolocation of the images.

The orientation refinement process starts with the automatic extraction and matching of image features through a SIFT [15] (Scale-Invariant Feature transform) algorithm which provides effectiveness against other feature detection algorithms. In addition, these features present optimal results for these type of aerial surveys where scale variations are minimal and perspective effects are almost nonexistent thanks to the gyro-stabilized platform. Next, taking as initial approximation the external orientation parameters provided by the GNSS system and as fixed the laboratory internal calibration parameters, an orientation refinement was performed through a global bundle adjustment between all images by means of the collinearity equations [16,17]. As a result, the spatial and angular geo-positioning of the RGB sensor was computed enabling the dense point cloud generation. Next, a dense matching method through the MicMac implementation [18] based on the semi global matching technique (SGM) [19,20] allows the generation of a dense and scaled 3D point cloud resulting from the determination of the 3D coordinates of each pixel. This process was performed using the RGB images because their higher resolution provides a 3D point cloud with higher point density than the point cloud generated from thermographic images. Regarding computation effort, the last step is the most expensive and time-consuming. All these photogrammetric tasks were performed using the Photogrammetry Workbench (PW) [21].

Last but not least, a dense 3D point cloud with thermographic information was obtained for the case study. This thermographic mapping was obtained thanks to the known baseline between both sensors which was previously calibrated in laboratory. In particular, this calibration consisted in solving the relative orientation of the thermographic camera regarding the RGB camera through the simultaneous visualization of a common pattern (the same used for the thermographic camera calibration). After the identification of homologous entities between both sensors, the calculation of the baseline (*bx, by, bz*) and the boresight (*Rx, Ry, Rz*) values was performed.

2.2.3. Automatic Planes Segmentation

Once the 3D point cloud is generated from the RGB images and the thermographic texture is applied, the following procedure is the segmentation of the roofs. This is performed in different steps using the Point Cloud Library (PCL) [22], open source and licensed under BSD (Berkeley Software Distribution) terms, which includes a collection of state-of-the-art algorithms and tools useful for 3D processing, computer vision and robotic perception. First, ground and vegetation are removed using a pass through filter with a *Z* coordinate restriction. A pass through filter performs a simple filtering along a specified dimension removing the elements that are either inside or outside a given range. In this case, the filter gets the minimum *Z* value of the point cloud and removes all points with a *Z* value close to the minimum Z. The distance threshold is established as a parameter set for the user, being 5 m a recommendable value, established experimentally by the authors after the segmentation of several point clouds. In addition, although the presence of points belonging to the facades is minimal due to the vertical configuration of the capture, a conditional filter based on the angle between the normal vector of each point and the vertical axis is applied to remove these points. Then, the point cloud including the roofs is segmented using a Euclidean cluster segmentation, providing better and faster results for the subsequent extraction of the different planes of each roof [23,24] by dividing the point cloud in individual roofs. This way, RANSAC (Random Sample Consensus) algorithm [25] is applied to each roof individually for the extraction of the composing planes.

Once each roof is clustered in different planes, and the coefficients that describe each surface in a Cartesian coordinate system by the general equation of the plane are known (Equation (2)), the geometric evaluation is performed, resulting in the orientation and tilt values for each surface from which we can calculate the percentage of solar energy productivity regarding the maximum productivity of the installed solar panels.

$$Ax + By + Cz = D \qquad (2)$$

where *A*, *B* and *C* are the components of the vector normal to the plane, whereas *D* is the independent term.

From the components A, B and C of the general equation of the plane that match with the values of the unit vector normal to the plane, denoted as $\vec{v}$ (Equation (3)) we can proceed to compute the orientation and tilt values for each roof. The orientation of the surface requires the computation of the angle between the projection of vector $\vec{v}$ on the horizontal plane and the cartesian Y-axis (Equation (4)). The quadrant to which the angle belongs, identified through the evaluation of the sign of A and B components of the unit vector (Figure 3 left), will allow us to get the orientation angle of the surface (Figure 3 right).

$$\vec{v} = (A, B, C) \tag{3}$$

$$\beta = arcTan\left(\frac{A}{B}\right) \tag{4}$$

where $\vec{v}$ is the vector normal to the plane; β is the angle between the horizontal projection of the normal vector and the cartesian Y-axis.

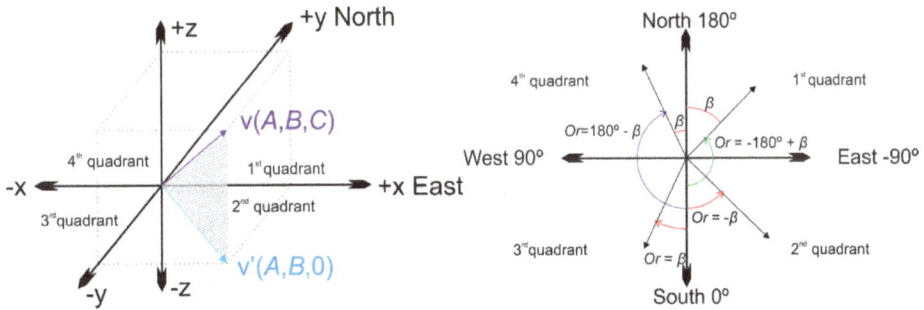

Figure 3. (Left) Quadrant determination from the components of the vector normal to the plane. (Right) Absolute orientation of the roof, denoted as "Or", from the angle between the projection of the vector on the horizontal plane and the cartesian Y-axis.

The tilt of the surface is calculated by a simple trigonometric process in a right-angled triangle where the normal vector is considered the hypotenuse and its projection on the horizontal plane ($Z = 0$) and C value the adjacent and opposite legs respectively (Figure 4) (Equations (5–7)).

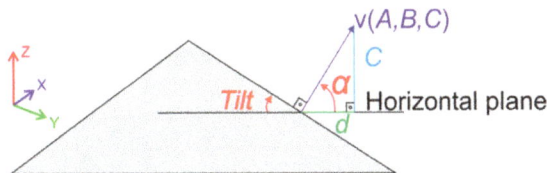

Figure 4. Tilt angle from the values of the normal vector of the plane.

$$d = \sqrt{A^2 + B^2} \tag{5}$$

$$\alpha = arcTan\left(\frac{C}{d}\right) \tag{6}$$

$$Tilt = 90° - \alpha \tag{7}$$

where d is the module of the vector $\vec{v}$ projected on the horizontal plane ($Z = 0$); α is the angle between the horizontal plane and the normal vector of the surface; *Tilt* represents the tilt of the plane.

2.2.4. Geometric Analysis and Classification

Once the different planes of the roofs are detected and segmented, their areas, tilts and orientations are analyzed in order to perform their geometric classification. The roofs with an area smaller than required for the installation of solar panels and those with North orientation are discarded due to their inadequacy. The remaining roofs are classified in different groups according to their theoretical productivity as a function of their tilt and orientation, also taking into account the possibilities of either integrating the solar panels in the roof or installing them in configurable supports. The CTE (Technical Edification Code) sets the South orientation as the optimal position for these elements, with a tilt equal to the latitude where they are installed. However, the limits on the tilt can be computed according to the minimum efficiency allowed for the orientation of the surface using the method explained by the IDAE (Institute for Diversification and Saving of Energy) (Equations (8) and (9)) [9].
If $Tilt > 15°$

$$Solar\ energy\ losses\ (\%) = 100 \cdot \left[1.2 \cdot 10^{-4} \cdot (Tilt - \varphi + 10)^2 + 3.5 \cdot 10^{-5} \cdot Or^2\right] \tag{8}$$

If $Tilt < 15°$

$$Solar\ energy\ losses\ (\%) = 100 \cdot \left[1.2 \cdot 10^{-4} \cdot (Tilt - \varphi + 10)^2\right] \tag{9}$$

where Or is the Orientation of the plane and φ its latitude.

These limits allow the evaluation of the suitability of the surface under study and the computation of the ideal geometry of the support platforms if needed, taking into account that productivity losses cannot exceed the 20% if solar panels are installed directly over the roofs, consequently keeping the angle of solar incidence of the roof. However, productivity losses should be under 10% using the general method which implies the installation of solar panels in supports to modify the angle of solar incidence regarding the tilt of the roof.

2.2.5. Thermographic Refinement of Surfaces

Once the geometric classification is performed, the thermographic data enables the location of elements that constitute an obstacle for the installation of solar panels in order to avoid protrusions or shadows that could reduce their productivity, produced by both the elements of the roof surface and by nearby buildings. This identification of obstacles allows the analysis to find the optimal location within each roof. The existence of obstacles and anomalies in the roofs involves the presence of different materials and surfaces with different emissivity values, so the temperature detected by the thermographic images will be different even if the solar radiation received was the same. In addition, the existence of obstacles that prevent direct sunlight will be manifested by changes in the temperatures of the roof surfaces. These facts allow the location of those surfaces of the roofs affected by anomalies through the performance of a statistical study of the mean and the standard deviation of temperatures. A point will be considered as an obstacle when the difference between its temperature value and the mean temperature of the roof is higher than the standard deviation of the temperatures of the surface that is being analyzed. Considering the fact that the shadow produced by an obstacle will move following the path of the Sun, a perimeter defined by the user around each obstacle will be considered for further analysis. Although a self-occlusion analysis would be appropriate for the determination of the perimeter, the performance of this processing for each obstacle in large study areas would be a high computationally demanding task. Experimental analysis performed by the authors establishes 0.5 m as a recommendable value.

2.2.6. Location of the Most Suitable Zones

The existence of obstacles, detected by the thermographic refinement step, prevents the installation of solar panels covering the whole surface of those roofs classified by the geometric analysis as having the optimal geometry (orientation and tilt) for this purpose. Therefore, the development of a procedure to locate the optimal zones for panel installation inside optimal surfaces (Figure 5) is necessary. This task is addressed by analyzing the spatial distribution of those 3D points belonging to the optimal surfaces and not considered as obstacles. This approach is focused on four steps individually applied to each surface. (i) Projection of the point cloud to the horizontal plane in order to evaluate the surfaces in a 2D environment, simplifying the process; (ii) Extraction of those points that describe the perimeter of the surface to evaluate (concave hull); (iii) Computation of the largest empty rectangles (no concave hull points inside); (iv) Re-projection of the largest empty rectangles to 3D. The concave hull can be defined as a set of points that enclose a concave region and define the perimeter of an unorganized set of points allowing any angle between consecutive edges. The area of the defined shape should be minimized without distorting the appearance of the point cloud

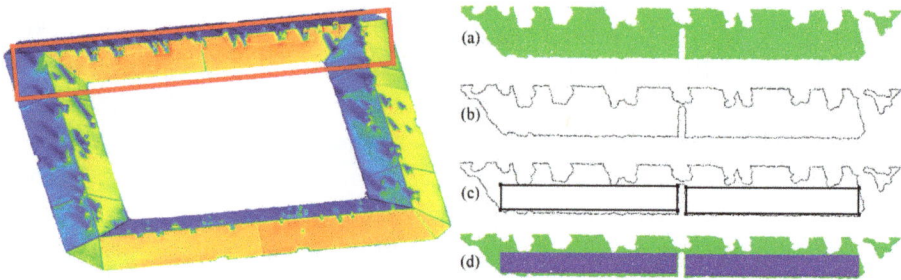

Figure 5. Procedure to locate optimal zones inside optimal surfaces. (**Left**) 3D surface selected (red rectangle) over the roof. (**Right**) From top to bottom: (**a**) surface under evaluation, without obstacles, projected to the horizontal plane; (**b**) concave hull that defines the perimeter of the surface; (**c**) largest empty rectangles inside the concave hull and (**d**) evaluated surface inliers of the largest empty rectangles re-projected to 3D.

The extraction of the points that describe the perimeter of the surface to evaluate is performed through the computation of their concave hull using an implementation of the "alpha shape" computational geometry approach based on the Delaunay triangulation [26]. The computation of the largest empty rectangles is performed through an iterative implementation presented in [27], restricting the orientation of the edges of the rectangles to the same orientation of the evaluated surface and its perpendicular vector on the horizontal plane. Finally, the largest empty rectangles are re-projected to the 3D space obtaining the georeferenced location of the optimal zones for the installation of solar panels in each roof, and their area.

2.2.7. Estimation of the Solar Irradiation

To introduce this section it will be useful to explain the differences between the terms "solar irradiance" and "solar irradiation". Solar irradiance describes the instantaneous radiant flux per unit area that is being delivered to a surface, usually expressed in W/m^2. It varies depending on the location of the surface, time and date, and atmospheric conditions, among other factors. Solar irradiation, also known as insolation and typically expressed in $kWh/m^2/a$, represents the amount of solar energy that can be collected on a surface per unit of area within a given time (*i.e.*, solar irradiance integrated over time). The use of georeferenced point clouds at this point is crucial given that solar irradiation is heavily influenced by day length and the position of the Sun (solar ephemeris), obtaining very different results for different locations on Earth. Solar irradiation may also significantly differ between building roofs in the same zone, depending on the orientation of the roof: a more favorable angle of a surface to the Sun means better exposure and more solar energy. Two roof surfaces of the same size at the same location but different orientations (azimuth) and tilts, may drastically differ in their solar potential [28,29]. This solar irradiation value, complemented with the area of the optimal surface obtained in the previous step, allows the computation of the solar irradiation of a surface in kWh/a [6,30]. Therefore, we can say that solar irradiation on a roof surface is mainly derived of four essential geographic parameters (area, tilt, azimuth and geographic location) that may be defined as a function (Equation (10)):

$$I = \int ([\varphi, \lambda], \beta, \theta, A, [\omega]) \tag{10}$$

where I is the solar irradiation on a roof; φ is the latitude; λ is the longitude; β is the tilt of the surface; θ is the azimuth; A is the area and ω other components such as cloud cover.

Solar irradiance can be decomposed into three components [31,32]: direct irradiance, diffuse irradiance and reflected irradiance. These three components are important and significantly influence the total irradiance [33], so they have been analyzed separately in this approach. Direct and reflected irradiance and their adjustment for a tilted and oriented surface have been calculated as described in [34]. The estimation of the diffuse solar irradiance is more complex and must be adjusted to one of the several existing empirical models [35–37].

Therefore, solar irradiation can be estimated by integrating the solar irradiance over a period of time (Equation (11))

$$I = \sum_{i=start}^{end} E_i + \Delta t_i \tag{11}$$

where I is the solar irradiation; Δt_i is the length of the ith time interval and E is the solar irradiance

The annual solar irradiation is calculated following the protocol used in the Solar3DCity [5,38] that includes the modeling of "real-sky" conditions through the integration of the historical EPW (Energy Plus Weather file format) weather data downloaded from the nearest weather station [39]. Solar ephemerides are from XEphem [40,41], which have been proven suitable for solar irradiation studies [42]. The computations use the empirical anisotropic irradiance model developed by [43], which was implemented in the solpy library [44]. The annual solar irradiation is then calculated for each roof surface by integrating the hourly irradiance values of the entire year. The computation of solar irradiation values can be extrapolated for every azimuth/tilt combination obtaining a function, usually called tilt-orientation-factors (TOF), which can be represented in a 3D diagram. This TOF function can be considered an additional product, since it represents the optimal tilt and orientation for a location, which is the main aim of several location-based studies [45,46].

3. Experimental Results

3.1. Study Case

The proposed methodology has been validated in a wide urban area in the city of Avila (Latitude 40°66'N, Longitude 4°70'W), with an extension of 237,250 m^2 in a rectangular shape of 365 m × 650 m, chosen due to the existence of roofs with different geometric characteristics, which is an interesting characteristic for testing the methodology.

Flight planning was performed considering the limitations of the thermographic sensor (Figure 6). Time between shots, in order to ensure properly overlapping, was stablished as 500 ms for the RGB sensor and 160 ms for the thermographic sensor, both for a flight speed of 50 km/h. Due to the speed flight, it was necessary to minimize the shutter opening time in order to avoid the motion blurring effect. This requirement needed several tests to find the optimal ISO, aperture and shutter speed configuration to ensure the correct exposure of the images minimizing the effect of any external factor that might damage the survey. As result, 409 images for the RGB sensor and 5312 images for the thermographic sensor were captured covering the whole study area. The flight altitude over the ground selected was 160 m, resulting on a GSD of 11 cm and 2 cm for the thermographic and RGB cameras, respectively.

The full resolution RGB images were processed according to the photogrammetric and computer vision methodologies obtaining as result a point cloud of 24,858,863 points, implying an average resolution of 100 points/m^2 ($\approx$1 point/GSD). The thermographic texture was projected over the point cloud obtaining a hybrid 3D point cloud (Figure 7) with thermographic texture mapped over the points of the roofs.

The survey was designed in order to fulfill the conditions established by PNOA (Spanish National Plan of Aerial Orthophotography) [47]. These conditions state the required accuracy to the GSD value for the X, Y and Z point coordinates. In order to check compliance with these requirements, a topographic GPS survey was performed, consisting of 26 check points homogeneously distributed over the whole study area (Figure 8).

This 3D point cloud with thermographic texture is the input of the algorithm developed. In the first step, ground, vegetation and façade points are removed using the pass through and conditional filters obtaining as result a point cloud of the roof with 8,930,030 points.

The point cloud of the roofs with thermographic texture is introduced into the segmentation process (Figure 9). The result of the Euclidean cluster segmentation is the extraction of 37 roofs that have been automatically segmented into 168 planar surfaces by the RANSAC algorithm and evaluated and classified regarding their geometrical suitability (location, area, orientation and tilt) for the installation of solar panels.

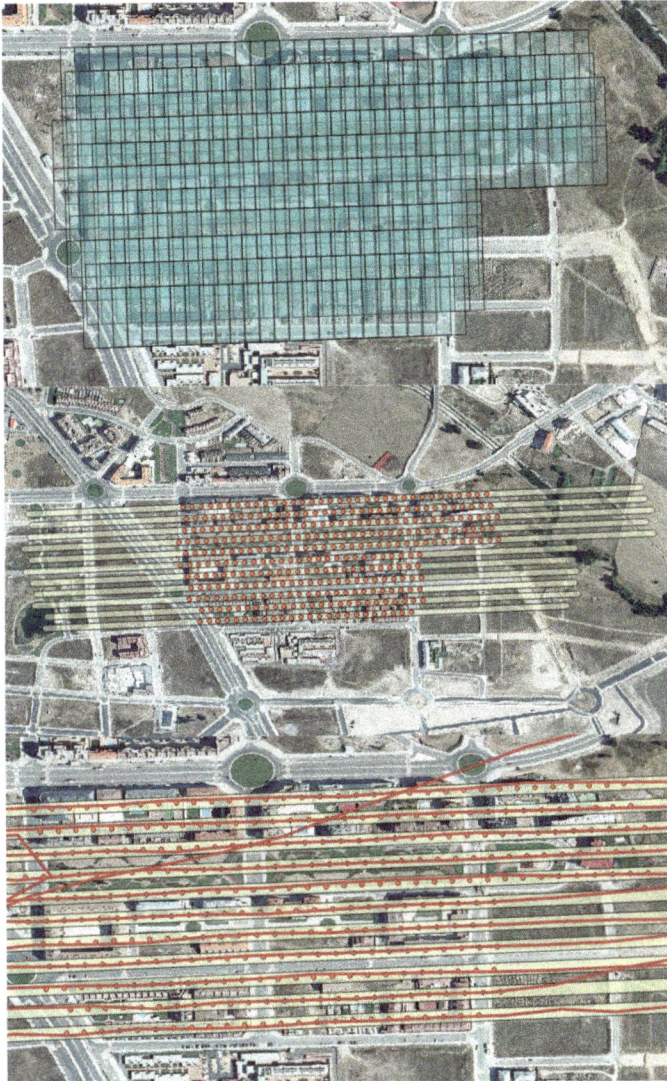

Figure 6. Flight planning: (**Top**) Areas captured by each image shot for the thermographic sensor. (**Middle**) Flight planning for the navigation of the aerial trike. (**Bottom**) GPS track after the aerial trike flight with the position of each image shot.

Figure 7. Hybrid 3D point cloud generated from images captured with the RGB camera integrating thermographic texture mapped over the roofs. Areas shown in detail in consecutive figures are remarked in red.

Figure 8. Check points homogeneously distributed over the whole study area.

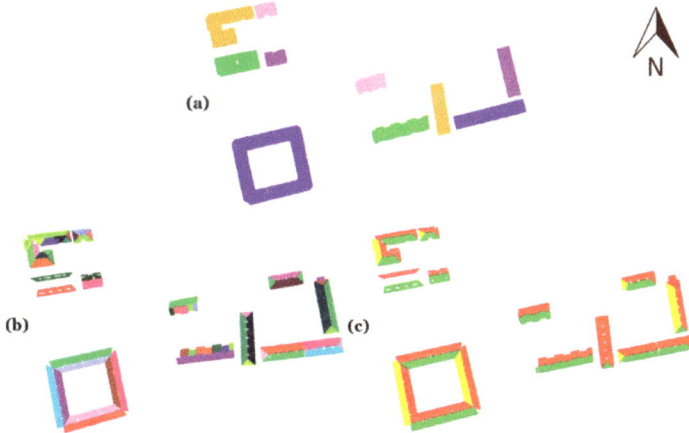

Figure 9. (a) Result of the extraction of each roof using the Euclidean cluster extraction algorithm. (b) Results of the extraction of each planar surface using the RANSAC algorithm. (c) Details of the previous point cloud segmented and classified by its suitability to install solar panels. (**Green**) Surfaces suitable to install solar panels without supports. (**Yellow**) Surfaces suitable to install solar panels using a support to modify the solar incidence angle. (**Red**) Surfaces not suitable for the installation of solar panels.

The radiometry of the thermographic images mapped on the 3D point cloud has been used as described in the methodology for the identification of possible obstacles (Figure 10). These obstacles have been taken into account for the location of the optimum zones for the installation of solar panels. This way, 67 zones (Figure 11) have been classified as optimal for the installation of solar panels, covering an area of 2230.77 m^2. Finally, all these data have been complemented with the estimation of the solar irradiation received by each optimal surface (Figure 12), quantifying an accumulated solar irradiation for the whole study area of 4.01×10^6 kWh/a.

The geometric and irradiation results are represented individually for each roof (Table 4). The geometric results show the azimuth (°) and tilt (°) of each surface and the area (m^2) of each optimal location. The irradiation results show the yearly solar irradiation for unit of area (kWh/m^2/a), computed from the azimuth and tilt values supported with the irradiation models and the atmospheric data, and the yearly total solar irradiation (kWh/a) of each optimal surface from its area. The area and the irradiation results of each optimal location for the installation of solar panels inside each roof surface are represented consecutively. The results of each segmented surface are recorded showing the total calculation of area and solar irradiation of the optimal areas in the whole surface.

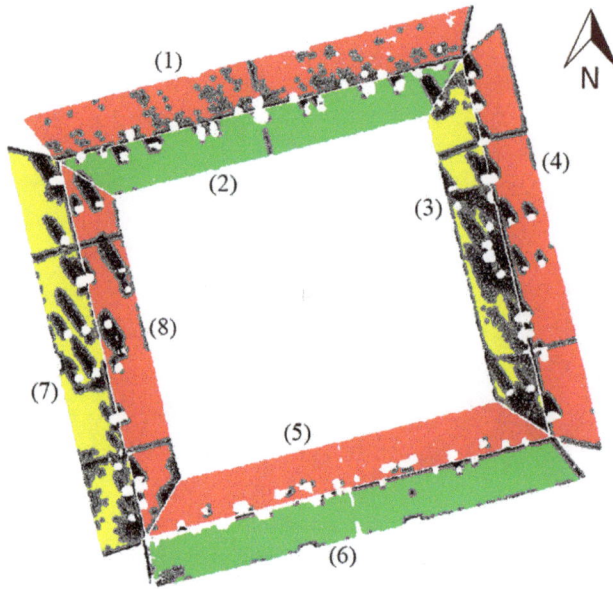

Figure 10. Detail of the point cloud segmented and classified after the thermographic analysis. (Black) Points removed by the statistical thermographic analysis. (Grey) Perimeter defined by a distance parameter around the obstacles detected to assist the optimal location of solar panels.

Figure 11. Georeferenced point cloud of the detail zones after the process. Optimal surfaces (rectangles) for the installation of solar panels are highlighted in blue. Roof ID's are used in Table 4.

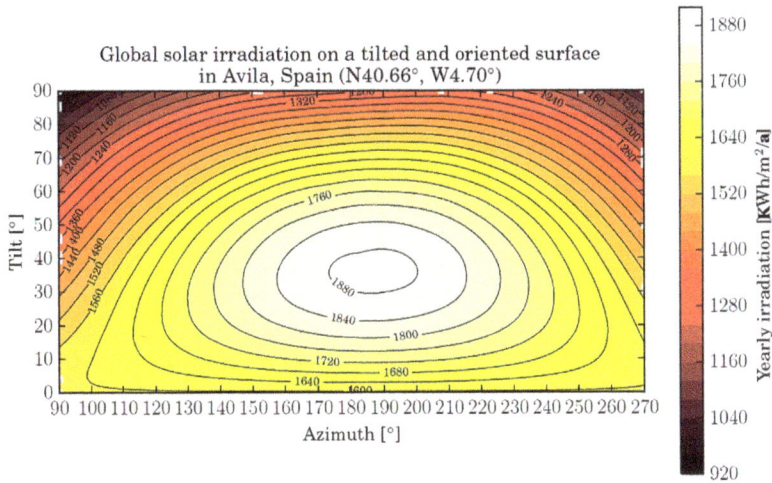

Figure 12. Tilt-orientation-factors (TOF) representation for Ávila (Spain) estimated by Solar 3D city.

Table 4. Example of the results after the processing of a roof of the dataset (roof and numbers in Figure 8).

Surface	Tilt (°)	Azimuth (°)	Yearly Solar Irradiation by m² (kWh/m²/a)	Area of Optimal Location (m²)	Yearly Total Solar Irradiation (kWh/a)
1	19.3568	346.0080	1295.011	–	–
2	19.4968	165.6463	1824.290	103.066	188,023
–	–	–	–	99.026	180,652
–	–	–	–	18.197	33,196.8
3	19.0947	156.1201	1651.910	22.938	37,891.8
4	19.3133	76.2060	1523.485	–	–
5	19.2929	346.1950	1295.834	–	–
6	19.5358	166.1320	1824.914	140.417	256,249.0
–	–	–	–	110.023	200,782.0
–	–	–	–	41.781	76,245.8
–	–	–	–	25.003	45,628.7
7	19.3667	156.0211	1652.508	38.410	63,472.9
8	19.2102	76.2400	1524.167	–	–
Total results for optimal locations of surface 2				220.289	401,871.8
Total results for optimal locations of surface 3				22.938	37,891.8
Total results for optimal locations of surface 6				317.224	578,905.5
Total results for optimal locations of surface 7				76.703	126,752.5
Total results for optimal locations of the roof				637.154	1,145,421.6

Those surfaces with North orientation have not been evaluated as not being candidates for the installation of solar panels by their geometry. For that reason, the "area of optimal location" and "yearly total solar irradiation" values of these surfaces are not available.

281

In order to evaluate the weaknesses and the reliability of the process, the roof planar surfaces of the study area have been analysed by visual inspection. Through this inspection 216 planar surfaces have been counted in the whole area under study. The difference of 48 planes existing between the reality and the result of the processing is distributed as follows. 42 planar surfaces have been correctly extracted by the algorithm but their evaluation has been discarded due to their minimum size that would make impossible the installation of solar panels. The 6 remaining surfaces match with roofs where the angle between their planar surfaces is next to 180° (Figure 13). In this case, RANSAC algorithm is not able to segment the roof in different planar surfaces obtaining as result of the interpolation a single horizontal plane that does not match with the reality of the roof. Surfaces whose tilt comes close to horizontal plane can be classified as non-optimal for installation of solar panels. For these reasons, it could be considered that the proposed approach is efficient enough for the location of the optimal zones to install solar panels, since it only misses 2.8% of the number of roofs of interest.

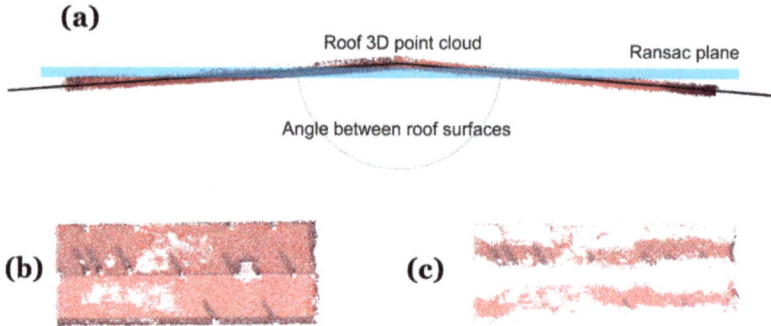

Figure 13. (**a**) Diagram of the RANSAC interpolation between two roof surfaces with an intersection angle next to 180°. (**b**) Top view of the evaluated roof. (**c**) Top view of the RANSAC algorithm results.

However, it must be emphasized that this algorithm only allows the automatic extraction of those surfaces which geometric approach can be interpreted and represented by a plane. Both curve and complex surfaces, where the installation of solar panels might be possible if complex support structures are used, are discarded.

3.2. Computing Efficiency Analysis

The machine used to perform the computation was a Microsoft Windows 8.1 workstation with 32 GB RAM, a 3.40 GHz Intel Core i5-3570K processor and an Nvidia Quadro 2000 GPU. The 3D dense point cloud generation was the most computationally demanding process investing a total of 13 h 20 min for the whole data set.

Regarding the algorithm developed for the automatic location of the best places for the installation of solar panels and the estimation of their incident solar irradiation, the execution time performed through a single thread algorithm was 26 min for the whole study area. However, multithreading computation is available, which would reduce the computation time proportionally to the number of concurrent running threads.

4. Conclusions

This article presents a methodology for the automatic processing of 3D point clouds with thermographic information for the automatic plane segmentation of the roof surfaces and their classification according to their theoretical productivity derived from their geometric characteristics (orientation, tilt and area). A refinement according to the thermographic radiometry allows the location of anomalies or obstacles that could prevent the installation of solar panels or reduce their productivity. This way, the location of the optimal zones for the installation of solar panels inside each surface is determined, avoiding obstacles detected in the thermographic refinement and allow the computation of the solar irradiation received on these zones. In particular, the hybrid product obtained provides complete thermographic and metric information of the different roof surfaces, which enables better detection, spatial location and interpretation of the suitability of each surface to harbor solar panels than traditional methods. In addition, the high level of automation of the procedure allows the evaluation of wide urban areas in a fast and accurate way, without the need to consult the technical documentation of each building.

The results of the application of the procedure are geo-referenced point clouds of the surfaces of the roofs with their geometric information (area, tilt, orientation), classified by their theoretical productivity regarding solar energy catchment. The computation has considered those elements that could reduce the productivity of the solar panels, as well as a perimeter zone potentially affected by the influence of the obstacles along the day. What is more, the procedure calculates the optimal tilt of the supports if they were necessary. In addition, results are combined in order to locate the optimal zones to install the solar panels inside each roof surface and complemented with the computation of the received solar irradiation. The latter, together with the technical characteristics of the panels to install, make possible the estimation of the productivity of the solar panels.

Thanks to that, the geometric information of the 3D point clouds has been preserved during all processing steps, enabling the performance of measurements directly on the point clouds of the segmented planes. As a result, the software developed could be used as a decision-making tool for those issues related with the dimension or type of solar panels that could be installed, calculating the actual productivity of the selected components.

In addition, the methodology proposed is valid for processing datasets captured with different airborne sensors such as LiDAR or any other RGB and thermographic sensors transported by any manned or unmanned aerial vehicle. This fact, supported with the existence of open access (public and free) LiDAR data [48] throughout the national territory, provides to this methodology and the developed software a direct applicability with minimum cost.

Regarding the efficiency of the approach developed, the methodology is applicable to the processing of entire towns or cities. Even assuming that the entire process of 3D reconstruction and the post-processing of point clouds for very large areas could imply some days of processing, the time required to perform the in-situ visual inspection (supported with technical documentation of the object) of large areas by human operators to achieve a similar result is not viable. It should also be noted that a human operator currently performs a subjective decision of the suitability of a roof to harbor solar panels (without generation of metric products and objective evaluation of obstacles). For this last reason, even in the case that a significant number of operators is available to perform the in-situ inspection in a "reasonable" period of time, the proposed approach is also an important breakthrough, especially since only an aerial trike pilot and an expert operator are required to perform the survey and to oversee the processing and analyze the results, respectively.

The main drawback of the methodology proposed is the possibility of finding particular homogeneous surfaces on the roofs where either the photogrammetric approach or the LiDAR measurement fails. These could lead to the appearance of areas without points or too noisy that prevent the segmentation process.

This project opens new trends for future work both from a sensorial and methodological point of view. Concerning the first, the evolution of multimodal matching techniques which can be applied to the matching of RGB and thermographic images would allow the automatic registration between both data sets, avoiding the tedious manual identification of homologous points between them, in those cases where the accurate positioning of both cameras is not possible. From the methodological point of view, the integration of the Sun path enables the most accurate determination of the location of the shadow zones that could reduce the productivity of the solar panels. The procedure considers shadows produced both by obstacles located on the same roof and by adjacent buildings. Improving the latter through a self-occlusion analysis directly from the 3D geometric information, together with the incorporation of the possible surrounding orography occlusion effects will be the next milestone towards the increase of the reliability of the decision-making processes using as input the processing results obtained from this methodology.

Regarding the possibility to take advantage of the large amount of information derived from this methodology, it would be interesting to advance in the automatic generation of BIM (Building Information Modelling) and CIM (City Information Modelling), using gbXML or CityGML languages, respectively, for the representation of the 3D point cloud geometry of the segmented roofs and the inclusion of the solar irradiation estimation. Progress toward these systems will allow to work with lighter and more agile information that we can include as an additional layer on a GIS enabling the integration of this information with other data sources (cadastral information, demographic information, urban parameters, *etc.*) to perform urban energy management tasks.

Acknowledgments: Authors would like to give thanks to the Ministerio de Educación. Cultura y Deporte (Gobierno de España) for the financial support given through human resources programs (FPDI-2013-17516). Also, we thank the support given by the Ministerio de Economía y Competitividad through projects ENE2013-48015-C3-1-R.

Author Contributions: All authors conceived and designed the study. Luis López-Fernández implemented the methodology. All authors discussed the basic structure of the manuscript. Luis López-Fernández wrote the document and all authors read and approved the final manuscript.

Conflicts of Interest: The authors declare no conflict of interest

References

1. Agugiaro, G.; Remondino, F.; Stevanato, G.; De Filippi, R.; Furlanello, C. Estimation of solar radiation on building roofs in mountainous areas. In Procceedings of the ISPRS Conference, Munich, Germany, 5–7 October 2011; pp. 155–160.
2. Agugiaro, G.; Nex, F.; Remondino, F.; de Filippi, R.; Droghetti, S.; Furlanello, C. Solar radiation estimation on building roofs and web-based solar cadastre. *ISPRS Ann. Photogramm. Remote Sens. Spat. Inf. Sci.* **2012**, *1*, 177–182.
3. Hofierka, J.; Suri, M. The solar radiation model for open source gis: Implementation and applications. In Proceedings of the Open Source GIS-GRASS Users Conference, Trento, Italy, 11–13 September 2002; pp. 1–19.
4. Nguyen, H.; Pearce, J.M. Estimating potential photovoltaic yield with r. Sun and the open source geographical resources analysis support system. *Sol. Energy* **2010**, *84*, 831–843.
5. Biljecki, F.; Heuvelink, G.B.M.; Ledoux, H.; Stoter, J. Propagation of positional error in 3D GIS: Estimation of the solar irradiation of building roofs. *Int. J. Geogr. Inf. Sci.* **2015**.
6. Strzalka, A.; Alam, N.; Duminil, E.; Coors, V.; Eicker, U. Large scale integration of photovoltaics in cities. *Appl. Energy* **2012**, *93*, 413–421.
7. Christensen, C.; Barker, G. Effects of tilt and azimuth on annual incident solar radiation for united states locations. In Procceedings of the Solar Forum 2001: Solar Energy: The Power to Choose, Washington, DC, USA, 21–25 April 2001; pp. 225–232.

8. López, L.; Lagüela, S.; Picon, I.; González-Aguilera, D. Automatic analysis and classification of the roof surfaces for the installation of solar panels using a multi-data source and multi-sensor aerial platform. *Int. Arch. Photogramm. Remote Sens. Spat. Inf. Sci.* **2015**, *1*, 171–178.

9. IDAE. Available online: http://www.Idae.Es/uploads/documentos/documentos_5654_fv_pliego_condiciones_tecnicas_instalaciones_conectadas_a_red_c20_julio_2011_3498eaaf.Pdf (accessed on 12 January 2015).

10. Microlight Navigation. Available online: http://www.boe.es/diario_boe/txt.php?id=BOE-A-1986-11068 (accessed on 12 January 2015).

11. Premerlani, W.; Bizard, P. *Direction Cosine Matrix IMU: Theory*; DIY Drone: Washington, DC, USA, 2009.

12. Lagüela, S.; González-Jorge, H.; Armesto, J.; Herráez, J. High performance grid for the metric calibration of thermographic cameras. *Meas. Sci. Technol.* **2012**, *23*.

13. Kraus, K.; Waldhäusl, P. *Photogrammetry: Fundamentals and Standard Processes*; Dümmmler Verlag: Bonn, Germany, 1993.

14. Agarwal, S.; Furukawa, Y.; Snavely, N.; Simon, I.; Curless, B.; Seitz, S.M.; Szeliski, R. Building rome in a day. *Commun. ACM* **2011**, *54*, 105–112.

15. Lowe, D.G. Object recognition from local scale-invariant features. In Proceedings of the Seventh IEEE International Conference on Computer Vision, Kerkyra, Greece, 20–27 September 1999; pp. 1150–1157.

16. Luhmann, T.; Robson, S.; Kyle, S.; Harley, I. *Close Range Photogrammetry: Principles, Methods and Applications*; Whittles: Dunbeath, UK, 2006.

17. Deseilligny, M.P.; Clery, I. Apero, an open source bundle adjusment software for automatic calibration and orientation of set of images. In Procceedings of the International Archives of the Photogrammetry, Remote Sensing and Spatial Information Sciences, Trento, Italy, 2–4 March 2011; pp. 269–276.

18. Apero-Micmac. Available online: http://www.tapenade.gamsau.archi.fr/TAPEnADe/Tools.html (accessed on 18 June 2015).

19. Gehrke, S.; Morin, K.; Downey, M.; Boehrer, N.; Fuchs, T. Semi-global matching: An alternative to lidar for DSM generation. In Proceedings of the 2010 Canadian Geomatics Conference and Symposium of Commission I, Calgary, AB, Canada, 15–18 June 2010.

20. Hirschmuller, H. Accurate and efficient stereo processing by semi-global matching and mutual information. In Procceedings of the IEEE Computer Society Conference on Computer Vision and Pattern Recognition, San Diego, CA, USA, 20–25 June 2005; pp. 807–814.

21. Gonzalez-Aguilera, D.; Guerrero, D.; Hernandez-Lopez, D.; Rodriguez-Gonzalvez, P.; Pierrot, M.; Fernandez-Hernandez, J. Silver CATCON award, technical commission WG VI/2. In Proceeding of the XXII ISPRS Congress, Melbourne, Australia, 25 August–1 September 2012.

22. Rusu, R.B.; Cousins, S. 3D is here: Point cloud library (pcl). In Proceedings of the 2011 IEEE International Conference on Robotics and Automation (ICRA), Shanghai, China, 9–13 May 2011.

23. Gallo, O.; Manduchi, R.; Rafii, A. Cc-ransac: Fitting planes in the presence of multiple surfaces in range data. *Pattern Recognit. Lett.* **2011**, *32*, 403–410.

24. Hulik, R.; Spanel, M.; Smrz, P.; Materna, Z. Continuous plane detection in point-cloud data based on 3D hough transform. *J. Vis. Commun. Image Represent.* **2014**, *25*, 86–97.

25. Fischler, M.A.; Bolles, R.C. Random sample consensus: A paradigm for model fitting with applications to image analysis and automated cartography. *Commun. ACM* **1981**, *24*, 381–395.

26. Edelsbrunner, H.; Kirkpatrick, D.G.; Seidel, R. On the shape of a set of points in the plane. *IEEE Trans. Inf. Theory* **1983**, *29*, 551–559.

27. Orlowski, M. A new algorithm for the largest empty rectangle problem. *Algorithmica* **1990**, *5*, 65–73.

28. Yang, H.; Lu, L. The optimum tilt angles and orientations of pv claddings for building-integrated photovoltaic (BIPV) applications. *J. Sol. Energy Eng.* **2007**, *129*, 253–255.

29. Santos, T.; Gomes, N.; Freire, S.; Brito, M.; Santos, L.; Tenedório, J. Applications of solar mapping in the urban environment. *Appl. Geogr.* **2014**, *51*, 48–57.

30. Li, Z.; Zhang, Z.; Davey, K. Estimating geographical pv potential using lidar data for buildings in downtown san francisco. *Trans. GIS* **2015**.

31. Šúri, M.; Hofierka, J. A new GIS-based solar radiation model and its application to photovoltaic assessments. *Trans. GIS* **2004**, *8*, 175–190.

32. Liang, J.; Gong, J.; Li, W.; Ibrahim, A.N. A visualization-oriented 3D method for efficient computation of urban solar radiation based on 3D–2D surface mapping. *Int. J. Geogr. Inf. Sci.* **2014**, *28*, 780–798.

33. Gulin, M.; Vašak, M.; Perić, N. Dynamical optimal positioning of a photovoltaic panel in all weather conditions. *Appl. Energy* **2013**, *108*, 429–438.

34. Masters, G.M. *Renewable and Efficient Electric Power Systems*; John Wiley & Sons Inc.: Hoboken, NJ, USA, 2013.

35. David, M.; Lauret, P.; Boland, J. Evaluating tilted plane models for solar radiation using comprehensive testing procedures, at a southern hemisphere location. *Renew. Energy* **2013**, *51*, 124–131.

36. Demain, C.; Journée, M.; Bertrand, C. Evaluation of different models to estimate the global solar radiation on inclined surfaces. *Renew. Energy* **2013**, *50*, 710–721.

37. Gulin, M.; Vašak, M.; Baotic, M. Estimation of the global solar irradiance on tilted surfaces. In Proceedings of the 17th International Conference on Electrical Drives and Power Electronics (EDPE 2013), Dubrovnik, Hrvatska, 2–4 October 2013; pp. 334–339.

38. Solar3Dcity. Available online: https://github.com/tudelft3d/Solar3Dcity (accessed on 5 August 2015).

39. Energy Plus weather data. Available online: http://apps1.eere.energy.gov/buildings/energyplus/weatherdata_about.cfm (accessed on 5 August 2015).

40. XEphem. Available online: http://www.clearskyinstitute.com/xephem/ (accessed on 5 August 2015).

41. Meeus, J.H. *Astronomical Algorithms*; Willmann-Bell, Incorporated: Richmond, VA, USA, 1991.

42. Reda, I.; Andreas, A. Solar position algorithm for solar radiation applications. *Sol. Energy* **2004**, *76*, 577–589.

43. Perez, R.; Ineichen, P.; Seals, R.; Michalsky, J.; Stewart, R. Modeling daylight availability and irradiance components from direct and global irradiance. *Sol. Energy* **1990**, *44*, 271–289.

44. Solpy. Available online: https://github.com/nrcharles/solpy (accessed on 5 August 2015).

45. Šúri, M.; Huld, T.A.; Dunlop, E.D. Pv-gis: A web-based solar radiation database for the calculation of pv potential in europe. *Int. J. Sustain. Energy* **2005**, *24*, 55–67.

46. Rowlands, I.H.; Kemery, B.P.; Beausoleil-Morrison, I. Optimal solar-pv tilt angle and azimuth: An Ontario (Canada) case-study. *Energy Policy* **2011**, *39*, 1397–1409.

47. PNOA. Available online: https://contrataciondelestado.es/wps/wcm/connect/7050f80e-1d0c-4231-afad-5c8262529580/DOC20090605131632pdf.pdf?MOD=AJPERES (accessed on 16 August 2015).

48. LiDar Data. Available online: http://centrodedescargas.cnig.es/CentroDescargas/buscadorCatalogo.do?codFamilia=LIDAR (accessed on 12 August 2015).

A New Approach to the Generation of Orthoimages of Cultural Heritage Objects—Integrating TLS and Image Data

Jakub Stefan Markiewicz, Piotr Podlasiak and Dorota Zawieska

Abstract: This paper discusses the issue of automation of orthoimage generation based on Terrestrial Laser Scanning (TLS) data and digital images. The following two problems are discussed: automatic generation of projection planes based on TLS data, and automatic orientation of digital images in relation to TLS data. The majority of popular software applications use manual definitions of projection planes. However, the authors propose an original software tool to address the first issue, which defines important planes based on a TLS point cloud utilizing different algorithms (RANdom SAmple Consensus–RANSAC, Hough transform, "region growing"). To address the second task, the authors present a series of algorithms for automated digital image orientation in relation to a point cloud. This is important in cases where scans and images are acquired from different places and at different times. The algorithms utilize Scale Invariant Feature Transform(SIFT) operators in order to find points that correspond in reflectance intensity between colour images (Red Green Blue—RGB) and orthoimages, based on TLS data. The paper also presents a verification method using SIFT and Speeded-Up Robust Features (SURF) operators. The research results in an original tool and applied Computer Vision(CV) algorithms that improve the process of orthoimage generation.

Reprinted from *Remote Sens.* Cite as: Markiewicz, J.S.; Podlasiak, P.; Zawieska, D. A New Approach to the Generation of Orthoimages of Cultural Heritage Objects—Integrating TLS and Image Data. *Remote Sens.* **2015**, *7*, 16963–16985.

1. Introduction

Developing geometric documentation is one of the most basic tasks in the fields of conservation policy and management of cultural heritage objects. 3D documentation is a prerequisite of conservation or restoration work on historical objects and sites. Development of specific documentation such as high-resolution orthoimages is necessary at all stages of conservation works.

Orthoimages are attractive for archaeological and architectural documentation, as they offer a combination of geometric accuracy and visual quality [1], and can also be applied to different measuring techniques.

Integration of Terrestrial Laser Scanning (TLS) and image-based data can lead to better results [2,3]. Despite many advantages, the TLS technique also has a lot

289

of limitations; for example,it is not possible to acquire sharp edges, poor quality or reflective surfaces, *etc.* Another issue is caused by areas with obstacles or hidden points. In this case, in order to fill in the blind spots, more stations are required. Laser scanners are currently integrated with low-resolution cameras. In many cases, it is recommended to use additional, high-resolution images in order to produce the required orthoimages. However, in many cases of TLS data utilization, the additional information is an intensity value, applied by conservators to evaluate the conditions of the surfaces of given historical objects [4].

3D modelling systems have now also been dynamically developed; they are based on digital images and utilize Computer Vision (CV) algorithms to automatically create 3D models [5–9]. Tools which utilize the Structure from Motion (SfM) approach are also being developed. CV algorithms allow for automatic creation of textured 3D surface models based on a sequence of images, without any prior knowledge concerning objects or the camera. Developed software tools also allow generation of orthoimages [5,7,10].

Integration of TLS and SfM technologies therefore becomes an alternative in many works on complicated architectural objects [6,11]. Although new imaging systems have been dynamically developed, many problems inherent in the integration of TLS and digital images are still waiting to be solved [12]. The possibilities of using digital technologies in the conservation process have been widely analysed. However, solutions to automate the generation of cultural-heritage-object documentation are still lacking. In many cases scanning data on cultural heritage objects exist already; it is only necessary to acquire high-resolution orthoimages with the use of additional, "free-hand" images.

The authors of this paper closely co-operate with the Museum of King John III's Palace in Wilanów in conservation works, as well as in other projects. Development of metric documentation of complicated historical objects requires implementation of an automated technological process. Based on this experience the orthoimage generation process has been automated with the use of terrestrial laser-scanning data and "arbitrary" high-resolution images.

2. Problem Statement

This paper focuses on generating orthoimages through automatic integration of TLS data with "free-hand" images. In the past, many cultural heritage objects were measured with the use of scanners without digital cameras; the results were point cloud with intensity values (near infrared) and transformed to the orthoimages. Due to the spectral range in which such an orthoimage is recorded, it seems to have a low visual contrast. Therefore, it is necessary, in many cases, to supplement the data with digital images in order to generate RGB orthoimages (in natural colours).

When orthoimages are generated it is important to settle two basic issues: accuracy, and resolution [13]. In this case integration of a TLS point cloud and digital images improves the geometric and radiometric quality of generated orthoimages [5–8,11,14,15].

Problems can also result from the nature of close-range measurements, *i.e.*, the proximity of objects of complex geometry [1]. Such difficulties have been widely discussed in the context of "true-ortho" generation for architectural objects, where orthoimages are the basic tool for documentation [16,17].

Many publications have demonstrated the role of TLS data in improving the accuracy and automation level [16,18]. Unfortunately, software for full automatic generation of orthoimages does not yet exist.

Modern terrestrial scanners are equipped with digital cameras of low resolution, and measurements are performed from the same position. Thus, scans and images are automatically integrated. However, in order to obtain RGB orthoimages of sufficient quality, it is necessary to use images of higher resolution, acquired independently of the scanner station [16]. Therefore, a separate issue concerns the integration of TLS data with "free-hand" images of higher resolution [19,20].

It is necessary to create procedures of integration of a TLS point cloud and "arbitrary" acquired images in order to generate a high-resolution orthoimage that will improve the quality of the point clouds and resolution texture of 3D models. Achieving geometric accuracy and visual quality is time consuming; the process needs to be performed manually or semi-automatically.

The first experiments testing the presented approach were discussed in [18]. These initial experiments were performed using the unmodified Hough transform for uncomplicated architectural objects, creating one plane of smaller depth.

The authors of this paper have developed original tools that automate the process of RGB image generation.Experiments performed to this end are presented in Figure 1.

The following issues are discussed in the paper:

- detection of horizontal and vertical planes from the so-called "raw point clouds" with the use of the processed Hough transform;
- checking the accuracy of plane matching and setting the points buffer for creation of the Digital Surface Model (DSM) of an object and the intensity image;
- DSM generation in the rectangular grid (GRID) form, and generation of intensity orthoimages with a depth map;
- orientation of terrestrial images based on orthoimages with the assigned depth map using the utilize Scale Invariant Feature Transform (SIFT) algorithm;
- quality control of matching images in relation to the point clouds/an orthoimage with the depth map;

- orthorectification and mosaicking of terrestrial images based on the previously generated DSM;
- quality control of the coloure orthoimages (Red Green Blue—RGB) based on the Speeded-Up Robust Features (SURF) algorithm and the intensity orthoimage.

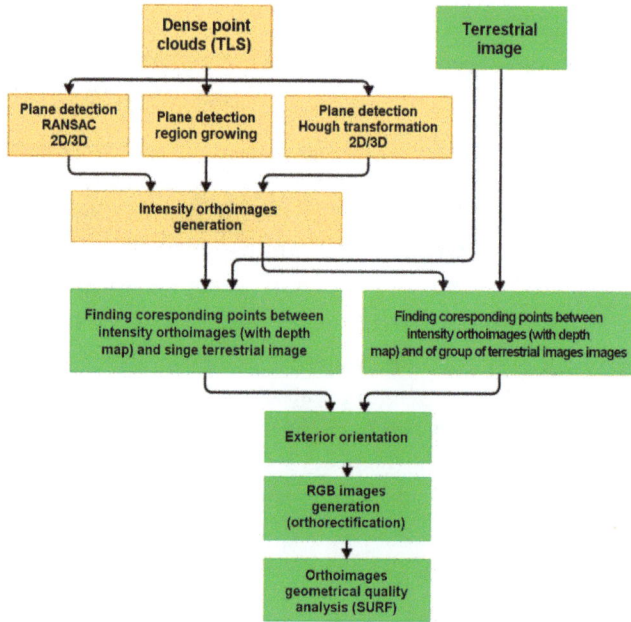

Figure 1. Diagram of the performed experiments.

3. Methodology

Utilization of terrestrial laser-scanning data requires consideration of the specifics of data recording.

3.1. Specifics of TLS Data

Data are recorded in the polar system, R, ω, φ, which causes deformation of straight lines (Figure 2). The point cloud recorded by the scanner contains points located on different planes (on the floor, on the ceiling, and, as in the discussed case, on the walls); therefore, it is necessary to filter points located on the plane of interest (again, the wall in the discussed case). Due to different disturbances, points determined by the scanner are not directly located on the reflecting surface. As shown by Figure 3, the distance measurement error (the scanner produces data in the polar system) as well as the laser's light incidence angle causes points to be located inside certain cuboids (not on the plane). The depth of these cuboids depends on the scanner's technical specifications (wavelength, modulation type, spot diameter, radiation power), as well as on the properties of the reflecting surface. Differences can emerge from scanner type (different companies have differing amounts of experience), as well as the surface structure: for example, Figure 3a shows a fabric, whose threads could cause multiple reflections; on the other hand, and Figure 3b shows a smoothly plastered and painted wall.

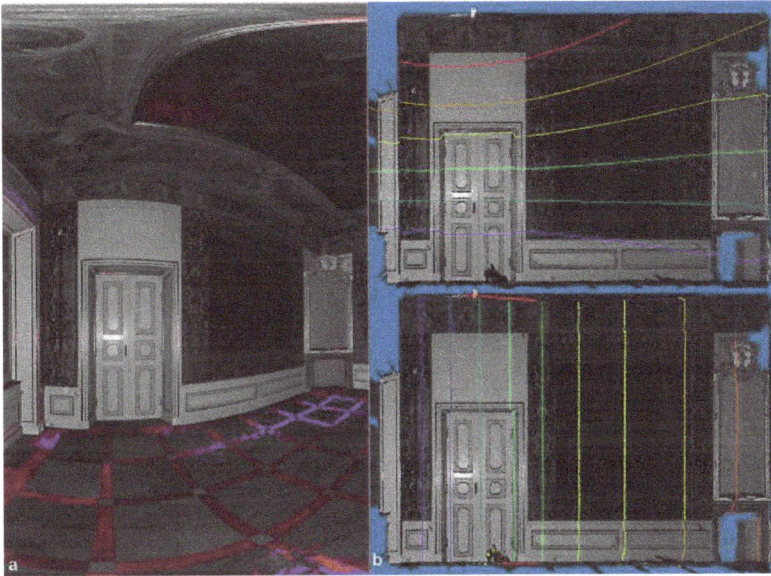

Figure 2. Data acquired by the ZFS 5003 scanner: (**a**) an image of the reflectance level in the ω, φ polar system; (**b**) in the rectangular XZ (rotated) system; colour lines present successive rows and columns of source data.

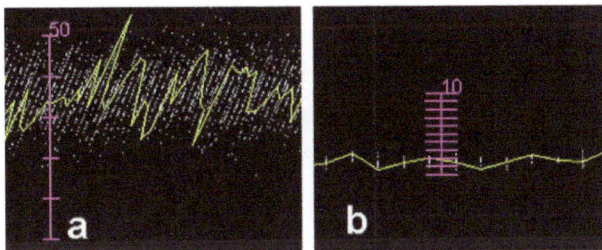

Figure 3. The "thickness" of the point cloud for data recorded by the (**a**) Z+F 5003 scanner for a wall covered with fabric and (**b**) by the Z+FS 5006 scanner for the painted wall. The yellow line presents the "row" of the scanner data (mm).

Even in the case of explicit classification of points, the thickness of the cloud and noises can influence the plane determination.

As can be seen, the properties of TLS point clouds require the use of methods resistant to gross errors. Three methods (with variants) were selected and tested: RANSAC, region-growing, and Hough transform.

3.2. Plane Equations

The plane equation in space can be described in different ways, but the general form of the equation is Equation (1):

$$ax + by + cz + d = 0 \tag{1}$$

or, in the vector form: unit length normal vector (n_x, n_y, n_z) and $r = $ distance from origin.

The normal vectorcan be determined from Equation (2):

$$Ax + By + Cz + 1 = 0 \tag{2}$$

and the distance from origin from Equation (3):

$$\begin{aligned}
n_x &= \cos(\varphi)\cos(\omega) \\
n_y &= \cos(\varphi)\sin(\omega) \\
n_z &= \sin(\varphi)
\end{aligned} \tag{3}$$

where X_p, Y_p, Z_p are the coordinates of a point lying on the plane.

Since vertical and horizontal planes are important for the purposes of documentation, the ranges of the permitted values of the coefficients in the above equations may be limited. For horizontal planes (*i.e.*, floors and ceilings) it can be assumed that $C \approx 0$ or $\varphi \approx \pm 90°$, and for vertical planes (*i.e.*, the walls) that $A \approx 0$ and $B \approx 0$ or $\varphi \approx 0°$. The presented equations describe the infinite plane. In order to utilize the results, it is necessary to limit the plane to a rectangular region and to determine four corners.

3.3. Determination of a Plane

Three different methods of plane determination were analysed. Each analysis searched for points located in the point cloud on (or close to) a plane. All algorithms are greedy algorithms, *i.e.*, they try to find the plane on which the most points are located. When the plane is determined, points located on this plane are eliminated and the operation is repeated for the remaining points. The process ends when certain number of points have been "used"; for the discussed algorithms, this was established a priori as 10% of the unclassified points.

3.3.1. The RANSAC Method

The RANSAC method was first described in [21]. It is an iterative method of estimation of parameters of a mathematical model, based on a set of observational data including outliers. Its algorithm is non-deterministic—the results are correct with a certain probability, which depends, among others, on the percentage of incorrect values in the data file. Due to the randomness of the procedure the results may differ slightly for successive algorithm iterations.

3.3.2. The Region-Growing Method

Region-growing methods consist of grouping points of certain, similar features. In the discussed case, the normal vector to a plane could become such a feature. It is not possible to determine the normal vector to an individual point; it needs to be calculated for certain groups of neighbouring points. As a result of the above-described noises the direction of the normal vector, determined only for three points, may be practically arbitrary, but using more points enables the calculation of an average. For each analysed point a certain number of neighbours are selected, the local plane is determined and the vector perpendicular to that plane is considered as the normal vector for the selected point. Two criteria for the selection of neighbours are applied in practice: "N"—nearest points, or points included in a certain sphere with a defined radius or a cube with a defined side length. Generally, it is not a trivial task to find such points in an unordered point cloud [22].

3.3.3. The Hough Transform

The Hough transform is a technique applied in image analysis [23]. Initially, it was designed to detect straight lines in a digital image. Later, the method was generalized for the detection of shapes that could be analytically described, such as circles [24,25]. The rule is based on the detection of objects from a given class by a voting procedure. This procedure is performed on parameter spaces that describe shapes; each point increases the counter (accumulator) value in the position corresponding to parameters it could be related to. After processing of all image points, the local maximum values in the parameter space describe the detected probable parameters of shapes. In practice, it is necessary to quantify the parameter values. In the simplest case of detection of straight lines, the line inclination angle φ ($0 \leqslant \varphi < \pi$) and the distance from the origin of the co-ordinate system $\mathbf{r}$ (the range depends on the image diagonal) could be used, as presented in Figure 4.

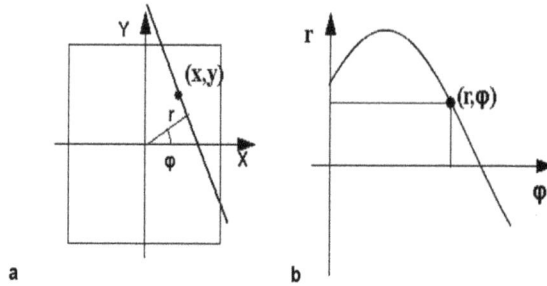

Figure 4. The image space (**a**) and the diagram (**b**) in the parameter space for an individual point.

For each point P(x,y), all cells of the accumulator that fit Equation (4) will be increased by a unit. It should be noted that, for each image point, it is necessary to determine the **r** value for each φ from the assumed range (Figure 4).

$$r = xcos\left(\varphi\right) + ysin\left(\varphi\right) \tag{4}$$

The Hough transform can be generalized for other cases [21,25,26]. Using the plane description by means of Equation (3), we can determine three parameters: r, φ and ω. Based on the angles $\varphi\left(-\pi/2 \leqslant \varphi < -\pi/2\right)$ and ω $\left(0 \leqslant \omega < \pi\right)$, we must determine the **r** value for each point X, Y and Z from the TLS data (Equation (4)). The counters need to be similarly increased. In this case the parameter space is three-dimensional, which causes difficulties for graphical representation. For computer processing, this means utilization of a three-dimensional instead of a two-dimensional array. This obviously requires more memory and computation time (reflected in Table 1).

Table 1. Comparison of different plane detection methods.

Algorithm	Sub-Sampling	Computation Time (s)		Main Wall Parameters			Number of Points in First Wall (% Total)
		First Wall	Next	ω (°)	φ (°)	R (m)	
RANSAC 3D	1:20	395	50–100	101.6	0.57	3.65	37
RANSAC 2D	1:20	85	60	102.2	-	3.74	34
Region growing	1:20	1000	-	100.6	0.99	3.44	30
Hough 2D	-	15	10	102.0	-	3.70	33
Hough 3D	-	820	100	102.0	1.01	3.72	34

3.4. Automatic Generation of Intensity Orthoimages

The hierarchical and ordered structure of data recorded is an advantage of acquisition by terrestrial laser scanning (Figure 2b). Such an approach, which is based on an assumed fixed scanning interval, explicitly defines the interpolation method that is applied for generation of the digital surface model (DSM) in the GRID form. Additionally, it allows us to recover the "missing points" (*i.e.*, those deleted in the data filtration process) based on the nearest points. This is why the *Triangular Irregular Network* (TIN) interpolation method would be unsuitable.

As a result, the orthoimages are obtained for use in further investigations into automation of the orientation of terrestrial digital images (Figure 5a). Additionally, depth is recorded for each orthoimage, as well as the intensity of the laser beam reflection (Figure 5b).

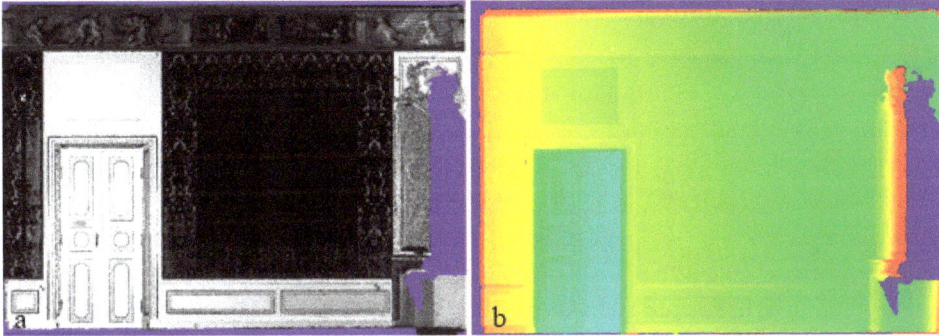

Figure 5. (**a**) Intensity orthoimage—Ground Sample Distance (GSD) = 2 mm; (**b**) map of depth.

3.5. Detection of Keypoints—SIFT and SURF

3.5.1. SIFT (Scale Invariant Feature Transform)

The SIFT algorithm allows for the detection (extraction) of characteristic points ("keypoints") in four main steps: scale-space extreme detection, keypoint localization, orientation assignment, and keypoint description. In the first step, the Gaussian difference function is used to detect potential characteristic points independently of the orientation scale. Areas located using the Difference of Gaussian (DoG) detector are described by the 128-dimensional vector, divided by the square root of the total squares of its elements, in order to achieve the invariance of changes in illumination [27].

3.5.2. SURF (Speeded-Up Robust Features)

The SURF algorithm was developed in response to SIFT in order to allow reception of similar results within a shorter time. Like SIFT, it is independent of scale and rotation, due to the use of the Hessian matrix during computation. The use of an integrated image in the SURF algorithm allows simple approximation of the Hessian matrix determinant using a rectangular filter (not DoG as in the case of SIFT). This reduces the computational complexity [28].

3.6. Orientation of Images in Relation to Intensity Orthoimages—SIFT

The objective of our experiments was to investigate the possibility of using "free-hand" terrestrial images for orthorectification or colouring of a point cloud. Although our scanners are equipped with built-in digital cameras that acquire images during scanning, these are characterized by low resolution and geometric quality [29].

In generating high-resolution orthoimages of cultural heritage objects, it is important to achieve high geometric and radiometric quality. For this reason, it is important to acquire "free-handed" images from positions different to those of the scanner stations. In the methodology proposed by the authors, orientation of images in relation to the point cloud is performed automatically, using the SIFT algorithm for the detector and the descriptor, which searches for tie points.

Common problems that occur in the acquisition of images of complex architectural objects relate to the maintenance of the reproducibility of the focal length and the limited possibility of framing appropriate parts when using fixed-focal-length lenses. The original application, based on processed algorithms from the OpenCV library,was used for image orientation in relation to scans. For determination of an image's exterior orientation elements, the perspective transformation algorithm was used, implemented in the OpenCV library [13,30].

The camera calibration matrix [13,30] contains information on changes of the camera focal length f_x, f_y, depending on the pixel dimension in the X and Y directions and the co-ordinates of the position of the camera's principal point (c_x and c_y, respectively).

In order to determine the camera orientation parameters, the solvePnP function is used, which requires the following input parameters:

- point co-ordinates on the image;
- point co-ordinates in the global system;
- selection of LM optimization as a minimization method. In this case the function finds a position that minimizes reprojection error, the sum of squared distances between the points of the image and of the projection [31].

Output parameters were matrices, which included information about exterior orientation elements, *i.e.*, the rotation matrix (rvec) and the translation vector (tvec).

The above approach requires knowledge of the camera calibration parameters (interior orientation). Another solution is possible, known as DLT (Direct Linear Transformation). Determined parameters include hidden parameters of interior and exterior orientation [32,33].

3.7. Terrestrial Images Orthorectification

A digital image is a faithful image of an object presented in the central projection [13]. Unlike in an orthoimage, as the result of an orthogonal projection (*i.e.*, a map), distortion of a digital image results from height differences of the object and the inclination of the image. In the theoretical approach, an orthoimage is an image whose projection plane is parallel to the reference plane, and where all rays are perpendicular to those two planes.

For the needs of the cultural heritage inventory, true orthophoto images are created [13,34]. Transformation of information acquired in the central projection for the orthogonal projection relies upon the processing of every pixel of the source image into the orthoimage, using knowledge of exterior orientation elements and the digital surface model. An image pixel would then be mathematically assigned to the generated orthoimage pixel as a result of superposition. This projection may be forward or backward [13]. In the case of forward projection, each subsequent pixel is processed and its corresponding pixel on the orthoimage.

In the case of backward projection, for each orthoimage pixel, the location of a corresponding pixels is found on the image.

4. Experiments

This section reports the experiments carried out on the influence of the quality and resolution of the point cloud on the quality of final products (Figure 1).

4.1. Determination of a Plane

Several experiments were performed to find the best algorithm for plane determination.

4.1.1. The RANSAC Method

The first experiments were performed with the use of Point Cloud Library algorithms [22]. The object pcl::SACSegmentation<pcl::PointXYZ> was selected from this library.After determination of appropriate data sources and parameters, plane coefficients were determined to meet the criteria. Additionally, the list of inliers was returned to allow a repeated search for successive planes with the use of unused points (outliers). Practical attempts to apply procedures for the complete source data, comprising 48 million points, proved that equipment resources did not allow sufficiently fast data processing. Only after 1:20 resampling were 16 regions obtained that corresponded to planes (see colours on Figure 6); some points remained unclassified by any plane (presented in grey).

The maximum deviation from the plane was assumed to be 5 cm (computed deviations are presented in Figure 6).This value was arbitrarily selected based on evaluation of the "depth" of the utilized point cloud (Figure 3). Points located further than 5 cm from the theoretical plane therefore were not considered in the plane determination, since these are either gross errors or belong to other objects, such as the floor or the ceiling. Most of the pseudo-3D drawingwas performed using the open-source 3D point cloud and the mesh-processing software CloudCompare [35].

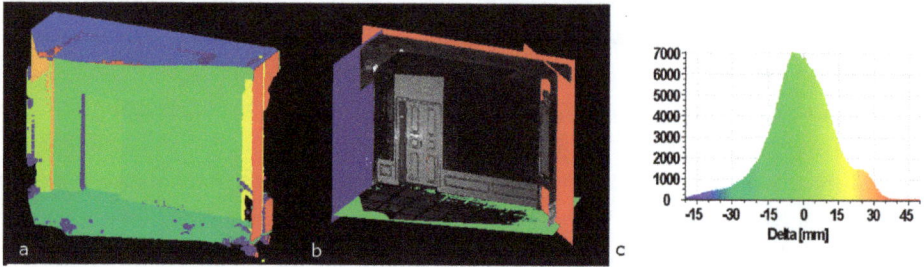

Figure 6. Planes detected by the PCL (RANSAC3D) algorithm: (**a**) Points on the same planes are marked with the same colour; (**b**) The rectangles describe the first four detected planes; (**c**) Histogram of the deviation.

The presented figures permit us to observe that the algorithm tends to separate points into groups, although the ω angle values are similar. The distribution of distance values from the theoretical wall surface is close to normal (Figure 6); this allows us to conclude that the plane was correctly determined.

Similar procedures can be found in the "openMVG" (open Multiple View Geometry) library [36,37]. These include Max-Consensus, RANSAC, and AC-RANSAC (model and precision estimation), which also allow testing of the 2D version (neglecting the Z co-ordinate). Similarity was found between the size of the problem and the obtained results (Figure 7).

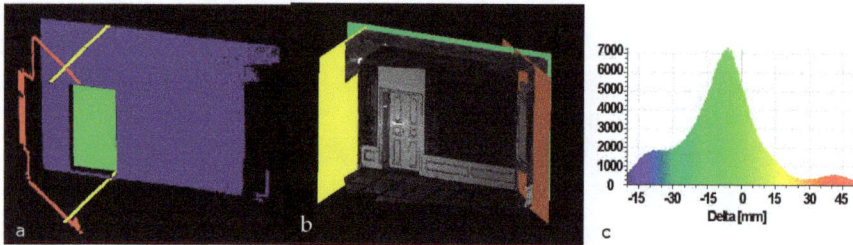

Figure 7. Planes detected by the 2D RANSACalgorithm: (**a**) Points on the same plane are marked with the same colour (the visible errors can be explained by inclusion of points on the floor and the ceiling); (**b**) The rectangles describe the first four detected planes; (**c**) Histogram of the deviation.

300

4.1.2. Region Growing

The PCL library contains functions used for the determination of normal vectors for point clouds within the defined environment. The first attempts were made for the 2 mm neighbouring area, resulting in a high dispersion of normal vectors.

Due to the limitations of the equipment, it was necessary to perform 1:10 resampling and select the neighbourhood area as 10 mm in order to reduce the dispersion value. Figures 8 and 9 present the components of the normal vector to neighbouring points in the cloud.

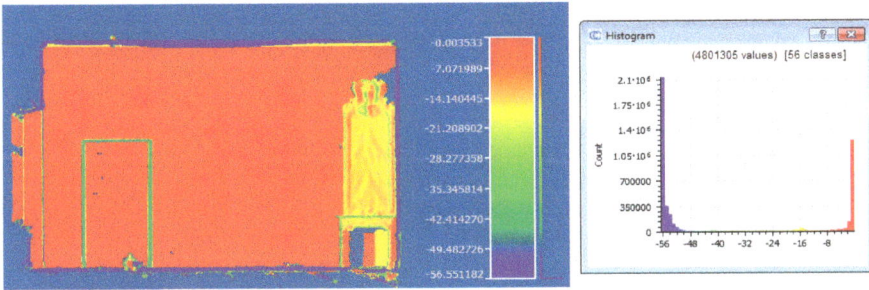

Figure 8. Distribution of the deviation of the normal vector (N_y component), expressed in degrees, determined for the neighbourhood of 10×10 mm. Three maximum values are visible in the histogram; they correspond to the floor (in blue), the mirror above the fireplace (in yellow) and the wall (in red). It can be seen that the niche area of the door was not separated, although it is located at a different depth.

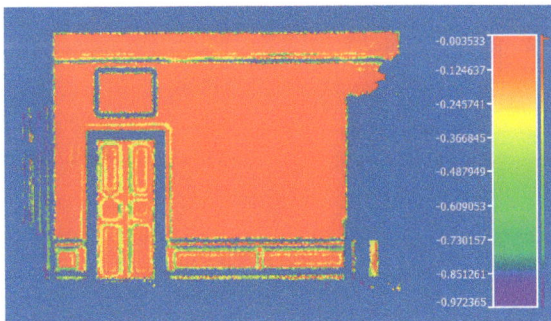

Figure 9. Points for which the deviation of the N_y normal vector does not exceed 1 degree.

4.1.3. The 2D Hough Transform

The direct utilization of the Hough transform requires an image to be generated on a 2D plane, *i.e.*, that the XY co-ordinates be quantified and the Z co-ordinate neglected. The first experiments proved that, the use of simple procedures from the OpenCV library [30,31],does not obtain satisfactory results. The image is disturbed by points on the floor and the ceiling, and therefore there are no clear maximum values (Figure 10b). Modification of this algorithm not by a unit, but by the number of cloud points that correspond to the given XY cell led to more promising results (Figure 10c) than were obtained using the Canny algorithm, as presented in our previous paper [18].

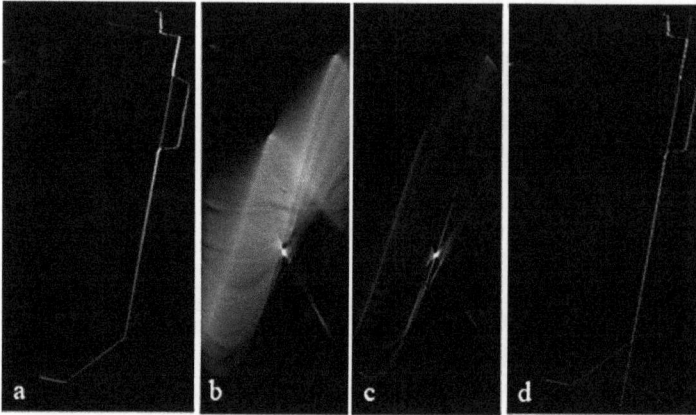

Figure 10. Example of 2D Hough transform: (**a**) The XY image—the brightness corresponds to the number of points included in the individual cell; the cell size has been assumed as 20 × 20 mm; (**b**) The parameter space when the counter is incremented by 1; (**c**) The parameter space when the counter is incremented by the number of points in a cell (the original algorithm)—the visible maximum corresponds to the direction of the wall; (**d**) The line corresponding to the maximum value from the image in (**c**), marked on the XY image.

Successive directions are determined after elimination of utilized points, zeroing of the counters, re-calculation of the parameter space, and selection of successive maximum values (Figure 11).

In order to avoid quantification of XY co-ordinates, a new version of the software application was developed that determines the image in the parameter space, directly on the basis of real XY co-ordinates from the point cloud. However, since the image must be quantified in the parameter space, it turns out to be practically identical with the image presented in Figure 11, albeit with considerably longer computing time—the value from Equation (4) must be determined for each of 48 million points, and in the previous version for 212 × 430 = 91,000 points only.

Figure 11. Planes detected by the 2D Hough transform: (**a**) Points on the same plane are marked with the same colour (the visible errors can be explained by inclusion of points on the floor and the ceiling); (**b**) The rectangles describe the first three detected planes; (**c**) Histogram of the deviation.

4.1.4. The 3D Hough Transform

Utilization of the 3D version requires the software developed by the authors. The quantified 3D image is created based on the XYZ point co-ordinates (the cell size is assumed as $2 \times 2 \times 2$ cm). Figure 12 illustrates the cross-sections.

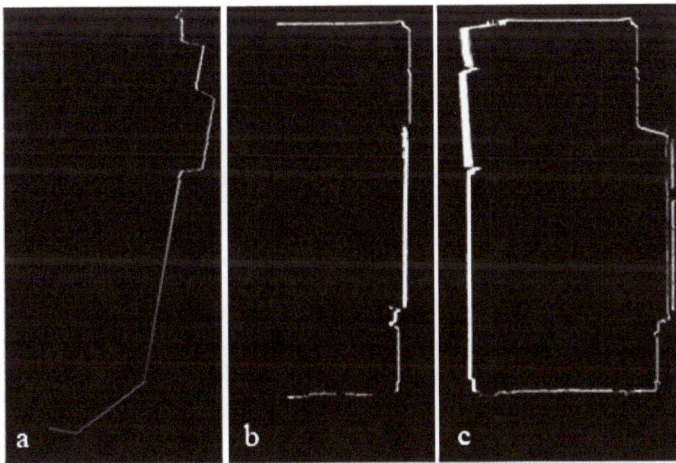

Figure 12. Example cross-sections of the 3D image utilized in the 3D Hough transform (the brightness corresponds to the number of points included in an individual cell): (**a**) XY image; (**b**) XZ image; (**c**) YZ image.

Then, the 3D image is generated in the parameter space by determining the r parameter as the ω, φ function and updating the counters by the number of points included in a corresponding XYZ cell. The angular resolution of 1 degree for $r = 2$ cm is assumed. As mentioned above,the parameter space is three-dimensional and its presentation on a plane is difficult. Figure 13 illustrates the parameter-space cross-sections.

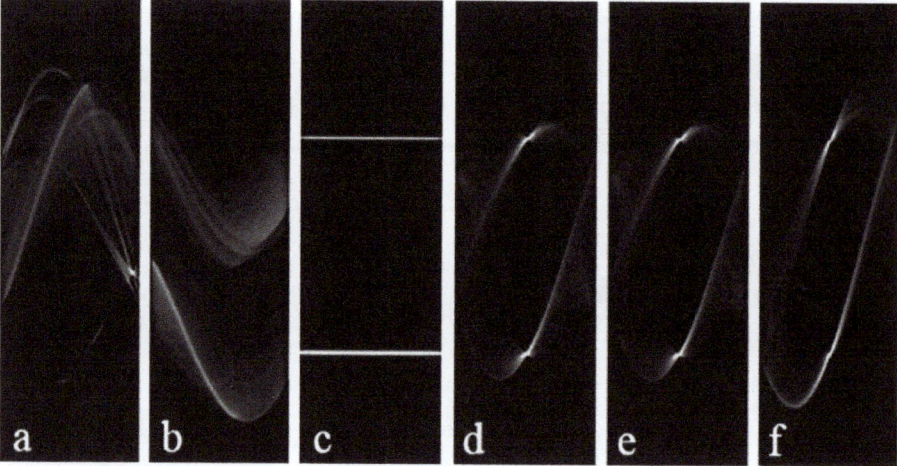

Figure 13. Example cross-sections of the domain of 3D parameters determined by the 3D Hough transform. The first three drawings show sections in the ω-r plane for the different parameters: φ: (**a**) $\varphi = 0°$; (**b**) $\varphi = 45°$; (**c**) $\varphi = 90°$. The next three show sections in the φ-r plane for the different parameters ω: (**d**) $\omega = 0°$; (**e**) $\omega = 45°$; (**f**) $\omega = 90°$.

The search planes illustrated in Figure 14 are almost identical to those used with the 3D RANSAC method because the calculated parameters are very similar (see Table 1).

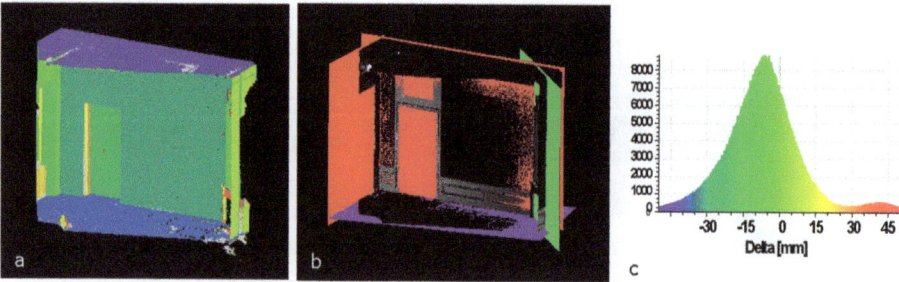

Figure 14. Planes detected by the 3D Hough algorithm: (**a**) Points on the same plane are marked with the same colour; (**b**) The rectangles describe the first four detected planes; (**c**) Histogram of the deviation.

Similarly to the 2D case, the algorithm utilizing the original XYZ values instead of 3D images was also tested. The calculation time was unacceptable; therefore the results of this case have not been presented in this paper.

4.1.5. Results

It can be observed that all the algorithms tend to divide points into groups, though ω values are similar (Table 1). In practice, the distribution of distance values from the theoretical surface of the wall is always close to normal, which allows us to consider the determined parameters as correct.

Source data comprised 48 million points with an average density of 10 points per cubic cm. Calculations were performed using an i5 2.8 GHz, 8 GB RAM PC. The original software was used, with the exception of the case of 3D RANSAC.

It can be seen that the computed plane normal vectors are similar. Significant differences in the values of R do not affect the further processing of orthoimages, as the projection is along the direction of R. The authors decided to use the 2D Hough algorithm because it gave the shortest computation time. Additionally, the 2D Hough algorithm not only detects planes but also allows similar planes to be connected that are not direct neighbours. Cultural heritage buildings have complicated structures and contain a lot of similar planes. Figure 15 shows an example of planes detected by the 2D Hough algorithm.

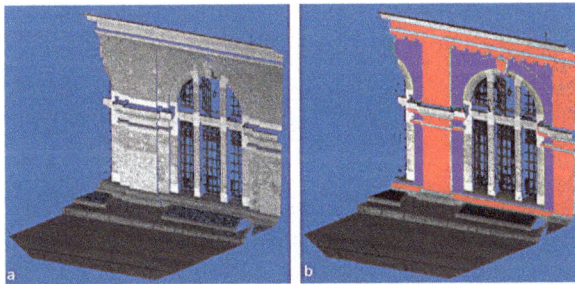

Figure 15. Results of 2D Hough algorithm for an object with many planes: (a) the raw point cloud; (b) detected points belonging to two different planes.

4.2. Automatic Generation of Intensity Orthoimages

The orthoimage generation process was performed automatically using the LupoScan software. A point cloud of 2 mm resolution was used. Thus, it was not possible to generate an orthoimage with a Ground Sample Distance (GSD) smaller than 2 mm. The following input parameters were assumed:

- NN interpolation;
- assumed deviation of automatically generated plane = 5 cm;

- buffer of projected points = $\pm 3 \times 5$ cm;
- resolution of generated object = 3 mm.

The resulting orthoimages (Figure 5a) were then used for further investigations concerning automation of terrestrial images. Additionally, depth information was recorded for each orthoimage, as well as information concerning the laser-beam reflectance intensity (Figure 5b).

4.3. Orientation of Images in Relation to Intensity Orthoimages—SIFT

Preliminary works were presented in [18]. This preliminary research was based on the utilization of two approaches to data orientation, with the use of cameras of determined interior orientation elements and images acquired with the use of a non-calibrated camera. The test site was the ruined castle in Iłża, which is characterized by relatively non-complicated surfaces whose shape could be considered as a plane. In the next stage, presented here, in order to test the methodology of the automatic generation of orthoimages some historical rooms at the Museum of King John III's Palace in Wilanów were selected as test sites. These rooms are characterized by numerous decorative elements and sculptural details. The terrestrial image orientation process was carried out based on TLS data, according to the two different approaches:

Variant I—the utilized image covered the entire area; the camera was calibrated in a test field; the exterior orientation elements were determined and considered further at the data processing stage (Figure 16a).

Variant II—a group of images was acquired that covered the analysed area. The horizontal coverage of the images was 70% and the vertical coverage up to 50%. Self-calibration was performed and the interior parameters were considered in further data processing (Figure 16b).

Figure 16. Examples of used images: (**a**) Variant I: one image covering the entire orthoimage area; (**b**) Variant II: part of multiple images covering the entire orthoimage area.

The images were processed in the following stages:

- SIFT keypoints detection on the image/images and the intensity orthoimage.

Use of this detector made it possible to detect evenly distributed keypoints, which were used as tie (homologous) points in the next step.

- SIFT keypoints description and matching.

During the first step the gradient value and orientation for each keypoint was calculated. In order to compare and match the described features, the "Best match" function was used [36].

- Division of points into control and check points;
- Bundle adjustment.

Due to the different textures and constituent materials of cultural heritage objects, incorrect distribution of tie points may occur. An automatic procedure was developed for this eventuality, which may be applied for controlling the distribution of points before the adjustment is performed. At the first stage the algorithm divides images into four areas and checks whether the points are distributed approximately evenly. If one group of points is 10 times greater than another, an incorrect distribution is signalled. In the next step the algorithm checks whether the number of points in each image quarter is smaller than two (*i.e.*, eight points in the entire image). If such a situation occurs, the process is terminated and the image is not oriented. After this stage the adjustment using the least-square method is applied (Gauss-Markov).

- Control of results

The obtained results were controlled based on "checkpoints".

4.3.1. Detection of Tie Points with the Use of the SIFT Algorithm (Authors' Application)

The SIFT algorithm [27] was applied to search for and match tie points, implemented in the OpenCV function library with default parameter values. Generated orthoimages were geo-referenced and the depth was recorded; this not only allowed the determination of field point co-ordinates in the orthoimage plane, but also the addition of the third dimension. These data allowed observations to be adjusted appropriately.

4.3.2. Orientation with the Use of Two Methods

Unfortunately, not all points that were automatically detected by the SIFT algorithm were correctly matched [18]. The criterion for correct point qualification related to the difference between the co-ordinates of a point determined on an image and the calculated value. The process was performed automatically using the software developed by the authors. In order to set a filtering threshold a series of empirical tests were performed.

Variant I

Eight single images of four walls were used for testing and analysis. This approach allowed implementation of the orthorectification process without inclusion of a mosaic step. The advantage of using the images in this way was the possibility of eliminating the influence of texture repeatability on the number and quality of detected homologous points (Figure 17).

Figure 17. Example of utilized image with control points in green and points eliminated from the adjustment process in red; a yellow "o"—a check point.

The proposed algorithm allows to detect about 2254 tie points but unfortunately, about 11% of them was incorrect (Appendix 1).

In order to enable an accurate analysis of image matching based on control points; deviations were also analysed. Due to the great number of tie points, the decision was made to present the results in the form of histograms showing deviations from the X and Y axes, respectively. Figure 18 shows four example diagrams for one of the images.

All analyses were performed using the authors' original software, based on the Matlab package.

It is assumed in geodesy that least-square adjustment methods have Gaussian distribution [13]. This confirms that only random errors occur, without gross errors. The possibility of evaluating the correctness of the results based on the normal distribution is influenced by the oscillation of the majority of deviations close to zero (the expected value). The control-point histograms show proximity to normal distribution, but are characterized by left- and right-hand obliquity (Figure 18). The histograms are not shifted, which means that systematic errors do not exist. The obliquity may be considered as having been influenced by the terrestrial laser-scanner data. The distribution of the control points shows deviations in an interval of ±3 pixels, which corresponds to a range of ±3 mm.

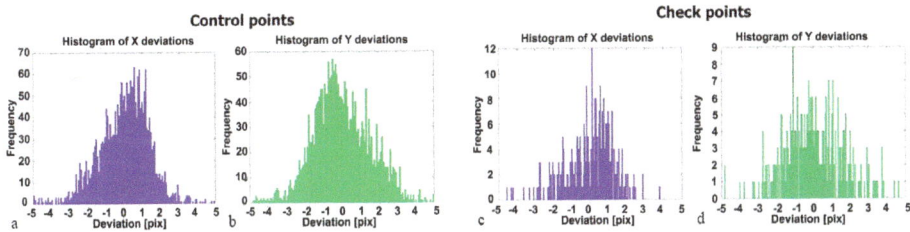

Figure 18. Histograms of control-and check-point deviation: (**a**) X direction in pixels—control; (**b**) Y direction in pixels—control; (**c**) X direction in pixels—check; (**d**) Y direction in pixels—check.

The check-point histograms also show an approximately normal distribution, where the majority of deviations are oscillating around zero. This confirms the lack of systematic errors. The deviations are in an interval of ±3.5 pixels, which corresponds to a range of ±3.5 mm. This image orientation accuracy is sufficient for orthoimage generation of GSD 3 mm.

The applied intensity orthoimage (including the depth map) was characterized by a GSD of 3 mm, and the image had a GSD of 1 mm. The obtained mean standard deviation for matching of a digital image was equal to approximately three pixels, which is equal to the size of the intensity orthoimage. Appendix 2 presents the extended statistical data.

Variant II

In the second variant, the group of images covering the entire wall was used for generation of the orthoimage.

Due to the repetitive texture of the analysed object (decorated fabrics), the tie points were incorrectly detected by the algorithm. During the process of filtration about 67% of points were eliminated from further processing. Examples of images with tie points are presented in Figure 19.

The analysed histograms show close-to-normal distribution, but are characterized by left- and right-hand obliquity (Figure 20a,b). In the distribution of control points, deviations are in the interval of ±3.5 pixels, which corresponds to a range of ±3.5 mm. The check-points histogram (Figure 20c,d) shows that Gaussian distribution was not achieved, due insufficient samples. Intervals of deviation values of between −2 and 2.5 pixels for the X coordinate and −2 and 5 pixels for the Y coordinate were taken to define accurate measurements; however, the majority of points oscillated between −2.5 and 2.5 pixels.

Figure 19. Distribution of points detected in the single image (from a group) and in the orthoimage: green "o"—control points; red "+"—wrong matched points; yellow "o"—check points.

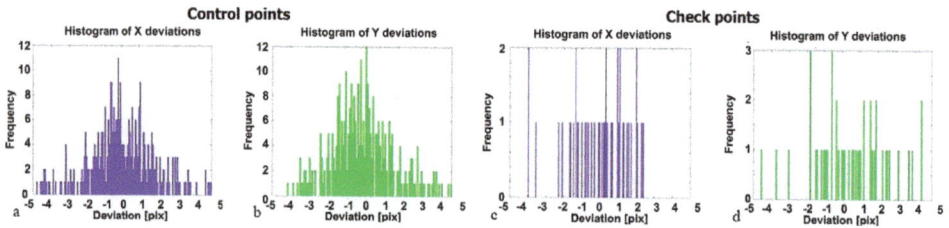

Figure 20. Histograms of control- and check-point deviations: (**a**) X direction in pixels-control; (**b**) Y direction in pixels control; (**c**) X direction in pixels-check; (**d**) Y direction in pixels-check.

Due to an excessive number of points being eliminated during tie-point-detection control, the decision was made to superimpose masks on orthoimages in order to improve the quality and the number of points detected in the combination of images and orthoimages. The mask generation process was performed automatically. After initial orientation of an image covering a part of an orthoimage, the projections of image coordinates on the orthoimage plane were calculated. Generation of such a mask allows generation of a new orthoimage for re-orientation of images.

Utilization of masks for tie-point detection improved the number of correctly detected points. The proposed algorithm allows to detect 690 tie points however about 19% of them was incorrect (Appendix 3).

Matching-point deviations were analysed and the results presented in the form of histograms. Figure 21 presents examples of four diagrams.

Figure 21a,b shows that Gaussian distribution was not achieved. An interval between −2 and 2.5 pixels for the X coordinate and −2 and 5 pixels for the Y coordinate were taken as measures of accuracy; however, the majority of points oscillated between −5 and 5 pixels, corresponding to a range of ±5 mm. In the case of the check points (Figure 21c,d), deviations are around ±5 pixels, corresponding to a range ±5 mm. Extended statistics are shown in Appendix 4.

The obtained maximum values of standard deviations for matching of a digital image were equal to approximately four pixels, close to the size of the intensity orthoimage.

These analyses present only the accuracy of image orientation in relation to TLS data.

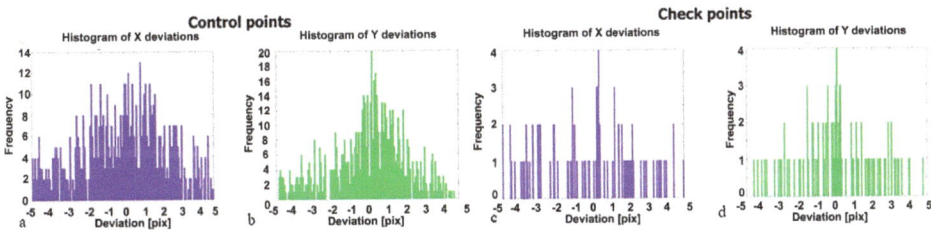

Figure 21. Histograms of control- and check points deviations: (**a**) X direction in pixels-control; (**b**) Y direction in pixels-control; (**c**) X direction in pixels-check; (**d**) Y direction in pixels-check.

4.3.3. Orientation Using 3D DLT Method

Another method of orientation of images in relation to terrestrial laser-scanning data is the DLT (Figure 22). In order to test the efficiency of this method, several images covering the investigated area were analysed to assess the quality of the generated orthoimage close to the area borders. When the DLT method is applied it is not possible to directly utilize interior orientation elements obtained from calibration in particular distortion. Image orientation using the DLT method takes 11 coefficients (or more when deformations are caused, for example, by distortion) into consideration. DLT is performed separately for each image; therefore, it is not possible to apply the camera calibration *a priori*; it is performed separately for each image. It is also impossible to determine DLT coefficients if all matching points are located on one plane, which is highly probable in the analysed cases.

Figure 22. Example of utilized image with points used for orientation (control points, in green) and yellow "o"—check points.

Control point deviations were also analysed, and the results presented as histograms showing deviations from the X and Y axes, respectively (Figure 23).

Compared to the independent bundle method, the orientation results obtained with the DLT method are characterized by significantly lower orientation accuracy for control points. Obtained deviations for check points reach as high as 20 pixels (GSD 3 mm).

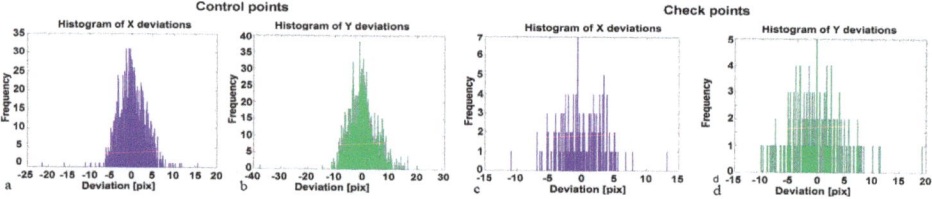

Figure 23. Histograms of deviations of control and check points: (**a**) X direction in pixels-control; (**b**) Y direction in pixels-control; (**c**) X direction in pixels-check; (**d**) Y direction in pixels-check.

4.4. RGB Orthoimages Generation and Inspection (SURF Algorithm)

The process of generation of colour orthoimages was performed automatically using the LupoScan package. Due to the nature of the regular DSM, the backward projection based on interpolation of the radiometric values for all nodes of the network was applied. This avoided deformations of radiometric values and allowed colour to be maintained, as required in conservation analyses.

The final orthoimage (after mosaicking) is shown in Figure 24 (Appendix 5). The red rectangle shows the places where two different images overlap. It can be seen that there are no parallaxes, only differences in colour.

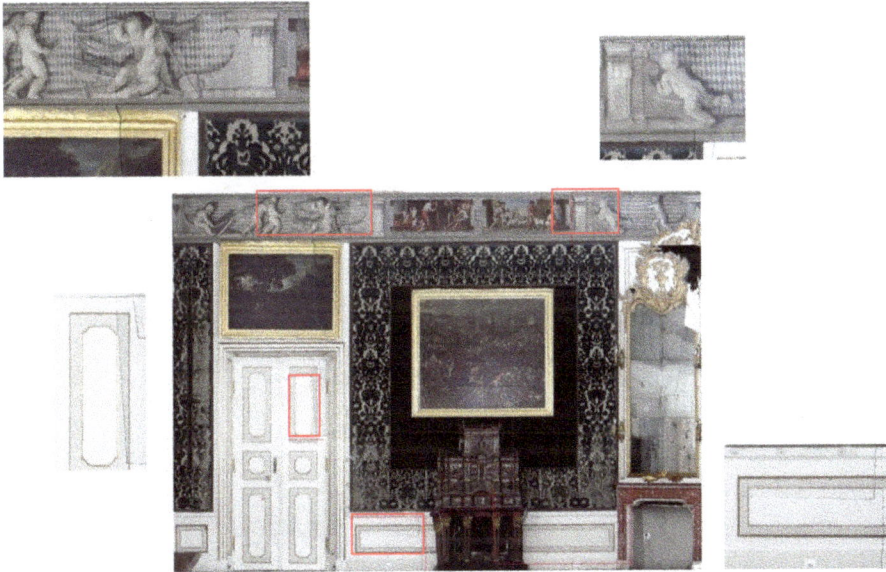

Figure 24. RGB orthoimage with marked overlaps and seam lines.

In order to analyse the accuracy of the generated RGB orthoimage, it was compared with the intensity orthoimage (Table 2).

Table 2. Deviation between check points on intensity orthoimages and RGB orthoimages (GSD—3 mm).

Id	RMSE		>2 RMSE	
	X (mm)	Y (mm)	X (%)	Y (%)
1	2.69	3.12	2.5	1.8
2	2.03	2.00	2.4	2.4
3	2.15	2.06	4.0	4.6
4	2.19	1.76	10.6	13.4

Figure 25 shows considerable displacements at the image borders. This results from uneven distribution of automatically detected matching points and from distortion, which has not been considered in the utilized 3D DLT method.

Figure 25. An RGB image with displacements resulting from inaccurately oriented images using the 3D DLT method.

In order to analyse the accuracy of the generated RGB orthoimage, it was compared with the intensity orthoimage. The obtained root mean square error (RMSE) values were: X—5.91 mm, Y—6.76 mm, Z—6.76 mm. For 28% of points, the deviation was larger than 2 RMS.

5. Discussion

The experiments proved that, depending on the scanner and the surface type of the analysed object, the "depth" of the cloud of points is often great; indeed, it can be greater than the scan resolution (see Section 3.1). Therefore, the quality of the point cloud has an important influence on the correctness of the determined projection plane, the quality of the image generated in the intensity, and the accuracy of the orientation of terrestrial images. The present authors propose that every orthoimage generation process, particularly in relation to cultural heritage objects, should be preceded by an initial analysis of the geometric quality of the point cloud. In the experiments presented in this paper, original software applications were used.

Five methods of detection of the projection plane of an orthoimage in the point cloud were tested (described in Section 4.1). Each of the search methods produced similar results, but some turned out to be highly time consuming [23–25,30]. Based on the performed analyses, it can be stated that the fastest implementation of automatic detection of a plane is with the modified Hough 2D algorithm (Table 1). In order to verify the operations of particular methods, the original software application was developed in C++; this will be modified in further experiments.

Due to the fact that the orientation process of images requires searching for homologous points, CV algorithms—the detector and descriptor SIFT—were used for that purpose. The experiments were performed in two variants: in the first variant, an individual image covering the entire investigated area was used, and in the second variant a group of images covering the entire area (see Section 4.3.2). Matched points were divided into two groups—control and check points—for orientation and checking. For the first variant, a detailed accuracy analysis is presented in Appendix 1; the maximum relative error is equal to 0.2%. For the second variant, the solution utilized the entire intensity orthoimage and the mask that limits the projected area. When the area is not limited by the mask, the points are incorrectly matched by the SIFT descriptor (about 67% incorrectly matched points) [27]. Tests were performed towards the automation of the process of defining the mask on the intensity image. The basic version of the 3D DLT method used here involves no interior orientation elements obtained from calibration. The accuracy of image orientation is therefore lower in relation to TLS data than when the independent bundle method is used, and generated images are characterized by considerable displacements in the common overlap areas.

A detailed geometric inspection of obtained orthoimages was performed for both variants. For this purpose, an independent CV algorithm—SURF—was applied. The intensity orthoimage and the RGB orthoimage were compared (see Section 4.4). The mean value of deviations was smaller than 2 pixels (GSD 3 mm).

6. Conclusions

The performed experiments proved that the orthoimage generation process can be automated, starting with generation of the reference plane and ending with automatic RGB orthoimage generation.

The proposed original approach, based on the modified Hough transform, allows us to considerably improve the process of searching for planes on entire scans, without the necessity of dividing them. Thus, the split-up scan with automatically generated planes simultaneously determines the number of orthoimages and the maximum deviations from the plane, which define the range (in front of and behind the plane) of points required for orthoimage generation. This new approach accelerates the orthoimage generation process and eliminates scanning noises from close-range scanners.

Utilization of orthoimages in the form of raster data, complemented by the third dimension, allows the application of CV algorithms to search for homologous points and perform spatial orientation of free-hand images. Such an approach eliminates the necessity to measure the control points.

This approach is not limited to images acquired by cameras integrated with scanners, which are characterized by lower geometric and radiometric quality. The solution is particularly recommended for generation of photogrammetric documentation of high resolution, when resolution and accuracy in the order of individual millimetres are required.

In conclusion, this approach with original software allows more complete automation of the generation of the high-resolution orthoimages used in the documentation of cultural heritage objects.

Acknowledgments: The authors would like to thank the reviewers and copy editors, whose helpful comments led to a better paper overall and also the Museum of King John III's Palace in Wilanów for cooperation.

Author Contributions: Dorota Zawieska came up with the concept for the paper and wrote the first draft. Piotr Podlasiak performed algorithms and applications of plane direction of point clouds. Jakub Stefan Markiewicz implemented the OpenCV SIFT and SURF algorithms and performed algorithms and applications of data orientation. All authors participated in the data analysis.

Conflicts of Interest: The authors declare no conflict of interest.

References

1. Mavromati, D.; Petsa, E.; Karras, G. Theoretical and practical aspects of archaeological orthoimaging. *Int. Arch. Photogram. Remote Sens.* **2002**, *34*, 413–418.
2. Guarnieri, A.; Remondino, F.; Vettore, A. Digital photogrammetry and TLS data fusion applied to Cultural Heritage 3D modeling. In Proceedings of the ISPRS Commission V Symposium Image Engineering and Vision Metrology, Dresden, Germany, 25–27 September 2006.
3. Grussenmeyer, P.; Landes, T.; Voegtle, T.; Ringle, K. Comparison methods of terrestrial laser scanning, photogrammetry and tacheometry data for recording of cultural heritage buildings. *Int. Arch. Photogram. Remote Sens. Spat. Inf. Sci.* **2008**, *37*, 213–218.

4.   Armesto-González, J.; Riveiro-Rodríguez, B.; González-Aguilera, D.; Rivas-Brea, M.T. Terrestrial laser scanning intensity data applied to damage detection for historical buildings. *J. Archaeol. Sci.* **2010**, *37*, 3037–3047.

5.   Giuliano, M.G. Cultural heritage: An example of graphical documentation with automated photogrammetric systems. *Int. Arch. Photogram. Remote Sens. Spat. Inf. Sci.* **2014**.

6.   Kersten, T.; Mechelke, K.; Maziull, L. 3D model of al zubarah fortress in Qatar—Terrestrial laser scanning *vs.* dense image matching. In Proceeding of the International Archives of the Photogrammetry, Remote Sensing Spatial Information Science, 2015 3D Virtual Reconstruction and Visualization of Complex Architectures, Avila, Spain, 25–27 February 2015.

7.   Ippoliti, E.; Meschini, A.; Sicuranza, F. Structure from motion systems for architectural heritage. A survey of the internal loggia courtyard of Palazzo Dei Capitani, Ascoli Piceno, Italy. In Proceeding of the International Archives of the Photogrammetry, Remote Sensing Spatial Information Science, 2015 3D Virtual Reconstruction and Visualization of Complex Architectures, Avila, Spain, 25–27 February 2015.

8.   Ballabeni, A.; Apollonio, F.I.; Gaiani, M.; Remondino, F. Advances in image pre-processing to improve automated 3D reconstruction. In Proceeding of the International Archives of the Photogrammetry, Remote Sensing Spatial Information Science, 2015 3D Virtual Reconstruction and Visualization of Complex Architectures, Avila, Spain, 25–27 February 2015.

9.   Alsadik, B.; Gerke, M.; Vosselman, G.; Daham, A.; Jasim, L. Minimal camera networks for 3D image based modeling of cultural heritage objects. *Sensors* **2014**, *14*, 5785–5804.

10.   Chiabrando, F.; Donadio, E.; Rinaudo, F. SfM for orthophoto generation: A winning approach for cultural heritage knowledge. *Int. Arch. Photogram. Remote Sens. Spat. Inf. Sci.* **2015**, *1*, 91–98.

11.   Markiewicz, J.S.; Zawieska, D. Terrestrial scanning or digital images in inventory of monumental objects? Case study. *Int. Arch. Photogram. Remote Sens. Spat. Inf. Sci.* **2014**, *5*, 395–400.

12.   Ramos, M.M.; Remondino, F. Data fusion in cultural heritage—A review. In Proceeding of the 25th International CIPA Symposium on The International Archives of the Photogrammetry, Remote Sensing and Spatial Information Sciences, Taipei, Taiwan, 31 August—4 September 2015; pp. 359–363.

13.   Luhmann, T.; Robson, S.; Kyle, S.; Boehm, J. *Close Range Photogrammetry and 3D Imaging*, 2nd ed.; Walter De Gruyter: Boston, MA, USA, 2013.

14.   González-Aguilera, D.; Rodríguez-Gonzálvez, P.; Gómez-Lahoz, J. An automatic procedure for co-registration of terrestrial laser scanners and digital cameras. *ISPRS J. Photogram. Remote Sens.* **2009**, *64*, 308–316.

15.   Remondino, F. Heritage recording and 3D modeling with photogrammetry and 3D scanning. *Remote Sens.* **2011**, *3*, 1104–1138.

16. Georgopoulos, A.; Tsakiri, M.; Ioannidis, C.; Kakli, A. Large scale orthophotography using DTM from terrestrial laser scanning. In Proceeding of the International Archives of the Photogrammetry, Remote Sensing Spatial Information Science, Istanbul, Turkey, 12–23 July 2004; Volume 35, pp. 467–472.

17. Georgopoulos, A.; Makris, G.N.; Dermentzopoulos, A. An alternative method for large scale orthophoto production. In Proceedings of the CIPA 2005 XX International Symposium, Torino, Italy, 26 September–1 October 2005.

18. Markiewicz, J.S.; Podlasiak, P.; Zawieska, D. Attempts to automate the process of generation of orthoimages of objects of cultural heritage. *Int. Arch. Photogram. Remote Sens. Spat. Inf. Sci.* **2015**.

19. Moussa, W.; Abdel-Wahab, M.; Fritsch, D. An automatic procedure for combining digital images and laser scanner data. *Int. Arch. Photogram. Remote Sens. Spat. Inf. Sci.* **2012**, *39*, 229–234.

20. Meierhold, N.; Spehr, M.; Schilling, A.; Gumhold, S.; Maas, H.G. Automatic feature matching between digital images and 2D representations of a 3D laser scanner point cloud. In Proceedings of the ISPRS Commission V Mid-Term Symposium on Close Range Image Measurement Techniques, Newcastle upon Tyne, UK, 21–24 June 2010.

21. Fischler, M.A.; Bolles, R.C. Random sample consensus: A paradigm for model fitting with applications to image analysis and automated cartography. *Comm. ACM* **1981**, *24*, 381–395.

22. PCL. Available online: http://pointclouds.org/ (accessed on 25 May 2015).

23. Hough, P.V.C. Method and Means for Recognizing Complex Patterns. U.S. Patent 3,069,654, 18 December 1962.

24. Duda, R.O.; Hart, P.E. Use of the Hough transformation to detect lines and curves in pictures. *Commun. ACM* **1972**, *15*, 11–15.

25. Ballard, D.H. Generalizing the Hough transform to detect arbitrary shapes. *Pattern Recognit.* **1981**, *13*, 111–122.

26. Overby, J.; Bodum, L.; Kjems, E.; Ilsoe, P.M. Automatic 3D building reconstruction from airborne laser scanning and cadastral data using Hough transform. *Int. Arch. Photogram. Remote Sens. Spat. Inf. Sci.* **2004**, *35*, 296–301.

27. Lowe, D. Distinctive image features from scale-invariant key points. *Int. J. Comput. Vis.* **2004**, *60*, 91–110.

28. Bay, H.; Tuytelaars, T.; van Gool, L. SURF: Speeded Up Robust Features. Available online: www.vision.ee.ethz.ch/~surf/eccv06.pdf (accessed on 1 September 2015).

29. Markiewicz, J.S.; Zawieska, D.; Kowalczyk, M.; Zapłata, R. Utilisation of laser scanning for inventory of an architectural object using the example of ruins of the Krakow Bishops' Castle in Ilza, Poland. In Proceeding of the 14th GeoConference on Informatics Geoinformatics and Remote Sensing, Ilza, Poland, 19–25 June 2014; pp. 391–396.

30. Bradski, G.; Kaehler, A. *Learning OpenCV Computer Vision with the OpenCV Library*; O'Reilly Media: Sebastopol, CA, USA, 2008.

31. OpenCV. Available online: http://opencv.org/ (accessed on 12 May 2015).

32. 3D Reconstruction Using the Direct Linear Transform with a Gabor Wavelet Based Correspondence Measure. Available online: http://bardsley.org.uk/wp-content/uploads/2007/02/3d-reconstruction-using-the-direct- linear-transform.pdf (accessed on 23 October 2015).

33. Direct Linear Transformation(DLT). Available online: https://me363.byu.edu/sites/me363.byu.edu/ files/userfiles/5/DLTNotes.pdf (accessed on 23 October 2015).

34. Stavropoulou, G.; Tzovla, G.; Georgopoulos, A. Can 3D point clouds replace GCPs. *ISPRS Ann. Photogram. Remote Sens. Spat. Inf. Sci.* **2014**, *5*, 347–354.

35. Cloud Compare. Available online: http://www.danielgm.net/cc/ (accessed on 14 May 2015).

36. OpenMVG. Available online: http://imagine.enpc.fr/~moulonp/openMVG/index.html (accessed on 30 May 2015).

37. OpenMVG. Available online: https://openmvg.readthedocs.org/en/latest/ (accessed on 30 May 2015).

A Multi-Data Source and Multi-Sensor Approach for the 3D Reconstruction and Web Visualization of a Complex Archaelogical Site: The Case Study of "Tolmo De Minateda"

Jose Alberto Torres-Martínez, Marcello Seddaiu, Pablo Rodríguez-Gonzálvez, David Hernández-López and Diego González-Aguilera

Abstract: The complexity of archaeological sites hinders creation of an integral model using the current Geomatic techniques (i.e., aerial, close-range photogrammetry and terrestrial laser scanner) individually. A multi-sensor approach is therefore proposed as the optimal solution to provide a 3D reconstruction and visualization of these complex sites. Sensor registration represents a riveting milestone when automation is required and when aerial and terrestrial datasets must be integrated. To this end, several problems must be solved: coordinate system definition, geo-referencing, co-registration of point clouds, geometric and radiometric homogeneity, etc. The proposed multi-data source and multi-sensor approach is applied to the study case of the "Tolmo de Minateda" archaeological site. A total extension of 9 ha is reconstructed, with an adapted level of detail, by an ultralight aerial platform (paratrike), an unmanned aerial vehicle, a terrestrial laser scanner and terrestrial photogrammetry. Finally, a mobile device (e.g., tablet or smartphone) has been used to integrate, optimize and visualize all this information, providing added value to archaeologists and heritage managers who want to use an efficient tool for their works at the site, and even for non-expert users who just want to know more about the archaeological settlement.

Reprinted from *Remote Sens.* Cite as: Torres-Martínez, J.A.; Seddaiu, M.; Rodríguez-Gonzálvez, P.; Hernández-López, D.; González-Aguilera, D. A Multi-Data Source and Multi-Sensor Approach for the 3D Reconstruction and Web Visualization of a Complex Archaelogical Site: The Case Study of "Tolmo De Minateda". *Remote Sens.* **2016**, *8*, 550.

1. Introduction

Several techniques have been applied thus far for the 3D reconstruction and visualization of archaeological settlements based on the use of close-range photogrammetry [1,2], terrestrial laser scanner (TLS) [3,4] or unmanned aerial vehicles (UAV) [5]. However, due to the inherent complexity of these sites, several problems arise when 3D reconstruction of these sites is mandatory and just one type of geotechnology is applied. For instance, aerial photogrammetry exhibits problems reconstructing vertical planes, common in archaeological sites, whereas terrestrial laser scanners or terrestrial photogrammetry could provide good results. However, these terrestrial techniques are subject to problems with horizontal surfaces or elevated areas. Other authors have explored multi-data and multi-sensor approaches to record and reconstruct complex archaeological sites. Recently in [6], the potential of this type of hybrid approach is shown for the analysis and interpretation of 3D/4D information applied to archaeological settlements. In order to guarantee geometric and radiometric quality, a combination of TLS and terrestrial photogrammetry is used and applied, enabling the monitoring of the settlement based on a volume analysis. However, this approach has problems related to the recording of more elevated areas. In order to solve this limitation, in [7], the authors use an aerial multi-sensor approach for the 3D reconstruction and visualization of archaeological settlements which provides a very good coverage of those elevated areas. The quality and precision of the TLS and UAV registration has been outlined in [8] where authors reconstruct the interior and exterior of the Church of Santa Barbara (Italy) after the earthquake it suffered in 2012. Trying to overcome the main UAV limitations, payload and autonomy, other authors have proved that low-cost manned platforms such as the paratrike can be an efficient solution for the recording of large sites [9], including archaeological sites [10], and allowing to put on board multiple sensors such as thermographic or multispectral cameras [11]. In those cases where the archeological site is complex and subterranean, other hybrid, dynamic and terrestrial approaches could be interesting. Unfortunately, there are not many mobile laser scanners, photogrammetric or hybrid systems for subterranean sites available in the market [12,13], even less specific for the field of archaeological recording. Some authors such as Canter et al. [14] have developed indoor mapping systems for the generation of indoor cartography from accurate geospatial information. The main advantage of these systems is that they integrate high-precision GNSS with advanced inertial technology (accelerometers and gyroscopes) for the geo-referencing of the site using measures from its exterior, apart from an Applanix POS system for positioning and orientation that provides uninterrupted measurements of the true position, roll, pitch and yaw of the vehicle moving indoors. Other authors have advanced more sophisticated systems based on a wheeled mobile robot and a multi-sensor global registration approach [15]. In particular, a geometric model to derive depth information is proposed based on a registration of heterogeneous 3D data arising from eight ultrasonic sonars, one TLS and three visual sensors.

It thus becomes clear that with the advances in multi-sensor and multi-data from different sources, data integration has become as a valuable tool in archaeological applications. The main objective of multi-sensor data integration is to register sensor data from different sources—with different characteristics, resolution, and quality in order to provide more reliable, accurate, and useful information required for diverse archaeological applications. In addition, this multi-data and multi-sensor integration can be improved through 3D geographical information systems (3DGIS) together with the final presentation of the products based on 3D Web.

The goal of this study is to propose a multi-data source and multi-sensor workflow in order to obtain high quality archaeological products, contributing to the robust interpretation of the observed objects/scenes and providing the basis of effective planning and decision making, essential in archaeology. To this end, terrestrial scans and images will be registered with aerial images acquired from a paratrike and an unmanned aerial vehicle (UAV). The methodology employed for data processing has been extensively tested by several authors. In [16], the authors employed computer vision algorithms for image-based modelling from aerial imagery and point cloud generation in urban environments, which are similar to those used in this work.

To carry out studies of characterization, measurement and analysis of the surface and the elements presented in the site, the geometrical information (acquired with geomatic sensors) is combined with other available thematic data such as photographs, sketches, restoration reports, schedules, etc. in a 3DGIS, Geoweb3D® [17]. In this way, a description of the site through time is possible. Also, the availability of the three-dimensional information of the settlement through the Web using mobile devices provides added value for archaeologists and heritage managers, simplifying the data acquisition, as well as its analysis in the field. Moreover, the centralization of the information and its external storage means it becomes available to different experts and organizations. Last but not least, the knowledge of the settlement is open to the general public based on an easy-to-use interface which integrates different 2D and 3D resources.

In this study, a specific simplification and optimization procedure was implemented to visualize the different 3D products through the Web using mobile devices and the Open Source library Cesium [18], developed by the company Analytical Graphics, Inc. (AGI, Greenbelt, MD, USA) [19]. Other authors [20] have developed similar works using their own system, *SGIS3D*, which uses VRML (Virtual Reality Modeling Language) format. However, one of the main limitations of this language is the mandatory use of plugins, as well as its lack of optimization through the graphical processing unit (GPU), crucial in those steps related to texture mapping. Another similar approach in the archaeological field is developed by [21], who implements a spatial data infrastructure known as *QuaeryArch3D* and applied it to the archaeological settlement of "Maya de Copan" in Honduras, a UNESCO World Heritage Site. In this case, they use the Open Source PostgreSQL and PostGIS for the integration and management of information in the database, whereas the 3D visualization is performed with Unity. However, one of the main limitations of Unity is dealing with huge 3D models coming from the point clouds. To overcome this problem, WebGL seems to provide an appropriate solution using similar developments to those presented in [22] or developing a new approach as the one presented in this paper. Furthermore, advancing a 3DGIS Web solution requires using different simplification and optimization processes, as well as different hierarchical (i.e., pyramidal) strategies for visualization which are not compatible with the Unity engine.

This paper is organized as follows: after this introduction, in Section 2, the different sensors employed and their characteristics are detailed. In Section 3, the proposed multi-data source and multi-sensor approach is described, as well as the simplification and optimization process. Experimental results are shown and discussed in Section 4. Conclusions and future directions are given in Section 5. Finally, two appendices are included; the first corresponding to the abbreviations used in the paper, whereas the second encompasses the explanation of the methodology employed for 3D Web visualization.

2. Materials

2.1. Paratrike

The main aerial platform employed to the documentation of the archaeological site was a paratrike (Figure 1a). It is a low-cost aircraft with more flexibility than conventional aircrafts, and more autonomy and payload capacity than the UAVs. This last characteristic allows the possibility of boarding better sensors than the UAVs, or even multiple sensors in a stabilized gimbal (MUSAS-MUltiSpectral Airborne Sensors). In particular, a tandem trike AIRGES (Table 1) was used to map the whole archaeological site following a vertical flight and using a full-frame reflex camera.

Table 1. Technical specifications of the paratrike.

Motor	Rotax 503 Two-Stroke Motor
Trike	Tandem Trike AIRGES
Tandem paraglide	MAC PARA Pasha 4 Trike 39 6 42
Emergency system	Ballistic parachutes GRS 350
Weight	110 kg
Weight capability	165–250 Kg
Air velocity range	30–60 km/h

The use of the gyro-stabilized camera platform, MUSAS (Figure 1b), guarantees the accurate orientation of the camera according to the flight planning by two servomotors arranged on the x and y axes, controlled by an Arduino board, which

323

incorporates an IMU with 6 degrees of freedom: three accelerometers, a double-shaft gyroscope (for pitch and roll) and an additional gyroscope for yaw.

For the paratrike, a full-frame reflex camera, Canon 5D MkII, with a fixed focal length to achieve a GSD of 3 cm, was used. This camera was also used for the terrestrial photogrammetry and photorealistic texture mapping due to its better image quality.

In contrast to the UAV platform which encloses an integrated navigation system (GPS, IMU and barometric altimeter), the paratrike requires an external set of sensors in order to provide navigation capabilities and thus fulfill photogrammetric constraints for data acquisition. In particular, the planimetric position is provided by a GPS antenna (Trimble Bullet III), installed in the camera platform close to the optical centre of the camera, connected to a mono-frequency receiver Ublox EVK-6T-0. This system yields an absolute precision of ± 9 m on the horizontal axis for 95% of the time [23]. During data acquisition the pilot follows the planned photogrammetric mission in a rugged table connected to the GPS system, where the real-time track is contrasted with the planned flight. The final component in the navigation system, the altimetry, which affects the GSD, is controlled by an altimetric barometer (DigiFly VL100) with an absolute precision of ± 8 m. This solution is chosen instead of GPS receiver, since the absolute altimetric GPS precision is just ± 15 m.

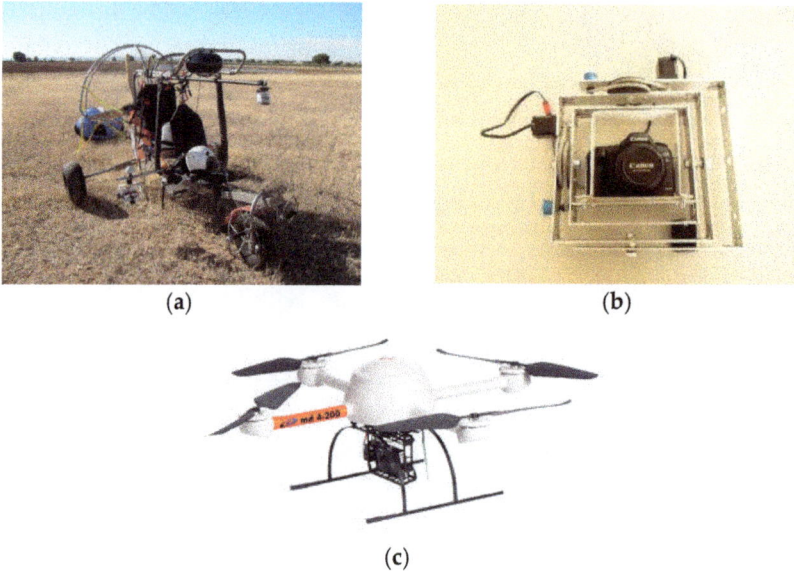

(a)

(b)

(c)

Figure 1. Paratrike employed (**a**); detail of the stabilization platform "MUSAS" (**b**) and UAV used in complex areas (**c**).

2.2. Unmanned Aerial Vehicle

For the aerial photogrammetric acquisition in the "El Reguerón" site, the paratrike platform was rejected due to the morphological characteristics of the terrain and the height of flight required to reach the high spatial resolution (GSD of 1 cm). The main drawback was the need to vary the flying height to keep the same scale, due to the presence of high reliefs along the strips. As an additional disadvantage, the walled constructions were occluded between walls of natural rock. Therefore, in order to complete the archaeological site documentation, a UAV was employed. Specifically, a Microdrone md4-200 (Table 2) was used (Figure 1c) to map the most challenging area following vertical and oblique flights through use of a compact camera.

In spite of the manoeuvrability provided by this UAV, it has a limited payload required to employ a compact camera for the photogrammetric flight.

For the UAV, an ultra-compact camera, Canon IXUS 115 HS, was chosen allowing a GSD of 1 cm.

Basically, the multi-data obtained by UAV and paratrike is geometrical, i.e., two point clouds with metric properties and texture information (RGB) which are homogenized under a common reference system based on a network of control and check points.

Table 2. Technical specifications of the Microdrone md4-200 platform.

UAV Weight	900 g
Payload	up to 200 g
Size	54 cm between rotors
Flight time	10 to 20 min
Operating temperature	−10 to 50 °C
Max. height flight	500 m
Max. wind	5 m/s

2.3. Terrestrial Laser Scanner

In those complex zones, a phase shift terrestrial laser scanner, Faro Focus 3D, was employed (Table 3).

According to the archaeological settlement characteristics and the TLS technical specifications, laser stations were established in a network that guaranteed an average spatial resolution of 5 mm for the whole scenario. The mean distance acquisition was 15 m.

Table 3. Technical specifications of the terrestrial laser scanner (TLS), Faro Focus 3D.

Model	Faro Focus 3D
Principle	Phase Shift
Wavelength	905 nm (Near infrared)
Field of view	360° H × 320° V
Range std. deviation	2 mm at 25 m
Measurement range	0.19 mrad
Beam divergence	8 mm at 50 m
Scanning speed	976,000 points/s

2.4. Geo-Referencing System

The establishment of the mapping frame in the study area is performed with two GNSS bi-frequency (L1, L2) receivers, Topcon manufacturer. The GNSS observation method was real-time kinematic (RTK) getting a relative and absolute precision of 1 cm and 3 cm, respectively.

The coordinate reference system was comprised of the official coordinate system established by Spanish law, a Compound Coordinate Reference System (CCRS) integrated by horizontal CRS referred to ETRS89 geodetic reference system and UTM Zone 30 mapping projection (EPSG: 25830), and vertical CRS with geoid's origin defined in Alicante (Spain) (EPSG: 5782). This mapping frame was materialized by a GNSS surveyal using natural features for "El Reguerón" area, which could be clearly identified on aerial images (e.g., corners of well-defined objects, small features with excellent contrast, etc.). In this way, we avoid artificial targets appearing in those more emblematic parts of the archaeological settlement. On the contrary and due to the large extension area, artificial targets were used for the full recording of the archaeological settlement (with the paratrike) in order to establish a network of control and checkpoints.

3. Methodology

Given the complexity of the archaeological site and the archaeological documentation requirements, the aerial and terrestrial data acquisition was planned in order to provide an integral and integrated recording of the site.

All the data from the different sensors and plataforms were processed according to the workflow outlined in Figure 2 and explained in the following subsections.

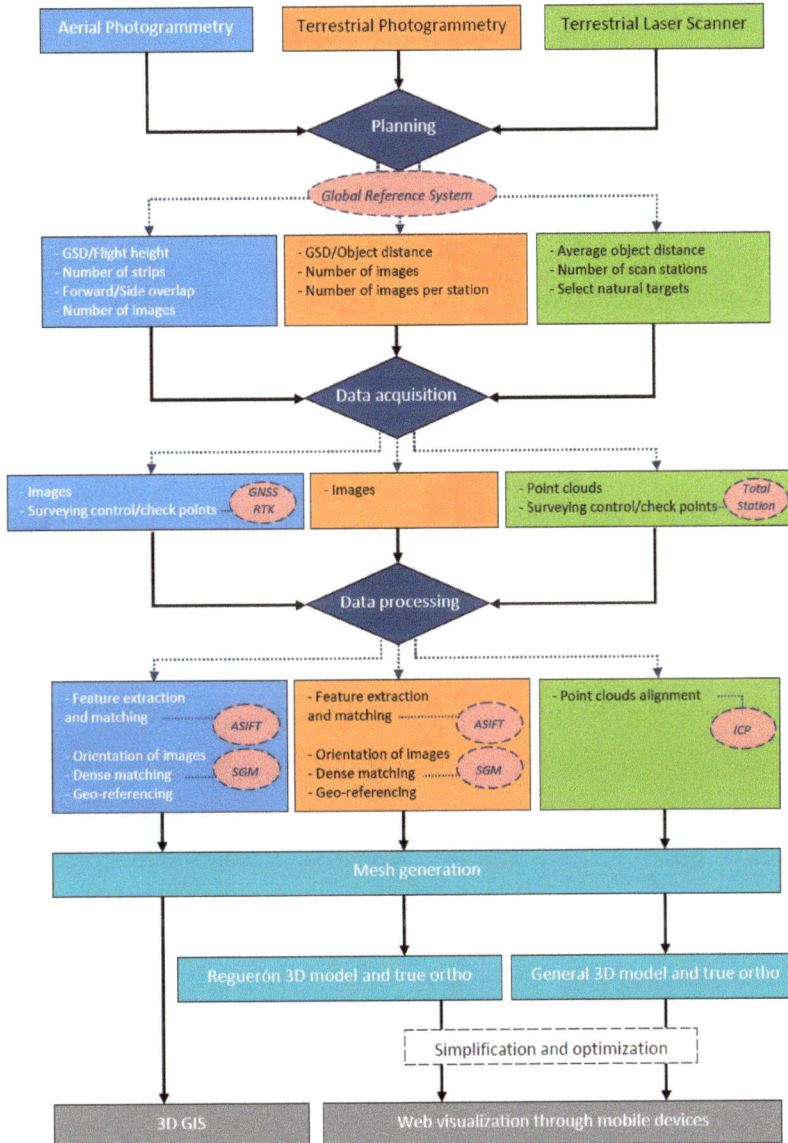

Figure 2. Multi-sensor and multi-data source workflow for the 3D reconstruction of complex archaeological sites.

3.1. Data Acquisition

Each technique (i.e., photogrammetry, laser scanning) requires different procedures and protocols for data acquisition.

In the case of aerial data acquisition, image projection centers and camera attitude must be previously defined according to the classical photogrammetric parameters. The flight planning was done using the in-house software, MFlip and PFlip, for the UAV and paratrike flights, respectively [24]. The main difference between both flights was the type of flight planning: for the paratrike, a standard stereoscopic photogrammetric flight was undertaken (Figure 3a), whereas for the case of UAV, oblique and vertical images were also considered. Therefore, data acquisition was completely planned in order to get better results. In particular, for the UAV, a script was prepared for an automatic photogrammetric flight; whereas for the paratrike, the flight axis and flight height were planned and followed by the pilot. Both flights were planned with higher overlaps in order to get better results in the dense matching process and both flights were planned considering the relief variation using a public digital terrain model (DTM) [24].

(a)   (b)

Figure 3. Example of image footprints projected over terrain (**white** line) and the different workspaces (**red** line). (a) Paratrike flight planned over the whole area and (b) UAV flight planned over "El Reguerón" area.

The result shown in Figure 3b corresponds to the UAV flight planning, which was complemented with oblique aerial images acquired manually. As can be seen through the image footprints, the relief effects make it difficult to maintain a constant GSD, so the data acquisition is complemented with terrestrial images. For its part, the paratrike was used for the whole recording of the settlement due to the autonomy and sensors limitations of the UAV.

For the "El Reguerón" area a more detailed flight with the UAV was designed using the ultra-compact camera.

An important issue for the aerial data acquisition was the image sharpness, which affects the final photogrammetric reconstruction. This issue is affected by the camera parameters (i.e., aperture, shutter time and sensibility), the platform performance (i.e., flight speed and efficiency of the stabilized gimbal to absorb the paratrike vibrations) and the scenario illumination conditions. The flight was executed on a cloudy day to avoid shadows being projected, and the camera sensibility and shutter time were set up for these conditions. The aperture and the focal length were fixed constant to avoid variations in the internal camera parameters.

As commented previously, the high vertical reliefs of the site and the level of detail required in some areas (e.g., those constructive elements that integrate the defensive system of the entrance, details of the walls, etc.) entailed that aerial images were not suitable; terrestrial images were thus acquired with the full-frame reflex camera. In addition, terrestrial laser scans were used in those complex areas where photogrammetry could entail problems requiring a lot of images to enclose the whole geometry or due to the presence of textureless objects or materials. A network of 13 TLS stations (Figure 4) was designed for an average distance of 15 m with an average spatial resolution of 5 mm.

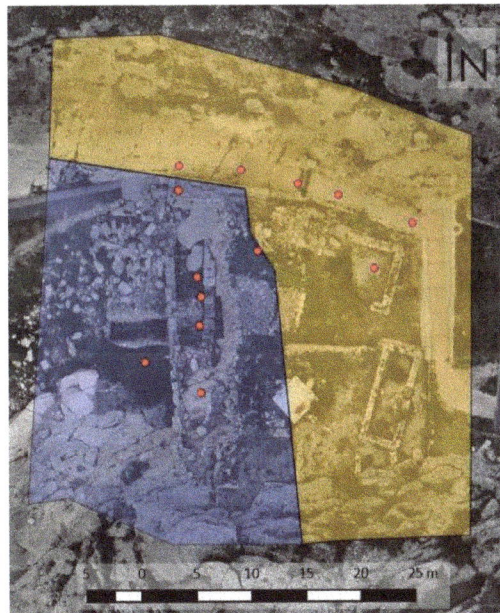

Figure 4. TLS stations distribution (**red** circles), including the area surveyed with TLS during the different campaigns performed (first campaign in **orange** colour and second campaign in **blue** colour).

From Figure 4, it can be observed that two laser scanning campaigns were planned for two different areas. The first campaign (orange colour) was designed for recording the existing constructions using six stations. The second campaign (blue colour) was designed for recording the walls of "El Reguerón" and the adjacent environment using seven stations.

3.2. Data Processing

Handling multiple sensors requires multi-data processing approaches that take the point cloud as the basic unit. However, since all data must be integrated to generate a single model, it is necessary to homogenize the information, establishing filtering, simplification and optimization algorithms. Thus, the result obtained by each method is the corresponding point cloud that will be registered, triangulated and textured for generating a single 3D model with metric and radiometric capabilities.

3.2.1. TLS Filtering and Alignment

Data acquired with TLS were processed with commercial software FARO SCENE 5.4 [25]. The raw TLS scans were filtered removing noise and undesired information which could affect the alignment process. Automatic filtering has been applied in those more conflictive areas. Firstly, each scan was filtered according to a distance threshold (20 m) in order to remove distant points. Afterwards, two specific filters (intense-based filter and outlier filter) were applied. The former applies a reflectance threshold to remove those points with the lowest intensity, whereas the latter analyses the point and its environment (3×3) using distances. For instance, if there is a distance variation of 1 cm between the point and its neighbourhood for more than the 50% of the neighbourhood, the point will be removed. These thresholds should be tested adaptively by the user depending on the type and geometry of the area. Finally, the more delicate areas (i.e., walls or vegetation) have been filtered manually. For instance, vegetation between the blocks of the wall was identified and removed manually.

Since artificial targets were not used, scan alignment was done by a solid rigid transformation of an iterative closest point (ICP) technique [26]. This iterative process was applied in pairwise stations, so the final a priori error was computed on the basis of the number of stations and the technical specifications of the TLS (Table 4) reaching 1.3 cm for the assumption of two consecutive overlapping point clouds. The reference system of the aligned TLS point cloud is defined through a local Cartesian system corresponding to the first scan station with coordinates (300,000, 4,000,000, 1000).

Table 4. Simplification and optimization results for "El Reguerón" area.

	Simplified	Simplified and Optimized
Number of points	11,267,122	2,816,853
Number of triangles	22,532,754	5,635,313
Spatial resolution * (Min, Avg, Max)	(60.5, 64.3, 69.1) mm	(69.2, 81.1, 98.9) mm

* Confidence Interval = 1σ.

3.2.2. Photogrammetric Processing

The generation of the dense point cloud, from both aerial and terrestrial images, was automatized through the Photogrammetry Workbench (PW) in-house software [27], following a three step workflow: 1. Image registration; 2. Camera orientation and 3. Dense matching.

The aerial images, coming from the paratrike and UAV, were processed to generate a hybrid model, with a total of 293 images and two different cameras. The dataset was checked to assure the sharpness of the images, which could be decreased by the motion blur.

1. The feature extraction has been carried out by the ASIFT (Affine Scale-Invariant Feature Transform) algorithm [28]. As its most remarkable improvement, ASIFT includes the consideration of two additional parameters that control the presence of images with different scales and rotations. In this manner, the ASIFT algorithm can cope with images displaying a high scale and rotation difference, common in oblique images. The result is an invariant algorithm that considers the scale, rotation, and movement between images. The main contribution in the adaptation of the ASIFT algorithm is its integration with robust strategies that allow us to avoid erroneous correspondences. These strategies are the Euclidean distance [29] and the Moisan-Stival ORSA (Optimized Random Sampling Algorithm) [30]. This algorithm is a variant of Random Sample Consensus (RANSAC) [31] with an adaptive criterion to filter erroneous correspondences by the employment of the epipolar geometry constraints. Once the feature points have been extracted and described, the final matching points are assessed based on their spatial distribution on the CCD. An asymmetric distribution (radial and angular) of matching points regarding the principal point will affect the correct determination of internal camera parameters and also the image orientation. Therefore, if the matching points do not cover an area more than two-thirds of the CCD format, the user will be alerted in order to modify the detector (ASIFT) and descriptor (SIFT) parameters. Through this quality control we try to minimize problems associated with the weakness and common deficiencies in the photogrammetric network geometry of both aerial flights (UAV and paratrike).

This result provides the next expression:

$$\mathbf{A}_F = \begin{bmatrix} a & b \\ c & d \end{bmatrix} = H_\lambda R_1(\kappa) T_1 R_2(\omega) = \lambda \begin{bmatrix} \cos\kappa & -\sin\kappa \\ \sin\kappa & \cos\kappa \end{bmatrix} \cdot \begin{bmatrix} t & 0 \\ 0 & 1 \end{bmatrix} \cdot \begin{bmatrix} \cos\omega & -\sin\omega \\ \sin\omega & \cos\omega \end{bmatrix} \quad (1)$$

where $\mathbf{A}_F$ is the affinity transformation that contains scale, λ, rotation, κ, around the optical axis (swing) and the perspective parameters that correspond to the inclination of the camera optical axis, φ (tilt) or the vertical angle between optical axis and the normal to the image plane; and ω (axis), the horizontal angle between the optical axis and the fixed vertical plan.

In order to accelerate the process, the overlapped aerial images were identified by their approximate camera orientations provided by the navigation system. In the case of terrestrial images, an all-to-all comparison was applied. This sub-step is a time-consuming process which increases exponentially with the number of images [32].

2. The multi-image protocol acquisition will require robust orientation procedures. For this purpose, a combination between computer vision and photogrammetric strategies was used. This combination is fed by the resulting keypoints extracted previously. In a first step, an approximation of the external orientation of the cameras was calculated following a fundamental matrix approach [33]. Later, these spatial (X,Y,Z) and angular (ω-omega, φ-phi, and χ-kappa) positions are refined by a bundle adjustment complemented with the collinearity condition [34]. In this field, several open source tools have been developed such as Bundler [35] and Apero [36]. For the present case study, both were combined and integrated. In particular, a specific converter has been developed for reading Bundler orientation files (*.out) and computing the three rotation angles and three translation coordinates of the camera in Apero. In addition, a coordinate system transformation has been implemented for passing from the Bundler to the Apero coordinate system. It is remarkable that at the same time, thanks to the reliability of the photogrammetric procedures used, it is possible to integrate as unknowns several internal camera parameters (focal length, principal point, and radial distortions). This possibility allows the use of non-calibrated cameras and guarantees acceptable results. For the present case study, a self-calibration strategy supported by a basic calibration model which encloses five internal parameters (focal length, principal point, and two radial distortion parameters) was used [37,38]. In order to provide metric capabilities to the model, manual identification of ground control points (GCPs) in the images were accomplished. Including these as an input in the bundle adjustment, the model is oriented according to the global coordinate system.

$$(x - x_0) + \Delta x = -f \frac{r_{11}(X-S_X)+r_{21}(Y-S_Y)+r_{31}(Z-S_Z)}{r_{13}(X-S_X)+r_{23}(Y-S_Y)+r_{33}(Z-S_Z)}$$
$$(y - y_0) + \Delta y = -f \frac{r_{12}(X-S_X)+r_{22}(Y-S_Y)+r_{32}(Z-S_Z)}{r_{13}(X-S_X)+r_{23}(Y-S_Y)+r_{33}(Z-S_Z)} \tag{2}$$

where x and y are the known image coordinates, X_i, Y_i and Z_i are the corresponding known GCPs, r_{ij} are the unknown 3×3 rotation matrix elements, S_X, S_Y and S_Z represent the unknown camera position, f is the principal distance, x_0 and y_0 are the principal point coordinates and Δx and Δy are the lens distortion parameters. These internal camera parameters may be known or unknown by the user and thus are introduced as equations or unknowns (self-calibration), respectively.

3.   One of the greatest breakthroughs in recent photogrammetry has been exploiting, from a geometric point of view, the image spatial resolution (size in pixels). This has made it possible to obtain a 3D object point of each of the image pixels. Different strategies have emerged in recent years, such as the Semi-Global Matching (SGM) approach [39] that allows the 3D reconstruction of the scene, in which an object point corresponds with a pixel in the image. These strategies, fed by the external and internal orientations and complemented by the epipolar geometry, are focused on the minimization of an energy function [39]. However, besides the classical SGM algorithm based on a stereo-matching strategy, multi-view approaches are incorporated in order to increase the reliability of the 3D results and to better cope with the case of complex archaeological sites (where the images are captured with different sensors). Considering the two types of flights performed (UAV and paratrike), two different multi-view algorithms were used. For the vertical flight (paratrike), the multi-view MicMac algorithm [40] was used. Meanwhile, for the oblique flight (UAV), the multi-view SURE algorithm [41] was used, which allows a complete reconstruction of the scene. Both strategies consist of minimizing an energy function throughout the eight basic directions that a pixel can take (each 45°). This function is composed of a function of cost, **M** (the pixel correspondence cost), that reflects the degree of the similarity of the pixels between two images, x and x', together with the incorporation of two restrictions, P_1 and P_2, to show the possible presence of gross errors in the process of SGM. In addition, a third constraint has been added to the process of SGM; it consists of the epipolar geometry derived from the photogrammetry, and it can enclose the search space of each pixel in order to reduce the enormous computational cost. In that case, it will generate a dense model with multiple images, obtaining more optimal processing times.

$$E(D) = \sum_x \left(M(x, D_x) + \sum_{x' \in N_x} P_1 T \left[|D_x - D_{x'}| = 1 \right] + \sum_{x' \in N_x} P_2 T \left[|D_x - D_{x'}| > 1 \right] \right) \tag{3}$$

where $E(D)$ is an energy function that must be minimized on the basis of the disparity (difference of correspondence) through the counterpart characteristics, the function M (the pixel correspondence cost) evaluates the levels of similarity between the pixel x and its counterpart x' through its disparity Dx, while the terms P_1 and P_2 correspond with two restrictions that allow for avoiding gross errors in the dense matching process for the disparity of 1 pixel or a larger number of them, respectively.

3.2.3. Data Fusion

Data fusion has been performed homogenizing the data provided for each sensor and generating a common product, a point cloud, with metric properties' multi-resolution and photorealistic texture. Concretely, in order to fuse data, both flights (UAV and paratrike) were solved under a combined photogrammetric bundle adjustment using common control points and two different cameras. Through this combined bundle adjustment, a better and more homogeneous aerial photogrammetric point cloud is obtained, avoiding errors that would be obtained and propagated using a solid rigid transformation. In particular, the combined bundle adjustment (UAV-*uav* and paratrike-*pt*) is solved through a least square adjustment based on collinearity condition Equation (2), as follows:

$$x = x\left(\bar{c}_{pt}, \bar{c}_{uav}, \overline{eo}_{pt\,i}, \overline{eo}_{uav\,j}, X_k\right) y = y\left(\bar{c}_{pt}, \bar{c}_{uav}, \overline{eo}_{pt\,i}, \overline{eo}_{uav\,j}, X_k\right) \qquad (4)$$

where:

- $\bar{c}_{pt}$ and $\bar{c}_{uav}$ are the camera vectors used for paratrike and UAV, respectively, and which include the internal camera parameters (principal point and focal length) and lens distortion coefficients (radial-K, decentering-P and affinity parameters-b). A total of ten unknowns were used for each camera vector, $\bar{c} = (x_0,\, y_0, f,\, K_1,\, K_2,\, K_3, P_1,\, P_2, b_{1,}, b_2)$.
- $\overline{eo}_{pt\,i}$ and $\overline{eo}_{uav\,j}$ correspond with the six unknowns of the external orientation for paratrike and UAV images, respectively. Being the external orientation vector, $\overline{eo} = (S_x,\, S_y,\, S_z,\, \omega,\, \varphi,\, \chi)$.
- X_k represents the spatial coordinates vector (X,Y,Z) of the unknown object points.

Therefore, the equation system is defined as follows:

$$\mathbf{A}x - \mathbf{K} = V \qquad (5)$$

where $\mathbf{A}$ corresponds with the design matrix based on collinearity equations and linearized through a first-order Taylor series, $\mathbf{K}$ is the observations matrix, V is the residual vector and x is the unknown's vector solved through a least squares adjustment as follows:

$$x = \left(A^T PA \right)^{-1} A^T PK \tag{6}$$

P is the weight matrix which corresponds with the inverse cofactor matrix of the observations. The equation system is solved through a twofold step: first by computing the exterior orientation parameters, $\overline{eo}$, and then computing the object points, $\overline{X}_k$, that represent the point cloud.

Next, the remaining step is the registration of the aerial point clouds derived from the different sensors under a global coordinate system. To this end, a GNSS campaign of 3 h based on three permanent ERGNSS stations was performed in order to provide precise coordinates to the GNSS base station used in the archaeological settlement. GNSS observations were processed guaranteeing an absolute error of 3 cm. The RTK surveying of the GCPs allowed us to obtain the aerial point cloud (i.e., coming from UAV and paratrike) under a global coordinate system (EPSG: 25830 and height EPSG: 5782), reaching a final relative precision of 1 cm.

Finally, regarding terrestrial laser scanner (TLS), the different scans were aligned and then co-registered with the photogrammetric point cloud coming from UAV and paratrike, using matching points defined manually as initial approximations. This was carried out using each dataset of coordinates in its coordinate system, that is, the local system in the case of the aligned TLS point cloud and the absolute system for the photogrammetric point cloud, and then applying a variation of the ICP technique, Least Squares Matching (LSM). A figure (Figure 5) to visually illustrate how this fusion has been done is included. A is the function that represents the point cloud coming from the aerial photogrammetry (UAV and paratrike) and B is the function that represents the aligned TLS point cloud, the registration of both point clouds will be obtained as follows:

$$A_i\,(x,y,z) - e_i\,(x,y,z) = B_j\,(x,y,z)\ i,j = 1,\ldots,n\,,i \neq j \tag{7}$$

The equation that relates both models is a solid rigid transformation with seven parameters:

$$\begin{bmatrix} X_A \\ Y_A \\ Z_A \end{bmatrix} = R \begin{bmatrix} X_B \\ Y_B \\ Z_B \end{bmatrix} + \begin{bmatrix} T_x \\ T_y \\ T_z \end{bmatrix} \tag{8}$$

where **R** represents the rotation matrix and T the translation vector. The correspondence between both point clouds is obtained through a minimisation, based on least squares adjustment, of the Euclidean distances between both point clouds. Since the rotation matrix is composed of non-linear functions, first-order Taylor series were used for the linearization of the Equation (8) as follows:

$$-e_i\left(x,y,z\right) = B_j^0\left(x,y,z\right) + \frac{B_j^0\left(x,y,z\right)}{\partial x_j}dx_j + \frac{B_j^0\left(x,y,z\right)}{\partial y_j}dy_j + \frac{B_j^0\left(x,y,z\right)}{\partial z_j}dz_j - A_i^0\left(x,y,z\right)$$
$$- \frac{A_i^0\left(x,y,z\right)}{\partial x_i}dx_i - \frac{A_i^0\left(x,y,z\right)}{\partial y_i}dy_i - \frac{A_i^0\left(x,y,z\right)}{\partial z_i}dz_i$$

$$(9)$$

where:

$$dx = dt_x + a_{10}dm + a_{11}d\omega + a_{12}d\varphi + a_{13}d\kappa$$
$$dy = dt_y + a_{20}dm + a_{21}d\omega + a_{22}d\varphi + a_{23}d\kappa \qquad (10)$$
$$dz = dt_z + a_{30}dm + a_{31}d\omega + a_{32}d\varphi + a_{33}d\kappa$$

Through an iterative process, the linear system $-e = Ax - l$ is solved, x being the unknown's vector encompassing the transformation parameters $(dt_x, dt_y, dt_z, dm, d\omega, d\varphi, d\kappa)$, whereas l is the observation vector enveloping the discrepancies of Euclidean distances between both point clouds, and e is the residual vector.

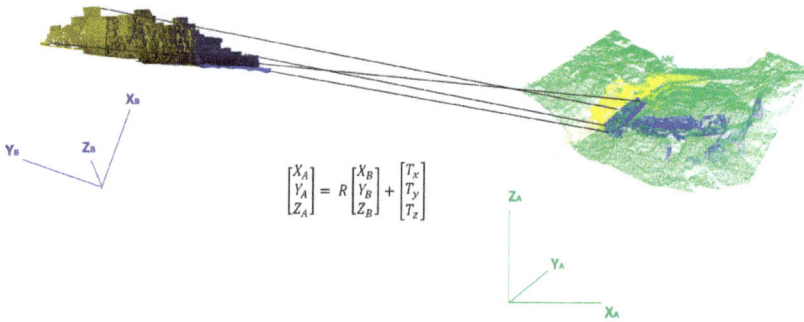

Figure 5. An example of data fusion using Least Squares Matching (LSM) between the laser (**Left**) and photogrammetric (**Right**) point clouds.

Figure 5 graphically outlines the process followed for data fusion. An example of the aligned TLS point clouds for two scans of the wall (yellow and blue point clouds) in a local coordinate system is outlined to the left side of the image. Subsequently, the aligned TLS point clouds are fused with the photogrammetric point cloud in a global coordinate system (green point cloud) by means of LSM.

3.2.4. Post-Processing

Once a metric and geo-referenced point cloud was generated, a mesh strategy was applied to generate the digital surface model (DSM). In this case, the incremental Delaunay triangulation algorithm was applied [42].

To improve the model quality, break lines were incorporated as geometric constraints. Break lines were manually restituted by the operator. Its use is relevant for the accurate representation of significant slope changes, as well as for correctly defining the defensive walls of the site.

In those cases where images come from different methodologies or acquisition time, a radiometric adjustment was necessary to improve the final model visualization and avoid abrupt radiometric changes in the texture.

Once the DSM was obtained, it was possible to generate a true orthoimage from the oriented images. Since the multi-data source and multi-sensor approach provide different DSM resolutions, it was possible to generate true orthoimages with different spatial resolutions. For instance, a true orthoimage that depicts the complete archaeological site was generated with a resolution of 10 cm; whereas a 2 cm spatial resolution was employed for the settlement entrance area in order to appreciate the construction features.

3.2.5. Simplification and Optimization

Different simplification and optimization strategies were applied to the different geomatics products (DSM and orthoimages) in order to access and analyse 2D and 3D information through a Web service using mobile devices such as tablets or smartphones. In particular, a pyramidal structure was generated for the Web visualization of the 2D orthoimages and 3D models, including "El Reguerón" as the area with the most resolution. The methodology used for simplifying the 3D models is based on the strategy "smooth looking" [43] which consists in a straightforward mesh generation procedure applying the Poisson algorithm [44] to the hybrid point cloud. The mesh resolution is determined by the octree level, chosen according to the user-defined spatial resolution. Unlike a direct mesh generation process, which usually requires mesh editing operations such as filtering and refinement [45,46], Poisson algorithm directly encloses a smoothing step and provides a continuous geometry. This process could be controlled by some computation parameters as the minimum number of sample points that should fall within a node of the resulting octree. This parameter controls the loss of detail by the smoothing process. Although a high number of sample points implies a decrease of the number of mesh vertices, its geometry could be affected, being the final mesh mildly shrunk. For this reason, and as the noise level of the hybrid point cloud is low, the threshold value was kept low in order to avoid redundant smoothing processes. Then, a mesh optimization strategy, based on reducing the final number of triangles through the "collapse" of non-relevant areas without losing significant level of detail, is performed. For this purpose, an iterative process is proposed where the mesh derived from the previous procedure (smooth-looking) is "collapsed" by 5% of the total number of triangles using the quadratic edge collapse algorithm [47]. The algorithm essentially removes edges by merging and regrouping nearby vertices. With the aim of minimising the distortion of the surface geometry, it is necessary to establish a precision threshold to stop the iterative process. So that, if the resulted mesh error against the original input point cloud remains unchanged (with respect to the previous iteration), the collapse process continues. An iterative procedure is required since the quadratic edge collapse algorithm implementation does not allow decimation at fixed spatial resolution. Although our approach is based on an edge-collapse algorithm for the mesh optimisation, others approaches could be applied, such as the remeshing with recursive resampling as shown in [48] based on the Marching Cubes algorithm [49]. However, this remeshing approach exhibits higher deviations from the original model, up to 20 times higher than the quadratic edge collapse algorithm [48], and is therefore inadequate for our final archaeological products.

Finally, a texture mapping of the simplified and optimised mesh was performed using the commercial software Agisoft Photoscan®.

4. Experimental Results

4.1. Area of Study

The city of "Tolmo de Minateda" (Figure 6) was a strategic settlement of great importance for several centuries, largely because of its peculiar topography and geographical location. It is placed on a plateau hill of about 50 m of height, located at the junction of the route from Complutum to Carthago Nova, one of the principal Roman routes connecting the interior of the plateau with the southeast coast, and a road connecting Castulo with Saetabis.

Figure 6. Aerial view of the study area "Tolmo de Minateda" (from [50]).

The archaeological research from the last 30 years has highlighted the importance of this site, revealing a history from the Middle Bronze Age, through the Iberian era, the Roman period, and the Middle Ages to the twentieth century. The Middle Age period provided most of the information by an important Visigoth settlement located in the upper part of the "El Tolmo", where an important Christian basilica was found between houses and cemeteries.

One of the most interesting areas of the archaeological site is "El Reguerón" (Figure 6), an area of natural drainage 12 m in width, with a main entrance at the top of the hill where the city was located. In the settlement, an important fortification system consisting of three walls of different chronology and architectural typologies has been documented [51]. The oldest fortification is represented by the so-called "embanked" wall and was built during the final phase of the Iberian period (4th−2nd century B.C.). Currently, only the remains of a wall 6 m high and 10 m wide at the top, embanked in the external front and built in irregular masonry work, was preserved in the soil. During the archaeological excavation of this ancient wall, an earlier phase, which dates back to the Middle Bronze Age, was discovered. The Iberian wall was successively used as a retaining wall for a new fortification building during Roman times when the "Tolmo de Minateda" probably received the title of municipality.

The last fortification found in "El Reguerón" comes from the period of peninsula occupation by the Visigoths (5th–6th century A.D.). This wall is presented as a solid, L-shaped bulwark that encloses the valley and flanks the main road access to the city. It is here at this point that a monumental gate, probably formed by two solid towers of blocks, was once located. Only partial remains have been preserved. The wall is formed by a line of blocks with inscriptions and architectural elements from older constructions (among which are examples from the Roman period).

The relevant archaeological stratification of this area with different structures and building types requires of a multi-data source and multi-sensor approach that allows us to properly record and classify archaeological surfaces, thereby establishing an integration of topographic data with documents of archaeological excavations.

In order to fulfill the archaeological documentation requirements, the aerial data gathering was performed with a paratrike which allows us to enclose the whole extension of the archaeological site using a full-frame reflex camera, assisted by a specific gyrostabilized platform (MUSAS-MUltiSpectral Airborne Sensors) with a ground sample distance (GSD) of 3 cm. According to the GSD desired for the whole archaeological site and considering the camera specifications, a maximum flight height of 224 m was established. As a result, a total of 268 images along seven strips (NW-SE direction) were required for guaranteeing a side overlap of >30% and a forward overlap of >70%.

To record with higher spatial resolution the area of interest of the "El Regueron" site (Figure 7), an UAV with an ultra-compact camera (Table 5) acquired the fortified walls with a GSD of 1 cm. Finally, 25 images and a flight height of 32 m guaranteed the desired spatial resolution. The overlap parameters were the same.

Figure 7. Location of the defensive area "El Reguerón".

Table 5. Geo-referencing errors.

Check Points			Discrepancies with 3D Model Coordinates			
X (m)	Y (m)	H (m)	ΔX (m)	ΔY (m)	ΔXY (m)	ΔH (m)
621,321.452	4,259,638.120	448.672	0.005	0.001	0.005	−0.013
621,375.933	4,259,646.226	454.775	−0.009	0.004	0.009	0.032
621,416.258	4,259,667.686	463.661	0.007	−0.011	0.013	0.030
621,419.022	4,259,658.232	464.618	0.001	0.008	0.008	0.046
621,345.185	4,259,659.272	454.267	0.002	0.018	0.006	0.024
621,321.298	4,259,660.311	452.106	−0.010	−0.026	0.028	0.036

In addition, to increase the level of detail in the fortification walls and thus avoid the occlusions due to the effects of terrain relief (i.e., areas occluded or without information in the model generated by aerial photogrammetry), a combination of terrestrial photogrammetry and terrestrial laser scanner (TLS) was employed for improving the final 3D hybrid model. As a result, a total of 36 terrestrial oblique images and 13 TLS point clouds were acquired in order to avoid areas without information, guaranteeing a subcentimeter resolution.

4.2. Workflow

The application of the multi-data source and multi-sensor fusion workflow was tested in the study area of "Tolmo de Minateda," which includes a total of 9 ha.

Firstly, the aerial images were processed by the automated photogrammetric approach and the cameras were self-calibrated according to Section 3.2.2. As a result of this process, the raw reconstructed point cloud is obtained which contains 4,271,354 points, while the TLS integrated point cloud reaches up to 1,314,136 points. Secondly, the geo-referencing of the 3D model was solved by the employment of 28 control points homogeneously distributed across the site (but with higher density in the interest area), while ten were used as checkpoints. Finally, the final multi-resolution 3D model after triangulation is obtained (Figure 8), encompassing the whole study area with a spatial resolution of 3 cm. A total of 10,542,505 triangles were obtained.

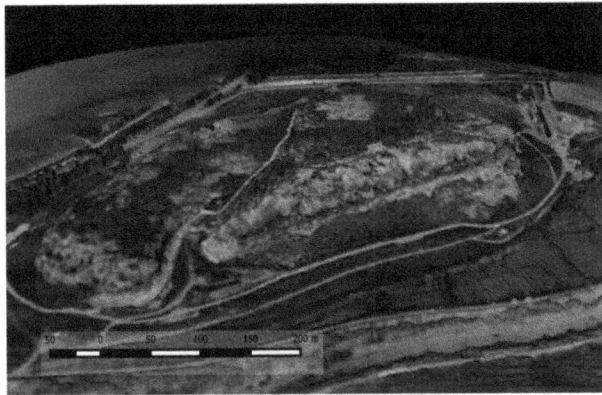

(a)

(b)

Figure 8. Multi-resolution 3D model of the archaeological settlement (a). Detailed 3D model of "El Reguerón" (b).

For the detailed area which integrates different data sources (i.e., terrestrial laser scanner and photogrammetry), a spatial filtering of 0.5 cm was applied to avoid areas with excessive point density caused by the different overlaps.

By this multi-resolution approach, the final inspection of the archaeological site could be adapted to any area, as illustrated with "El Reguerón", where the subcentimete resolution achieved for the walls and construction (Figure 9b) was integrated with the rest of the archaeological site (Figure 9a). This multi-resolution capability opens a range of possibilities for a spatial analysis, settlement interpretation, pattern recognition, and the establishing of relationships among the elements (sites, artefacts locations, etc.).

(a)

(b)

Figure 9. Multi-resolution model of "El Reguerón" area (**a**) and detail for the wall (**b**).

In order to validate the final integration of the different data sources, a series of checkpoints were used. The different error components are shown in Table 4. The average vertical error (2.6 cm) was higher than the horizontal error (1.4 cm), as was expected for the GNSS technique. An average precision of 3 cm for the 3D vector error was obtained, which is statistical compatible with the expected a priori error of 3.1 cm composed by the model and GNSS check point errors.

342

To manage more efficiently the available information of the archaeological settlement, true orthoimages were generated as derived products (Figure 10), since they combined the photorealistic texture with the metric properties in an easy-to-use document for non-experts.

However, to allow a more complete interpretation and to make use of the potential of the generated 3D product, an integration with a 3DGIS tool, GeoWeb3D, was performed. In particular, the 3D geometry provided by the multi-sensor approach has been integrated with 2D archaeological archives such as sketches, pictures, part details, etc., thus enhancing the subsequent analysis and providing problem solving and decision-making capabilities. Figure 11 shows the integration of different historical events in the reconstructed 3D model.

The integration of 2D information and 3D models allows us to extract intangible information that improves the analysis capabilities of the archaeological settlement. For instance, Figure 11 integrates an archaeological sketch of defensive constructions (2012) with the generated hybrid 3D model (2015). It should be noted that the high accuracy and proportion of the sketch is a perfect coincidence of the main homologous entities. However, some detached blocks belonging to the wall of the 3D model do not appear in the sketch. Analysing the position of blocks in the 3D model, it seems that they were spread along the natural drainage bed of the Tolmo, possibly suggesting that a runoff flow took place from the upper part of the Tolmo, providing this current block distribution.

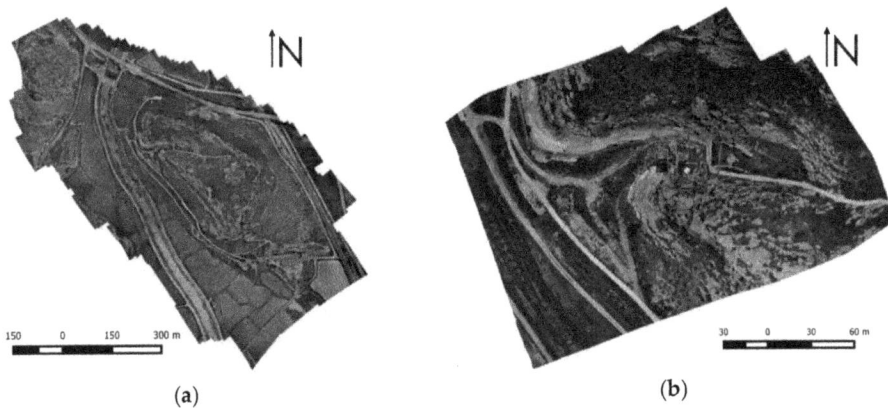

(a)   (b)

Figure 10. True orthoimages of the whole settlement (**a**) and defensive area "El Reguerón" (**b**).

Next, with the aim of showing the potential of analysing in situ the archaeological settlement, "El Reguerón" was tested using a mobile device (smartphone). The simplification and optimisation level was fixed considering different aspects: (i) the amount of 2D information which should be integrated; (ii) the minimum size of the interest elements and (iii) the level of reduction. A minimum resolution of 10 cm was established for the "El Reguerón" area.

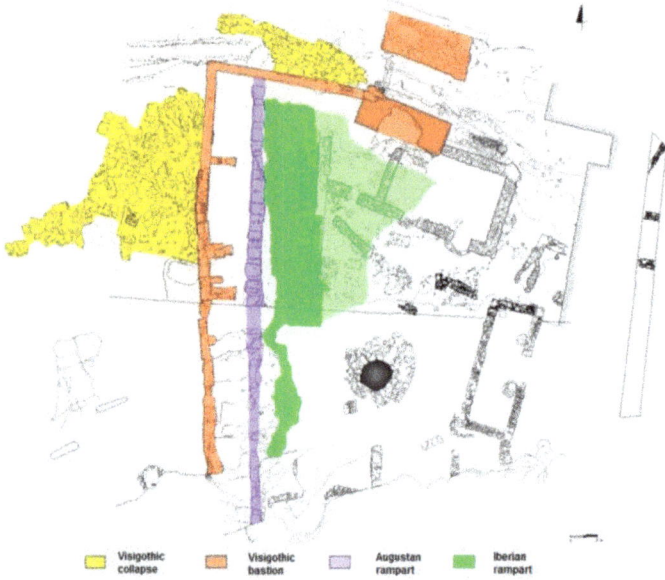

(a)

(b)

Figure 11. Archaeological sketch (a) overlapped with the 3D model (b).

Looking to the smallest and most emblematic elements of "El Reguerón," it can be confirmed that these correspond to the blocks of the wall, which are bigger than the minimum resolution, meaning that they should be correctly represented. Subsequently, an iterative optimization process of the mesh was performed maintaining relevant information and stopping when the threshold of resolution (10 cm) was surpassed. As we can see in Table 5, the simplified and optimised models maintain the initial resolution (10 cm), with a level of reduction of 75% in comparison with the simplified model (without optimisation).

The simplification and optimisation procedures have removed 31,039,445 points of the 33,856,298 initial points. However, in order to assess the final error of the simplification and optimisation strategy, we have compared the simplified and optimised model with the original point cloud, obtaining an average error of 0.2 mm and a standard deviation of ±22 mm.

As we can see in Figure 12, higher discrepancies are located in peripheral areas due to the presence of vegetation with an irregular typology. Conversely, the walled area exhibits minimum discrepancies always less than ±5 mm.

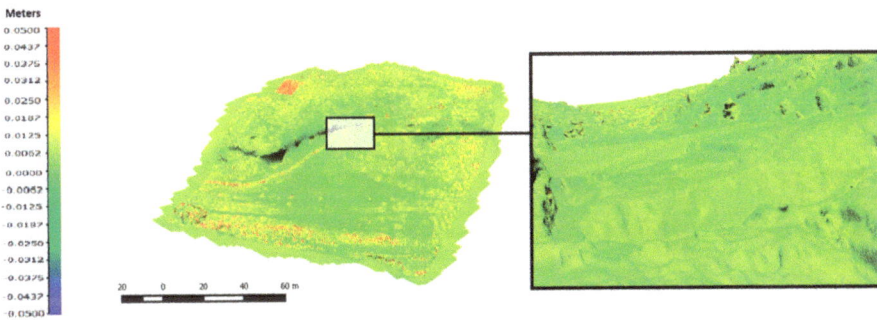

Figure 12. Analysis of discrepancies between the simplified-optimised 3D model and the original point cloud (**Left**) and a detailed comparison over the walled area (**Right**).

Finally, results were presented through the Web based on the Open Source library Cesium. A specific template based on HTML language was prepared to show the 3D models and additional information using external geospatial services such as WMS, MapServer, Google Earth or Bing, among others. Thanks to the flexibility and portability of the mobile devices, it was possible for the archaeologist to interact directly with the platform at the field: recording data and adding new information with corresponding attributes and descriptions. An example is outlined in Figure 13, where the application is loaded in a smartphone BQ Aquaris E4.5 using Android 4.4.2 and Google Chrome 45. Results were incorporated to the archaeological information system and the spatial data infrastructure of the archaeological cultural heritage of Castilla La Mancha (ideARQ + SIA) [50].

Additionally, the optimal value of points and triangles for the proper management of 3D models in smartphones was examined. For this aim, three levels of simplification were analysed using three, two and one million(s) triangles, respectively. Afterwards, loading and operation times together with RAM were monitored in order to see the best simplification level for a conventional smartphone. Table 6 outlines the results obtained.

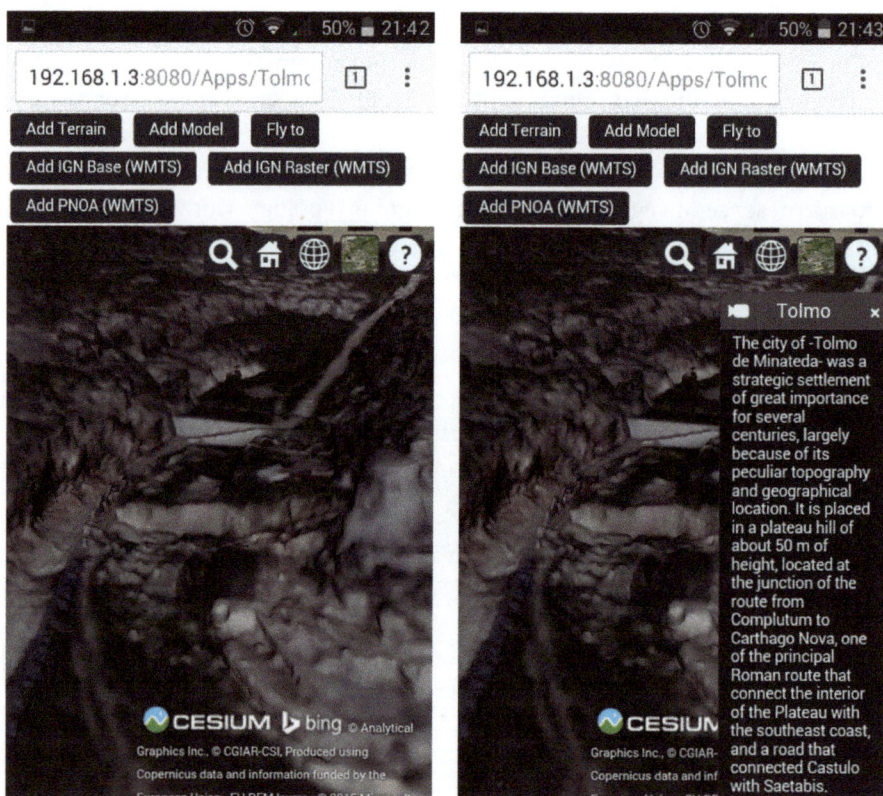

Figure 13. "El Regueron" 3D model and its additional information loaded in a smartphone (Base model from: © Analytical Graphics, Inc., © CGIAR-CSI, Produced using Copernicus data and information funded by the European Union—EU-DEM layers, © Commonwealth of Australia (Geoscience Australia) 2012) [19].

As we can see, with less than one million triangles, very good loading and operation times (instantaneous) are obtained, as well as an affordable RAM consumption for the majority of smartphones. Therefore, for optimal user experience and navigation for each visualization level (defined in terms of spatial resolution; see

Section 3.2.5), the maximum display area is cropped accordingly to this empirical number of triangles.

Table 6. Times and resources required with lighter models using mobile devices.

Model	Number of Points	Number of Triangles	
3M	1,408,636	2,816,156	
	Load	Operation	RAM consumption
	10 s	2–4 s	2.39 GB
2M	986,150	1,971,309	
	Load	Operation	RAM consumption
	8 s	1–3 s	2.24 GB
1M	493,198	985,653	
	Load	Operation	RAM consumption
	3 s	Instantaneous	1.79 GB

5. Conclusions

This paper presents a methodology based on a combination of multiple sensors, platforms and techniques, which has been tested in a complex archaeological site. As is shown in the experimental results, the automation provided by the photogrammetric and laser techniques, along with the versatility of the aerial platforms (paratrike and Unmanned Aerial Vehicle UAV), provide the suitability of this methodology for complex archaeological sites.

The potential of ultralight aerial platforms (paratrike) is highlighted, due to its payload and flight autonomy, which overpasses the UAVs capabilities. Regarding the usual archaeological surveys, one of the main differential factors in this work has been the integration of aerial images at different resolutions, terrestrial images and terrestrial laser scans.

The generation of high resolution products based on photogrammetry requires high computational costs. To this end, the presented algorithms, such as Affine Scale Invariant Feature Transform ASIFT, have been developed to take advantage of Graphical Processing Unit GPU capabilities, reduce time operations and consequently improve workflow efficiency.

The management of the final models through a 3DGIS tool opens new analysis capabilities for the archaeologist (e.g., analysis though time, archaeological investigations, integration of historical and geometric models).

Finally, it has been demonstrated that through the simplification and optimization strategies, complex hybrid 3D modes which enclose 2D information can be flexibly shown through the Web and even embedded in smartphones. As a result, we can interact directly with hybrid 3D models using mobile devices, allowing the recording, storage and analysis of data in situ. The implemented methodology to optimize the visualization and interaction of geomatics products has proven to be effective. However, it lacks a fully complete user interaction, since the final navigation is constrained by the definition of the visualization level. To overcome this limitation, a future work line is aimed to a more efficient visualization management by the 3DGIS tool, avoiding the definition of level of visualization by the final user.

The technical case outlined in this paper could be of great interest for different stakeholders:

- *Researchers* for the interpretation, spatial and temporal analysis of the archeological settlement thanks to the integration capabilities and portability of the system.
- *Managers* through the monitoring of the archaeological settlement through time, the diffusion of the site using videos, documents, etc.
- *Students* who could exploit the didactical possibilities of the 3D inspection, interaction and superposition of thematic information.
- *General public* allowing a flexible and enjoyable accessibility to the archaeological settlement which complements and provides added value to a visit to an historical site.

The data acquisition and processing methodologies from Geomatic science broaden the possibilities of sensors, configurations and/or combination as future perspectives. Concretely, the resolution of the external orientation of the images directly, just using the integrated Global Navigation Satellites System GNSS/IMU Inertial Measurement Unit of the aerial platform, would allow us to speed up the field work and processing time. Furthermore, in order to obtain realistic textured models and orthoimages, the use of techniques such BRDF (Bidirectional Reflectance Distribution Function) to improve the radiometric parameters will be studied.

Acknowledgments: This research has been developed in the framework of the research project *"Infraestructura de datos espaciales de patrimonio arqueológico de Castilla-La Mancha"* (POII-2014-004-P) of the 2014-2017 Scientific Research Projects cofounded by the European Regional Development Fund. The authors would like to thank the Ministry of Education, Culture and Sport of Castilla-La Mancha, especially to the Directorate-General for Universities, Research and Innovation, the Directorate-General for Cultural and the Museum of Albacete. We also specially thank Lorenzo Abad Casal (University of Alicante), Sonia Gutiérrez Lloret (University of Alicante), Rubí Sanz Gamo (Museum of Albacete) and Blanca Gamo Parras (Museum of Albacete) for their support during the course of this study.

Author Contributions: All authors conceived, designed and performed the experimental campaign and the implemented methodology. José Alberto Torres processed and analysed the results. All authors wrote the manuscript.

Conflicts of Interest: The authors declare no conflict of interest.

Abbreviations

AGI	Analytical Graphics, Inc.
ASIFT	Affine Scale Invariant Feature Transform
BRDF	Bidirectional Reflectance Distribution Function
CCRS	Compound Coordinate Reference System
CRS	Coordinate Reference System
DSM	Digital Surface Model
DTM	Digital Terrain Model
ETRS89	European Terrestrial Reference System 1989
EPSG	European Petroleum Survey Group
GIS	Geographical Information System
GNSS	Global Navigation Satellites System
GPS	Global Positioning System
GPU	Graphical Processing Unit
GSD	Ground Sample Distance
HTML	HyperText Markup Language
ICP	Iterative Closest Point
IMU	Inertial Measurement Unit
LSM	Least Squares Matching
MI	Mutual Information
MUSAS	MUltiSpectral Airborne Sensors
ORSA	Optimized Random Sampling Algorithm
RAM	Random Access Memory
RANSAC	Random Sample Consensus
RTK	Real-Time Kinematic
SGM	Semi-Global Matching
SIFT	Scale Invariant Feature Transform
TLS	Terrestrial Laser Scanner
UAV	Unmanned Aerial Vehicle
UTM	Universal Transverse Mercator
VRML	Virtual Reality Modeling Language
WMS	Web Map Service

Appendix: 3D Web Visualization

The management of a 3D textured model is a key part of any spatial data infrastructure. However, it requires an efficient visualization system to visualize the geometric and semantic information. In spite of the mesh simplification and optimization process mentioned in Section 3.2.5, it is possible to optimize the visualization of 3D models in terms of WebGL framework Cesium. Some authors [52] propose an approach based on a custom geometry loader implemented in the Cesium APIs. By this approach any 3D element is rendered and handled by the Cesium rendering code. However, these authors remark some drawbacks in terms of memory and performance. Another alternative is the employment of the GL Transmission Format (glTF) proposed in [53] which allows directly loading the data to the WebGL buffers. This format aims to be a standard for data exchange and rendering. Although this glTF format is able to deliver an arbitrary number of mesh data buffers within a single file, it completely lacks any mechanism for progressive data transmission. Trying to overcome this limitation, new file formats are being developed [54]. The basic idea would be to improve the data streaming by means of a Level of Detail (LoD) visualization in order to reduce the number of WebGL draw calls.

Following this line, the implemented solution is based on the optimization of visualization levels. For each user call, a specific 3D sub-model is loaded accordingly to the user position, view orientation and zoom level. This is a trade-off between the visualized area and the model resolution, keeping an optimal number of mesh triangles for an easy navigation in terms of loading time and user experience. The main drawback is that the final user navigation has to be constrained into a set of positions to avoid the movement outside the active visualization level. The generation of the 3D sub-models for visualization purposes implies a spatial resolution simplification, where the visualization areas are extracted according to the optimal number of triangles mentioned above. The 3D sub-models are created in a discrete number of levels, where the minimum allowed resolution change is established at 5%.

References

1. Gomez-Lahoz, J.; Gonzalez-Aguilera, D. Recovering traditions in the digital era: The use of blimps for modelling the archaeological cultural heritage. *J. Archaeol. Sci.* **2009**, *36*, 100–109.

2. Bevan, A.; Li, X.; Martinón-Torres, M.; Green, S.; Xia, Y.; Zhao, K.; Zhao, Z.; Ma, S.; Cao, W.; Rehren, T. Computer vision, archaeological classification and China's Terracotta Warriors. *J. Archaeol. Sci.* **2014**, *49*, 249–254.

3. González-Aguilera, D.; Muñoz-Nieto, Á.; Rodríguez-Gonzálvez, P.; Menéndez, M. New tools for rock art modelling: Automated sensor integration in Pindal Cave. *J. Archaeol. Sci.* **2011**, *38*, 120–128.

4. Hakonen, A.; Kuusela, J.-M.; Okkonen, J. Assessing the application of laser scanning and 3D inspection in the study of prehistoric cairn sites: The case study of Tahkokangas, Northern Finland. *J. Archaeol. Sci.* **2015**, *2*, 227–234.

5. Themistocleous, K.; Ioannides, M.; Agapiou, A.; Hadjimitsis, D.G. The methodology of documenting cultural heritage sites using photogrammetry, UAV and 3D printing techniques: The case study of Asinou Church in Cyprus. In Proceedings of the Third International Conference on Remote Sensing and Geoinformation of Environment, Cyprus, Nicosia, 16 March 2015; pp. 16–19.

6. Lai, L.; Sordini, M.; Campana, S.; Usai, L.; Condò, F. 4D recording and analysis: The case study of Nuraghe Oes (Giave, Sardinia). *Digit. Appl. Archaeol. Cult. Herit.* **2015**, *2*, 233–239.

7. Torres, J.A.; Hernandez-Lopez, D.; Gonzalez-Aguilera, D.; Moreno Hidalgo, M.A. A hybrid measurement approach for archaeological site modelling and monitoring: The case study of Mas D'is, Penàguila. *J. Archaeol. Sci.* **2014**, *50*, 475–483.

8. Achille, C.; Adami, A.; Chiarini, S.; Cremonesi, S.; Fassi, F.; Fregonese, L.; Taffurelli, L. UAV-based photogrammetry and integrated technologies for architectural applications—Methodological strategies for the after-quake survey of vertical structures in Mantua (Italy). *Sensors* **2015**, *15*, 15520–15539. PubMed]

9. Ortega-Terol, D.; Moreno, M.A.; Hernández-López, D.; Rodríguez-Gonzálvez, P. Survey and classification of Large Woody Debris (LWD) in streams using generated low-cost geomatic products. *Remote Sens.* **2014**, *6*, 11770–11790.

10. Hailey, T. The powered parachute as an archaeological reconnaissance vehicle. *Archaeol. Prospect.* **2005**, *12*, 69–78.

11. Herrero-Huerta, M.; Hernández-López, D.; Rodriguez-Gonzalvez, P.; González-Aguilera, D.; González-Piqueras, J. Vicarious radiometric calibration of a multispectral sensor from an aerial trike applied to precision agriculture. *Comput. Electron. Agric.* **2014**, *108*, 28–38.

12. Trimble Indoor Mobile Mapping Solution, TIMMS. Available online: http://www.applanix.com/solutions/land/timms.html (accesed on 15 February 2016).

13. Viametris, Indoor Mobile Mapping System. Available online: http://viametris.fr/Produits_IMMS.php (accesses on 15 February 2016).

14. Canter, P.; Stott, A.; Rich, S.; Querry, J. Creating georeferenced indoor maps, images and 3D models: Indoor mapping for high accuracy and productivity. *J. Chart. Instit. Civ. Eng. Surv.* **2010**, *1*, 20–22.

15. Rivero-Juárez, J.; Martínez-García, E.A.; Torres-Mendez, A.; Mohan, R.E. 3D heterogeneous multi-sensor global registration. *Procedia Eng.* **2013**, *64*, 1552–1561.

16. Nilosek, D.; Sun, S.; Salvaggio, C. Geo-Accurate model extraction from three-dimensional image-derived point clouds. In Proceedings of the Algorithms and Technologies for Multispectral, Hyperspectral, and Ultraspectral Imagery XVIII, Baltimore, MD, USA, 23 Aril 2012.

17. Geoweb3d—3D GIS Visualization. Available online: http://www.geoweb3d.com/ (accessed on 15 February 2016).

18. WebGL Virtual Globe and Map Engine. Available online: http://cesiumjs.org/ (accessed on 15 February 2016).

19. AGI—Software to Model, Analyze and Visualize Space, Defense and Intelligence Systems. Available online: http://www.agi.com/ (accessed on 15 February 2016).

20. Le, H.S. An approach to construct SGIS-3D: A three dimensional WebGIS system based on DEM, GeoVRML and spatial analysis operations. In Proceedings of the 2nd IADIS International Conference Web Virtual Reality and Three-Dimensional Worlds, Freiburg, Germany, 27–29 July 2010; pp. 317–326.

21. Von Schwerin, J.; Richards-Rissetto, H.; Remondino, F.; Agugiaro, G.; Girardi, G. The MayaArch3D project: A 3D WebGIS for analyzing ancient architecture and landscapes. *Lit. Linguist. Comput.* **2013**, *28*, 736–753.

22. Auer, M.; Agugiaro, G.; Billen, N.; Loos, L.; Zipf, A. Web-based visualization and query of semantically segmented multiresolution 3D models in the field of cultural heritage. *Proc. ISPRS* **2014**, *II-5*, 33–39.

23. Takasu, T. RTKLIB: Open source program package for RTK-GPS. In Proceedings of the Free and Open Source Software for Geospatial (FOSS4G), Tokyo, Japan, 1–2 November 2009.

24. Hernandez-Lopez, D.; Felipe-Garcia, B.; Gonzalez-Aguilera, D.; Arias-Perez, B. An automatic approach to UAV flight planning and control for photogrammetric applications: A test case in the Asturias region (Spain). *Photogramm. Eng. Remote Sens.* **2013**, *79*, 87–98.

25. 3D Measurement Technology from FARO. Available online: http://www.faro.com/ (accessed on 15 February 2016).

26. Besl, P.; McKay, N. A method for registration of 3-D Shapes. *IEEE Trans. Pattern Anal. Mach. Intell.* **1992**, *14*, 239–256.

27. Fernández-Hernandez, J.; González-Aguilera, D.; Rodríguez-Gonzálvez, P.; Mancera-Taboada, J. Image-based modelling from Unmanned Aerial Vehicle (UAV) photogrammetry: An effective, low-cost tool for archaeological applications. *Archaeometry* **2015**, *57*, 128–145.

28. Morel, J.M.; Yu, G. ASIFT: A new framework for fully affine invariant image comparison. *SIAM J. Imaging Sci.* **2009**, *2*, 438–469.

29. Gruen, A. Adaptive least squares correlation: A powerful image matching technique. *S. Afr. J. Photogramm. Remote Sens. Cartogr.* **1985**, *14*, 175–187.

30. Moisan, L.; Stival, B. A probabilistic criterion to detect rigid point matches between two images and estimate the fundamental matrix. *Int. J. Comput. Vis.* **2004**, *57*, 201–218.

31. Fischler, M.A.; Bolles, R.C. Random sample consensus: A paradigm for model fitting with applications to image analysis and automated cartography. *Commun. ACM* **1981**, *24*, 381–395.

32. Rieke-Zapp, D.H.; Nearing, M.A. Digital close range photogrammetry for measurement of soil erosion. *Photogram. Rec.* **2005**, *20*, 69–87.

33. Hartley, R.; Zisserman, A. *Multiple View Geometry in Computer Vision*; Cambridge University Press: New York, NY, USA, 2003.

34. Kraus, K.; Jansa, J.; Kager, H. *Advanced Methods and Applications Volume 2. Fundamentals and Standard Processes Volume 1*; Institute for Photogrammetry Vienna University of Technology: Bonn, Germany, 1997.

35. Snavely, N.; Seitz, S.M.; Szeliski, R. Modeling the world from Internet photo collections. *Int. J. Comput. Vis.* **2008**, *80*, 189–210.

36. Deseilligny, M.P.; Clery, I. Apero, an open source bundle adjustment software for automatic calibration and orientation of set of images. *ISPRS Int. Arch. Photogramm. Remote Sens. Spat. Inf. Sci.* **2011**, *38*, 269–277.

37. Kukelova, Z.; Pajdla, T. A minimal solution to the autocalibration of radial distortion. In Proceedings of the IEEE Conference on Computer Vision and Pattern Recognition, Minneapolis, MN, USA, 17–22 June 2007.

38. Sturm, P.; Ramalingam, S.; Tardif, J.P.; Gasparini, S.; Barreto, J. Camera models and fundamental concepts used in geometric computer vision. *Found. Trends® Comput. Graph. Vis.* **2011**, *6*, 1–183.

39. Hirschmuller, H. Stereo processing by semiglobal matching and mutual information. *IEEE Trans. Pattern Anal. Mach. Intell.* **2008**, *30*, 328–341.

40. Micmac Website. Available online: http://www.tapenade.gamsau.archi.fr/TAPEnADe/Tools.html (accessed on 15 February 2016).

41. Rothermel, M.; Wenzel, K.; Fritsch, D.; Haala, N. SURE: Photogrammetric surface reconstruction from imagery. In Proceedings of the LC3D Workshop, Berlin, Germany, 4–5 December 2012; pp. 1–9.

42. Bourke, P. An algorithm for interpolating irregularly-spaced data with applications in terrain modelling. In Proceedings of the Pan Pacific Computer Conference, Beijing, China, 1 January 1989.

43. Rodríguez-Gonzálvez, P.; Nocerino, E.; Menna, F.; Minto, S.; Remondino, F. 3D surveying & modeling of underground passages in WWI fortifications. *Int. Arch. Photogramm. Remote Sens. Spat. Inf. Sci.* **2015**.

44. Kazhdan, M.; Bolitho, M.; Hoppe, H. Poisson surface reconstruction. In Proceedings of the 4th Eurographics Symposium on Geometry, Sardinia, Italy, 26–28 June 2006; pp. 61–70.

45. Attene, M. A lightweight approach to repairing digitalized polygon meshes. *Vis. Comput.* **2010**, *26*, 1393–1406.

46. Varnuška, M.; Parus, J.; Kolingerová, I. Simple holes triangulation in surface reconstruction. In Proceedings of the ALGORITMY 2005, Vysoke Tatry, Podbanske, 13–18 March 2005.

47. Garland, M.; Heckbert, P. Surface simplification using quadric error metrics. In Proceedings of the Special Interest Group on Computer Graphics and Interactive Techniques (SIGGRAPH), Los Angeles, CA, USA, 3 August 1997; pp. 209–216.

48. Minto, S.; Remondino, F. Online access and sharing of reality-based 3D models. *Sci. Res. Inf. Technol.* **2014**, *4*, 17–28.

49. Lorensen, W.E.; Cline, H.E. Marching cubes: A high resolution 3D surface construction algorithm. *ACM Siggraph Comput. Graph.* **1987**, *21*, 163–169.

50. Abad Casal, L.; Lloret, S.G.; Parras, B.G.; Guillen, P.C. El Tolmo de Minateda (Hellín, Albacete, España): Un proyecto de investigación y puesta en valor del patrimonio. *Debates Arqueol. Mediev.* **2012**, *2*, 351–381.

51. IdeARQ + SIA. Available online: http://161.67.130.146/test/potree/puntos_fot_tolmo_2015/examples/puntos_fot_tolmo_2015.html (accessed on 15 February 2016).

52. Prandi, F.; Devigili, F.; Soave, M.; Di Staso, U.; De Amicis, R. 3D web visualization of huge CityGML models. *Proc. ISPRS* **2015**, *40*, 601–603.

53. Robinet, F.; Cozzi, P. Gltf—The Runtime Asset Format for WebGL, OpenGL ES, and OpenGL. Available online: https://github.com/KhronosGroup/glTF/blob/master/README.md (accessed on 30 May 2016).
54. Limper, M.; Thöner, M.; Behr, J.; Fellner, D.W. SRC-a streamable format for generalized web-based 3D data transmission. In Proceedings of the 19th International ACM Conference on 3D Web Technologies, Vancouver, BC, Canada, 8–10 August 2014; pp. 35–43.

Multi-Sensor As-Built Models of Complex Industrial Architectures

Jean-François Hullo, Guillaume Thibault, Christian Boucheny, Fabien Dory and Arnaud Mas

Abstract: In the context of increased maintenance operations and generational renewal work, a nuclear owner and operator, like Electricité de France (EDF), is invested in the scaling-up of tools and methods of "as-built virtual reality" for whole buildings and large audiences. In this paper, we first present the state of the art of scanning tools and methods used to represent a very complex architecture. Then, we propose a methodology and assess it in a large experiment carried out on the most complex building of a 1300-megawatt power plant, an 11-floor reactor building. We also present several developments that made possible the acquisition, processing and georeferencing of multiple data sources (1000+ 3D laser scans and RGB panoramic, total-station surveying, 2D floor plans and the 3D reconstruction of CAD as-built models). In addition, we introduce new concepts for user interaction with complex architecture, elaborated during the development of an application that allows a painless exploration of the whole dataset by professionals, unfamiliar with such data types. Finally, we discuss the main feedback items from this large experiment, the remaining issues for the generalization of such large-scale surveys and the future technical and scientific challenges in the field of industrial "virtual reality".

Reprinted from *Remote Sens.* Cite as: Hullo, J.-F.; Thibault, G.; Boucheny, C.; Dory, F.; Mas, A. Multi-Sensor As-Built Models of Complex Industrial Architectures. *Remote Sens.* **2015**, *7*, 16339–16362.

1. Introduction

1.1. Industrial Context

In order to fulfill the need for as-built datasets to help workers in complex buildings in their daily jobs, large and multi-sensor surveys now have to be considered at the whole building size. Unfortunately, many current tools (including sensors, processing programs and visualization applications) have not been designed for such large surveys of complex indoor facilities.

Until today, the major uses with as-built data in the industry are related to the description of the actual shape of only a part of the facility, with its obstacles and free spaces with centimeter accuracy, to help maintenance planning, handling, storage, replacement or changing important components in that specific part of the plant; see Figure 1. The next step in the field of 3D surveying of facilities consists both of scaling up the current state of the art, without compromising data quality, and in dedicating as-built datasets to new users, who are not experts in CAD or terrestrial laser scanner (TLS) data.

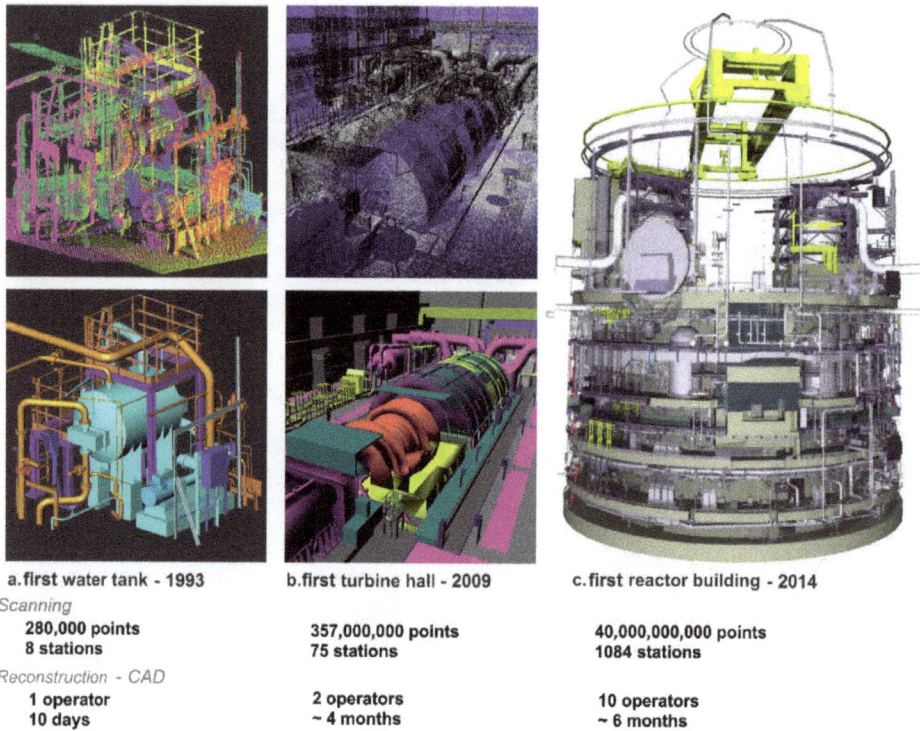

a. first water tank - 1993	b. first turbine hall - 2009	c. first reactor building - 2014
Scanning		
280,000 points 8 stations	357,000,000 points 75 stations	40,000,000,000 points 1084 stations
Reconstruction - CAD		
1 operator 10 days	2 operators ~ 4 months	10 operators ~ 6 months

Figure 1. Review of three breakthrough projects (1993–2014) of as-built reconstruction from laser scanning data of industrial facilities at Electricité de France (EDF) [1]: (**a**) first water tank (1993); (**b**) first turbine hall (2009); (**c**) first reactor building (2014).

1.2. Contributions

The contributions of this paper, which is an enhanced and detailed version of [1], are the following:

- The state of the art of tools and methods for the acquisition, processing and georeferencing of as-built datasets dedicated to the specific conditions of complex indoor facilities, Subsection 2.1 and Subsection 2.2;
- The proposition of a global method for multi-sensor acquisition and processing to represent complex architecture, Subsection 2.3 and Figure 2;
- Settings, processes and feedback from a large-scale multi-sensor scanning survey experiment on a whole 1300-megawatt nuclear reactor building (1000+ stations

of both TLS and panoramic images) with a highlight of the role of the human beings in the process, Section 3;

- A new interactive tool for pose estimation of panoramic images, Section 4;
- Recommendations and examples for developing dedicated applications for virtual tours of complex architectures using multiple data types in order to increase the value of the dataset and answer users' requirements, Section 5;
- An overview of the remaining bottlenecks and challenges in view of the generalization of large, dense, multi-sensor scanning surveys, Section 6.

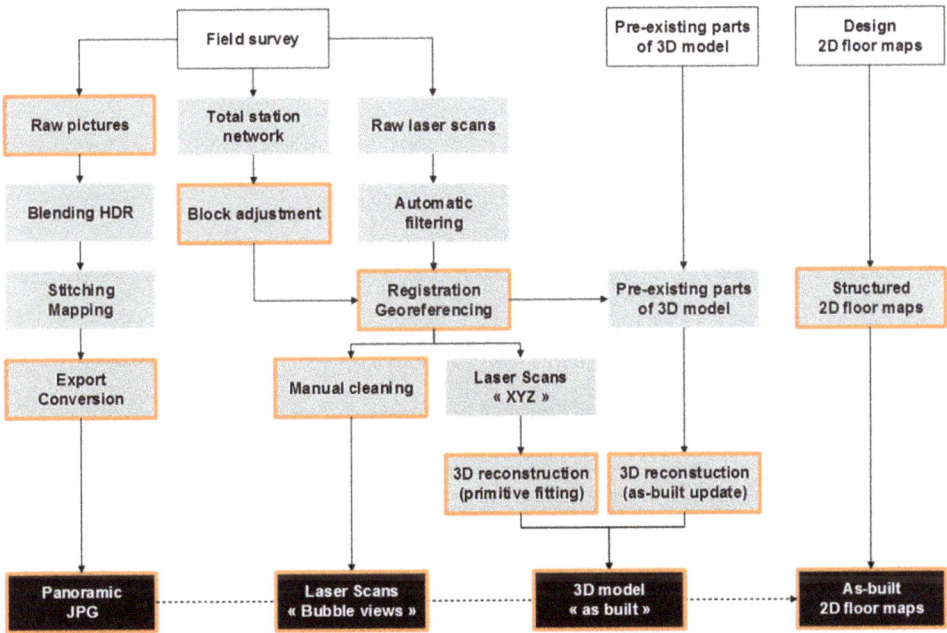

Figure 2. Global framework of as-built data production in the experiment. White boxes are the initial data sources; grey boxes are intermediate data versions; and black boxes are the datasets that composed the virtual replica of a complex building. Orange outlines represent stronger control procedures in the data production process.

2. State of the Art in 3D Surveying for the Industry

2.1. Scanning Industrial Interiors, a 20-Year-Old Challenge

As shown in Figure 3, power facilities are a specific subset of industrial environments, in their shapes (congested, with a huge number of objects, whose manufactured geometries correspond to a set of 3D primitives, like cylinders, boxes, *etc.*), their surfaces (objects sometimes very reflective or with very dark albedos) and accessibility (on-site restricted access time, difficult temperature and vibration conditions). The geometric complexity of these scenes led to the development of dedicated acquisition and processing tools, such as terrestrial laser scanning in the 1990s [2], used very often since then, see Figure 1a. These methods have been used until today to help maintenance workers, by providing a description of the geometry of key areas of the facilities.

Unfortunately, and because of the specific nature of industrial environments, many developments intended for the 3D laser scanning of non-industrial objects have not solved all of the issues arising in the large-scale dense scanning survey of whole industrial scenes with multiple floor heights. These main issues are the speed of surveying (since the plant is only accessible a couple of weeks in the year), the accuracy of the raw points for a large spectrum of albedos (with both mirror-like and dark surfaces in the same scene), the accuracy of registration and referencing (the 2.58σ global geometric tolerance is ±2 cm) and, also, the productivity of 3D reconstruction (up to 100.000 objects for one single building).

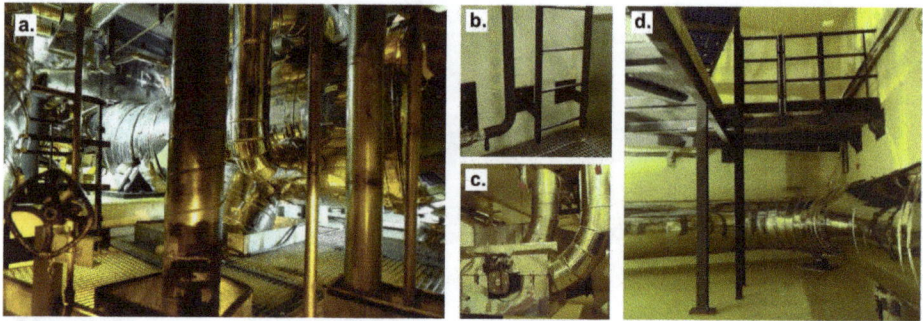

Figure 3. Specific issues of a power facility: (**a**) rooms are often very congested; (**b**) objects can be very dark and (**c**) stainless objects can be clean and then very reflective; (**d**) often coexisting in one place.

In an industrial context, many users are not experienced users of laser scanner or CAD data. Furthermore, geometry alone is not sufficient to meet all maintenance needs and to represent all of the complexity of the architecture and the details of a power plant. These observations lead to the following requirements for the tools and methods used in these new standards in the production and use of as-built datasets for the maintenance of facilities:

(1) Multi-sensor datasets that describe both the geometry and appearance of a facility;

(2) Fast and accurate sensors, responsive to harsh albedos or very reflective surfaces;

(3)   Automated pipelines for conversions and formatting;

(4)   Large-scale efficient tools for registration and processing;

(5)   User-friendly diagnostic tools to achieve high quality requirements on large and complex datasets;

(6)   Dedicated solutions for the intuitive exploration and use of multi-sensor datasets.

2.2. Requirement Engineering for As-Built Datasets of Complex Architectures

In order to represent a complex facility, the requirement engineering of several jobs, through meetings and discussions, to estimate the optimal value of the dataset produced, has led to the conclusion that the following data types must be produced together; see Figure 4:

(1)   Panoramic images, in order to be able to read each equipment tag that might be visible up to 5 m from the main pathways in the facility, as well as to understand the overall architecture of the building;

(2)   Georeferenced terrestrial laser scans, in order to deliver local sub-centimeter geometric accuracy on distance measurements, as well as global 3D locations with less than 2 cm of deviation from the ground truth (2.58σ tolerance);

(3)   2D floor maps, with all elements relative to the structure of the building (walls, floors, ladders, stairs, *etc.*), as well as landmarks to help pedestrians navigate the plant;

(4)   3D, as-built CAD models with categories, to capture empty areas or open walking spaces with less than 5 cm of error (2.58σ tolerance), as well as the shape of the main equipment of the facility.

In order to match these requirements, one must define relevant specification criteria for each type of data, as presented in Table 1.

Then, the tools and methods that might meet all of these criteria were selected using the experience of survey teams that should submit sample datasets to prove their expertise (from either previous surveys or small-scale experiments in similar conditions). The initial estimates of costs and delays were extrapolated to the building scale thanks to a deep analysis of these previous acquisitions.

Finally, to ensure the consistency of results for initial user needs, the dataset was fully checked, using a Level 2 control procedure and then reworked until quality standards were achieved. Finally, user feedback showed the validity of such approaches.

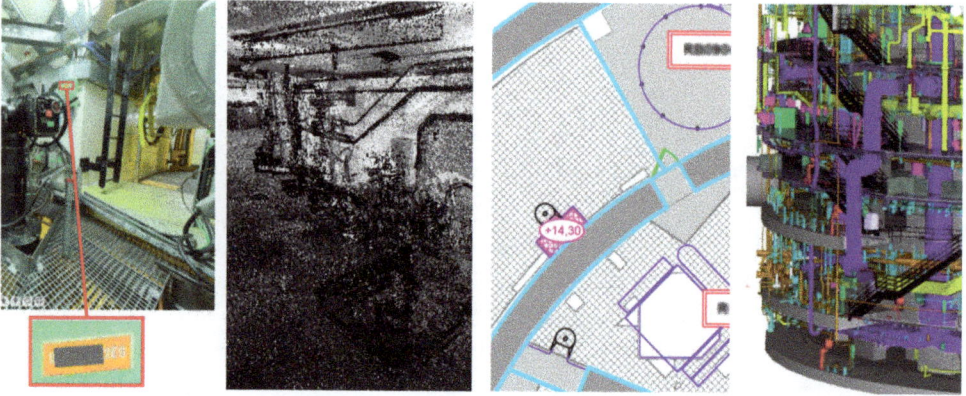

Figure 4. Four types of data are required to represent a power plant facility with a high degree of confidence (geometry and appearance): panoramic images, laser scans, floor maps and 3D CAD model.

Table 1. Criteria of a multi-sensor survey of a building with complex architecture.

Type of Data	Criteria
Panoramic images	completeness, location of stations, field of view, resolution and noise level of digital images, white balance and high dynamic range (HDR), naming, approximate poses, vertical estimation and mapping
Georeferenced and cleaned laser scans	completeness, location of stations, block structure of sub-networks, targets for registration, reference targets for georeferencing, field of view, resolution, signal and total station traverse survey
2D floor maps	completeness, geometry, naming and formatting
3D as-built CAD model	completeness, categories of objects, type of shapes, geometric accuracies by category, names and structure

2.3. State of the Art of Large-Scale Dense Scanning Surveys

Recent breakthroughs offer the prospect of large-scale and multi-sensor scanning surveys that comply with the previously defined technical requirements and tolerances, from acquisition and processing to data integration in dedicated software, developed specifically for maintenance work in a power plant. As will be shown in this section, new developments in data acquisition in the past five years have mainly improved the speed rate, with almost no gain in accuracy. The real benefit for scaling up dense scanning surveys consists of the improvement in data storage, visualization and processing.

2.3.1. Panorama Acquisition

Regarding panoramic image acquisition, current 360° video cameras allow fast acquisition of equirectangular images; the state-of-the-art highest resolution panoramic systems use 44 sensors to produce videos with a resolution of 9000 × 2400 pixels at a rate of 30 frames per second [3]. However, static tripod-mounted motorized heads with full frame digital single lens cameras (DLSR) can generate higher resolutions (gigapixels) panoramas, for example half a billion pixels in five minutes [4], with a high automation rate in stitching. This resolution allows reading labels up to 5 m, and the high dynamic range (HDR) rendering can deal with high contrast and poor lightning conditions.

2.3.2. 3D Point Cloud Acquisition and Registration

Regarding 3D georeferenced point cloud acquisition, some noticeable improvements have come from robotic and computer vision research programs on fast 3D scanning indoors, such as range sensors [5], backpack scanning systems [6], handheld [7] or mounted on mobile platforms in various shapes and sizes [8,9]. However, these systems, either for local precision [5–7] or global georeferencing errors [9], deal mostly with decimeter accuracy. Moreover, many indoor localization and mapping systems require planar objects and/or straight corridors to reduce drift using elaborate algorithms, such as "iterate closest point + inertial measurement unit + planar" proposed in [6]. For labyrinthine and crowded indoors, a fusion of several techniques may be available in a few years, including indoor localization systems based on inertial sensors (the state of the art may be found in the EVAAL indoor positioning competition—Evaluating Ambient Assisted Living Systems Through Competitive Benchmarking—won in 2015 by [10]), graph-SLAM (simultaneous localization and mapping) sensor positioning [11], efficient loop closure [12] and robust detection of features for geo-referencing [13].

Currently, phase-based terrestrial laser scanners are better sensors for small distances, and acquisition rate and can for instance capture 50 million points of mid-range measurements in approximately five minutes (<1 m–50 m). Regarding local accuracy, errors in depth measurements are under 1 cm in that range on most object surfaces [14]. However, the surfaces of some industrial objects do not correspond to the calibration standard ranges (very short distances, low albedos, high reflectance properties and low incidence angles), leading to significant consumption of the error budget [15]. A proportion of these errors can be filtered out automatically [16] (comet-tailed effect, outliers), although another proportion, often quite considerable, can only be removed through manual segmentation (multi-reflections on specular surfaces, in particular).

Regarding global accuracy, the fine registration of large laser scanner datasets is an open and active research topic. For complex buildings, cloud-to-cloud automatic approaches, such as ICP and variants, cannot be used for several reasons. First, due to the proximity of objects, a small relative change in scanner position induces strong differences on point clouds. Thus, cloud-to-cloud constraints cannot, by definition, lead to accurate and trustworthy results in complex scenes. Secondly, these approaches cannot take into account datum constraints for data referencing without a great loss of automation. Finally, these approaches are neither efficient nor robust for very large datasets (many hundreds of stations) [17]. Currently, only target-based registration paired with a total station survey for datum referencing allows the fine registration of hundreds of laser scans ($3\sigma = \pm 2$ cm) on the scale of 10-floor buildings.

2.3.3. 3D CAD Reconstruction

Recent improvements in processing tools for 3D reconstructions from large point clouds (tens of billions) have made possible the "as-built" reconstruction of a full mock-up. First, the data storage cost has increased by a factor of 25 in the last decade [18], while the data USB external transfer rate has increased by a factor of 20 [19], and internal SSD now reaches 500 Mb/s. However, if dozens of laser scans requires gigabytes, thousands of laser scans require terabytes. To improve both file storage, versioning and processing, data structures have been developed [20] and implemented in many commercial software. Other great improvements have been made for visualizing and manipulating a billion point clouds, for example using QuadTrees [21], such as implemented in the WebGL renderer PoTree [22], with real-time shaders, like Eye Dome Lightning [23], also implemented in the open source software CloudCompare.

Looking back, 3D CAD models have been used for planning maintenance operations of industrial installations since the 1990s [2]. Depending on the requirements, several formats can be used depending on the requirements of the industry. In the architectural, engineering and construction (AEC) industry, building information models (BIM) have recently attained widespread attention. These BIM models can be really valuable when they are used in the whole lifecycle of a building [24], from construction monitoring [25] to decommissioning [26], but are used for new rather than for existing buildings [27]. In the power plant industry, PDMS from Aveva has been a leading plant design model for more than 30 years [28]. For power plants designed before the 1980s, only 2D plans were created.

As detailed in [29], there is an obvious need for automated or semi-automated methods for the production of as-built BIMs; the current process for creating parametric BIM from a point cloud is largely a manual procedure, which is time consuming and lacks quality controls. Indeed, primitive-based 3D CAD models can still be reconstructed from point clouds with a better productivity than BIM [30], and most objects of a power plant can be considered as a combination of primitive geometries (due to a series effect of the design and part manufacturing process, except for molding large metal equipment). Though recent automatic algorithms for primitive fitting perform better with increasing sampling resolution, they are still far from a 2.58σ tolerance of detection and fitting [31]. To achieve centimeter accuracy on more than 99% of the reconstructed objects, human interaction is required for either initial segmentation or picking initial points for region growing. To assess this accuracy, several tools must be used: visual inspection and a cloud to shape distance computation. Finally, the primitive-based CAD models can also be displayed in the usual rendering engines of virtual reality [32].

3. Experiment of a Multi-Sensor Survey in a 1300-MW Nuclear Reactor

3.1. Goals and Context of the Experiment

In the wake of several research projects that contributed to the development of tools and methods for dense scanning [33], reconstruction [34] or registration [17] and also tools for the visualization of complex datasets [23], EDF carried out the first very large-scale experiment in the most complex building in a 1300-MW nuclear power plant facility: a reactor building.

The goal was to assess the effective performance (quality, cost and speed) of current tools and best practices through a major research project launched in 2013, in order to make a decision on the generalization of the multi-sensor scanning of power plants and highlight remaining bottlenecks to target lock ups. Another goal was to maximize the benefits of this dataset to the company, especially by sharing it with as many users as possible, to assist them in their daily work.

Therefore, to meet the needs of many maintenance procedures and operations, a dense multi-sensor survey (total stations, laser scanning and panoramic RGB) was carried out during the summer of 2013, using the technical specifications detailed above; Subsection 2.2; see Figure 5.

3.2. A Level 2 Procedure for the Quality Control of Large Datasets

We mentioned several definitions of tolerances and quality requirements to reach high-quality data production. In order to reach these requirements, it is necessary to use specific procedures for the quality control of large, as-built datasets (laser scans, RGB panoramic or 3D CAD models). These Level 2 procedures are a standard for quality monitoring and for detecting non-conforming materials. Our experiment demonstrated their usefulness. To enhance dataset quality, we therefore implemented it in the following manner:

- Level 1: the data creator checks 100% of the dataset and documents it;
- Level 2: the project data manager performs spot checks both on areas of interest and randomly on the whole dataset.

Due to the significant number of human actions involved in the processes and the size of the dataset, at least two iterations are generally needed to reach quality levels.

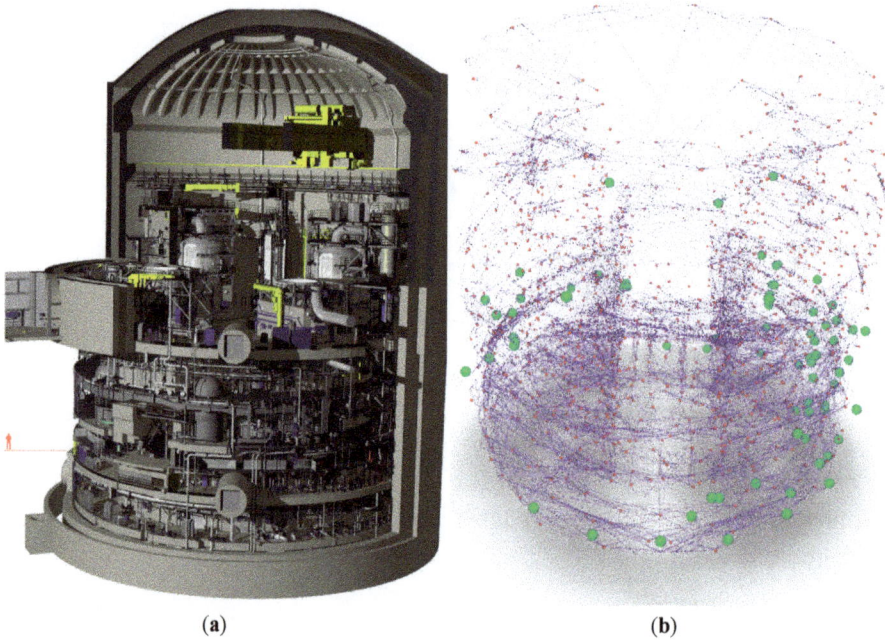

(a)   (b)

Figure 5. (a) Scale of the reactor in a power plant facility, a cylindrical building with 11 floors; (b) 1000+ TLS stations and targets for referencing appear as a very complex 3D network: green spheres are reference points; red spheres are TLS stations; and blue lines are the sights of registration targets from TLS stations.

3.3. Description of the Data Acquisition

The reactor building, whose shape is cylindrical, consists of eleven floors with additional floor heights in each of them. Moreover, the majority of the 100+ rooms in the building is particularly congested, and many of them are only accessible through ladders. Regarding the environmental conditions, exploring and scanning such environments is constrained by poor lighting, access restrictions and congestion due to the equipment and the civil works of the plant itself. During this specific experiment, due to the many maintenance operations occurring in the shutdown period, the building was exceptionally accessible for five weeks (35 days), but we expect the available survey time to be reduced by 50% in the future (17 days). Key numbers for the multi-sensor scanning survey are shown in Figure 6. The global framework for data production is shown in Figure 2.

Acquisition

5 weeks on site

1100 stations

Co-stationning

Laser Dataset

40.10⁹ 3D points

2 cm accuracy (3σ)

800 Gb laser data

RGB Dataset

located on maps

labels readable to 5 m

90 Gb RGB data

GLOBAL TIME ALLOCATION

Laser acquisition
Laser processing
RGB acquisition
RGB processing
CAD reconstruction

Figure 6. Key numbers in the multi-sensor scanning survey of a nuclear reactor building [1]; the main order of magnitude that may be highlighted is the number of stations: more than 1000 stations (for more details, see [35]). The relative time allocation is shown on the graph; CAD reconstruction is the main budget.

Discussion on on-site acquisition: As detailed in [35], this acquisition phase demonstrated the feasibility of such surveys thanks to some specific evolutions and developments of the tools and methods to take into account the up scaling. To succeed, a large number of sensors and operators (five laser-scanning surveyors with Leica HDS 6100 and Leica HDS 7000 and three photographers with Canon EOS 5D Mark III with motorized panoramic heads) is required. However, as mentioned above, we expect to have only half the time to carry out the survey, *i.e.*, in the future, twice the number of sensors and operators will have to be sent to a site, creating additional stress on current methodologies and synchronization between operators.

Discussion on processing TLS data: This experiment also underlined some constraints on scaling up the current post-processing tools and methods in terms of quality, time or cost (fine registration of large laser scanner datasets, quality monitoring and validation tools, issues for formats, storage and data sharing). To illustrate the complexity of the fine registration of laser scans in complex architectures, we show in Figure 5a view of the 3D network used in the adjustment. This experiment, mostly performed with a Leica Cyclone, underlined the lack of robustness of weighted least squares in such large blocks and the need to work with 3D topographic networks instead of 2.5D. Unfortunately, even state-of-the-art registration tools implement a basic version of the least squares algorithm and lack tools for blunder detection, error propagation and quality monitoring.

The main issues regarding the processing of TLS data are the following:

- Several tests on TLS sensors have shown that, even today, the scanning of dark surfaces is not efficient, and the reflective surfaces lead to billions of erroneous 3D points that cannot be filtered by current firmware or software. These points are certainly an issue for 3D reconstructions, but mostly for taking measurements of so-called "bubble views" or station-based views where no feedback on the real position of 3D points is given. Our solution consisted of developing an efficient interface to perform a manual segmentation of these outliers Figure 7.

- Without better algorithms for the fine referencing of TLS dataset and large 3D networks, the only way to comply with a global 2.58σ tolerance of 2 cm accuracy overall, the dataset consists of creating sub-networks with fewer than 200 stations, independently referenced to the external reference system. This constraint implies thorough planning on site for placing and surveying targets. Recent developments have shown that we can expect better results in automation and quality by bundler snooping (moving targets) and the use of robust estimation well known from geodesists [36].

Figure 7. An example of blunder removal in a 3D scan. Blue: 3D points filtered automatically. Red: 3D points manually segmented (the average segmentation time is 6.5 minutes per station).

Discussion on processing panoramic images: The level of automation and the quality of output using modern stitching software, such as the one we used, Kolor AutoPanoPro, are very good. However, some issues still have to be addressed:

- When dealing with 450 Mpixel images of indoor scenes, the field of view of a single picture is quite small and can lead to a lack of feature points when only uniform objects, such as a painted wall, are visible in the picture. The Level 2 control procedure has led to the reopening of 20% of the panoramic stitches for editing;
- Another recurrent error in panoramic images consists of estimating verticality, based on the images. In congested environments, default settings can lead to errors of up to $20°$ in that estimate. We recommend either the use of leveled panoramic heads with custom settings for estimating the relative position of unit images or the use of vertical definition tools through the software interface;

- A major issue is the registration of the panoramic images on the external reference frame; see Figure 8. As will be discussed later on, the fine registration of panoramic images is valuable in assisting with the quality control of 3D reconstructions and offers a better user experience when browsing the dataset. We will address this issue later in this paper.

Discussion on restructuring and updating 2D floor maps: A global model of the plant is key data for designing the structure of the dataset, unifying the names of the objects and performing an analysis on it. By constructing a graph of the various objects and their relationships (adjacency, verticality, inclusion, *etc.*), it makes it possible to answer questions like "What are the panoramic images taken in the rooms adjacent to a specific one?" or "Can I access a specific location in the building without taking ladders?"

This entire graph database was built using structured 2D floor plans, updated using the 3D dataset: walls, grounds, ladders, stairs and many others; *cf.* Figure 9a. These as-built floor plans were then automatically processed to extract all of the required information and to build a "topological graph", describing several properties of the installation (shape, location, names, types, navigation, access, *etc.*), as well as rendering specific maps using style sheets. A previous example of this kind of approach can be seen in [37]. This graph is then used in the software applications that we developed, in order to answer queries on the semantic, geometric or topological properties of the building and its components; *cf.* Figure 9b.

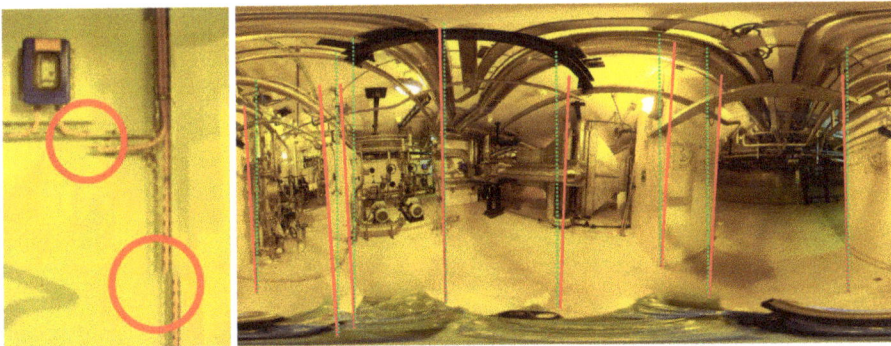

Figure 8. Typical issues in panoramic images. A lack of texture can lead to glitches in the final image (**Left**). The estimated vertical may have to be refined manually because horizon estimation algorithms can fail up to 15° (**Right**).

Regarding the processing steps for the floor maps, the pipeline is as follows:

(1)   Redraw existing floor maps in AutoCAD, following specific drawing rules and using only two types of objects: polylines and blocks (manual);

(2)   Compare the 3D as-built model and panoramic images to floor plans in a specific tool developed in Unity3D, to update them in AutoCAD (manual);

(3)   Convert polygons of floor maps to Scalable Vector Graphics (SVG) files (automatic) and export the blocks to XLS using EXTATTBE in AutoCAD (automatic);

(4)   In a dedicated C# tool, instantiate a class model of objects of the building:

    a   For each floor:

        i   import SVG and XLS to instantiate a relational model of the building (floors, ladders, walls, *etc.*), including controls to check with respect to the drawing rules,

        ii   using a Clipper Library clipping algorithm [38] and threshold, compute relations between objects, such as adjacency, inclusion, *etc.*

    b   For each pair of floors:

        i   Connect objects, such as elevators, ladders, stairs,

        ii   Using Clipper Library, compute vertical relationships between objects.

(5)   Export the instantiated building model to a topological graph in XML, with a description of the building model in XML Schema Definition (XSD).

Discussion on the 3D CAD reconstruction based on the TLS point cloud: A large dataset of 40 billion 3D points was used to reconstruct an as-built 3D CAD model of the facility (see Figure 10). The main part of the 3D reconstruction was produced using Trimble RealWorks Version 8.x (80% of the final CAD model); some objects from a pre-existing CAD model were adjusted in Dassault Systems SolidWorks, as well as other equipment that could not be modeled as a combination of primitive shapes (see Subsection 2.3.3). As illustrated in Figure 6, the 3D reconstruction was the main line in the budget of the data production. In order to meet end users' needs, the quality levels of the 3D data produced had to be defined in accordance with their intended uses: maintenance task planning (including the associated logistics: access, scaffolds, handling areas), worker safety and virtual tour for low accessibility rooms by inexperienced professionals.

The consequences of this multiplicity of needs led to a detailed specification for 3D reconstruction tasks depending on the object type among the 25 categories used: fitting tolerances, naming and methods (cloud least-square fitting, region growing cloud-shape fitting, cloud snapping and copies). The use of precisely georeferenced panoramic RGB images was very valuable in the reconstruction process to help with understanding complex areas.

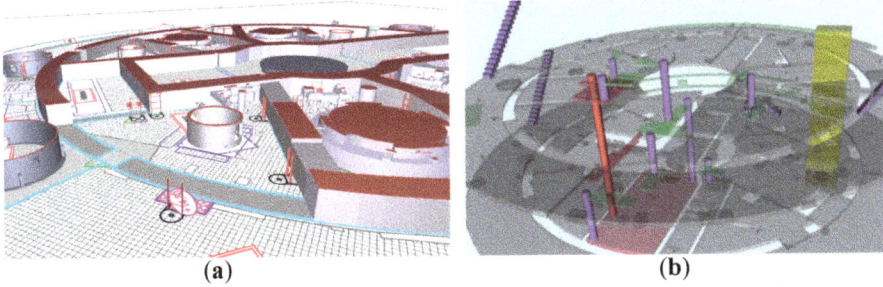

(a) **(b)**

Figure 9. 2D as-built floor maps (**a**) updated using the 3D as-built dataset, laser scans and panoramic images; (**b**) An interactive Virtual Reality Modeling Language (VRML) visualization of a semantic, geometric and topological representation of the building, based on the as-built floor maps [1].

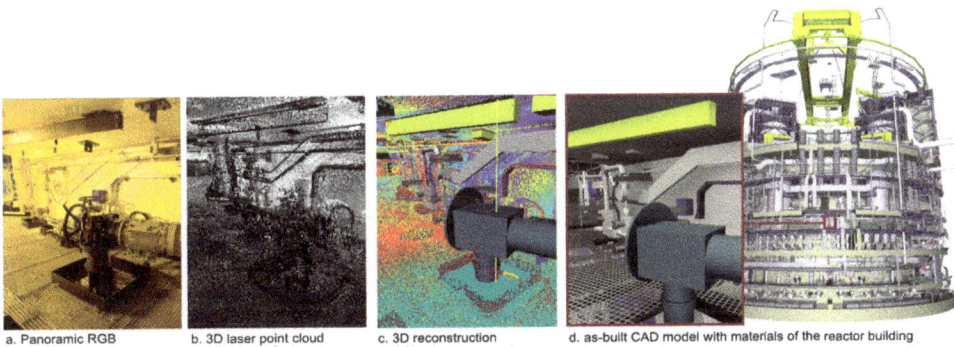

a. Panoramic RGB   b. 3D laser point cloud   c. 3D reconstruction   d. as-built CAD model with materials of the reactor building

Figure 10. The dense data sources used for as-built reconstruction of the nuclear reactor building from point clouds: (**a**) 1025 panoramic images with 450 million of pixels; (**b**) 1085 laser scans with 40 million 3D points; (**c**) 3D reconstruction of 25 types of objects with specific rules (fitting tolerances and naming encodings) and (**d**) as-built CAD model with 100 rooms and 100,000 3D objects.

In order to deal with the huge amount of work and to reduce the time of data delivery, the 3D CAD reconstruction had to be split and parallelized by sectors and then allocated to half a dozen CAD operators for almost a year (10 months). To achieve and certify the quality of the reconstructed as-built model, a Level 2 check procedure was carried out by two independent operators and led to further iterations and re-working (approximately 10% of the total effort). This validation was performed using mesh-cloud distance computations, by using software originally developed by EDF R&D and Telecom ParisTech (CloudCompare [33]) and "out-of-core" technologies to display a maximum number of points for the visual inspection in Trimble RealWorks 8.x, guided by the analysis

369

of an SQL database storing the standard deviation of each reconstructed geometry (~80,000 objects).

3.4. Summary of Tool Development during the Project

Many tools (hardware and software) were used to produce both the data and end-user applications. Some of them were satisfactory, but many had to be improved during the project.

Acquisition and stitching of RGB panoramic images: Canon EOS 5D cameras with a Dr Clauss motorized head were used. For stitching, mapping and blending, Kolor AutoPano was used, and XNview MP was very useful for batch conversions, resampling and renaming. None of the tools were improved.

Acquisition and processing of the TLS dataset: Leica HDS 6000 and 7000 with total stations were used. The Z+F software was used to convert and filter scans. Specific settings were added to that software for improve the results. For registering scans, Leica Cyclone was used. For manual segmentation, we had to add new features in Trimble RealWorks, used to remove noise that could not be filtered out (mirrors).

Floor plan editing and processing: Floor plans were edited in AutoCAD, and a specific tool had to be implemented in Unity3D to compare them to the as-built 3D CAD model. All of the processing, converting and analysis steps required to create the "topological graph" were made in a custom tool developed in C#, using Clipper Library, based on a Vatti clipping algorithm [38].

3D as-built CAD reconstruction and conversions: For main walls and civils works, Dassault Systems SolidWorks was used. For all of the other objects, Trimble Realworks was used and had to be improved to increase productivity (shortcuts, color palette, debugging fitting tools, etc.), as well as quality, by developing a dedicated SQL plug-in to store metadata for each 3D object ("was the object fitted or manually adjusted?", "is the object a copy?", etc.). For two-level control procedures, Hexagon 3DReshaper and CloudCompare were used. CloudCompare was specially enhanced for this project. In addition to the edition and conversion features offered by SAP Visual Enterprise Authors, we developed advanced scripts using Windows PowerShell.

4. Example of Specific Development to Reach Quality Expectations for Referencing Panoramic Images Precisely

During acquisition, panoramic images were roughly located on a map and oriented to the north. However, many usage examples have underlined the need for estimated correct pose in these images (dataset navigation, 3D reconstruction quality check through overlay, etc.). We propose several solutions, using either the TLS dataset, the 2D floor maps or the 3D CAD reconstruction model.

4.1. Camera Model for Panoramic Images

Using an equirectangular projection where the z axis of the reference frame is aligned with the meridians of the panorama (see Figure 11), the spherical coordinates of a pixel can be given using the following relations:

$$\theta = \frac{W - 2u}{W} \pi \in [-\pi, \pi[\quad \phi = \frac{H - 2v}{2H} \pi \in \left[-\frac{\pi}{2}, \frac{\pi}{2}\right[\tag{1}$$

If we represent these pixels on the surface of the sphere, the point $m_i = (x_i, y_i, z_i)$ can be calculated from its coordinates (θ_i, ϕ_i) using the following relations:

$$m_i = (cos\,(\phi)\,.cos\,(\theta)\,, cos\,(\phi)\,.sin\,(\theta)\,, sin\,(\phi)) \tag{2}$$

The epipolar constraints between two images I_1 and I_2, with a relative pose composed of a translation three-vector t and a 3×3 rotation matrix R, are given for a corresponding pair of points from the two images:

$$m_2^T E m_1 = 0 \text{ where } E = [t]_\times R \text{ is the essential matrix} \tag{3}$$

$[t]_\times$ is the anti-symmetric matrix induced by the vector product. The locus of epipolar constraint transforms from lines (pinhole cameras) to circles in the case of spherical panoramas.

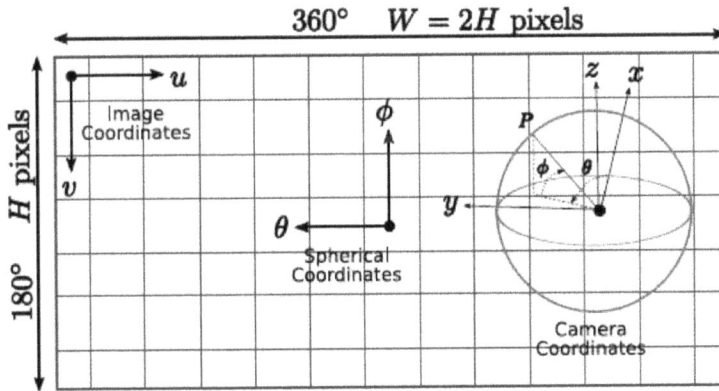

Figure 11. Spherical image and coordinate axes.

4.2. Pose Estimate Using Constraints between Image Pixels and 3D Points from Laser Scans

For the external pose estimate, *i.e.*, the absolute pose of the panoramic image in the global reference frame, we can use matching feature points between pixels in the image and real world points (for example, from a 3D laser scan of a close station). Thus, the problem is the following, given points m_i, the coordinates of the corresponding pixels on the sphere, and M_i, the homogeneous coordinates of 3D:

$$a_i m_i = R M_i + t = (R|t) \begin{pmatrix} M_i \\ 1 \end{pmatrix} = P M_{i(homogen.)} \text{ where } a_i \text{ is a scale factor for each point } i \quad (4)$$

This can be turned to a linear system problem $Ah = 0$ as follows, where p_j is the j-th row of:

$$Ah = \begin{pmatrix} M_1^T & 0 & 0 & -x_1 & 0 & 0 & 0 & \cdots \\ 0 & M_1^T & 0 & -y_1 & 0 & 0 & 0 & \cdots \\ 0 & 0 & M_1^T & -z_1 & 0 & 0 & 0 & \cdots \\ M_2^T & 0 & 0 & 0 & -x_1 & 0 & 0 & \cdots \\ 0 & M_2^T & 0 & 0 & -y_1 & 0 & 0 & \cdots \\ 0 & 0 & M_0^T & 0 & 0 & -x_3 & 0 & \cdots \\ \cdots & \cdots & \cdots & \cdots & \cdots & \cdots & \cdots & \cdots \end{pmatrix} \begin{pmatrix} p_1^T \\ p_2^T \\ p_3^T \\ a_{11} \\ a_{22} \\ \cdots \end{pmatrix} = 0 \quad (5)$$

We solve this linear system by choosing h as the last column of V where $A + UDV^T$ is the singular value decomposition of A. Taking the 12 entries of h, we obtain P_{est} (estimated P), which differs from P by a scale factor k, that can be solved through:

$$k = \frac{||P_{est} M_i||}{a_{2i}} \quad (6)$$

we can choose k as the average value computed from n points.

In our case, panoramas are "vertical", which simplifies the extraction of the only non-zero rotation angle κ (using Tait–Bryan angles) and t values from P. Thus, to solve pose estimation, we need a few reliable and well-distributed matching pairs. This can be done manually or automatically using a feature point extraction algorithm and more robust estimating schemes. However, achieving stable results in matching and pose estimates for the whole dataset is tedious, and manual editing requires a dedicated interface. In the next two sections, we propose interfaces for using either 2D maps or 3D as-built CAD models instead of feature point-based methods.

4.3. Partial Pose Estimate of Panoramic Images Using 2D Floor Maps

A user-oriented approach was implemented to get a first "independent" estimate of parameters (t_x, t_y) and (κ) of a vertical panoramic image by using 2D floor maps. Using a user interface with a synchronization of the panoramic view and its position with the camera orientation on the 2D floor map, the procedure followed by the operator is:

(1) Move (t_x, t_y) panoramic using near and identifiable details of the floor maps by estimating the ratio of distances (doors, holes in floors, *etc.*);

(2) Orient (κ) using far landmarks of the floor maps;

(3) Check at $+90°$ and $+180°$ and iterate the first two steps until the best estimate.

This method was applied to the entire dataset (1025 panoramic images), and 95% of the images were moved or oriented (970); see Figure 12. The average processing and control time for one panoramic image is 1 min 50 s.

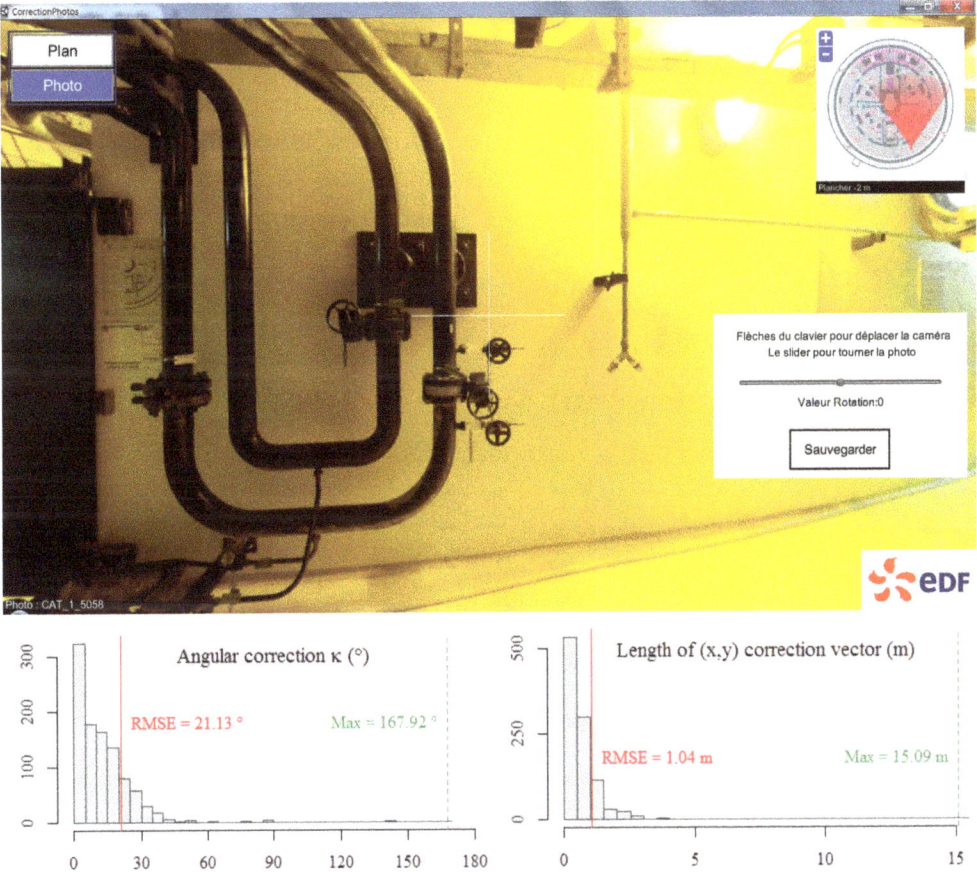

Figure 12. User interface of the partial pose estimation tool. The camera position, orientation and field of view of the panoramic image are synchronized with the floor map (**Top**). A central cross helps the sighting of landmarks; Key numbers of the corrections using 2D maps (1025 images) (**Bottom**).

4.4. Full Pose Estimate of Panoramic Images Using the 3D As-Built Dataset

Another user-oriented approach was developed to refine the whole pose (including t_z) of the panoramic images (t_x, t_y, t_z) and (κ) given the hypothesis of verticality ($\omega = 0$, $\phi = 0$). Using a user interface with a synchronization of the panoramic view and its position with the camera orientation in the 3D model and using switches for overlay and transparency, the procedure followed by the operator consists of independently correcting the four parameters of the pose; see Figure 13:

(1) Move (t_z) panoramic using landmarks of the 3D model that cross the equator of the panoramic in the image (boxes, stairs, guardrail, *etc.*); front view;

(2) Align (κ) panoramic using parallel objects of the 3D model to any meridian of the panoramic in the image (pipes, beams, *etc.*); zenith view;

(3) Move (t_x, t_y) panoramic using landmarks on the ceiling of the 3D model to align them to the zenith of the image; zenith view;

(4) Check horizontally for any issue at +90° and +180° and iterate the first three steps until the best estimate; an incorrect estimate of the vertical when mapping the panorama is the main source of a poor estimate of the pose of the panoramic, and such a panoramic should be corrected.

This method was applied to the entire dataset after the first estimate on 2D floor maps (1025 panoramic images). During this process, we moved and oriented 100% of the images (1025). The average processing and control time for one panoramic image is 30 s for low corrections and 1 min 30 s–2 min for more difficult images.

4.5. Overall Feedback on the Experiment and Discussion on Future Large Scanning Surveys

Even for a company with over 20 years' experience in the field of as-built documentation, a multi-sensor and dense scanning survey on the scale of an entire 11-floor building remains a challenge, for which tools and methods have not yet been designed.

In addition to the several recommendations and choices that were set out in the previous paragraphs, our experiment highlighted some specific aspects of large and multi-sensor scanning surveys in complex architectures. For these projects that respond to the growing need for as-built data for professionals, we recommend:

- specifying needs, requirements and constraints in detail. On the scale of a building, every misunderstanding or fuzzy specification may have a severe impact on costs, quality or durations;
- documenting every step in the process and performing quality monitoring from the beginning, to help both fixing non-conforming data and enriching the dataset for future use;
- parallelizing the tasks as much as possible (acquisition and post-processing); when the data production time increases, the number of non-qualities increases significantly.

Regarding remaining bottlenecks, we have mentioned several technical brakes on the generalization of current tools and methods for large and complex buildings, regarding raw data acquisition and processing. We have also discussed the three areas of 3D reconstruction: costs, duration and quality. Since the real requirement is the third one, the first two areas must be viewed as secondary in achieving the dataset in industrial processes. The low level of automation and the low quality of acquisition sensors and tools are the actual brakes on generalization. However, by performing rigorous quality monitoring and control, each dataset can be produced with a high degree of confidence. In such large projects, traceability is one of the keys to quality management. The Figure 2 gives a global overview of the whole framework of as-built data production; we can see that many steps are required, including many tools and human interactions. As shown, also, many control procedures are required to reach and certify a quality level. Indeed, the main drawback for achieving expected quality levels is the difficult inter-validation of data sources. To perform such a quality control policy, the dataset has to be merged, compared and visualized altogether. Nevertheless, due to the limitations of current tools and formats, this fusion and exhaustive checking of multi-source datasets remains very tedious.

1. Interactive pose determination procedure using a 3D reconstructed model

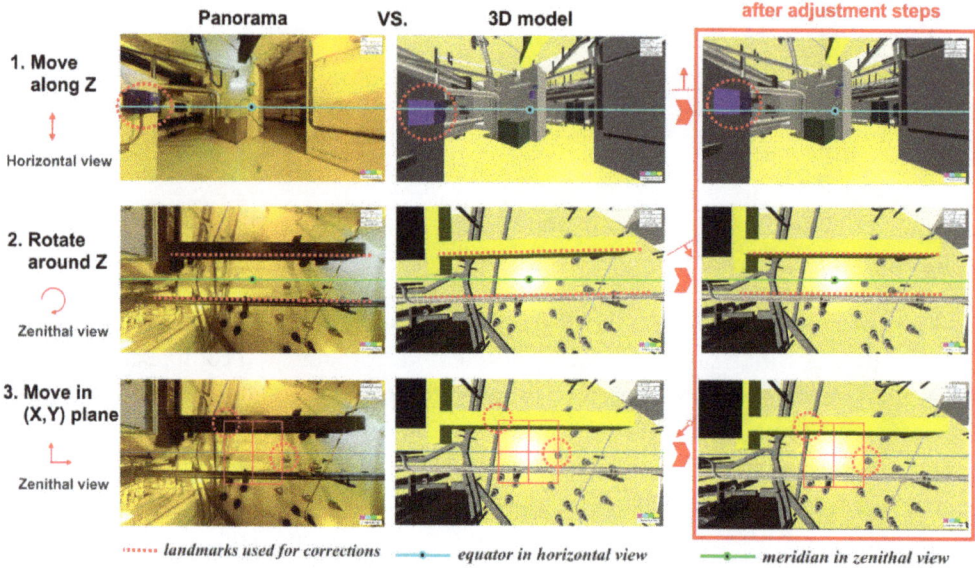

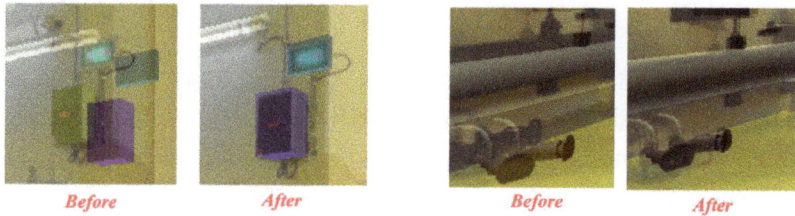

2. Zoom on details

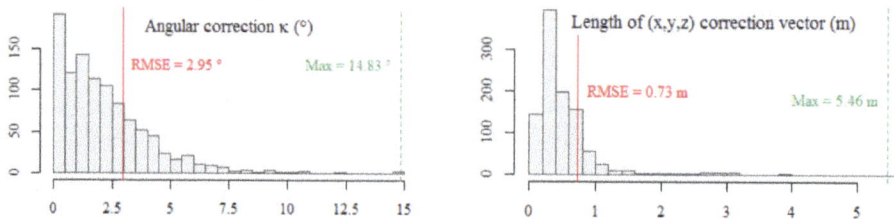

3. Key numbers in the experiment

Figure 13. User interface and procedure for fully estimating the pose of panoramic images in reference to a 3D model (**Top**); Equator, meridians and central dots help with the alignment and correction of parameters. Zoom on specific details (before/results) (**Middle**); Key numbers in the corrections using the 3D model (1025 images) (**Bottom**).

376

5. Developing New Software Applications to Increase the Value of the Dataset

In the previous sections, we have detailed the challenges in producing data to represent a complex building. Due to that complexity and the intended uses, multiple data types are required: high resolution panoramic RGB, laser scans, 3D as-built CAD models and 2D floor plans. Once the dataset is complete, issues remain for visualization and exploration, taking into account that users may not be familiar with these various data types.

5.1. State of the Art of Multi-Data Visualization for Complex Architectures

By its nature, each data source represents only one aspect of the reality of the plant and only meets some needs among many. The co-visualization of the multiple data sources is therefore required to assist and help workers with finding answers in their daily jobs. In recent years, several solutions have been proposed for the problem of multi-source, as-built data visualization; see Figure 14:

- virtual tours with floor plans and panoramic RGB [39];
- navigation through several spherical laser views;
- navigation and path calculation in 3D environments, for instance in 3D video games or 2.5D cartography services, such as Google Maps indoors.

Figure 14. Examples of user interfaces that can be found today for virtual tours: Leica TrueView, CSA VirtualTours, Kolor PanoTour, Faro WebShare.

However, none of these solutions integrates all of the data sources required to represent some complex buildings, such as nuclear reactor buildings, and easily navigate the large dataset. Indeed, apart from the constraints of technical integration, the complexity of the plant itself is an issue for virtual navigation (multiple levels with dozens of rooms and vertical junctions) and requires a specific interaction design to handle it.

5.2. Rules for the Development of Virtual Tours of a Complex Building

Our first goal is to develop an application that can be of value to many people working in nuclear plants, taking into account that the targeted building is rarely accessible. Potential values include several scenarios. The first scenario

377

consists of improving the productivity of maintenance operations through a virtual preparation stage that takes into account the spatial constraints of the environment; this requires the collaboration of different teams based on a shared representation of the environment (e.g., mechanical workers defining their scaffolding needs). The second scenario aims to improve accuracy and to reduce delays in engineering studies to prepare for modifications and revamping to the plant, with reduced on-site time for the teams. The third and last scenario consists of helping many recently-hired workers to become familiar with their working environment more rapidly, through dedicated training sessions, including tutoring courses.

These few scenarios illustrate the variety of user profiles that should be considered for the common view, which we wish to design. Many of them are not familiar with the handling of as-built datasets, and their use of the tool we developed may be very occasional, which reinforces the need for user-friendly and simple interfaces, as well as taking into account human perception in the design [40,41].

To address this problem, several principles have been selected for the development of navigation interaction within our applications, developed using the game engine Unity 3D. The first principle consists of the multi-view synchronization across a multi-source dataset, with the possibility at any time of moving from one view to another without losing user position/orientation (shared, unique position of the user in the environment), as shown in Figure 15. The second principle consists of the ability to explore the building in its horizontal and vertical dimensions (through stairs and ladders, as well as trapdoors and removable slatted floors), which is required by tasks, such as handling heavy equipment. The last principle consists of developing all tools and functionalities in a project interface for a better user data management: marks, snapshots, distance measurements with guidance instruction and path computations.

For instance, we estimate that in the synchronized, multi-view experience, spherical photographs need to be positioned in the virtual environment with a 2.58σ tolerance of 50 cm and 5°. This accuracy range seems sufficient to help users to focus on their task, by not being disturbed by some inaccuracies in the transitions when switching between views. Furthermore, the verticality of the panoramic images must be known with an error below 3°; otherwise, navigation becomes very uncomfortable.

In addition to the navigation features, we provide users with specific interactions with the dataset. Each feature has been designed according to the users' needs. To implement them, we carried out several iterations with users to overcome technical feasibility issues and create genuinely powerful and smart tools. Among these, we can mention distance measurements on 3D CAD models (perpendicular to the normal unit vectors, vertically constrained or free), annotations and snapshots in every data view, interactive cutaways of walls and user data management and sharing. In all of these developments, a large amount of feedback has been collected in a quick response development process to achieve optimum user benefit.

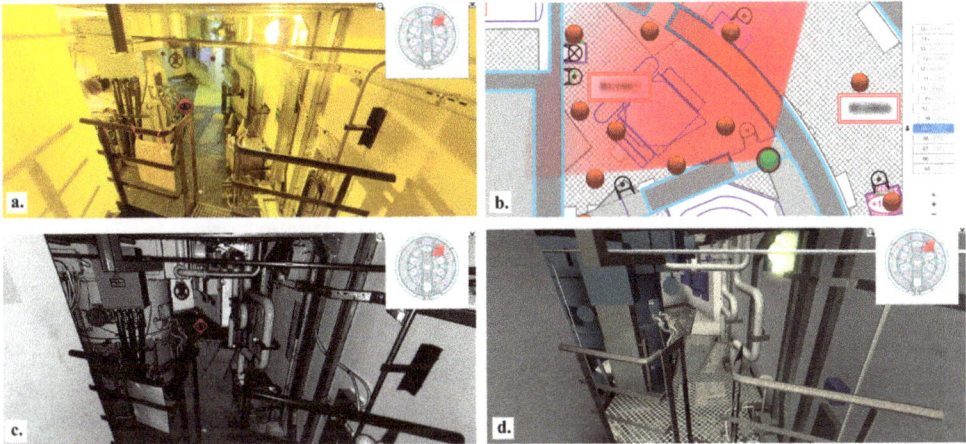

Figure 15. Multi-source data exploration and navigation in a complex building: visualization and station-based transitions of (**a**) panoramic images and (**b**) laser scans; (**c**) 2D map view of all stations and synchronized mini-maps and (**d**) first-person pedestrian navigation in the 3D model, including climbing ladders, taking stairs and crouching.

5.3. A Framework for Taking Measurements on Laser Scans for Non-Expert Users

So-called "bubble views" of laser scans are often used to measure distances. However, for non-familiar users, two main issues should be avoided. Firstly, errors in range measurement (reflective surfaces, for instance) impact the distance value; hence, many distance measurements should be checked more than once and interpreted with caution, although most users ignore the typical error budget of a laser scan. A manual segmentation process is highly recommended to remove as many noisy points as possible, to comply with the definition of tolerances. Secondly, wrong picks on edges can lead to wrong distance measurements, whereas in spherical view, no feedback can be returned. To decrease the rate of wrong measurements, we have developed a specific procedure, as detailed below.

To reduce the occurrence of false measurements in a laser scan, even after removing outliers, we propose the following procedure (see Figure 16):

(1)   choose a type of measurement from "almost horizontal", "almost vertical" and "oblique"

(2)   pick two points in the laser scan to define the distance to be measured.

→   If the picked segment is "consistent" with the type of measurement, then:

379

(3) pick the two points in the laser scan again.

→ If the measured distance is "similar" to the previous one:

(4) the measurement is displayed on the laser scan with centimeter accuracy; it can be checked against the reconstructed 3D model.

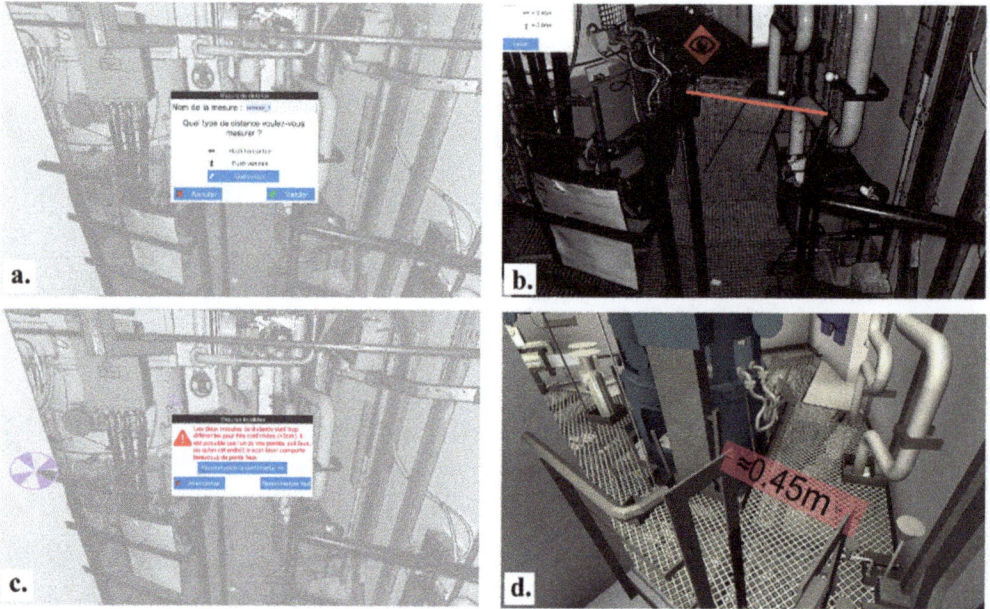

Figure 16. A procedure for improving the quality of distance measurement in laser scans. The user must (**a**) choose between measurement types and (**b**) control his/her own measurement by a double picking; then (**c**) check the result in "bubble view" and finally (**d**) in the 3D model.

5.4. An Example of Advanced Technological Features Using a Georeferenced Dataset

The quality of the entire dataset allows the development of advanced features, such as path overlay on panoramic images. Paths are computed on navmeshes, using Autodesk GameWare, and can be overlaid in real-time on panoramic images; see Figure 17.

Path computation on 3D model using navmeshes.

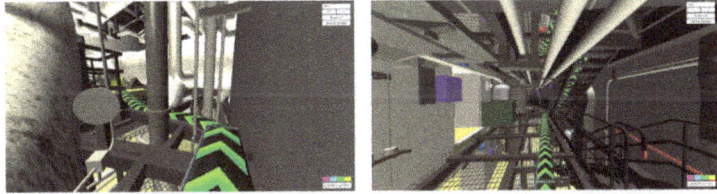

Overlay of path on referenced panoramic images

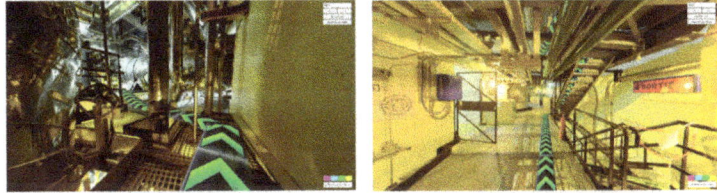

These details highlight the need of both accurate 3D reconstruction and precise referencing of panoramic images.

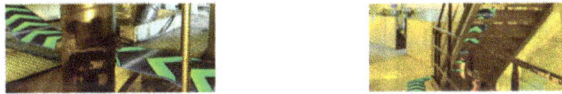

Figure 17. 3D pedestrian path (using navmeshes) overlay on panoramic images.

6. Conclusions and Future Work

The professional use of as-built models has recently increased significantly, resulting in two major challenges: scaling up dense and multi-sensor scanning surveys to a whole building and sharing this dataset with a very large audience.

In this paper, we reviewed the state of the art of scanning tools and methods for industrial installations. Then, we proposed a global methodology for acquisition and processing of multi-sensor datasets to represent complex architectures. To face the lack of automation to reach high quality in modelling multi-floor plants, we detailed the optimal contribution of human beings in the process. We then assessed this methodology in a large-scale experiment of modelling the 11 floors of a 1300-megawatt nuclear reactor building. From the acquisition of 1000+ co-stations (RGB panoramic and laser scans) to a 3D as-built reconstruction, the project has involved many contributors for almost a year and a half. At every step, quality requirements have put stress on tools and methods and led to the implementation of Level 2 quality controls. We finally presented new tools that have been developed to help many professionals in their daily jobs by allowing them to explore a complete digital plant easily using different types of data.

We can summarize the contribution made by our work in three points:

- We assessed the feasibility, as well as underlined the current complexity of tools and methods used in a multi-sensor scanning of industrial environments (1000+ stations). Issues remain to reduce the effort in the production of such models.

381

- We have highlighted human beings' contribution in the data production (from interactive tools to quality management and control).
- We have shown how the value of such datasets can be increased by developing multi-source data visualization and navigation applications in multi-floor plants with recommendations for general principles applied to virtual tours of complex architectures.

Based on these findings and experiments, future work is already planned, with a view toward reducing the significance of the remaining issues:

- Improve data referencing procedures and algorithms to ensure confidence in quality levels, across all of the datasets.
- Develop new tools for data cross-validation and consistency across a multi-source dataset.

However, some serious bottlenecks seem farther removed from the current state of the art:

- How could we significantly increase (benefit >50% of current cost) the productivity of 3D as-built CAD (or BIM) models from very large point clouds (hundreds of billions)?
- Which procedures and resources should be employed to update a large multi-sensor dataset and detect errors or inconsistencies between two epochs?

Acknowledgments: The authors would like to thank the reviewers and copy editors, whose helpful comments led to a better paper overall. The authors also want to thank all the involved teams in this project, with a special mention to survey teams (photographers and surveyors), CAD operators and software developers.

Author Contributions: Jean-François Hullo and Guillaume Thibault conceived and designed the tools and the methods of data acquisition and processing. Christian Boucheny and Arnaud Mas conceived and designed the end-user application. Fabien Dory developed tools for 3D path computations and panoramic pose estimation.

Conflicts of Interest: The authors declare no conflict of interest.

References

1. Hullo, J.-F.; Thibault, G.; Boucheny, C. Advances in multi-sensor scanning and visualization of complex plants: The utmost case of a reactor building. *ISPRS Arch. Photogramm.* **2015**, *1*, 163–169.
2. Pot, J.; Thibault, G.; Levesque, P. Techniques for CAD reconstruction of "as-built" environments and application to preparing for dismantling of plants. *Nucl. Eng. Des.* **1997**, *178*, 135–143.
3. Depraz, F.; Popovic, V.; Ott, B.; Wellig, P.; Leblebici, Y. Real-time object detection and tracking in omni-directional surveillance using GPU. In Proceedings of the 2015 SPIE Target and Background Signatures, Toulouse, France, 21 September 2015.

4. Kopf, J.; Uyttendaele, M.; Deussen, O.; Cohen, M. Capturing and viewing gigapixel images. *ACM Trans. Gr.* **2007**, *26*, 93.

5. Khoshelham, K.; Elberink, S.O. Accuracy and resolution of kinect depth data for indoor mapping applications. *Sensors* **2012**, *12*, 1437–1454.

6. Chen, G.; Kua, J.; Shum, S.; Naikal, N.; Carlberg, M.; Zakhor, A. Indoor localization algorithms for a human-operated backpack system. In Proceedings of the 5th International Symposium 3D Data Conference on Visualization, and Transmission, Paris, France, 17–20 May 2010.

7. Bosse, M.; Zlot, R.; Flick, P. Zebedee: Design of a spring-mounted 3-d range sensor with application to mobile mapping. *IEEE Trans. Robot.* **2012**, *28*, 1104–1119.

8. Adán, A.; Quintana, B.; Vázquez, A.S.; Olivares, A.; Parra, E.; Prieto, S. Towards the automatic scanning of indoors with robots. *Sensors* **2015**, *15*, 11551–11574.

9. Zlot, R.; Bosse, M. Efficient large-scale three-dimensional mobile mapping for underground mines. *J. Field Robot.* **2014**, *31*, 758–779.

10. Li, Y.; Zhang, P.; Lan, H.; Yuan, Z.; Niu, X.; El-Sheimy, N. Real-time indoor navigation using smartphones. In Proceedings of IPIN 2015 Conference on Evaal Indoor Navigation Competition, Banff, AB, Canada, 13–16 October 2015.

11. De la Puente, P.; Rodriguez-Losada, D. Feature based graph-SLAM in structured environments. *Auton. Robot.* **2014**, *37*, 243–260.

12. Labbe, M.; Michaud, F. Online global loop closure detection for large-scale multi-session graph-based slam. In Proceedings of THE 2014 IEEE/RSJ International Conference on Intelligent Robots and Systems, Chicago, IL, USA, 14–18 September 2014.

13. Ji, S.; Shi, Y.; Shan, J.; Shao, X.; Shi, Z.; Yuan, X.; Ynag, P.; Wu, W.; Tang, H.; Shibasaki, R. Particle filtering methods for georeferencing panoramic image sequence in complex urban scenes. *ISPRS J. Photogramm.* **2015**, *105*, 1–12.

14. Barras, V.; Delley, N.; Chapotte, G. Analyses aux limites des scanners laser terrestres. *XYZ Rev. Assoc. Fr. Topogr.* **2013**, *137*, 19–26.

15. Voegtle, T.; Wakaluk, S. Effects on the measurements of the terrestrial laser scanner HDS 6000 (Leica) caused by different object materials. In Proceedings of the 2009 ISPRS Work on Laser Scanning, Paris, France, 1–2 September 2009.

16. Tang, P.; Huber, D.; Akinci, B. A comparative analysis of depth-discontinuity and mixed-pixel detection algorithms. In Proceedings of the IEEE Sixth International Conference on 3-D Digital Imaging and Modeling, Montreal, QC, Canada, 21–23 August 2007.

17. Hullo, J.-F.; Thibault, G.; Grussenmeyer, P.; Landes, T.; Bennequin, D. Probabilistic feature matching applied to primitive based registration of TLS data. In Proceedings of the XXII ISPRS Congress on Annals of the Photogrammetry, Remote Sensing and Spatial Information Sciences, Melbourne, Australia, 25 August–1 September 2012.

18. Komorowski, M. A History of Storage Cost. Available online: http://www.mkomo.com/cost-per-gigabyte-update (accessed 30 October 2015).

19. Wikipedia. List of Device Bit Rates. Available online: https://en.wikipedia.org/wiki/List_of_device_bit_rates (accessed 4 October 2015).

20. Elseberg, J.; Borrmann, D.; Nüchter, A. One billion points in the cloud—An octree for efficient processing of 3D laser scans. *ISPRS J. Photogramm.* **2013**, *76*, 76–88.

21. Scheiblauer, C. Interactions with Gigantic Point Clouds. Ph.D. Thesis, Institute of Computer Graphics and Algorithms, Vienna, Austria, 2014.

22. Schütz, M. Potree, A Free Open-Source WebGL Based Point Cloud Renderer for Large Point Clouds. SCANOPY Project. Available online: https://github.com/potree/potree (accessed 30 October 2015).

23. Boucheny, C. Interactive Scientific Visualization of Large Datasets: Towards a Perceptive-Based Approach. Ph.D. Thesis, Université Joseph Fourier, Grenoble, France, 2009.

24. Azhar, S. Building information modeling (BIM): Trends, benefits, risks, and challenges for the AEC industry. *Leadersh. Manag. Eng.* **2011**, *11*, 241–252.

25. Golparvar-Fard, M.; Peña-Mora, F.; Savarese, S. Automated progress monitoring using unordered daily construction photographs and IFC-based building information models. *J. Comput. Civil. Eng.* **2012**, *29*, 04014025.

26. Jung, I.; Kim, W. Analysis of the possibility of required resources estimation for nuclear power plant decommissioning applying BIM. In Proceedings of the KNS 2014 Spring Meeting, Jeju, Korea, 28–30 May 2014.

27. Volk, R.; Stengel, J.; Schultmann, F. Building Information Modeling (BIM) for existing buildings—Literature review and future needs. *Automat. Constr.* **2014**, *38*, 109–127.

28. Trickett, K.G.; Chaney, J.C. *PDMS: Plant Layout and Piping Design*; Gulf Publishing Company: Houston, TX, USA, 1982.

29. Tang, P.; Huber, D.; Akinci, B.; Lipman, R.; Lytle, A. Automatic reconstruction of as-built building information models from laser-scanned point clouds: A review of related techniques. *Automat. Constr.* **2010**, *19*, 829–843.

30. Son, H.; Kim, C.; Kim, C. 3D reconstruction of as-built industrial instrumentation models from laser-scan data and a 3D CAD database based on prior knowledge. *Automat. Constr.* **2015**, *49*, 193–200.

31. Kang, Z.; Li, Z. Primitive fitting based on the efficient multiBaySAC algorithm. *PLoS ONE* **2015**, *10*, e0117341.

32. Whyte, J.; Bouchlaghem, N.; Thorpe, A.; McCaffer, R. From CAD to virtual reality: Modelling approaches, data exchange and interactive 3D building design tools. *Automat. Constr.* **2000**, *10*, 43–55.

33. Girardeau-Montaut, D.; Roux, M.; Marc, R.; Thibault, G. Change detection on points cloud data acquired with a ground laser scanner. *ISPRS Arch. Photogramm.* **2005**, *36*, 30–35.

34. Bey, A.; Chaine, R.; Marc, R.; Thibault, G.; Akkouche, S. Reconstruction of consistent 3D CAD models from point cloud data using a priori CAD models. In Proceedings of the 2011 ISPRS Workshop on Laser Scanning, Calgary, AB, Canada, 29–31 August 2011.

35. Hullo, J.-F.; Thibault, G. Scaling up close-range surveys: A challenge for the generalization of as-built data in industrial applications. *ISPRS Arch. Photogramm.* **2014**, *1*, 293–299.

36. Wicki, F. Robust estimator for the adjustment of geodetic networks. In Proceedings of the First International Symposium on Robust Statistics and Fuzzy Techniques in Geodesy and GIS, Zurich, Switzerland, 12–16 March 2001.

37. Whiting, E. Geometric, Topological & Semantic Analysis of Multi-Building Floor Plan Data. Master's Thesis, Massachusetts Institute of Technology, Deptment of Architecture, Cambridge, MA, USA, 2006.

38. Vatti, B.R. A generic solution to polygon clipping. *Commun. ACM* **1992**, *35*, 56–63.

39. Koehl, M.; Scheider, A.; Fritsch, E.; Fritsch, F.; Rachebi, A.; Guillemin, S. Documentation of historical building via virtual tour: the complex building of baths in Strasbourg. In Proceedings of the XXIV International CIPA Symposium on Archives of the Photogrammetry, Remote Sensing and Spatial Information Sciences, Strasbourg, France, 2–6 September 2013.

40. Thibault, G.; Pasqualotto, A.; Vidal, M.; Droulez, J.; Berthoz, A. How does horizontal and vertical navigation influence spatial memory of multifloored environments? *Atten. Percept. Psychophys.* **2013**, *75*, 10–15.

41. Dollé, L.; Droulez, J.; Bennequin, D.; Berthoz, A.; Thibault, G. How the learning path and the very structure of a multifloored environment influence human spatial memory? *Adv. Cogniti. Psychol.* **2015**, in press.

Chapter 4:
3D Modeling

Fine Surveying and 3D Modeling Approach for Wooden Ancient Architecture via Multiple Laser Scanner Integration

Qingwu Hu, Shaohua Wang, Caiwu Fu, Mingyao Ai, Dengbo Yu and Wende Wang

Abstract: A multiple terrestrial laser scanner (TLS) integration approach is proposed for the fine surveying and 3D modeling of ancient wooden architecture in an ancient building complex of Wudang Mountains, which is located in very steep surroundings making it difficult to access. Three-level TLS with a scalable measurement distance and accuracy is presented for data collection to compensate for data missed because of mutual sheltering and scanning view limitations. A multi-scale data fusion approach is proposed for data registration and filtering of the different scales and separated 3D data. A point projection algorithm together with point cloud slice tools is designed for fine surveying to generate all types of architecture maps, such as plan drawings, facade drawings, section drawings, and doors and windows drawings. The section drawings together with slicing point cloud are presented for the deformation analysis of the building structure. Along with fine drawings and laser scanning data, the 3D models of the ancient architecture components are built for digital management and visualization. Results show that the proposed approach can achieve fine surveying and 3D documentation of the ancient architecture within 3 mm accuracy. In addition, the defects of scanning view and mutual sheltering can overcome to obtain the complete and exact structure in detail.

Reprinted from *Remote Sens*. Cite as: Hu, Q.; Wang, S.; Fu, C.; Ai, M.; Yu, D.; Wang, W. Fine Surveying and 3D Modeling Approach for Wooden Ancient Architecture via Multiple Laser Scanner Integration. *Remote Sens*. **2016**, *8*, 270.

1. Introduction

The architecture of China is as old as its civilization. Together with European and Arabian architecture, ancient Chinese architecture is an important component of the world architectural system [1,2]. Since the Tang Dynasty, Chinese architecture has significantly influenced the architectural styles of Korea, Vietnam, and Japan. Ancient Chinese architecture mainly involved timberwork. Wooden posts, beams, lintels, and joists make up the framework of a house [3,4]. The specialty of woodworking requires the adoption of antisepsis methods, which eventually developed into the respective architectural painting decorations of China. Colored glaze roofs, windows

389

with exquisite applique designs, and beautiful flower patterns on wooden pillars reflect the high level of craftsmanship and rich imagination.

Unlike other building construction materials, old wooden structures often do not survive because they are vulnerable to weathering and fires and are naturally subjected to rotting over time. Consequently, only a few examples of ancient wooden architecture exist today [5–7]. YuZhen Palace, which was included in the World Heritage List, was a famous ancient wooden architecture complex in Wudang Mountain. It was built in 1417 to represent the Taoism culture. On 19 January 2003, a fire turned YuZhen Palace into ashes. Chinese traditional ancient wooden architecture is an important cultural heritage; thus, its protection, repair, maintenance, and restoration is a significant challenge [8,9].

Fine surveying and mapping to generate 3D models of historical buildings by obtaining all types of maps and drawings is an important task for digital protection projects and continuous monitoring of ancient wooden architecture [1–12]. Fine maps, drawings, and photorealistic models can provide high geometric accuracy and detail of the architectural layout and structure. For the past two decades, total station surveying and close-range photogrammetry were traditionally used for the required data collection of historical buildings [13–16]. The 3D data collection based on laser scanning has made great progress in the quick and reliable fine surveying and 3D documentation of heritage sites by obtaining millions of 3D points to effectively generate a dense representation of the respective surface geometry [16,17].

Laser scanning techniques can create the 3D nature of archaeological objects that vary in size and shape, ranging from something as small as a human molar to objects as large as a building. Terrestrial laser scanning has been widely adopted in archaeology site surveys and 3D documentation, particularly in projects dealing with ancient architecture [18–21]. Several surveying and digital heritage projects focusing on the use of laser scanning technologies for 3D modeling of heritage sites can be seen in literature [22–28]. Recent studies on terrestrial laser scanning techniques have achieved great progress in 3D archaeological documentation. However, some limitations remain. The restrictions of scanning view and mutual sheltering lead to data being missed in terms of structural details, particularly for complex ancient architecture, thereby affecting the quality of the final 3D model [29,30]. Establishing proper approaches to TLS data collection and processing for the fine surveying and 3D modeling of complex archeological sites is challenging and valuable work, specifically in transforming these approaches from research to generally accepted techniques.

The work presented in this paper aims to overcome the scanning view restriction and mutual sheltering in difficult surveying surroundings to obtain fine surveying and 3D documentation of Chinese wooden architecture through the combination of different levels of terrestrial laser scanners. Three-level TLS with a scalable measurement distance and accuracy is presented for the data collection of the architecture to solve the problems of fine surveying and 3D documentation of the ancient building complex, which is constructed on the cliffs of Wudang Mountains. A multi-scale data fusion approach is proposed for laser scanning data registration and filtering. A laser scanning point projection algorithm together with point cloud slice tools is designed for the fine surveying of the ancient architecture to generate all types of architecture maps, such as plan drawings, facade drawings, section drawings, and doors and windows drawings. The section drawings together with slicing point cloud are presented for the deformation analysis of the building structure. The 3D models of the ancient architecture components are built for digital management, research, and visualization through fine drawings and laser scanning data.

2. Background

2.1. Liangyi Temple in Wudang Mountains

Wudang Mountains are located in Shiyan, Hubei Province in Central China and are renowned for the practice of Tai Chi and Taoism. Liangyi Temple is the Buddhist counterpart of the Shaolin Monastery, which is affiliated with Chinese Chan Buddhism. A total of 53 ancient buildings and nine architectural sites were constructed from the early Tang Dynasty to the Yuan, Ming, and Qing dynasties. These structures are listed as World Heritage Sites because they include many Taoist monasteries and secular buildings, all of which have a profound influence on Chinese art and architecture.

The existing Liangyi Temple, which was built on the South Cliff during the Yongle Period (1403–1424) and undertaken by Emperor Zhu Di in memory of his parents, is one of the extremely precious official wooden buildings in Wudang Mountains. It has maintained its original structure since its construction during the Ming Dynasty. It is located in the middle of a cliff, as presented in Figure 1a,c. The gate of Liangyi Temple is masterly built at its side by ancient designers, and its external veranda extends straight to the Tianyi Zhenqing Stone Hall, which was built during the Yuan Dynasty; its front Longtou Incense (as shown in Figure 1d), which faces the Golden Dome, shows the superb idea of ancient Chinese craftsmen and fully reflects the Taoist "Imitation of Nature" [31,32].

As shown in Figure 1, Liangyi Temple was built on a mountain cliff. It was one of the important royal Taoist architectural complexes in Wudang Mountains. The temple, which includes a gable and hip roof, a wooden structure, and a single building, was the best preserved Ming Dynasty architecture in Wudang Mountains. The construction of Liangyi Temple represents the outstanding achievement of Chinese architectural technology and art, which has great historical, scientific, and artistic value.

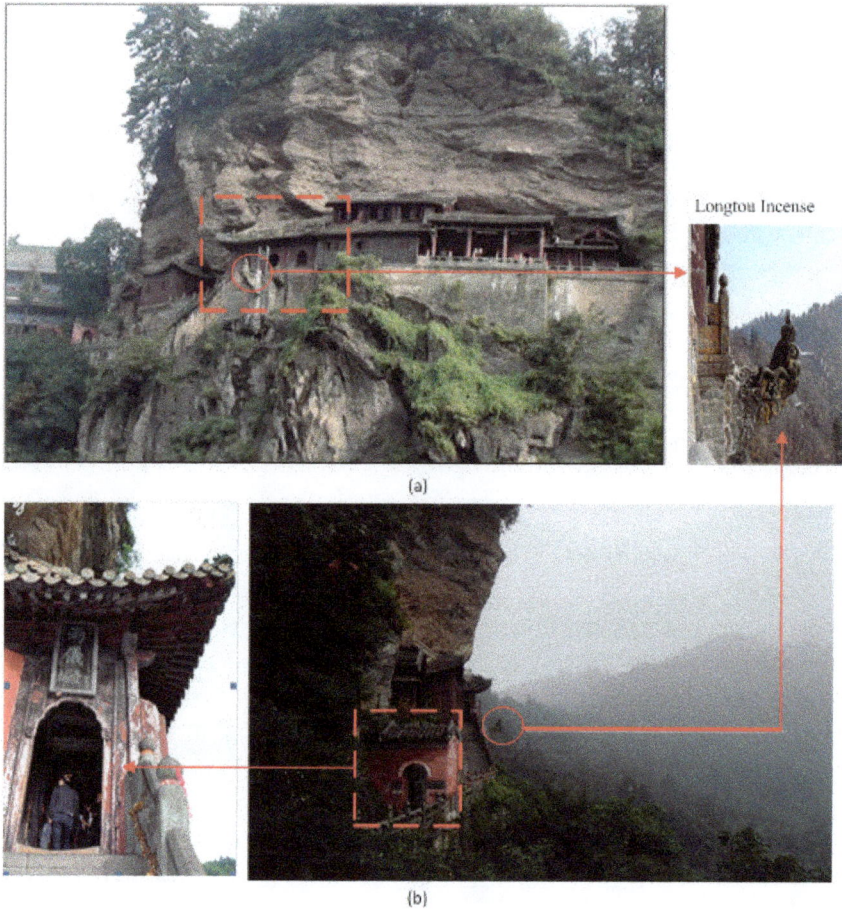

Figure 1. Ancient architecture of Liangyi Temple: (**a**) front side, (**b**) side face. The red dashed rectangle in (**a**) is the Liangyi Temple. The red dashed rectangle in (**b**) is the gate door of Liangyi Temple. The red circle is Longtou Incense.

2.2. Threats on Liangyi Temple

As a world cultural heritage, Liangyi Temple faces huge challenges for its protection. Wudang Mountains are located in a subtropical monsoon climate zone. The mild and humid weather provides a suitable living environment for white ants. The wood components of Liangyi Temple have no resistance against moths. The wooden columns serve as the load-bearing structure and are easily infested by white ants, which cause structural deformation or collapse. Thus, the structure of the building should be monitored to protect the wooden Liangyi Temple by preventing its deformation. Another significant threat is fire. On 19 January 2003, a fire broke out in Wudang Mountains, causing another wooden historical building, namely, Yuzhen Palace, to burn to ashes [9,33,34]. The Longtou Incense in front of Liangyi Temple is a fire source, and frequent Taoist activities with joss candles and electrical lightings are potential fire risks. As an internationally renowned area of cultural heritage, Wudang Mountains were visited by more than 3,689,000 tourists by the end of July 2015. The large number of visitors and narrow visiting places present another threat to the building structure and body.

Given the aforementioned threats, the fine surveying, mapping, 3D documentation, and modeling for the digital protection of the wooden Liangyi Temple are necessary and valuable. The Chinese government started the "Compass Plan" through fine surveying and 3D documentation projects to prevent accidents with ancient architecture. Liangyi Temple was selected as a representation of ancient wooden architecture.

2.3. Problems and Solution: Fine Surveying and 3D Documentation for Liangyi Temple

In previous literature, terrestrial laser scanning technologies are widely used for the surveying and documentation of all types of archaeological sites, such as caves, natural landscapes, buildings, and tombs. The basic principle of the laser scanner is that it rapidly captures the shapes of objects through distance measurement at every pointing direction with a rotation mirror. The object should be in the direction of the laser point without any sheltering. Thus, scanning view restriction and mutual sheltering, which can affect the quality of 3D data, are more serious factors than weather and fogs [28,29]. Liangyi Temples only consist of two houses with an area of less than 50 m². Except for the statues and tables, the standing place is less than 20 m² with no more than a 1 m corridor. The complicated setting and complex building structure lead to the following problems in surveying and documentation via laser scanning:

(1) **Missed corner data**: The small and narrow standing place requires many stations of TLS, leading to serious missed 3D data with the restriction of scanning view, as shown in Figure 2.

Figure 2 shows that the intersection of ceiling and cantilever causes blind scanning of the two stations, and 3D data of the blind corner are missed. Numerous close scans should be set up to collect the missed 3D data of the blind corner.

(2) **Incomplete 3D data of the structure**: As a Taoism worship site, Liangyi Temple has many statues, tables, boxes, and ritual supplies. Mutual sheltering in the small space also causes another kind of incomplete 3D data for the structure measurement. Changing the scanning view or using a handheld laser scanner can limit the missed 3D data for the detailed architecture structure surveying. A special point cloud data processing approach should be designed to achieve precise structure

surveying behind the sheltering objects with the adjacent 3D data, for example, the slicing of point cloud from a different angle to intersect the corner.

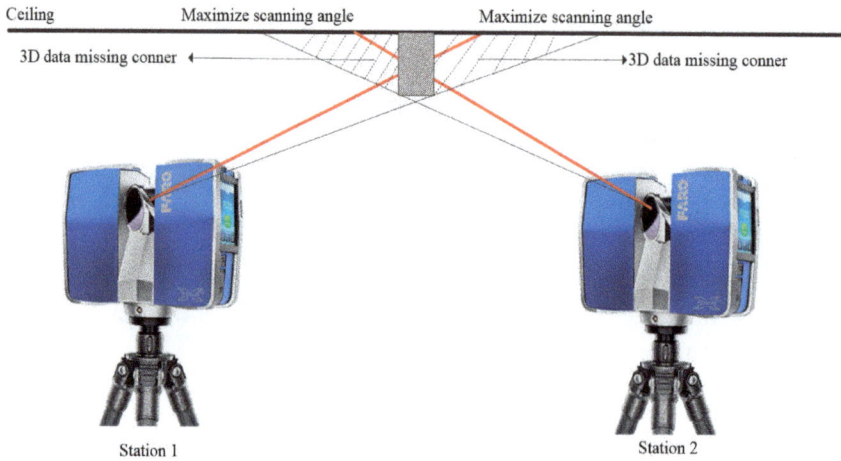

Figure 2. Blind corner in scanning of two stations.

(3) Separated 3D data of the difficult scene: As introduced in Section 2.2, Liangyi Temple was constructed on a cliff. Thus, collecting the whole 3D data in one scene is impossible. The roof surface in the scene of the building and the inside of the building from the roof of the building cannot be seen. The front and back sides of the cliff have no space for TLS data collection. The only position from the gate to obtain 3D data of the building with its roof is also restricted by the scanning view. Thus, the whole scene of 3D data of the cliff terrain can only be collected from the front side of the cliff, which should be at the other peak opposite to the cliff. The 3D data of the roof must be collected at the top of the building. Data registration and fusion of the separated 3D data are other challenges for 3D documentation and fine surveying.

The unique and steep location of Liangyi Temple with its complicated layout and structure makes fine surveying and 3D documentation difficult using the current terrestrial laser scanning technique. The scalable and difficult scene must use different types of laser scanners to collect 3D data with minimal loss. Our solution is to integrate long-distance TLS, middle-range laser scanner, and a handheld scanner. The long-distance TLS (Riegl VZ 1000) will collect 3D data of the cliff terrain and the whole building. The middle-range laser scanner (Faro Focus 3D) will cover the building body and roof. The handheld scanner (Handyscan 3D) will be adopted for the architectural components and the scattered blind corner. All these separated 3D data should be registered well by their spatial relationship for further mapping and 3D modeling. The proposed idea is explained in Figure 3.

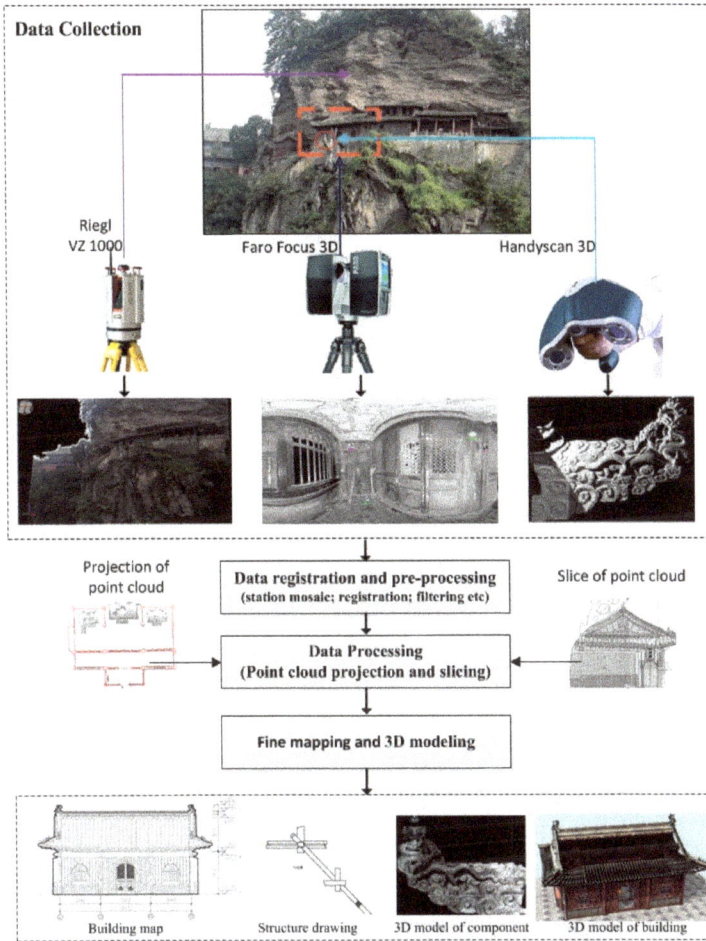

Figure 3. Proposed idea for the fine mapping and 3D documentation of Liangyi Temple.

The proposed idea for the fine mapping and 3D modeling of Liangyi Temple includes data collection, 3D data registration and preprocessing, 3D data processing, fine mapping, and 3D modeling. The rest of this paper is organized as follows. Section 3 introduces the proposed idea in detail. Section 4 discusses the final results. Finally, Section 5 presents the conclusions of the proposed approach and future work.

3. Methodology

The proposed approach defines an optimized terrestrial laser scanning flowchart for the fine surveying and 3D documentation of Liangyi Temple when compared with

395

the current method used for archaeological sites. Multiple laser scanner integration also brings related data processing methods, which are presented in Sections 3.2–4.

3.1. 3D Data Collection Based on Multiple Laser Scanner Integration

The different types of terrestrial laser scanning were applied according to local conditions. Differential GPS (DGPS) and total station were adopted to ensure the orientation and registration of different 3D datasets in a common coordinate system by measuring the ground control points (GCPs) and target points of TLS. The field data collection workflow is shown in Figure 4.

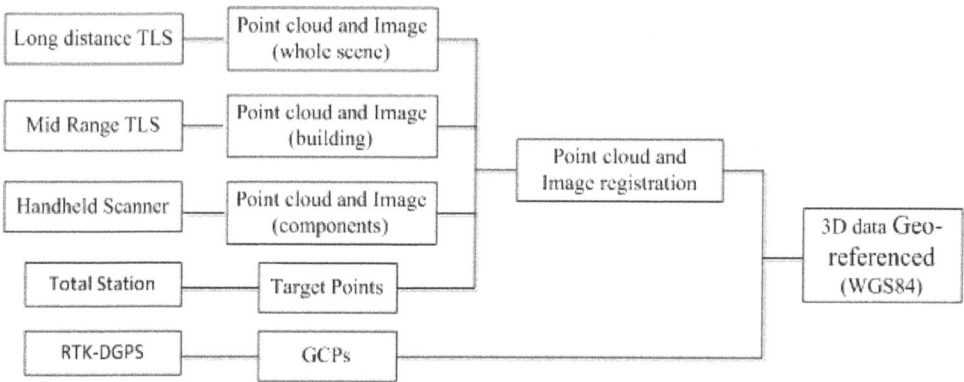

Figure 4. Workflow of field data collection.

(1) Control points and target point measurements

The GCPs were measured with DGPS in real-time kinematic mode (DGPS-RTK) in the WGS84 coordinate system with respective 3D accuracies of 10 and 15 mm in the horizontal and vertical directions after net adjustment. Three types of targets were adopted for the laser scanning data registration. One was a retro-reflecting circle provided by Riegl GmbH. The second was a reference sphere from Faro Focus 3D. The last one was a rectangle white–black cardboard. The target points with reference sphere were used for point cloud registration station by station, and their coordinates do not need to be measured. The target points with retro-reflecting circle and rectangle white–black cardboard were used for the long-distance Riegl point cloud registration together with the middle-distance point cloud from Faro Focus 3D. Thus, they needed absolute coordinates. Total station was adopted to survey all these target points based on GCPs measured with DGPS-RTK. A final total of five GCPs via DGPS-RTK and 35 target points were measured by the total station. The target points were all transformed to WGS84 with 3D accuracies of 21.2 and 26.8 mm in the horizontal and vertical directions, respectively, after triangulation net adjustment. With the target points, we combined and registered the point clouds of the cliff terrain via Riegl and those of the building via Faro Focus 3D.

(2) Data acquisition of multiple laser scanner integration

Three types of laser scanners with different distance ranges (*i.e.*, 1200 m with Riegl VZ 1000, 120 m with Faro Focus 3D, and 0.5 m with Handyscan) were adopted for the 3D data acquisition in difficult environmental conditions. Equipment choices, logistics, scanning stations, and environmental conditions must be considered in survey planning to ensure the quality of 3D data. The whole survey plan included three steps, and the complete data acquisition took 11 days of field work.

In the first step, the long-distance laser scanner with Riegl VZ 1000 was positioned on the three sites opposite Liangyi Temple, as shown in Figure 5. These sites have a platform and pedestrian steps that link to the Temple. The distance between the platform and Liangyi Temple is less than 1000 m. In each station, high-density scanning with the highest angle resolution of 0.0024° in both vertical and horizontal scan lines was taken first in the specific sight view angle to Liangyi Temple. Then, the fine and detailed scanning of the target point was taken point by point after the targets were recognized automatically with RiSCAN PRO software. Riegl VZ 1000 is composed of a high-performance long-range laser scanner with a wide field of view and a calibrated digital camera firmly mounted on the scanning head of the laser scanner. After 3D point cloud data acquisition, the system then took images with the top-mounted camera in certain positions to obtain colorful visual information of the cliff and building surface, as presented in Figure 5b. Each station took one day of field work, including 3 h for density scanning, 1 hour for target fine scanning, and 3 min for image capture. Scanning from the three stations finally acquired the 3D data of the cliff terrain and the whole building with ground sampled distances of 2 cm, as shown in Figure 5c.

In the second step, the middle-distance laser scanner with Faro Focus 3D was presented to collect 3D data of the building body in detail. The building was divided into five parts, namely, gate, corridor, attic, stair, and roof, and scanned from 53 positions (Figure 6) during five days of field work. Each position had more than five reference sphere targets, and no less than three targets that should be scanned in the neighboring position. For the narrow space, where the positions between stair and roof lacked target points, some feature points in the point cloud of the long-distance scanner with Riegl VZ1000 were used as target points for data registration. All these positions also collected 360° panoramic images with the built-in camera to obtain the visible information.

Figure 5. Long-distance laser scanner was used to acquire 3D data of the cliff terrain, and the scanning stations were located throughout the whole building (**a**); (**b**) scanning work scene on the platform; (**c**) 3D point cloud data with color information.

In the last phase, the handheld laser scanner with Handyscan was adopted to acquire 3D data of some important components of the building, such as Longtou Incense, status, decoration of the building, horned beast, and Vatan on the roof (shown in Figure 7). The windows with special engraving were also documented with the handheld scanner. In this phase, another two days of field work were taken to collect 3D data of some concealed regions in the corner and some missed important components after we merged all the data of the previous steps and checked the data quality.

Figure 6. 3D data collection for the building. (**a**) Scanning for the gate; (**b**) scanning for the corridor; (**c**) scanning for the Longtou Incense; (**d**) scanning for the roof; (**e**) scanning for the statues; (**f**) point cloud of the whole building; (**g**) scanning for the attic.

Figure 7. 3D data collection for some components with the handheld scanner. (**a**) Small statue scanning; (**b**) rows of horned beast scanning; (**c**) large statue scanning; (**d**) corner horned beast scanning.

Although the whole area of Liangyi Temple is very small, less than 50 m², 11 days of field work was spent for 3D data acquisition with the three-level laser scanner. The total data volume was approximately 2 TB, including 83 positions of point cloud, 53 panoramic images with a resolution of 1 cm, and 11 high-resolution images with a resolution of 5 mm.

3.2. 3D Data Registration and Preprocessing

The 3D data obtained from different laser scanners focused on different parts of Liangyi Temple. In accordance with our data collection plan, a combined registration approach based on different models in different software was presented for the 3D data registration, as shown in Figure 8. The foundation of 3D data registration is a unique WGS84 coordinate system. For the three-position dataset via Rigel VZ1000, we used RiSCAN PRO software for data registration. After directly adding WGS84 coordinates of the target points, all the points from three positions could be registered and geo-referenced in the WGS84 coordinate system. The 3D data registration of the building acquired with Faro Focus 3D included two steps. In the first step, 53 position data were registered through the reference sphere target points as indicated in the flowchart provided by Faro Scene software via an independent relative coordinate system. In the second step, the whole dataset was registered by the extracted target points for geo-referencing in the WGS 84 coordinate system. A surface-based registration method developed by the chair of Photogrammetry and Remote Sensing at ETH Zurich, called least squares 3D surface matching [35], was applied to register the 3D data of the scattered point cloud datasets from Handyscan by integrating the point cloud dataset of the components and the building with a strict geometric constraint. After performing all the registration processes, the absolute reference system for the 3D data was WGS84.

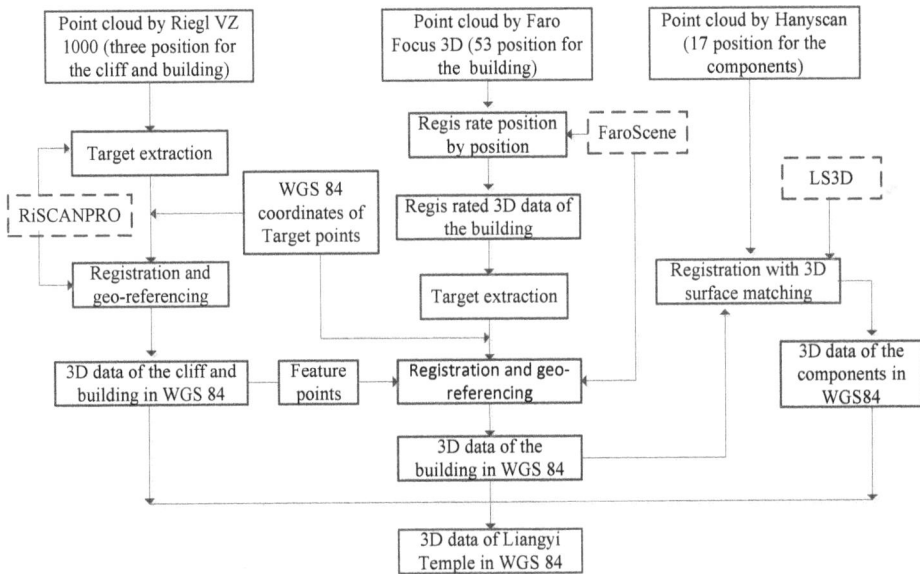

Figure 8. 3D Data registration approach for the point cloud of a different scanner.

A globally registered point cloud with high accuracy was generated through the proposed 3D data registration approach for further processing and application. Before the data processing for the fine map drawing and modeling, some data preprocesses were presented to remove the moving objects in the point cloud dataset and filter out the noise points. The preprocessing of moving object removal was implemented with Terrosolid software both semi-automatically and manually. For the point cloud of the components, both noise points and data holes exist, as revealed in Figure 9a,c. Geomagic Studio software was introduced for noise filtering and hole repair to achieve a high-quality point cloud dataset. Figure 9d,e show the result of one component dataset after filtering and repair.

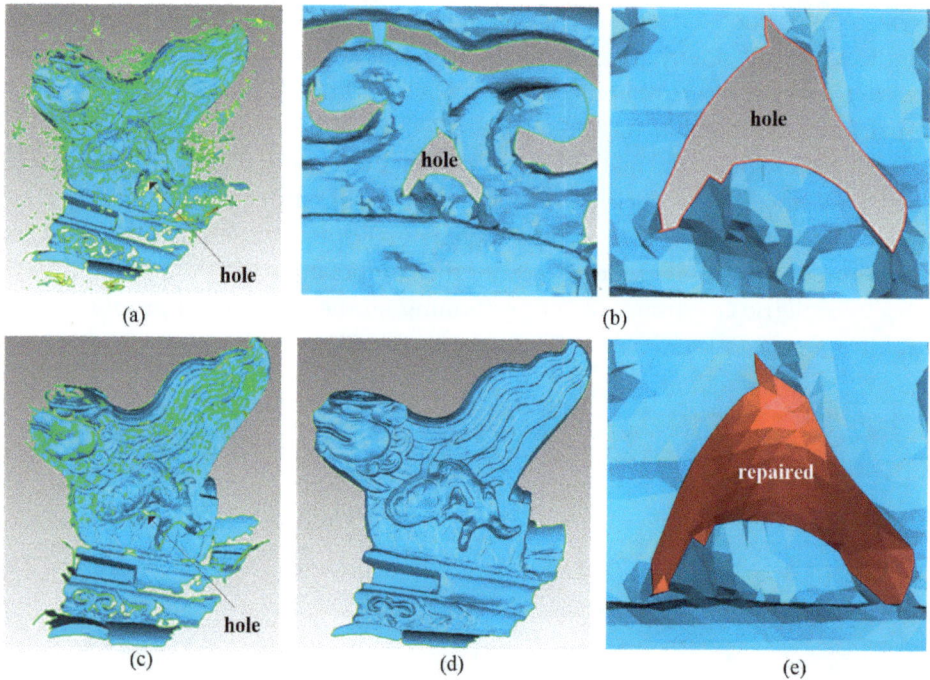

Figure 9. Noise filtering and data repair of the component point cloud: (**a**) original point cloud; (**b**) hole in detail; (**c**) result of noise filtering; (**d**) result after filtering and repairing; (**e**) result of data repair.

3.3. Data Processing: Point Cloud Projection and Slicing

The 3D data processing aims to achieve fine map drawing and 3D modeling with the registered point cloud. The complex architecture is composed of various geometric objects and is represented by three-view drawings in different planes. The main work of fine surveying aims to generate all types of three-view drawings and plane drawings of the geometric components of the building for use in 3D modeling of the architecture [36]. The projection and slicing of point cloud are usually adopted to simplify the data and highlight the geometric feature by drawing all these plane maps with a large volume of point cloud. In this section, point cloud projection and point cloud slicing approaches are proposed in detail for the fine surveying of complex architecture.

(1) Point cloud projection

The method of point cloud projection was first suggested for automatic building model reconstruction. Projection simulates the intuitive data handling of a user who tries to recognize the structure of a 3D point cloud by rotating it into specific projections [37,38]. In the specific projection, the geometric aspect, such as the

402

building layout, can be formalized and structured. The projection can also simplify 3D data into the 2D place with minimal data volume. The fine drawing of Liangyi Temple benefits from point cloud projection because the arbitrary view can be defined with the drawing of a user. AutoCAD 2011 can support point cloud data. The point cloud data should be converted in the specific format of *. pcg to obtain a high-efficiency index before importing the point cloud data into AutoCAD 2011 software. Once the point cloud data are loaded via AutoCAD 2011, the point cloud can easily be handled with a user-defined projection. Figure 10 shows some typical point cloud projections of the building. The point cloud projection converts 3D data into 2D space. In the 2D space, the point cloud can be rendered as a gray image (as Figure 10a), in which the gray value is mapped as the linear model based on the distance from the point to the projection plane. All the three-view drawings of the building can be generated with point cloud projection. The vertical projection is used for the layout map. The side view projection can generate the gate map, while the front view projection is used for the face map of the building.

(2) Point cloud slicing

The slicing method for mass 3D point data to construct a CAD model is commonly used in reverse engineering technology. Slicing data are automatically or manually grouped from the 3D point dataset according to the single principle feature to recognize the border, which is useful for surface modeling of complex objects [39,40]. The slicing method of point cloud always combines with the projection approach, as shown in Figure 11. First, the cloud data are segmented into several layers by slicing the point cloud along a user-specified direction. Next, the sliced data points in each layer are projected onto an appropriate plane and then used to reconstruct geometric objects.

The most important work for point cloud slicing is to determine the layer thickness, which is sensitive to accuracy and shape extraction. A small layer thickness leads to missed shape features, whereas a large layer thickness causes the shape feature to submerge in the heavy point cloud data. Thus, layer thickness should be defined as a different type of architectural component. The point cloud slicing function is provided by Kubit software from Faro Company. Kubit is a point cloud processing software that combines sophisticated measuring technologies with the established CAD-functionality seamless integration into AutoCAD for surveying, construction, and architecture. Kubit plug-in support can be used to define an arbitrary direction and layer with a specific thickness, and thus obtain the slicing data for the fine drawing of component. Figure 12 illustrates the fine surveying and drawing results of some typical components, such as ridge tie beam, crescent beam, hip rafter, and architrave, through the point cloud slicing method.

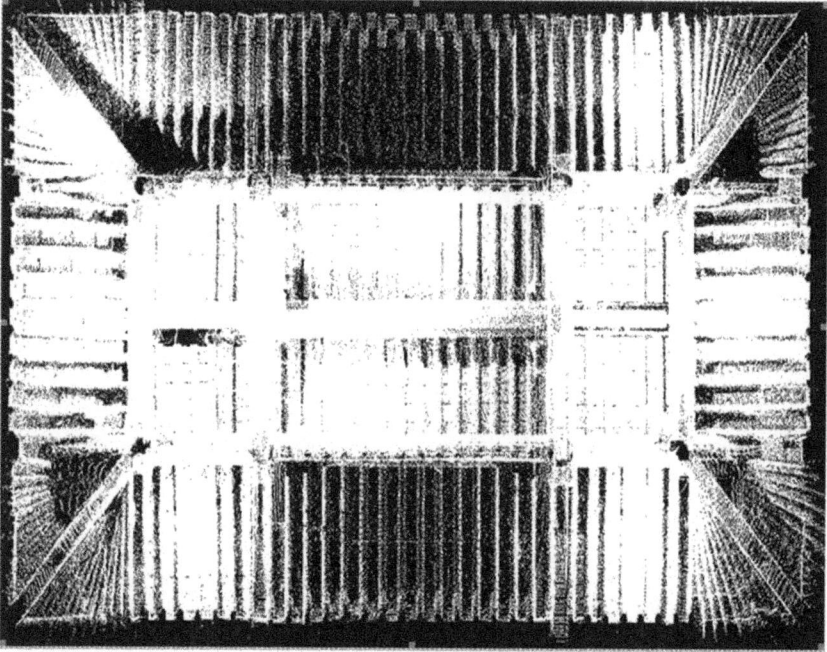

(a)

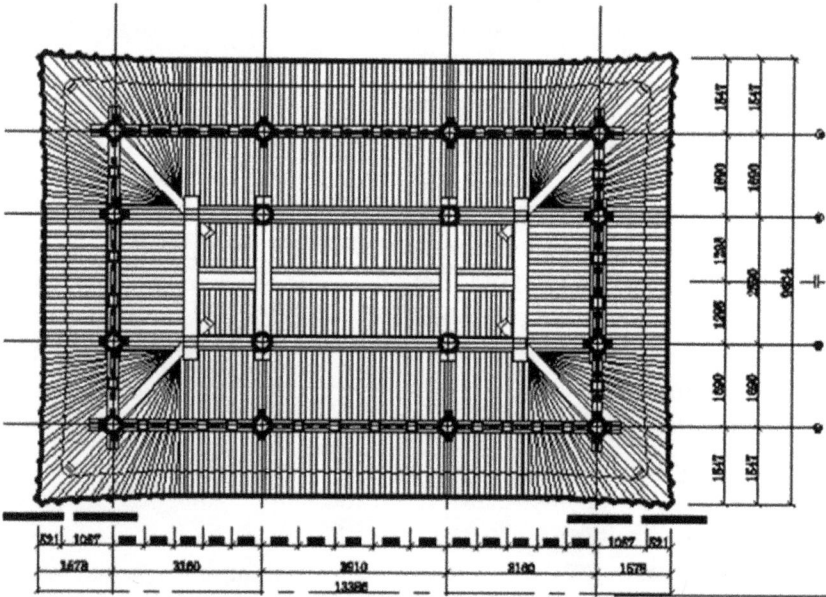

(b)

Figure 10. Point cloud projection for the plane drawing of the building roof. (**a**) Point cloud projection; (**b**) plane drawing based on point cloud projection.

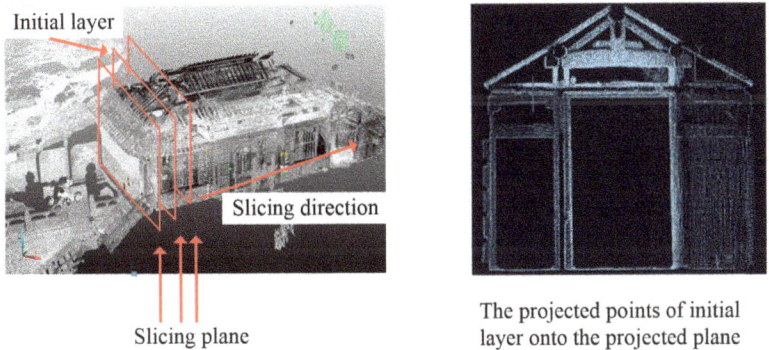

Figure 11. Point cloud slicing and projection.

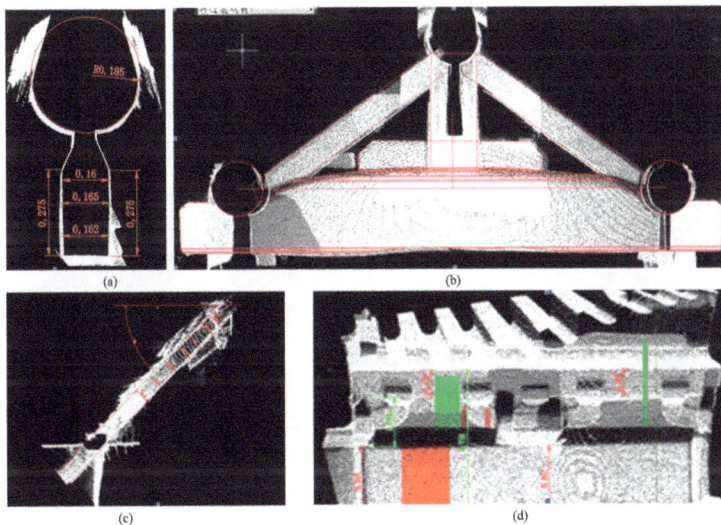

Figure 12. Fine drawings of typical components: (a) ridge tie beam; (b) crescent beam; (c) hip rafter; (d) architrave.

3.4. 3D Modeling

The 3D modeling of architecture has two levels. One is reality-based 3D modeling, and the other is reconstructive 3D modeling [41]. Reality-based modeling uses registered point clouds to generate high-resolution polygons based on 3D triangulation, called a polygonal model. The polygonal model can be directly rendered as a 3D scene, which can also map the image as visible information. Reality-based modeling presents several topological errors, which are caused by residual errors that survived the cleaning phase, and a huge numbers of holes, which are related to the shadow effects of the complex geometry. A limitation exists for the 3D representation of the complex architecture, particularly the internal composition [42,43]. Thus, we use reality-based modeling to represent the attached historical relics in Liangyi Temple, such as the famous Longtou Incense and Taoism statues shown in Figure 13. The procedure of reality-based polygon modeling and rendering is partly shown in Figure 9 together with the preprocessing for these types of components via Handyscan.

Figure 13. Reality-based 3D modeling of historical relics. (**a**) 3D model of horned beast; (**b**) 3D model of statue; (**c**) 3D model of sculpture; (**d**) texture-mapped 3D model of Longtou Incense; (**e**) texture-mapped 3D model of horned beast.

Compared with reality-based modeling, reconstructive 3D modeling aims to structure the architectural elements. It is also object oriented along with topological information, which can be spatially indexed for further query, highlighting, and selection. All these properties are very important for architectural management and reconstruction in the digital architecture preservation projection.

Reconstructive 3D modeling starts from the fine surveying drawings in Section 3.3, which contain the geometrical information in reality-based models of both single buildings and historical relics in the whole aligned scene. The three-view drawings, such as plan drawings, facade drawings, and section drawings, are adopted for the 3D modeling of the building in AutoCAD 2011 software, as shown in Figure 14. The accurate building footprints, height of architectural elements, and positioning of the relics are determined according to their actual configuration in the 3D reconstruction phase. The 3D models of the architectural components are also generated from their fine drawings and objective geometrical constraints in AutoCAD 2011, as presented in Figure 14e,f. Texture mapping is also implemented in AutoCAD 2011 with high-resolution images from the field data collection.

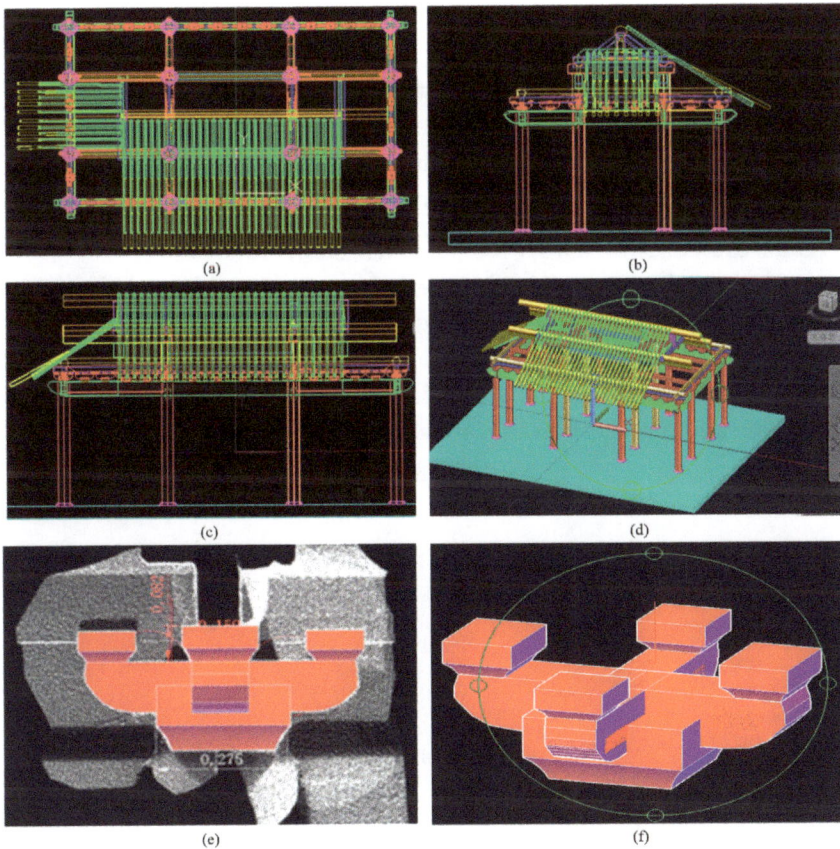

Figure 14. Reconstructive 3D modeling of the architecture. (**a**) Plan drawing of the building; (**b**) section drawing of the building; (**c**) facade drawing of the building; (**d**) reconstructive 3D model of the building; (**e**) architectural component modeling; (**f**) reconstructive 3D model of the component.

We also use the approach introduced by Gabriele to ensure the quality of 3D modeling by double checking the 3D model of the complex architecture [44]. First, the reality-based digital model is used to check the coherence of each reconstructive step. The digital reconstructed models are continuously compared with the reality-based models to verify possible incoherencies introduced during the reconstruction procedure. Second, the high-resolution images, including the 360 panoramic images from the field data collection together with the architectural archaeologist, are introduced to check the ancient architecture structure. These architectural structure checks include the architectural structure, architecture with decorations, relics in the architecture, and final texturized architecture.

4. Results and Discussions

4.1. Results and Analysis

The proposed approach aims to maintain and achieve the digital preservation of the wooden architecture of Liangyi Temple through fine surveying and 3D modeling. We introduce the final results, which include three parts: (1) fine drawings; (2) 3D models; and (3) deformation analysis of the architectural structure.

(1) Fine drawings of Liangyi Temple and accuracy analysis

A total of 148 drawings of Liangyi Temple are obtained from the multiple TLS integration point cloud data. The fine drawings include a series of scale maps. The details of these drawings are shown Table 1. Figure 15 shows some typical fine drawings of Liangyi Temple; these drawings are the foundation of the digital reconstruction and preservation of Liangyi Temple.

Table 1. Fine drawings of Liangyi Temple.

Type	Components	Description	Map Scale	Number
Plane drawing	building	two floors	1:50, 1:100	3
Facade drawing	building	front and side views	1:50	2
Section drawing	building	cross and vertical sections	1:50	9
Detail design	bracket set	four corners	1:10, 1:20	8
Bottom view	beam	hall and corridor	1:50	2
Detail design	components	entablature, column base, hip rafter, architrave, tile end, inverted V-shaped brace, camel hump-shaped support, partition door	1:5 1:10 1:20	124

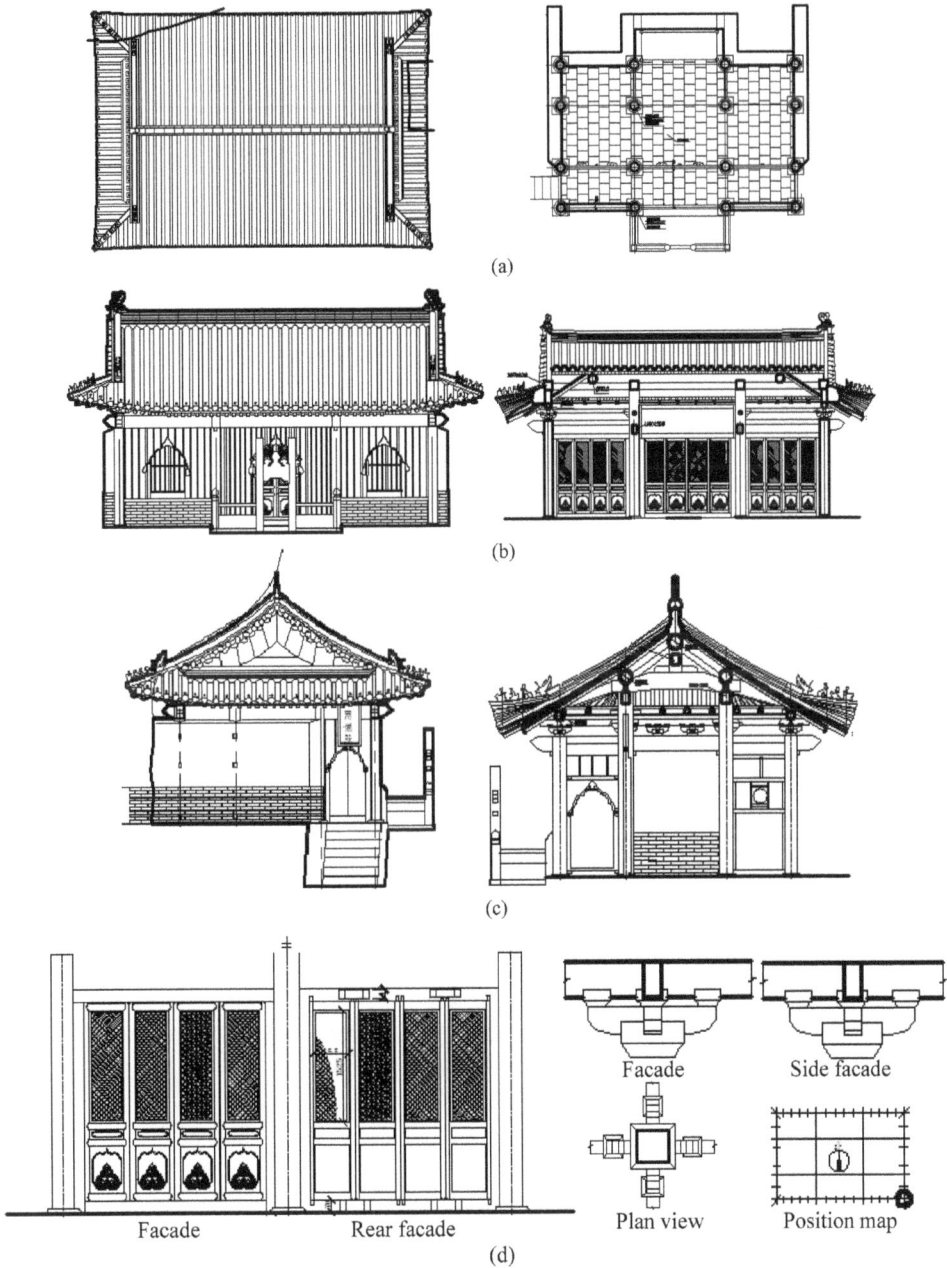

(a)

(b)

(c)

Facade Rear facade

Facade Side facade

Plan view Position map

(d)

Figure 15. Fine drawings of Liangyi Temple. (**a**) Plane drawings (layout); (**b**) facade drawings; (**c**) section drawings; (**d**) window drawings and bracket set drawings.

According to the fine surveying technique requirements of the wooden architecture, each component has a large-scale drawing from the point cloud with

409

precise geometric size. We randomly select some components of the building to evaluate the accuracy of the fine surveying results with the proposed approach in this study by measuring the geometric parameters with a ruler or a total station. These parameters were compared with the geometric size from the fine drawings. The accuracy result is reported in Table 2.

Table 2. Accuracy of fine surveying with the proposed approach.

Components	Type	View	Measurement [a] (mm)	Measurement [b] (mm)	Error (mm)
Narrow tie beam under ridged purlin	Length	Left	122	119	−3
wide tie beam under ridged purlin	Width	Left	168	169	1
narrow tie beam under south Quan	Length	Left	211	212	1
wide tie beam under south Quan	Width	Left	76	78	2
narrow tie beam under north Quan	Length	Left	205	204	−1
wide tie beam under north Quan	Width	Left	178	183	5
inverted V-shaped brace	Width	Left	143	142	−1
inverted V-shaped brace	Thickness	Plane	77	81	4
bracing of crescent beam	Length	Plane	301	302	1
bracing of crescent beam	Width	Plane	256	258	2
bracing of hip rafter	Length	Plane	213	213	0
bracing of hip rafter	Width	Plane	275	274	−1
door of front hall	Width	Back	3585	3588	3
crossbar of front hall door	Width	Back	113	111	−2
main door	Width	Front	2835	2838	3
right side of the second door	Width	Front	164	162	−2
bottom side of the second door	Height	Front	193	195	2
height of the wall	Height	Back	795	798	3
window	Width	Back	2862	2863	1
half of window	Width	Back	894	891	−3
baluster	Width	Back	1265	1263	−2
inside baluster	Width	Back	1078	1076	−2
Mean square error (MSE)					2.3

[a] measurement result by ruler or total station(mm); [b] measurement result from point cloud.

As Table 2 indicates, the accuracy of the proposed approach reaches 2.3 mm in MSE, which can meet the fine surveying requirement of the ancient wooden architecture. These fine drawings can be used to reconstruct the architecture and obtain the 3D model for digital architecture documentation.

(2) 3D models of Liangyi Temple

The majority of 3D models of Liangyi Temple are 3D reconstructive-based models. After texture mapping and 3D rendering of the high-resolution images, we can obtain the true 3D representation and documentation of Liangyi Temple (as displayed in Figure 16). The environment of the cliff terrain is finally generated based on the reality-based scene. The 3D models of vegetation are also added to the true Digital Terrain Model (DTM) to build a panoramic view of the true surroundings on a sphere, resulting in the final reconstructive panoramic view shown in Figure 17.

(a)

(b)

Figure 16. Reconstructive 3D model of Liangyi Temple. (**a**) Rendered with gray image; (**b**) rendered with texture mapping.

Figure 17a shows the front view of Liangyi Temple, which adds the surrounding 3D models of terrain and vegetables. Figure 17b shows the top view of the entire wooden architecture. The camera viewpoint roams and turns around the temple to help obtain a bird's-eye view of Liangyi Temple. Figure 17c shows the roaming scene of the panoramic view in the reconstructed temple, in which the corridor and hall have been passed through for a detailed observation.

(a)

(b)

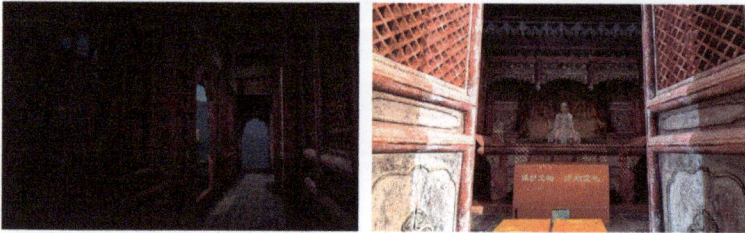

(c)

Figure 17. 3D scene of Liangyi Temple. (**a**) Front panoramic view; (**b**) top view; (**c**) inner view: corridor and hall.

(3) Deformation analysis of the architectural structure

The fine surveying of Liangyi Temple has another important goal, which is to determine and measure structural deformation for further monitoring and preservation projects. The proposed multiple TLS integration approach provides a well-registered global point cloud that can be used for the deformation analysis. We can obtain the slicing section in the column bases where structural deformation tends to occur through the slicing and projection method of the point cloud. Figure 18 shows one section of the building. The crooked column base is illustrated in the drawings of this section (Figure 18a).

Figure 18 shows that the wooden column base is sloped at the top, which may be caused by the extrusion of the cliff. The crooked column can be observed clearly from the section point cloud by slicing. The section drawing is measured from the sliced point cloud. The maximized crooked gap of the column base at the top is 102.1 mm. This gap is a very large displacement between the roof and the building foundation and threatens the integrity of the entire architecture. The deformation analysis results are given to the Heritage Management Department of Wudang Mountains for further maintenance and monitoring.

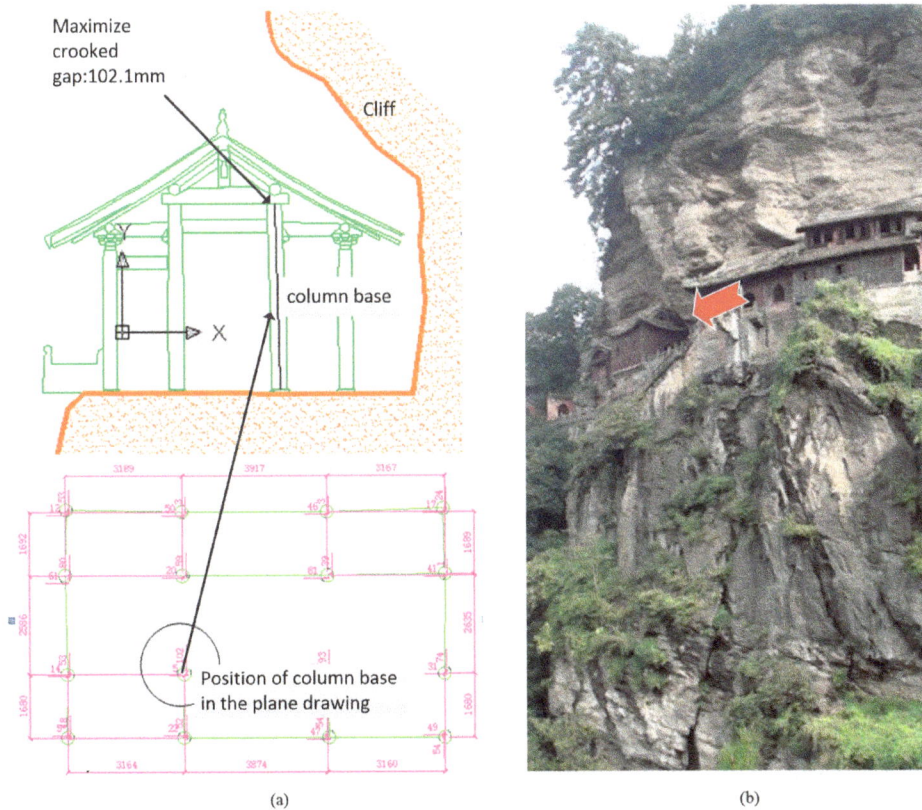

(a)   (b)

Figure 18. Deformation analysis of the column base. (**a**) Crooked column on the section drawing; (**b**) crooked direction of Liangyi Temple.

413

4.2. Discussions

As previously stated, our proposed approach aims to use multiple level terrestrial laser scanner integration for the fine surveying and 3D modeling of complex architecture in a steep and difficult environment. Although the proposed approach attempts to make up for the restrictions of mutual sheltering and reduce missed corner data, some objects in the multiple cross parts remain; these objects cannot be determined by the geometric relationship among the neighboring objects. The complex structure of the architecture with complicated elements requires both architectural knowledge and archaeological experience for the fine surveying and 3D modeling with the proposed slicing and projection method of 3D point clouds. In the 3D modeling with the point cloud and fine drawings, architectural and archaeological knowledge is also very important to the quality and accuracy of the 3D model of the architecture. All these factors are important for the reconstruction and preservation of the architecture. Hence, technical scientists and archaeologists should work together in laser scanning to conduct interdisciplinary research for digital ancient architecture projects [45].

The specific location terrain and climate environment of Liangyi Temple threaten the survival of this wooden architecture. Although we present a deformation analysis method based on the 3D point cloud, many destructive insect pests, such as the aforementioned white ants that often appear in this district, threaten the wooden components. The white ants can hurt the wooden component from the inside, which is the main factor behind the crooked column bases. Laser scanning can only measure the geometric deformation of the architectural structure. Microwave and X-ray technology together with high-density geometric shape measurement approaches via laser scanning have great potential in this field [46–48].

5. Conclusions

The work presented in this study is intended to be a valuable aid for the digital reconstruction, visualization, and preservation of ancient wooden architecture. A multiple level terrestrial laser scanning integration approach is proposed for complex architecture in steep terrain surroundings. The data processing approach with point cloud from a three-level terrestrial laser scanner is proven to be effective for fine surveying and 3D modeling. Liangyi Temple in the Wudang Tourism architecture complex is taken as a typical wooden ancient architecture to realize fine surveying and 3D documentation via the proposed approach. A total of 148 drawings of the architecture and 3D models with high-resolution texture are obtained. All these results are useful for the digital reconstruction, visualization, and preservation of the architecture. We also find a structural deformation in Liangyi Temple, which should be a subject of long-term monitoring for architectural heritage, having more than 600 years of history, in further preservation and maintenance projects. Along with future development, combining microwave and laser scanning technology for full 3D modeling of the wooden components is significant for both the surface shape and the inside structure, such as the hole infected by insect pests.

Acknowledgments: The authors would like to thank "The Compass Plan" supported by State Administration of Cultural Heritage. This research is also supported by National Natural Science Foundation of China (Grand No: 41271452) and Key Technologies R&D Program of China (Grand No. 2015BAK03B04).

Author Contributions: Qingwu Hu and Shaohua Wang conceived the study and wrote the paper; Caiwu Fu conceived the study; Mingyao Ai did the field data collection and data processing; Dengbo Yu and Wende Wang implemented 3D modeling and accuracy analysis.

Conflicts of Interest: The authors declare no conflict of interest.

References

1. Zhang, Q. A Scientific approach to the origins of Chinese civilization. In *An Introduction to Chinese History and Culture*; Springer: Berlin, Germany, 2015; pp. 1–21.
2. Wibowo, A.S. Conservation of wooden architecture: Willingness, support and tradition. *Procedia Soc. Behav. Sci.* **2015**, *184*, 388–393.
3. Rujivacharakul, V.; Hahn, H.H.; Oshima, K.T.; Christensen, P. *Architecturalized. Asia: Mapping a Continent through History*; Hong Kong University Press: Hong Kong, China, 2013.
4. Choi, J.; Kim, Y.; Kang, J.; Choi, Y. Comparative analysis of the spatial structure of apartment unit plans in Asia-apartments in Korea, Vietnam, and Kazakhstan. *J. Asian Archit. Build. Eng.* **2014**, *13*, 563–569.
5. Chun, Q.; Van Balen, K.; Pan, J.; Sun, L. Structural performance and repair methodology of the Wenxing lounge bridge in China. *Int. J. Archit. Herit.* **2015**, *9*, 730–743.
6. Mazzeo, R.; Cam, D.; Chiavari, G.; Fabbri, D.; Ling, H.; Prati, S. Analytical study of traditional decorative materials and techniques used in Ming Dynasty wooden architecture. The case of the Drum Tower in Xi'an, PR of China. *J. Cult. Herit.* **2004**, *5*, 273–283.
7. Bridge, M. Locating the origins of wood resources: A review of dendroprovenancing. *J. Archaeol. Sci.* **2012**, *39*, 2828–2834.
8. Li, H.Q.; Yu, Y.; Yu, X. On fire protection problems and its countermeasures about Chinese ancient architecture. *Appl. Mech. Mater.* **2012**, *204*, 3365–3368.
9. Tang, Z. Does the institution of property rights matter for heritage preservation? Evidence from China. In *Cultural Heritage Politics in China*; Springer: New York, NY, USA, 2013; pp. 23–30.
10. Fregonese, L.; Barbieri, G.; Biolzi, L.; Bocciarelli, M.; Frigeri, A.; Taffurelli, L. Surveying and monitoring for vulnerability assessment of an ancient building. *Sensors* **2013**, *13*, 9747–9773.

11. Al-Kheder, S.; Al-Shawabkeh, Y.; Haala, N. Developing a documentation system for desert palaces in Jordan using 3D laser scanning and digital photogrammetry. *J. Archaeol. Sci.* **2009**, *36*, 537–546.

12. Oreni, D.; Cuca, B.; Brumana, R. Three-dimensional virtual models for better comprehension of architectural heritage construction techniques and its maintenance over time. In *Progress in Cultural Heritage Preservation*; Springer: Berlin, Germany; Heidelberg, Germany, 2012; pp. 533–542.

13. Brumana, R.; Oreni, D.; Cuca, B.; Binda, L.; Condoleo, P.; Triggiani, M. Strategy for integrated surveying techniques finalized to interpretive models in a byzantine church, Mesopotam, Albania. *Int. J. Archit. Herit.* **2014**, *8*, 886–924.

14. McCarthy, J. Multi-image photogrammetry as a practical tool for cultural heritage survey and community engagement. *J. Archaeol. Sci.* **2014**, *43*, 175–185.

15. Martínez, S.; Ortiz, J.; Gil, M.L.; Rego, M.T. Recording complex structures using close range photogrammetry: The cathedral of Santiago De Compostela. *Photogramm. Rec.* **2013**, *28*, 375–395.

16. Remondino, F. Heritage recording and 3D modeling with photogrammetry and 3D scanning. *Remote Sens.* **2011**, *3*, 1104–1138.

17. Guarnieri, A.; Milan, N.; Vettore, A. Monitoring of complex structure for structural control using terrestrial laser scanning (TLS) and photogrammetry. *Int. J. Archit. Herit.* **2013**, *7*, 54–67.

18. Ercoli, L.; Megna, B.; Nocilla, A.; Zimbardo, M. Measure of a limestone weathering degree using laser scanner. *Int. J. Archit. Herit.* **2013**, *7*, 591–607.

19. Lambers, K.; Eisenbeiss, H.; Sauerbier, M.; Denise Kupferschmidt, D.; Gaisecker, T.; Sotoodeh, S.; Hanusch, T. Combining photogrammetry and laser scanning for the recording and modelling of the Late Intermediate Period site of Pinchango Alto, Palpa, Peru. *J. Archaeol. Sci.* **2007**, *34*, 1702–1712.

20. Rüther, H.; Chazan, M.; Schroeder, R.; Neeser, R.; Held, C.; Walker, S.J.; Matmon, N.; Horwitz, L.K. Laser scanning for conservation and research of African cultural heritage sites: The case study of Wonderwerk Cave, South Africa. *J. Archaeol. Sci.* **2009**, *36*, 1847–1856.

21. Lezzerini, M.; Antonelli, F.; Columbu, S. The documentation and conservation of the cultural heritage: 3D laser scanning and GIS techniques for thematic mapping of the stonework of the facade of St. Nicholas church (Pisa, Italy). *Int. J. Archit. Herit.* **2014**.

22. Chellini, G.; Nardini, L.; Pucci, B. Evaluation of seismic vulnerability of Santa Maria del Mar in Barcelona by an integrated approach based on terrestrial laser scanner and finite element modeling. *Int. J. Archit. Herit.* **2014**, *8*, 795–819.

23. Pesci, A.; Casula, G.; Boschi, E. Laser scanning the Garisenda and Asinelli towers in Bologna (Italy): Detailed deformation patterns of two ancient leaning buildings. *J. Cult. Herit.* **2011**, *12*, 117–127.

24. Kuzminsky, S.C.; Gardiner, M.S. Three-dimensional laser scanning: potential uses for museum conservation and scientific research. *J. Archaeol. Sci.* **2012**, *39*, 2744–2751.

25. Pesci, A.; Bonali, E.; Galli, C. Laser scanning and digital imaging for the investigation of an ancient building: Palazzo d'Accursio study case (Bologna, Italy). *J. Cult. Herit.* **2012**, *13*, 215–220.

26. Domingo, I.; Villaverde, V.; López-Montalvo, E. Latest developments in rock art recording: Towards an integral documentation of Levantine rock art sites combining 2D and 3D recording techniques. *J. Archaeol. Sci.* **2013**, *40*, 1879–1889.

27. Hinzen, K.G.; Schreiber, S.; Rosellen, S. A high resolution laser scanning model of the Roman theater in Pinara, Turkey—Comparison to previous measurements and search for the causes of damage. *J. Cult. Herit.* **2013**, *14*, 424–430.

28. Guidi, G.; Russo, M.; Angheleddu, D. 3D survey and virtual reconstruction of archeological sites. *Digit. Appl. Archaeol. Cult. Herit.* **2014**, *1*, 55–69.

29. Remondino, F.; ElHakim, S. Image-based 3D modelling: A review. *Photogramm. Rec.* **2006**, *21*, 269–291.

30. Soudarissanane, S.; Lindenbergh, R.; Menenti, M. Scanning geometry: Influencing factor on the quality of terrestrial laser scanning points. *ISPRS J. Photogramm. Remote Sens.* **2011**, *66*, 389–399.

31. Lei, Z.K.; Zhou, L.; Zheng, J.H. Preliminary study of building pathological information system for wall painting-in the case of Liang-Yi Temple at Wudang Mountain. *Appl. Mech. Mater.* **2012**, *174*, 1645–1650.

32. Jun, Y.; Shaohua, W.; Jiayuan, L. Research on fine management and visualization of ancient architectures based on integration of 2D and 3D GIS technology. *IOP Publ.* **2014**, *17*, 012168.

33. Li, X.F.; Tao, H. Study on the digitalization and virtual restoration of the hall of Yuzhen Palace in Wudang Mountain. *Archit. J.* **2004**, *12*, 66–68.

34. Zhang, K.; Hu, C.S. *World Heritage in China*; South China University of Technology Press: Guangdong, China, 2006.

35. Gruen, A.; Akca, D. Least squares 3D surface and curve matching. *ISPRS J. Photogramm. Remote Sens.* **2005**, *59*, 151–174.

36. Yin, X.; Wonka, P.; Razdan, A. Generating 3D building models from architectural drawings: A survey. *IEEE Comput. Graph. Appl.* **2009**, *29*, 20–30.

37. Peng, T.; Gupta, S.K. Model and algorithms for point cloud construction using digital projection patterns. *J. Comput. Inf. Sci. Eng.* **2007**, *7*, 372–381.

38. Yoo, D.J. Rapid surface reconstruction from a point cloud using the least-squares projection. *Int. J. Precis. Eng. Manuf.* **2010**, *11*, 273–283.

39. Wu, Y.F.; Wong, Y.S.; Loh, H.T. Modelling cloud data using an adaptive slicing approach. *Comput. Aided Des.* **2004**, *36*, 231–240.

40. Zhang, Y.F.; Wong, Y.S.; Loh, H.T. An adaptive slicing approach to modelling cloud data for rapid prototyping. *J. Mater. Process. Technol.* **2003**, *140*, 105–109.

41. Manferdini, A.; Remondino, F. A review of reality-based 3D model generation, segmentation and web-based visualization methods. *Int. J. Herit. Digit. Era* **2012**, *1*, 103–124.

42. Remondino, F.; Rizzi, A. Reality-based 3D documentation of natural and cultural heritage sites—Techniques, problems, and examples. *Appl. Geomat.* **2010**, *2*, 85–100.

43. Apollonio, F.I.; Gaiani, M.; Benedetti, B. 3D reality-based artefact models for the management of archaeological sites using 3D Gis: A framework starting from the case study of the Pompeii Archaeological area. *J. Archaeol. Sci.* **2012**, *39*, 1271–1287.

44. Guidi, G.; Remondino, F.; Russo, M. A multi-resolution methodology for the 3D modeling of large and complex archeological areas. *Int. J. Archit. Comput.* **2009**, *7*, 39–55.

45. Toniolo, L.; Boriani, M.; Guidi, G. *Built Heritage: Monitoring Conservation Management*; Springer: Berlin, Germany, 2015.

46. Pieraccini, M.; Fratini, M.; Dei, D. Structural testing of Historical Heritage Site Towers by microwave remote sensing. *J. Cult. Herit.* **2009**, *10*, 174–182.

47. Weritz, F.; Kruschwitz, S.; Maierhofer, C. Assessment of moisture and salt contents in brick masonry with microwave transmission, spectral-induced polarization, and laser-induced breakdown spectroscopy. *Int. J. Archit. Herit.* **2009**, *3*, 126–144.

48. Mannes, D.; Schmid, F.; Frey, J. Combined neutron and X-ray imaging for non-invasive investigations of cultural heritage objects. *Phys. Procedia* **2015**, *69*, 653–660.

Frescoed Vaults: Accuracy Controlled Simplified Methodology for Planar Development of Three-Dimensional Textured Models

Marco Giorgio Bevilacqua, Gabriella Caroti, Isabel Martínez-Espejo Zaragoza and Andrea Piemonte

Abstract: In the field of documentation and preservation of cultural heritage, there is keen interest in 3D metric viewing and rendering of architecture for both formal appearance and color. On the other hand, operative steps of restoration interventions still require full-scale, 2D metric surface representations. The transition from 3D to 2D representation, with the related geometric transformations, has not yet been fully formalized for planar development of frescoed vaults. Methodologies proposed so far on this subject provide transitioning from point cloud models to ideal mathematical surfaces and projecting textures using software tools. The methodology used for geometry and texture development in the present work does not require any dedicated software. The different processing steps can be individually checked for any error introduced, which can be then quantified. A direct accuracy check of the planar development of the frescoed surface has been carried out by qualified restorers, yielding a result of 3 mm. The proposed methodology, although requiring further studies to improve automation of the different processing steps, allowed extracting 2D drafts fully usable by operators restoring the vault frescoes.

Reprinted from *Remote Sens.* Cite as: Bevilacqua, M.G.; Caroti, G.; Zaragoza, I.M.-E.; Piemonte, A. Frescoed Vaults: Accuracy Controlled Simplified Methodology for Planar Development of Three-Dimensional Textured Models. *Remote Sens.* **2016**, *8*, 239.

1. Introduction

Planar development of more or less complex curved surfaces, also possibly decorated, is of great interest for experts working in the field of documentation and preservation of cultural heritage. The scientific community has addressed the problem in a rigorous way, with both analytical [1] and digital [2–6] methodologies. These investigations have led to the development of software applications that have only partially automated the procedures required for planar development of curved surfaces. Studies carried out by the authors have not found as yet any commercial tool allowing automated and controlled development of geometry and texture.

The goal of this research was to study a simplified methodology enabling both planar development of geometry and texture of frescoed vaults (surveyed with geomatics techniques) and checking of the errors related to the different operating steps. In particular, this methodology has been developed in the case of the "a schifo" vault typology [7,8] and does not provide any kind of cartographic projection. It uses very common commercial software and includes some processing steps requiring user operation.

2. State of the Art

3D photorealistic environments allow engineers, historians and restorers to research, investigate, and simulate outcomes of restoration projects before these are executed. For all these aspects, 3D-textured metric modeling is currently the most sought after tool for cognitive evaluation and operating approach in the field of cultural heritage [9]. Creation of a 3D-textured model includes three steps: geometry modeling, parameterization and texture creation.

2.1. Geometry Survey

Laser scanning is a well-established surveying methodology, whose output is readily usable for representing historical and architectural heritage [10–17]. Accuracy and resolution attainable in comparatively short times (last-generation scanners for architectural surveys can acquire millions of points per second with sub-centimeter accuracy) are the main strengths of these systems although prices are still quite high.

The new approach to softcopy photogrammetry realized by Structure from Motion (SfM) and MVS (Multi-View Stereo) algorithms generates very dense 3D color point clouds quite similar in size to those produced from laser scanning surveys [18–23]. However, even if software evolution in this field is very fast and performance is good in terms of processing time, and the amount of manageable data and obtainable precisions are gradually improving, these procedures may not always be considered reliable. In fact, matching algorithms can be very sensitive to recording and illumination differences and not reliable in poorly textured or homogeneous regions. This can result in noisy point clouds and/or difficulties in feature extraction [24]. These matching algorithms could suffer from variable precision, strongly dependent on the pattern present on surveyed objects, as well as the difficulty of having control of the achievable accuracy at the geometric and morphological levels [20,25].

2.2. Texture Mapping

In large-scale 3-D models used as supporting documentation in restoration works, textures are not a mere aesthetic complement. In fact, besides supporting construction, material and chromatic studies, they also act as metric surveying tools, providing, once applied to the models, a guideline for measurements. Therefore, if textures have to meet these requirements, their positioning accuracy must be consistent with the scale used, besides having the necessary chromatic precision [26].

Many laser scanners have built-in cameras, whose relative orientation is calibrated by the manufacturer, which allow direct true coloring of point clouds. These textures are characterized by a high geometric accuracy, but the systems used for the photographic takes usually do not achieve good results in terms of resolution and color fidelity [27].

Simplified, realistic-looking models may not suffice for restorers, who require rigorous texture mapping for both morphology and color information. In these cases, it is essential to resort to a dedicated photographic campaign, performed with high quality cameras as regards optics, sensor size and image post-processing.

In the case of SfM software, the creation of models and textures is almost contextual, and the procedure usually involves self-calibration of the camera, which also takes account of characteristic distortion parameters. In these models, although textures usually have good photographic quality, it is necessary to check the overall morphological reliability.

2.3. Vault Development

Commercial and open-source software currently available are capable to render architectures in 3D as regards formal appearance and color. On the other hand, operative steps of restoration interventions still require large-scale, 2D metric surface representations. The transition from 3D to 2D representation, with the related geometric transformations, has not yet been fully formalized and still features open issues, e.g., in the case of planar development of frescoed vaults [28]. Methodologies proposed so far on this subject provide transitioning from point cloud models to known mathematical surfaces (developed on plane, or not), and afterwards seeking an ideal representation of the actual surface, losing some architectural and building details in this process [29–33]. To the best knowledge of the authors, modeling and reverse engineering software commonly used do not have dedicated tools that enable automatic development of geometry and textures. Moreover, the tools that only partially solve the problem do not take account of the introduced deformations.

3. Materials and Methods

In order to achieve planar development of frescoed vaults, textured 3D models of the vaults are required. These should feature good geometric accuracy to identify the ideal geometric root that best fits the actual vault. Textures associated with the models will have good radiometric quality and true colors, and will also faithfully reproduce position and dimension of any fresco detail.

3.1. The Case Study

The object of this investigation is a vault in Palazzo Roncioni (Pisa, Italy). Its entire surface is covered by a XVIII century fresco, painted by Tuscan painter G.B. Tempesti, which has over time undergone extensive damage (cracks, plaster gaps, *etc.*). Currently, the vault is the subject of safety and restoration work. A laser scanning survey of the vault has been performed with the pulse shift-based laser scanner Leica Geosystems C10 ScanStation, with a point cloud density of about 70 pts/cm^2 on average. Use of a phase-based laser scanner would have allowed for more accurate results at short distances and therefore less noisy reference data. The photogrammetric survey was performed with a Nikon D700 SLR camera (f = 20 mm lens) at about 4.5 m range, ensuring a roughly 2-mm pixel covering on the vault surface. It has been subjected to image processing via SfM algorithms, granting an overlap ⩾70%.

3.2. 3D Modeling and Texturing

A separate 3D TIN model has been built by means of each surveying methodology. Both models have been rigorously registered in the same reference system due to the extrapolation, from the colored point cloud, of the coordinates of 12 Ground Control Points (GCPs). These have been chosen as easily identifiable fresco details, spread evenly across the entire vault, and have been used in the processes of scaling and rototranslation of the photogrammetric model.

The model obtained via laser scanning survey, henceforth referred to as "model LASER" (Figure 1c), features homogenous geometric precision and a high enough resolution to show elements in the sub-centimeter range (cracks, plaster displacements, *etc.*). As regards textures, images collected via the on-board camera (single image 17° × 17°, 1920 × 1920 pixel) do not grant adequate radiometric quality. The model obtained by means of SfM/MVS methodology, henceforth referred to as "model SfM/MVS", features uneven geometric precision, mostly in the areas where radiometric uniformity reduces the performance of the SfM algorithms (Figure 1b). On the other hand, it has been obtained by a photographic campaign executed with a high quality camera, so that despite some local errors in detail rendering, image orientation is substantially fit to the needs of the application, as detailed later. A new textured model ("model SfM/LASER") has been generated via SfM software from the geometry of model LASER and the images orientated for creation of model SfM/MVS (Figure 1a).

Model SfM/LASER is the best result for geometry and texture quality, starting from collected data. As regards processing time, this texturing process definitely has lower requirements than single image orientation based on GCPs [26].

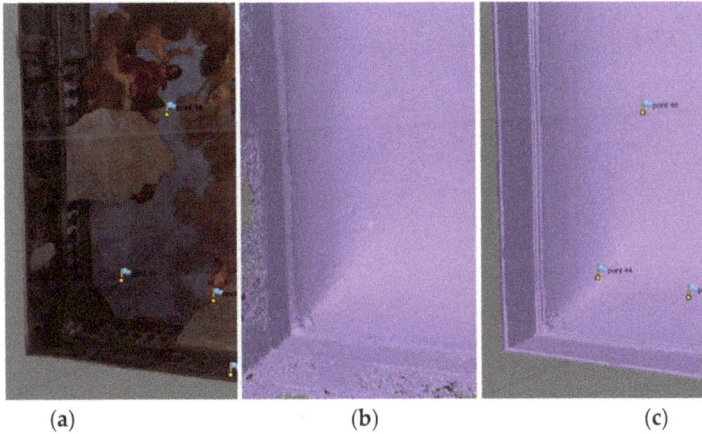

(a)   (b)   (c)

Figure 1. (a) Model SfM/LASER; (b) Model SfM/MVS; (c) Model LASER.

3.3. Vault Development

Point clouds, either collected by laser scanning (Cloud LASER) or obtained by photogrammetry (Cloud SfM/MVS), have been framed in a single reference system. The X- and Y-axes lie in the speculated vault impost plan, which is not horizontal (axes origin in a corner, X-axis on the long side and Y-axis on the short side) and the Z-axis completes the orthogonal triplet. In order to proceed with the 2D vault development, the 3D model has been analyzed.

3.3.1. Analysis and Preliminary Processing of Laser Data

In order to define the geometric components that constitute the vault, a dense contour (step = 2 cm) representation of model LASER has been generated according the three coordinate planes (Figure 2).

Figure 2. Dense contour model (isometric view).

The study of this contour representation (Figure 3) has allowed identification of nine discontinuity directions that divide the vault in 6 areas, each featuring its own section profiles with almost constant radius: areas 2, 4 and 6, close to the vault impost, have greater section profile radii than those in the upper part of the vault (areas 1, 3 and 5). Separation between the lower and upper parts of the vault is located at about one third of the vault height above the impost plane.

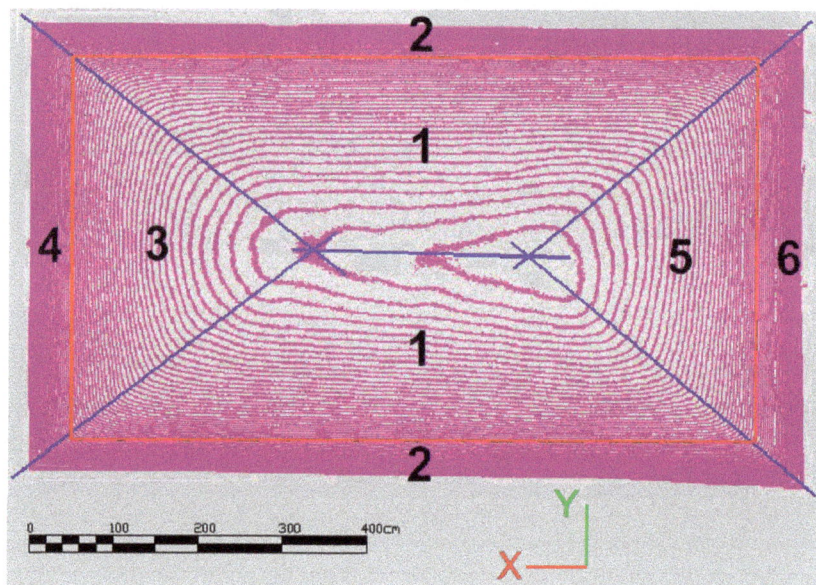

Figure 3. Discontinuity directions and vault areas (bottom view).

The interpretation of these results suggested that the vault could be generated from the combination of several elements belonging to different cylindrical surfaces, and could be part of the "a schifo" type. This includes a lower portion, similar to a section of a pavilion vault, and an upper one, named "specchio" (mirror), which features so wide a curvature to appear almost planar. This vault type has been widely used in architecture since Renaissance exactly in the case of fresco decorations.

Once the hypotheses about the building type of the vault have been substantiated, the values of the geometric parameters (axis and radius) of the elementary cylindrical surfaces that best fit the point clouds of each of the six areas detected were computed by means of approximation algorithms.

As an example, approximation by cylindrical surfaces of the long side (Figure 4) yielded the following results (Table 1).

Figure 4. Approximation by cylindrical surface.

Table 1. Approximation by cylindrical surface—radius.

	Area 1	Area 2
Radius (m)	7.349	5.280
STDV (m)	0.007	0.008

Analysis of Standard Deviation (STDV) should take in account that the vault does not actually show neat transitions between contiguous cylindrical surfaces, but rather the curvature radius changes gradually. In fact, the higher values of the difference between ideal and actual surfaces are found in these transition areas (Figure 5).

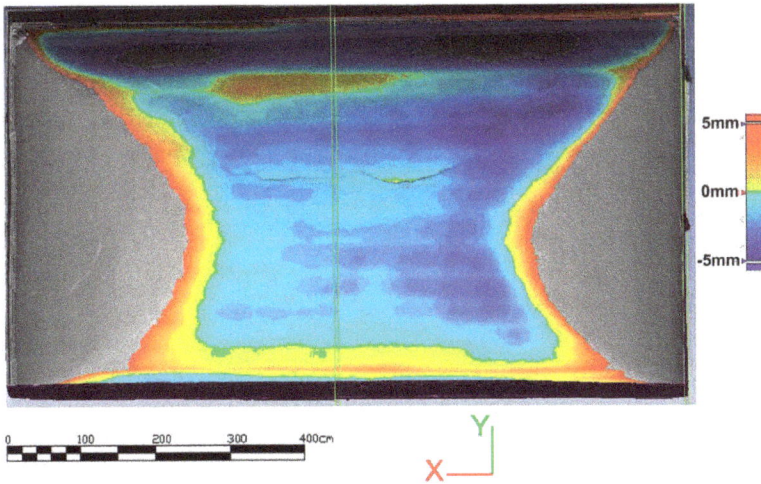

Figure 5. Actual 3D model—Cylindrical surface error.

425

3.3.2. Analysis of the Development Methodology

In order to achieve a planar development of vault geometry and texture using well known and easily available software tools, a methodology using model representation by contours, rather than by ideal shapes such as cylinders, has been investigated and applied to this case study.

For this purpose, just the contour lines lying in the XY coordinate plane have been used in a CAD environment, with a 20-cm step (Figure 6).

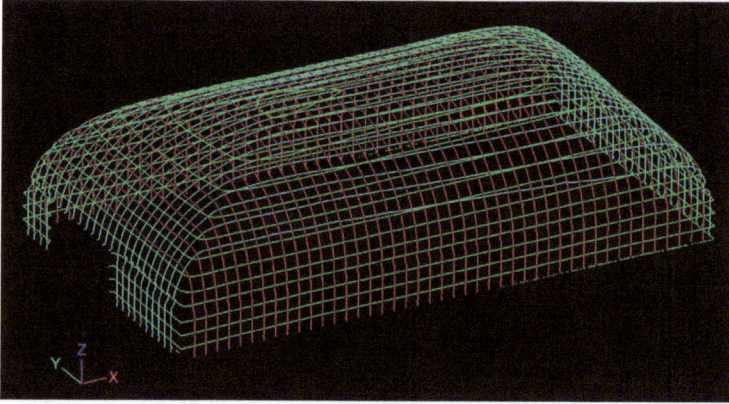

Figure 6. 20-cm step contour lines model.

It has been assumed that vault sections between adjacent contours were planar. Contours have been assumed as connections between adjacent planes. In order to appraise the error introduced by this assumption, an orthogonal section of the interpolating cylinders has been checked. In the most unfavorable situation (Figure 7), the difference between the section arc Equation (1) and the related chord Equation (2), bounded by two adjacent contours, has been computed.

$$\widehat{AB} = \beta \cdot R \tag{1}$$

$$\overline{AB} = 2 \cdot R \cdot \sin\left(\beta/2\right) \tag{2}$$

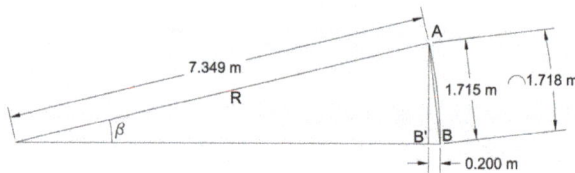

Figure 7. Section arc—chord comparison.

On an arc length = 1.718 m, the maximum difference was 3 mm, with a sub-millimeter relative error. This approximation has been deemed acceptable. For the planar development of the XY contour lines model with a 20-cm step (the same used to detect the different portions of the vault), the crown of the vault has been outlined in CAD at 1:1 scale (line AB in Figure 8).

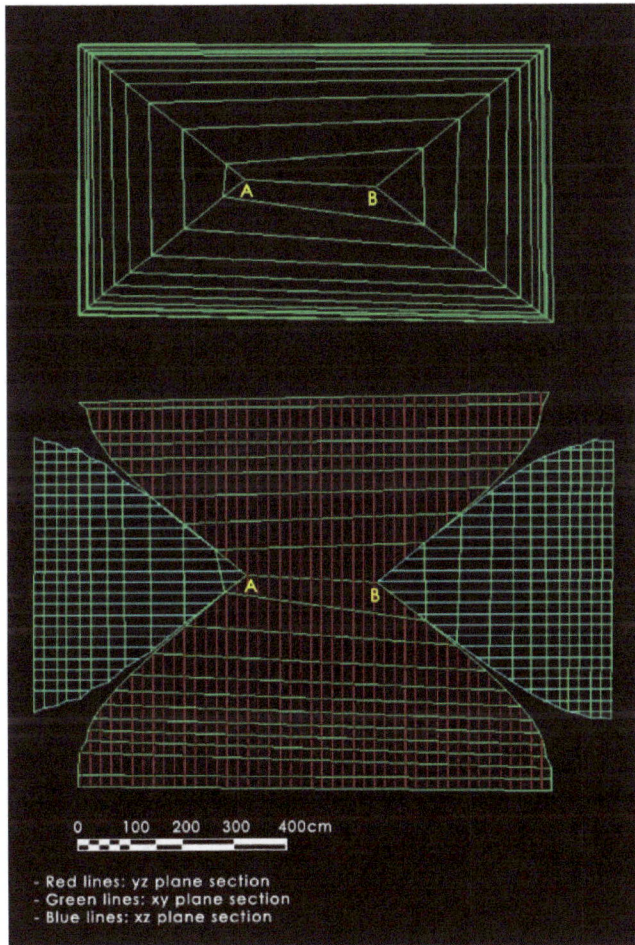

Figure 8. Geometric development of contours model.

Subsequent contour lines have been separately developed by trilateration for each surface of the vault. Assuming the extremes of the previous contour as fixed points, the extremes of the next contour have been plotted. Figure 8 shows how planar geometric development of the actual vault lacks the regular course found in the development of an ideal surface constituted just by portions of cylinder with parallel axes.

Sections defined by ZX (blue lines) and ZY (red lines) planes have then been superimposed on the development of the XY sections (green lines).

Besides geometric vault development, sections are also required as a reference for correct texture placement on the developed model.

In order to apply textures to the developed geometric model, the following methodology has been chosen.

For each surface, eight directions have been identified to set orthogonal views of 3D model SfM/LASER. These directions are orthogonal to the axis of the theoretical cylinders and, starting from the horizontal view, tilted by 15° relative to the previous view.

For each viewing direction, images of the model of vault portion bounded by a 15° cylindrical arc have been collected. These are an orthogonal projection of the vault texture on an orthogonal plane relative to the viewing direction [34]. Fresco elements projected in this way are obviously deformed.

Accepting the simplification that the vault is represented by the surfaces of the interpolating cylinders, it is possible to quantify this deformation. As for the orthogonal projection on a plane tangent to the cylinder, there is no deformation along parallel directions relative to the tangency line, while deformation is highest along orthogonal directions. Linear deformation module (m_l) at the extremes of the orthogonally projected area is defined by Equation (3).

$$m_l = \frac{\overline{AB\prime}}{\overarc{AB}} = \frac{2 \cdot R \cdot sin\,(\beta/2)}{R \cdot \beta} \tag{3}$$

In the projection used, deformation is highest in the furthest point relative to the tangency line (for breadth = 15°, distance is about 1 m), where $m_l = 0.9970$ and deformation = 3 mm (Figure 9).

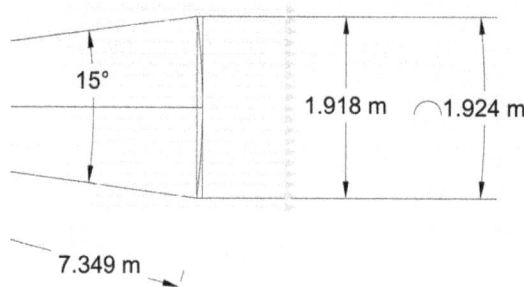

Figure 9. Projection deformation.

In accordance with the operators dealing with the restoration of the fresco, this deformation has been deemed acceptable.

Each orthogonal view of model SfM/LASER has been performed in two configurations and saved in two separate image files. Configuration 1 provided for superimposing the section lines to the model (Figure 10a). Configuration 2 viewed the model with just the high quality texture applied (Figure 10b).

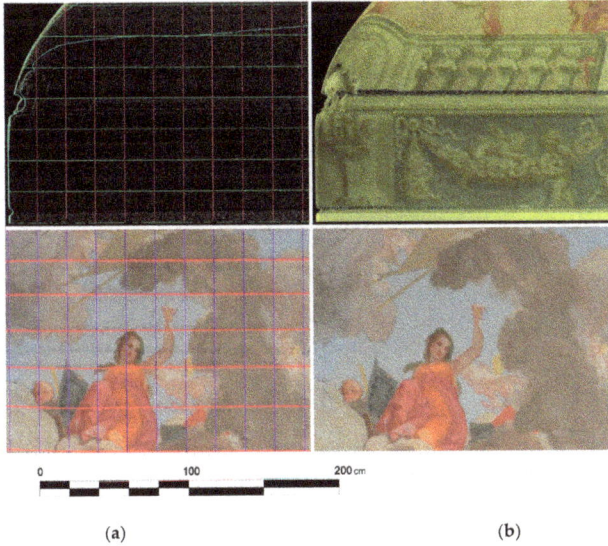

(a)   (b)

Figure 10. (a) Orthogonal view with section lines; (b) Orthogonal view with high quality texture.

The following processing steps have therefore been run for each image pair:

- Image pairs and the geometric vault development frame (Figure 8) have been imported in the same photo editing software environment.
- A single block has been created with both images, so that any transformation applied to any one image was similarly applied to the other.
- The layer containing the image with just the texture has been turned off, leaving visible just the image with the section lines.
- The image has been scaled and moved on the geometric frame, assuming the section lines obtained with XZ and YZ planes (vertical lines in Figure 10) as reference.

As proof of the small deformation of the images, it has been noticed that after scaling the image in the direction of the axis of the interpolating cylinders for a single projection direction, it aligns with images derived from other projection directions at less than the computed deformation (Figure 11).

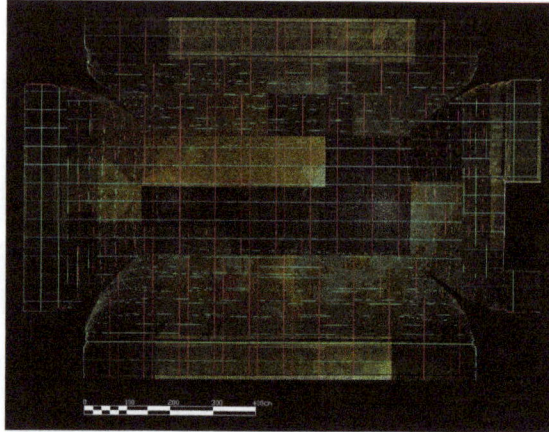

Figure 11. Superimposition of orthogonal view on sections model development.

4. Results and Discussion

In order to validate the methodology used, results must be checked for both geometric precision of the different models obtained and precision of placement, dimension and shape of the applied textures. Finally, the quality of the planar development of the vault was assessed (Figure 12).

Figure 12. High quality textured model development.

4.1. Assessment of Model Geometrical Accuracy

The laser scanning colored point cloud model (Cloud LASER) can be assumed as the absolute geometric reference in this application. It has very high point density, and allows extraction of coordinates of features for both geometry (cracks, gaps, *etc.*) and painting (boundary lines, color transitions, *etc.*) with a sub-centimeter resolution.

Standard deviation obtained by comparing Cloud LASER with Model LASER is 1 mm, with peaks in the 3 mm range. These results highlight that the transition from point cloud to surface model entails a small decay of geometric precision.

A second check has been performed comparing Cloud LASER with Cloud SfM/MVS; the standard deviation averaged at 3 mm, with peaks of about 6 mm.

Finally, Cloud LASER has been compared against Model SfM/MVS; the standard deviation was 3 mm on average, peaking at about 10 mm.

These results show that image orientation in SfM is substantially accurate and confirm the mean reprojection error involved with orientating each image via SfM to be 0.70 pixel with an average of 9000 tie points per image. On the other hand, maximum deviation values are in the range of 7–10 mm and refer to cracks and plaster collapse borders. Figures 13–17 show an overview of the fresco and some details on local deviations.

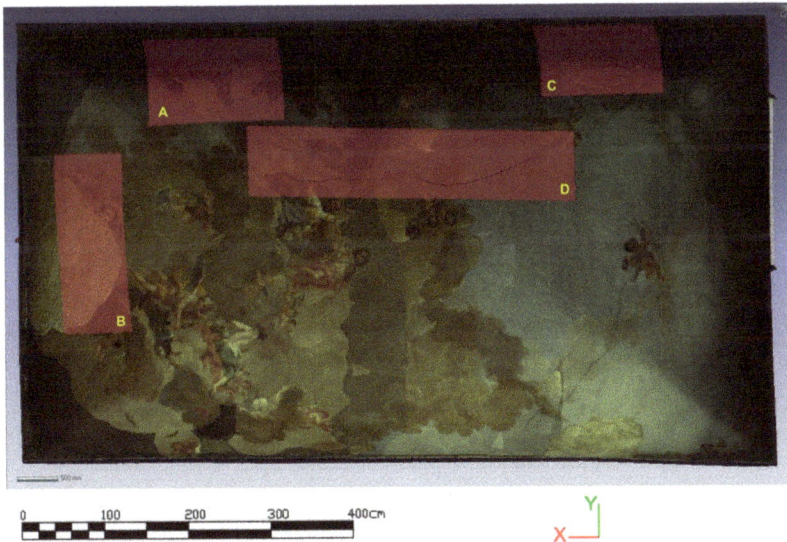

Figure 13. Regions checked for deviations between cloud LASER and model SfM/MVS.

Taking into account all these cases, greater deviations are found when surveyed surfaces are orthogonal to the vault. SfM/MVS methodology does not correctly represent the transitions typical of deep cracks and delamination. This result is in the authors' opinion due to the fact that these surface regions are acquired by inclined views with different inclinations and sometimes with the camera axis parallel to the surface. This fact, reported in the literature, leads to worse performance of the matching algorithms [24].

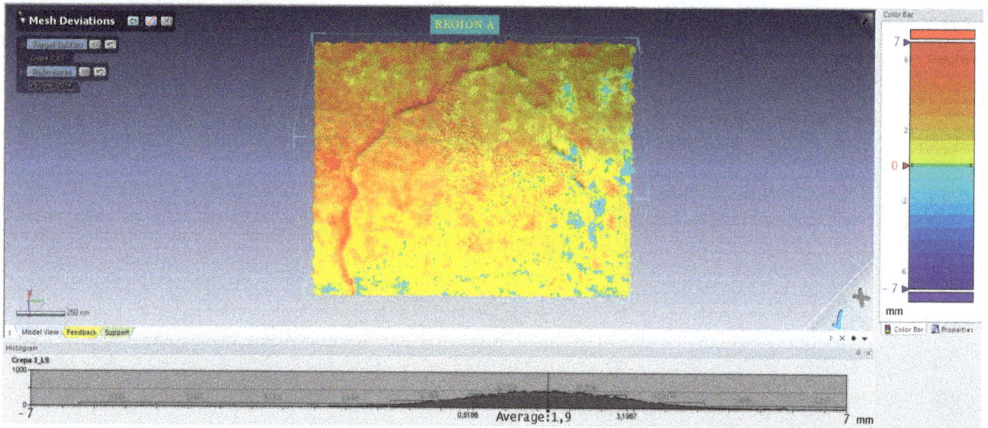

Figure 14. Region A: total plaster collapse borders.

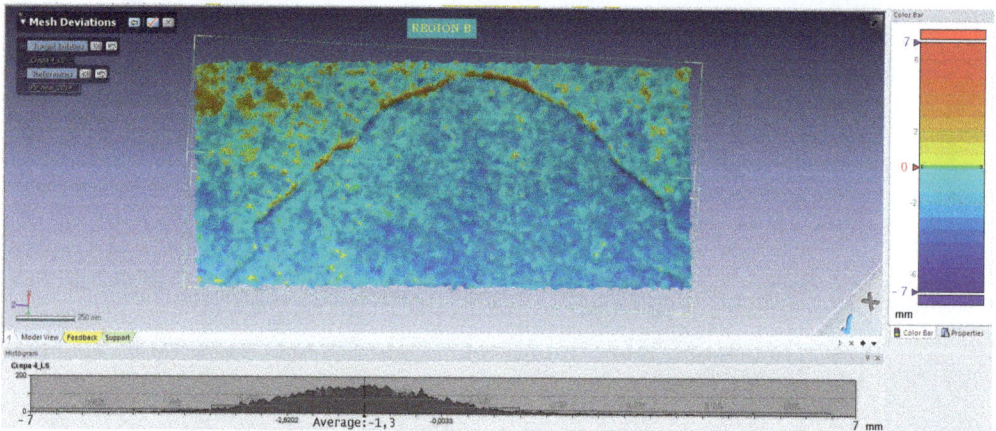

Figure 15. Region B: total plaster collapse borders.

Hence, the overall accuracy of the SfM-derived model is good (3 mm), but shows some flaws precisely in the regions of most interest to restorers. This processing methodology, on the other hand, has the advantage of significantly lower resource requirements: manual intervention is limited to inputting the support points to orient and scale the model.

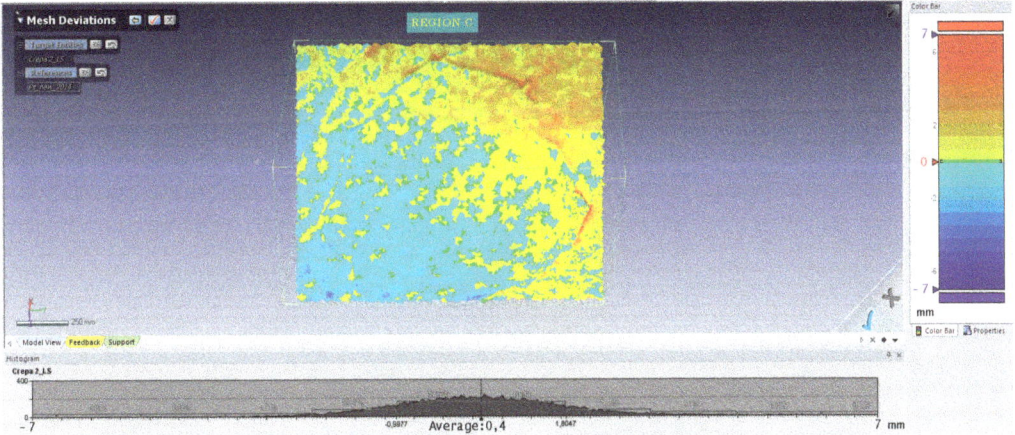

Figure 16. Region C: gap in the fresco.

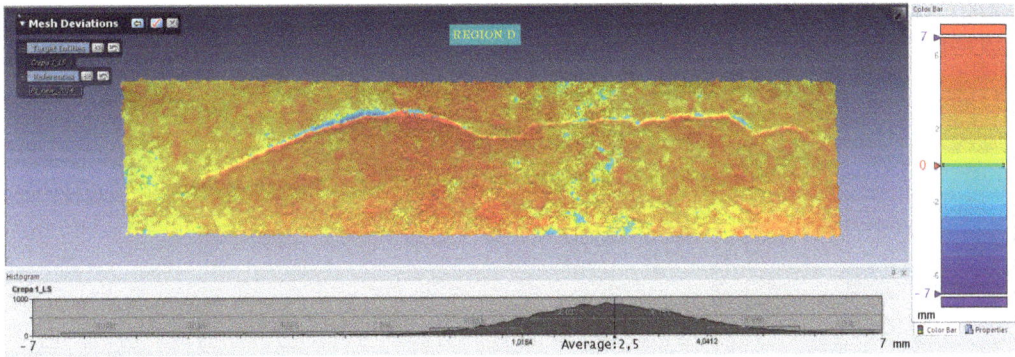

Figure 17. Region D: crack in the topmost region of the vault.

These considerations on geometric precision led to the choice of Model SfM/LASER as a starting model for vault development.

4.2. Texture Dimension and Positioning Accuracy Assessment

After geometric accuracy of the models has been checked, texturing precision has also been monitored. For this purpose, the coordinates of 36 Control Points (CPs) have been extracted by Cloud LASER. These coordinates have been firstly compared with those obtained by digitizing the points on the images and obtaining their 3D position in Cloud SfM/MVS (Figure 18). The comparison provided the statistics displayed in Table 2.

433

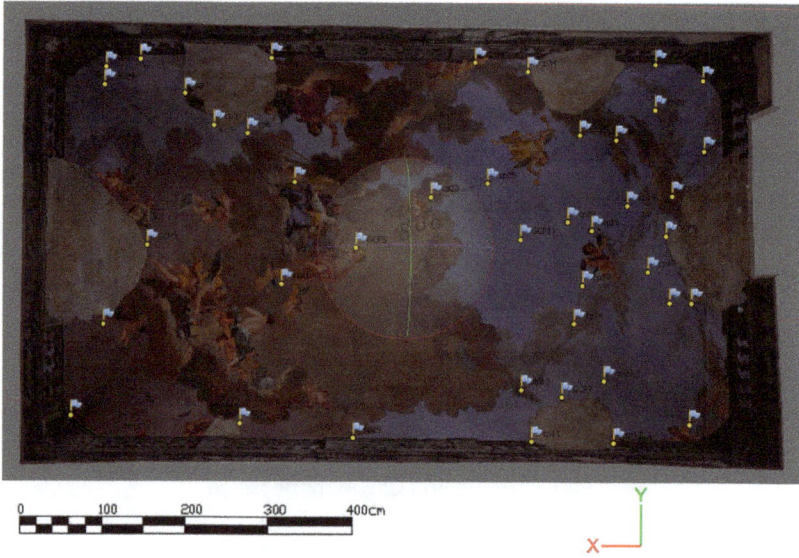

Figure 18. CPs on the vault.

Table 2. CP coordinates comparison Cloud LASER—Cloud SfM/MVS.

	X	Y	Z
mean (m)	0.000	0.000	0.000
max (m)	0.005	0.005	0.010
STDV (m)	0.002	0.003	0.004

Note: The values are in line with the geometric comparison between point clouds LASER and SfM/MVS.

Subsequently, the same points have been digitized directly on Model SfM/LASER. A comparison with the reference CP coordinates yielded the results displayed in Table 3.

Table 3. CP coordinates comparison Cloud LASER—Model SfM/LASER.

	X	Y	Z
mean (m)	0.000	0.000	0.000
max (m)	0.019	0.023	0.026
STDV (m)	0.007	0.007	0.008

This comparison shows that precision checks on texture yielded a slightly worse result relative to those on geometry. Such an outcome was predictable, assuming the addition of errors for geometry with those for image orientation and texture projection, as well as those for direct CP collimation on Model SfM/LASER.

4.3. Vault Development Accuracy Assessment

Besides the 3D comparison between Model SfM/LASER and Cloud LASER, planar development has also been validated at actual scale. Some portions of the image, representing the vault development, have been printed at 1:1 scale on A0 tracing paper. Subsequently, restorers checked the prints directly with the represented fresco portions (Figure 19), noticing the accordance of shapes and dimension of the checked portions in line with the deformations already expected and accepted in the processing steps. On the same tracing paper sheet, restorers have drafted the outlines of the actual fresco paintings; the resulting accuracy is 3 mm.

Figure 19. Development accuracy assessment at 1:1 scale.

5. Conclusions

The methodology discussed has proposed a simplified solution for the problem of a metrically correct planar representation of a frescoed "a schifo" vault. The processing steps shown can be carried out even by relatively inexperienced users and do not require specific software.

A peculiar feature of this methodology is the creation of collages of several orthogonal views of the textured 3-D model, thanks to geometrical references provided by the section lines of the model. These lines are visible in the three-dimensional model, its geometric development and on the images used for the collage.

The methodology proposed for modeling, texturing and planar development was verified by both calculating the theoretical error introduced by the single processing step and by comparing the final products with a reference survey and then directly with the surveyed object.

The theoretical development accuracy is 3 mm. The comparison between the laser scanner model textured with oriented images through SFM and the original laser scanning point cloud yielded a 3-mm accuracy. Finally, the direct verification of the development of the model confirmed an accuracy of 3 mm, which allowed drafts to be obtained that are fully usable by restorers for 3D fresco reconstruction on a vaulted surface.

In particular, these are most useful for faithful reconstruction of the geometry in damaged fresco portions, for which a photographic documentation suitable for 3-D modeling is available.

The same methodology can also be applied to domes and vaults of different types. The authors are currently planning further testing on barrel and pavilion vaults and on elliptical and spherical domes.

The present research will be prosecuted with the aim of automating the different processing steps, particularly as regards monitoring of deformations and errors introduced in the final representation.

Further interest also lies in investigating differences between developments obtained by extracting contours by actual surfaces or by approximating them to ideal surfaces.

Acknowledgments: This research work was funded by the University of Pisa internal research funds and by PRIN (Programmi di Ricerca Scientifica di Rilevante Interesse Nazionale—Scientific Research Programs of Relevant National Interest) 2010–2011 funds (coordinator: Raffaele Santamaria—Pisa local research unit).

Author Contributions: This research was performed by Marco Giorgio Bevilacqua, Gabriella Caroti, Isabel Martínez-Espejo Zaragoza and Andrea Piemonte. Each author contributed extensively and equally to prepare this paper.

Conflicts of Interest: The authors declare no conflict of interest.

References

1. Dequal, S. An unconventional application of analytical plotters to architectural photogrammetry: Projection, plotting and digitizing on non plane surfaces. *Int. Arch. Photogramm. Remote Sens. Spat. Inf. Sci.* **1988**, *27*, 140–146.
2. Karras, G.E.; Petsa, E.; Dimarogona, A.; Kouroupis, S. Photo-textured rendering of developable surfaces in architectural photogrammetry. In *Virtual and Augmented Architecture (VAA'01)*; Springer: London, UK, 2001; pp. 147–158.
3. Karras, G.E.; Patias, P.; Petsa, E. Digital monoplotting and photo-unwrapping of developable surfaces in architectural photogrammetry. *Int. Arch. Photogramm. Remote Sens.* **1996**, *31*, 290–294.
4. Georgopoulos, A.; Makris, G.N.; Tournas, E.; Ioannidis, C.H. Digitally developing works of art. In Proceedings of the CIPA 2001 International Symposium, Potsdam, Germany, 18–21 September 2001.

5. Valanis, A. Fitting, portrayal and mapping for the production of 2nd order surfaces photomosaics. *Int. Arch. Photogramm. Remote Sens.* **2002**, *34*, 463–467.

6. Lerma, J.L.; Tortosa, R.V. Digital development of a small Valencian tower. *Int. Arch. Photogramm. Remote Sens. Spat. Inf. Sci.* **2004**, *35*, 451–454.

7. Pera, L. *Tecnica Dell'architettura. Tipologia Strutturale*; Goliardica: Pisa, Italy, 1987.

8. De la Plaza Escudero, L. *Diccionario Visual de Términos Arquitectónicos*; Grandes Temas Cátedra: Madrid, Spain, 2008.

9. Lerma, J.L.; Pérez, C. 3D Photorealistic and interactive reconstruction of covered up frescoes. In *Recording, Modeling and Visualization of Cultural Heritage*; Baltasavias, E., Gruen, A., Van Gool, L., Pateraki, M., Eds.; Taylor & Francis Group: London, UK, 2006; pp. 485–491.

10. Bevilacqua, M.G.; Caroti, G.; Piemonte, A. Rilievi integrati della Basilica romanica di San Gavino a Porto Torres. In Proceedings of the 16ª Conferenza Nazionale ASITA, Vicenza, Italy, 6–9 November 2012; pp. 391–396.

11. Caroti, G.; Piemonte, A. An integrated survey for knowledge and preservation of a cultural heritage: The Albanian fortified citadel of Elbasan. *Int. Arch. Photogramm. Remote Sens. Spat. Inf. Sci.* **2008**, *37*, 373–378.

12. Caroti, G.; Piemonte, A. Studio relativo al rilevamento di dettaglio di una porzione del transetto nord della chiesa di San Paolo a Ripa d'Arno al fine di procedere al consolidamento strutturale. In Proceedings of the Un incontro informale per i 70 anni del Prof. Carlo Monti, Milano, Italy, 3 May 2012; pp. 1–6.

13. Dore, C.; Murphy, M. Integration of Historic Building Information Modelling (HBIM) and 3D GIS for recording and managing cultural heritage sites. In Proceedings of the IEEE 18th International Conference on Virtual Systems and Multimedia (VSMM), Milan, Italia, 2–5 September 2012; pp. 369–376.

14. Juan Vidal, F.; Martínez-Espejo Zaragoza, I. Hypotesisforthe virtual anastilosis of themainchapel of the Iglesia de Los Desamparados de Les Coves de Vinromá. In Proceedings of the 16th International Conference on Cultural Heritage and New Technologies, Vienna, Austria, 14–16 November 2011; pp. 201–212.

15. Lerma, J.L.; Cabrelles, M.; Navarro, S.; Seguí, A.E. La documentación patrimonial mediante sensores de imagen o de barrido láser. In *Documentación Gráfica del Patrimonio*; Ministerio de Cultura: Madrid, Spain, 2011; pp. 108–117.

16. Marambio, A.; Pucci, B.; Nuñez, M.A.; Buill, F. La aplicación del escáner láser terrestre en la catalogación del patrimonio arquitectónico del casco histórico de Castellfollit de la Roca. In Proceedings of the VIII Semana Geomática Internacional, Barcelona, Spain, 3–5 March 2009; pp. 1–8.

17. Martínez-Espejo Zaragoza, I.; Juan Vidal, F. Gestión gráfica avanzada de edificaciones antiguas. In Proceedings of the XI Congreso Internacional de Expresión Gráfica Aplicada a la Edificación, Universidad Politécnica de Valencia, Valencia, Spain, 29 November–1 December 2012; pp. 710–714.

18. Alsadik, B.; Remondino, F.; Menna, F.; Gerke, M.; Vosselman, G. Robust extraction of image correspondences exploiting the image scene geometry and approximate camera orientation. *Int. Arch. Photogramm. Remote Sens. Spat. Inf. Sci.* **2013**, *40*, 1–7.

19. Guidi, G.; Russo, M.; Beraldin, J.A. *Acquisizione 3D e Modellazione Poligonale*; McGraw-Hill: Milano, Italy, 2010.

20. Guidi, G.; Remondino, F. 3D Modelling from real data. In *Modelling and Simulation in Engineering*; Catalin Alexandru, C., Ed.; InTech: Vienna, Austria, 2012; pp. 69–102.

21. Koska, B.; Křemen, T. The combination of laser scanning and structure from motion technology for creation of accurate exterior and interior orthophotos of St. Nicholas Baroque church. *Int. Arch. Photogramm. Remote Sens. Spat. Inf. Sci.* **2013**, *40*, 133–138.

22. Lerma, J.L.; Navarro, S.; Cabrelles, M.; Villaverde, V. Terrestrial laser scanning and close range photogrammetry for 3D archaeological documentation: The upper Palaeolithic Cave of Parpalló as a case study. *J. Archaeol. Sci.* **2010**, *37*, 499–507.

23. Wenzel, K.; Rothermel, M.; Fritsch, D.; Haala, N. Image acquisition and model selection for multi-view stereo. *Int. Arch. Photogramm. Remote Sens. Spat. Inf. Sci.* **2013**, *40*, 251–258.

24. Remondino, F.; Spera, M.G.; Nocerino, E.; Menna, F.; Nex, F. State of the art in high density image matching. *Photogramm. Rec.* **2014**, *29*, 144–166.

25. Apollonio, F.I.; Ballabeni, A.; Gaiani, M.; Remondino, F. Evaluation of feature-based methods for automated network orientation. *Int. Arch. Photogramm. Remote Sens. Spat. Inf. Sci.* **2014**, *40*, 47–54.

26. Caroti, G.; Martínez-Espejo Zaragoza, I.; Piemonte, A. Accuracy assessment in structure from motion 3D reconstruction from UAV-born images: The influence of the data processing methods. *Int. Arch. Photogramm. Remote Sens. Spat. Inf. Sci.* **2015**, *40*, 103–109.

27. Apollonio, F.I.; Remondino, F. Modellazione 3D da sensori attivi Pipeline con laser scanner. In *Modelli Digitali 3D in Archeologia: Il Caso di Pompei*; Scuola Normale Superiore Pisa: Pisa, Italy, 2010; pp. 94–117.

28. Carpiceci, M. Survey problems and representation of architectural painted surfaces. *Int. Arch. Photogramm. Remote Sens. Spat. Inf. Sci.* **2011**, *38*, 523–528.

29. Menna, F.; Rizzi, A.; Nocerino, E.; Remondino, F.; Gruen, A. High resolution 3D modeling of the behaim globe. *Int. Arch. Photogramm. Remote Sens. Spat. Inf. Sci.* **2012**, *39*, 115–120.

30. Cipriani, L.; Fantini, F.; Bertacchi, S. Survey and representation of vaults and cupolas: An overview on some relevant Italian UNESCO Sites. In Proceedings of the IEEE International Conference on Virtual Systems & Multimedia (VSMM), Hong Kong, China, 9–12 December 2014; pp. 50–57.

31. Chiabrando, F.; Rinaudo, F. Recovering a collapsed medieval fresco by using 3D modeling techniques. *ISPRS Ann. Photogramm. Remote Sens. Spat. Inf. Sci.* **2014**, *2*, 105–112.

32. Pancani, G. Lo svolgimento in vera grandezza delle volte affrescate delle sale dei quartieri al piano terreno di Palazzo Pitti a Firenze. In Proceedings of the Giornate di Studio Il Disegno delle trasformazioni, Napoli, Italy, 1–2 December 2011; pp. 1–11.

33. Cannella, M. Sviluppo e Rappresentazione Digitale di Superfici Architettoniche Complesse per la Documentazione e il Restauro. Available online: http://hdl.handle.net/10447/130394 (accessed on 25 November 2015).

34. Meyer, E.; Parisel, C.; Grussenmeyer, P.; Revez, J.; Tidafi, T. A computerized solution for the epigraphic survey in Egyptian Temples. *J. Archaeol. Sci.* **2006**, *33*, 1605–1616.

Assessment and Calibration of a RGB-D Camera (Kinect v2 Sensor) Towards a Potential Use for Close-Range 3D Modeling

Elise Lachat, Hélène Macher, Tania Landes and Pierre Grussenmeyer

Abstract: In the last decade, RGB-D cameras - also called range imaging cameras - have known a permanent evolution. Because of their limited cost and their ability to measure distances at a high frame rate, such sensors are especially appreciated for applications in robotics or computer vision. The Kinect v1 (Microsoft) release in November 2010 promoted the use of RGB-D cameras, so that a second version of the sensor arrived on the market in July 2014. Since it is possible to obtain point clouds of an observed scene with a high frequency, one could imagine applying this type of sensors to answer to the need for 3D acquisition. However, due to the technology involved, some questions have to be considered such as, for example, the suitability and accuracy of RGB-D cameras for close range 3D modeling. In that way, the quality of the acquired data represents a major axis. In this paper, the use of a recent Kinect v2 sensor to reconstruct small objects in three dimensions has been investigated. To achieve this goal, a survey of the sensor characteristics as well as a calibration approach are relevant. After an accuracy assessment of the produced models, the benefits and drawbacks of Kinect v2 compared to the first version of the sensor and then to photogrammetry are discussed.

Reprinted from *Remote Sens.* Cite as: Lachat, E.; Macher, H.; Landes, T.; Grussenmeyer, P. Assessment and Calibration of a RGB-D Camera (Kinect v2 Sensor) Towards a Potential Use for Close-Range 3D Modeling. *Remote Sens.* **2015**, *7*, 13070–13093.

1. Introduction

Saving three-dimensional information about geometry of objects or scenes tends to be increasingly applied in the conventional workflow for documentation and analysis, of cultural heritage and archaeological objects or sites, for example. In this particular field of study, the needs in terms of restoration, conservation, digital documentation, reconstruction or museum exhibitions can be mentioned [1,2]. The digitization process is nowadays greatly simplified thanks to several techniques available that provide 3D data [3]. In the case of large spaces or objects, terrestrial laser scanners (TLS) are preferred because this technology allows collecting a large amount of accurate data very quickly. While trying to reduce costs and working on smaller pieces, on the contrary, digital cameras are commonly used. They have the advantage of being rather easy to use, through image-based 3D reconstruction techniques [4]. Besides, both methodologies can also be merged in order to overcome their respective limitations and to provide more complete models [5,6].

In this context, regarding some aspects like price or computation time, RGB-D cameras offer new possibilities for the modeling of complex structures, such as indoor environments [7]. Indeed, these sensors enable acquiring a scene in real-time with its corresponding colorimetric information. Among them, the Kinect sensor developed by Microsoft in 2010 and the Asus Xtion Pro in 2011, based on the PrimeSense technology [8,9], have encountered a great success in developer and scientific communities. Since July 2014, a second version of Microsoft's Kinect sensor has been available, based on another measurement technique. Such range imaging devices make use of optical properties, and are referred to as active sensors since they use their own light source for the active illumination of the scene. However, whereas Kinect v1 was based on active triangulation through the projection of structured light (speckle pattern), Kinect v2 sensor has been designed as a time-of-flight (ToF) camera [10]. Thanks to this technology, a depth measurement to the nearest objects is provided for every single pixel of the acquired depth maps. To illustrate this change of distance measurement principle, Figure 1 offers an overview of non-contact 3D measuring methods.

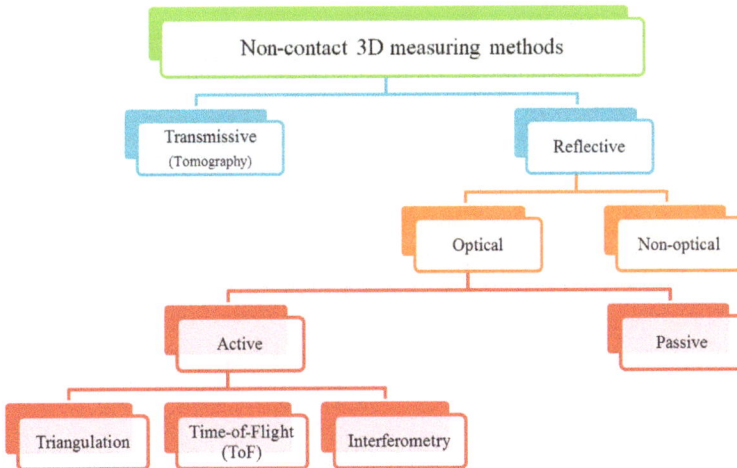

Figure 1. Time-of-Flight technology among non-contact 3D measuring methods. A more detailed taxonomy of active vision techniques can be found in [11].

In the following sections, an overview of application fields and calibration methods related to range imaging devices is presented. Then, characteristics of the recent Kinect v2 sensor are summarized, through a set of investigations carried out according to different aspects. The specific calibration process is also reported, in order to analyse its influence on the metric performances of the camera. Once all these parameters are known, the question of close range 3D modeling approaches can be addressed. To make a conclusion on the ability of Kinect v2 to provide reliable 3D models, the accuracy of the produced data has to be assessed. Considering not only the accuracy obtained but also the computation time or ease of use of the proposed methodology, finally a discussion about improvements and possible use of the sensor is reported.

2. Related Works

Because of their attractiveness and imaging capacities, lots of works have been dedicated to RGB-D cameras during the last decade. The aim of this section is to outline the state-of-the-art related to this technology, considering aspects such as fields of application, calibration methods or metrological approaches.

2.1. Fields of Application of RGB-D Cameras

A wide range of applications can be explored while considering RGB-D cameras. The main advantages are the cost, which is low for most of them compared to laser scanners, but also their high portability which enables a use on board of mobile platforms. Moreover, due to a powerful combination of 3D and colorimetric information as well as a high framerate, RGB-D cameras represent an attractive sensor well-suited for applications notably in the robotic field. Simultaneous localization and mapping (SLAM) tasks [12] or dense 3D mapping of indoor environments [13,14] are good examples of this problematic. Besides, industrial applications can be mentioned for real-time change detections [15] or for detections on automotive systems [16]. Some forensics studies are also reported, for crime scene documentation [17].

The Kinect sensors marketed by Microsoft are motion sensing devices, initially designed for gaming and entertainment purposes. They enable a contactless interaction between the user and a games console by gesturing. Since the release of the Kinect v1 sensor in 2010, lots of studies involving the device in alternative applications have been released. The capacities of localization [18] and SLAM systems [19] with Kinect devices have also been investigated. Moreover, pose recognition and estimation of human body [20] or hand articulations [21,22] have been considered. To answer to the need for 3D printing, [23] suggests the design of a scanning system based on a Kinect-type consumer-grade 3D camera. Finally, with the more recent Kinect v2 sensor, applications for face tracking [24], coastal mapping [25] or cyclical deformations analysis [26] appear in the literature.

2.2. Towards 3D Modeling of Objects with a RGB-D Camera

The creation of 3D models represents a common and interesting solution for the documentation and visualization of heritage and archaeological materials. Because of its remarkable results and its affordability, the probably most used technique by the archaeological community remains photogrammetry. However, the emergence in the past decades of active scanning devices to acquire close-range 3D models has provided new possibilities [27]. Given their characteristics, RGB-D cameras belong to the most recent generation of sensors that could be investigated for such applications. The original use of Kinect v1 device has already been modified for the creation of a hand-held 3D scanner [28] or for a virtual reality project [29]. Since the sensor can also provide 3D information in the form of point clouds, the geometrical quality of data acquired on small objects such as a statue [30], or on bigger scenes such as archaeological walls [31] has also been assessed. More recently, the use of a Kinect v2 sensor with the Kinect Fusion tool [32] has shown good results for the direct reconstruction of 3D meshes [33].

2.3. Error Sources and Calibration Methods

The main problem while working with ToF cameras is due to the fact that the measurements realized are distorted by several phenomena. For guarantying the reliability of the acquired point clouds, especially for an accurate 3D modeling purpose, a prior removal of these distortions must be carried out. To do that, a good knowledge of the multiple error sources that affect the measurements is useful. A description of these sources of measuring errors is summarized in [34,35]. Considering especially the Kinect gaming sensors, analysis of related error sources are reported for example in [36,37].

First of all, a systematic depth-related deformation, also depicted as systematic wiggling error by some authors [38] can be reported. This deformation is partially due to inhomogeneities within the modulation process of the optical beam, which is not perfectly sinusoidal. Lots of works have been devoted to the understanding of this particular error. As a matter of fact, calibration of time-of-flight sensors was a major issue in many studies in the last decade. Most of the time, the introduced approaches are based either on Look-Up Tables [39] to store and interpolate the related deviations, or on curve approximation with B-splines [38] to model the distance deviations. In other methods of curve approximation, polynomials have been used [40] to fit the deviations. These models require a smaller number of initial values than B-splines, but are also less representative of the actual deviations. Whatever the method, the aim of this calibration step is the storage of depth residuals as a function of the measured distance.

Since object reflectivity varies with the range and can also cause a distance shift depending on the distance to the observed object, an intensity-related (or amplitude-related) error must be mentioned. Specific calibration steps for its correction have been investigated, for example on a PMD time-of-flight camera [41]. Furthermore, the depth calibration needs to be extended to the whole sensor array, because a time delay of the signal propagation is observed as a function of the pixel position on the sensor array. A per-pixel distance calibration [42] can be considered. Besides, [39] suggests the computation of a Fixed Pattern Noise (FPN) matrix containing an individual deviation value for each pixel, that allows the diminution of the global surface deformations. Sometimes, this global correction is performed together with the wiggling error correction, by considering in the same mathematical model the depth-related error and the location of the pixel on the array [40]. Finally, some non-systematic depth deformations which rather correspond to noise are also reported. One should notice the existence of denoising and filtering methods for their reduction [43], as well as methods related to the multiple returns issue [44].

3. Survey of Kinect v2 Specifications

Since Kinect v2 is initially a motion sensing device produced to be used with a gaming console, it seems obvious that the measurements provided by it will be affected by some unavoidable error sources. Indeed, the environment in which the acquisitions are performed has an influence (e.g., temperature, brightness, humidity), as well as the characteristics of the observed object (reflectivity or texture, among others). Once the camera is on its tripod, the user intervention is limited to choosing settings (time interval, types of output data), and therefore, the automation level seems to be relatively high. However, errors can occur due to the internal workings of the sensor itself.

To quantify the accuracy of the acquired data, a good knowledge of these potential sources of errors is required. For the Kinect v1, some of these aspects have been investigated, such as pre-heating time [9], or the influence of radiometry and ambient light [45]. Because of the change of the distance measurement principle in the Kinect v2, the phenomena observed with this new sensor might be different. In spite of this change, it is important to investigate their influence on the produced data. After a review of the way the sensor works and the data it provides, a few experimentations are presented in this section. They deal with pre-heating time, noise reduction, and some environment or object-related criteria.

3.1. Sensor Characteristics and Data Acquisition

Kinect v2 sensor is composed of two cameras, namely a RGB and an infrared (IR) camera. The active illumination of the observed scene is insured by three IR projectors. Some features of the sensor are summarized in Table 1. For example, given the specifications, at 2 m range a pixel size of 5 mm can be obtained.

As mentioned on the Microsoft website [46], the Kinect v2 sensor is based on ToF principle. Even though time-of-flight range imaging is a quite recent technology, many books deal with its principles and its applications [47–50]. The basic principle

444

is as follows: knowing the speed of light, the distance to be measured is proportional to the time needed by the active illumination source to travel from emitter to target. Thus, matricial ToF cameras enable the acquisition of a distance-to-object measurement, for each pixel of its output data. It should be noted that, unlike other ToF cameras, it is impossible to act on the modulation frequency or within the integration time of the input parameters of the Kinect v2 sensor.

Table 1. Technical features of Kinect v2 sensor.

Infrared (IR) camera resolution	512 × 424 pixels
RGB camera resolution	1920 × 1080 pixels
Field of view	70 × 60 degrees
Framerate	30 frames per second
Operative measuring range	from 0.5 to 4.5 m
Object pixel size (GSD)	between 1.4 mm (@ 0.5 m range) and 12 mm (@ 4.5 m range)

As shown in Figure 2, three different output streams arise from the two lenses of the Kinect v2 device: infrared data and depthmaps come from one lens and have the same resolution. The depthmaps are 2D images 16 bits encoded in which measurement information is stored for each pixel. The color images of higher resolution come from the second lens. The point cloud is calculated thanks to the depthmap, because of the distance measurements it contains. From depth values stored in the pixel matrix, two possible ways can be considered to infer 3D coordinates from the 2D data. Either a mapping function provided by the SDK (Software Development Kit) is applied between depth and camera space, or self-implemented solutions can be used applying the perspective projection relationships. The result is a list of (X, Y, Z) coordinates that can be displayed as a point cloud.

If colorimetric information is required, a transformation of the color frame has to be performed because of its higher resolution compared to the depthmap. This transformation is achieved based on a mapping function from the SDK, which enables to map the corresponding color to a corresponding pixel in depth space. The three dimensional coordinates coming from the depthmap are combined with the corresponding color information (as RGB values) and constitute a colorized point cloud.

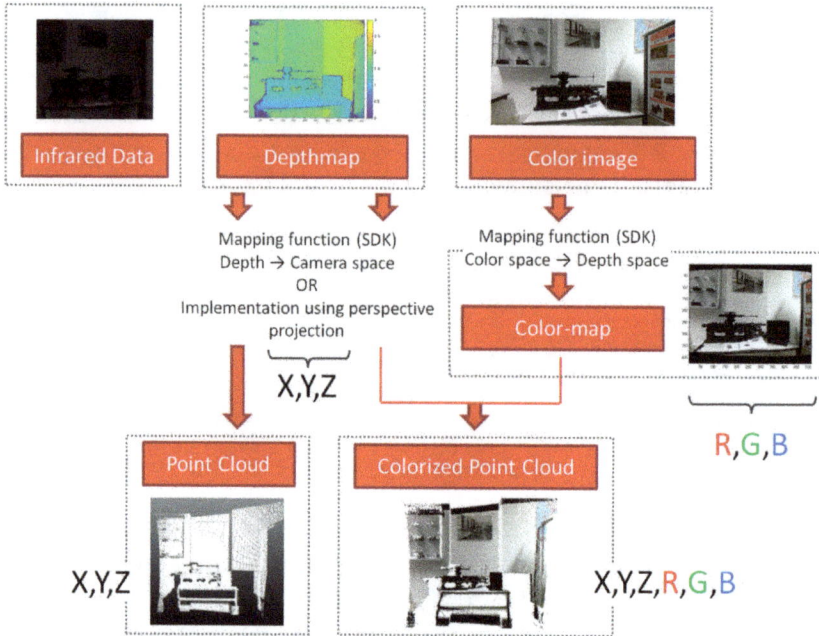

Figure 2. Schematic representation of the output data of Kinect v2 and summary of point cloud computation.

3.2. Pre-Heating Time and Noise Reduction

The first study involved the necessary pre-heating time. Indeed, as shown in [9,51], some RGB-D sensors need a time delay before providing reliable range measurements. In order to determine this time delay which is necessary for the Kinect v2 to reach constant measurement, the sensor was placed parallel to a white planar wall at a range of about 1.36 m. Three hundred and sixty depthmaps were recorded during one and a half hours, which is one depthmap recorded each 15 seconds. An area of 10×10 pixels corresponding to the maximal intensities in IR data was then selected in each recorded depthmap. Indeed, the sensor is not placed perfectly parallel to the wall, thus it cannot be assumed that the shortest measured distances are located at the center of the depthmaps. Mean measured distances can be calculated and represented as a function of the operating time. The distance varies from 5 mm up to 30 minutes and becomes then almost constant, around 1 mm. It can be noticed that the sensor's cooling fan starts after 20 minutes. For future tests, the pre-heating time of 30 minutes will be respected even if the distance variation is quite low. This experiment is related in [52].

Secondly, the principle of averaging several depthmaps has also been investigated in order to reduce the measurement noise inherent to the sensor and its technology [45]. With this process, the influence of the frame number on the final measurement precision is studied. For this purpose, the sensor was placed in front of a planar wall, at a distance of 1.15 m. To investigate the influence of the frames averaging, different sizes of samples have been considered. Figure 3a,b presents the standard deviation computed for each pixel on datasets of respectively 10 successive frames and 100 successive frames. The standard deviation values obtained with 10 frames are lower than 1 cm, except on the corners. Considering the averaging of 100 successive frames, these values are not lower, but a smoothing effect is observed especially on the corners. The variation of the standard deviations along a horizontal line of pixels in the middle of the frame (see red line) is presented in Figure 3c, where the case of 10 frames and 100 frames are plotted together. Regarding this graph it is clearly visible that there is no gain in terms of precision when more frames are averaged, because the mean standard deviation values are almost the same in both cases. Only the variations of the standard deviations are reduced for the higher number of frames. The smoothing effect observed while increasing the number of frames appears through the diminution of the peaks in Figure 3c, expressing a reduction of the noise. In conclusion, the use of a larger sample does not really enhance the precision but it makes it more uniform. That is why for the future acquisitions, datasets of about 10–50 frames will be considered. Besides, similar results obtained for the case of real scenes are reported in [52].

3.3. Influences of the Environment and Acquired Scene

The previous section described phenomena which are rather related to the sensor itself. However, some characteristics of the environment in which the acquisitions are realized also have an influence on the measured distances.

3.3.1. Color and Texture of the Observed Items

The effects of different materials on intensity as well as on distance measurements have first been assessed using samples characterized by different reflectivities and roughness. The sensor was placed parallel to a board with samples. It appears that reflective as well as dark materials stand out among all the samples. Indeed, very reflective as well as very dark surfaces display intensity values which are lower than for other samples. In the depthmaps, this phenomenon results in measured distances which are larger than expected. The more important effect is observed for a compact disk, *i.e.*, a highly reflective material. For this sample, distance measurements vary up to 6 cm in an inhomogeneous way. Always considering very reflective materials, experiments performed on mirror surfaces have confirmed the previous findings with measurement deviations that can reach up to 15 cm.

447

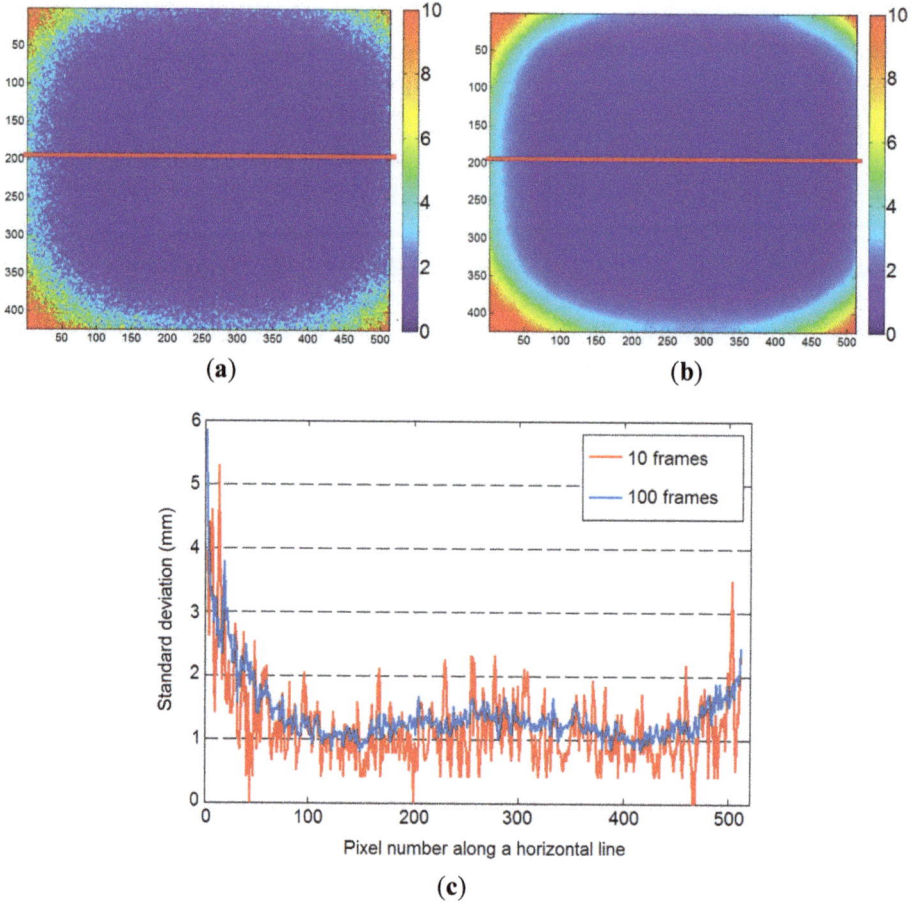

Figure 3. Visualization of color-coded standard deviation calculated for each pixel over (**a**) 10 successive frames, and (**b**) 100 successive frames; (**c**) Observation of standard deviations along a line of pixels (see red line) in the set of 10 frames (orange) and 100 frames (blue).

To complete these investigations, observations have been realized on a checkerboard of grey levels inspired from [38] (Figure 4). As illustrated in Figure 4b, the displayed intensity increases with the brightness of the patterns. Intensity values displayed by black squares are almost six times lower than the ones displayed by white squares (2000 against 12,000). Besides, considering the values stored in the depthmaps (Figure 4c), one can conclude that the lower the intensities are, the longer the measured distances are. Deviations of the distance up to 1.2 cm are observed for the black patterns. As mentioned in Subsection 2.3, the reduction of intensity-related errors can be considered in a specific calibration process described in [41] for a

PMD ToF camera. This correction technique is not discussed here. Considering the modeling approach developed in Section 5 of this paper, the reconstructed object does not show significant color changes. Hence, in that case this error source does not really affect the measurements and is safely negligible.

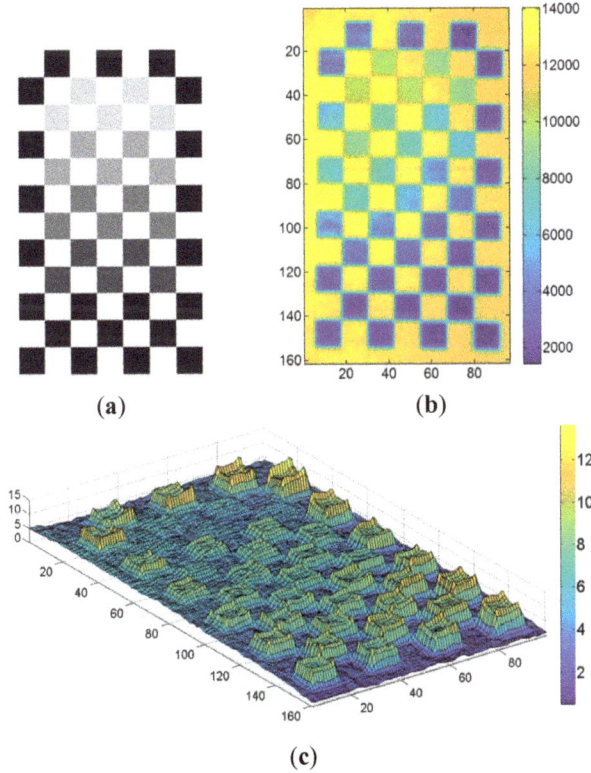

(a)　　　　　　　　　　　(b)

(c)

Figure 4. (a) Checkerboard of grey levels inspired from [38], (b) intensity values measured on this checkerboard, and (c) corresponding 3D representation of distance measurements variations (in mm).

3.3.2. Behavior in High Brightness Conditions

Because the previous Kinect device was not suited for sunny outdoor acquisitions, the influence of the brightness conditions on the measurements realized with a Kinect v2 was finally studied. Some applications of the sensor could require a proved efficiency in outdoor conditions. Hence, acquisitions were performed with the sensor during a sunny day. The observed scene composed of a table and a squared object is presented on Figure 5a, as well as the resulting point cloud on Figure 5b. It appears that parts of the scene are missing, such as the table stands. As a matter of fact, about 35,000 points are visible on this point cloud over the 217,088 maximal points of a single point cloud, which is about 20%. Regarding Figure 5c which represents the entire point cloud, about 2% of "flying pixels" (in red on Figure 5c) are included in the total number of points.

(a)   (b)   (c)

Figure 5. Outdoor acquisitions: (**a**) Picture of the observed scene; (**b**) Corresponding point cloud acquired with the sensor, (**c**) and the same entire point cloud in a profile view (scene in a black rectangle) without removal of the "flying pixels" (red).

Despite a clear lack of points to depict the whole scene, the results show that the sensor is able to work during a sunny day provided that the light does not directly illuminate the sensor. Indeed, strong backlighting conditions cause the sensor's disconnections from the computer. Two phenomena summarize the previous observations: the number of "flying pixels" is clearly visible particularly on the edges of the sensor field of view, and the number of acquired points decreases when the light intensity raises.

4. Calibration Method

A major drawback of 3D ToF cameras is the significant number of errors that can influence the acquired data. These errors, which affect both the overall performances of the system and its metrological performances, were emphasized in the related work section as well as in the section related to the survey of the sensor.

This section deals essentially with the assessment and the correction of the systematic distance measurement errors. To do that, a calibration method divided into several steps is proposed. Considering the fact that range imaging cameras combine two technologies, two types of calibrations have to be considered. Firstly, the geometric calibration of the lens is performed in order to correct its geometric distortions. Secondly, a depth calibration is suggested in order to assess and correct depth-related errors.

4.1. Geometric Calibration

Since lenses are involved in the Kinect device acquisitions, geometric distortions as reported with DSLR cameras can be observed. For the first version of the Kinect sensor, some computer vision based calibration algorithms have been developed [53,54]. Unfortunately, these algorithms have not yet been adapted for the Kinect v2 sensor. Nevertheless, it appears in lots of works, e.g., [55], that time-of-flight cameras can be geometrically calibrated with standard methods.

As for common 2D sensors, several images of a planar checkerboard have been taken under different points of view. It is worth noting that the infrared data have been used to handle this geometric calibration, because depth and infrared output streams result from the same lens. To determine the necessary intrinsic parameters, our dataset was treated with the "Camera Calibration Toolbox" proposed by [56] under the Matlab software. In this tool, the Brown distortion model is implemented. This model is largely used for photogrammetric applications because of its efficiency. The "Camera Calibration Toolbox" allows the determination of radial as well as tangential distortion parameters, and also internal parameters such as focal length and principal point coordinates.

With this technique applied, one should underline the fact that changing or removing some of the images causes the computed results to vary by a few pixels. This phenomenon is only due to the low sensor resolution. As a matter of fact, the best calibration results are obtained regarding the lower uncertainties calculated on the parameters. An overview of results obtained for these parameters is available in [52], as well as in [57]. The direct integration of the distortion coefficients into our point clouds computation algorithm allows the obtainment of a calibrated three dimensional dataset, which is no longer affected by the actual distortions of the initial depthmap.

4.2. Depth Calibration

As mentioned in Section 2.3, errors related to the measurement principle systematically affect the distances stored in the depth images. Accordingly, a depth-related calibration model needs to be defined to correct the measured distances and thus enhance the quality of the acquired data. This step is particularly significant since the aim of this study is to produce 3D models in an as accurate as possible way. After a description of the method that has been set up for this purpose, the processing of the data and the results that they provide are analyzed in this section.

4.2.1. Experimental Setup

A common way to assess the distance inhomogeneity consists on positioning the camera parallel to a white planar wall at different well-known ranges. In our study, the wall has been surveyed beforehand with a terrestrial laser scanner (FARO Focus^{3D}, precision 2 mm). The standard deviation computed on plane adjustment in the TLS point cloud is 0.5 mm. It allows confirming the reference wall planarity assuming that the TLS provides higher accuracy than the investigated sensor. After that, a line has been implanted perpendicularly to the wall by tachometry. Conventional approaches make use of track lines [39,42] to control the reference distances between plane and sensor, or rather try to estimate the camera position with respect to the reference plane through a prior calibration of an additional combined CCD camera [38,40]. In our experimental setup, marks have been determined by tachometry along the previously implanted line at predetermined ranges. They represent stations from which the acquisitions are realized with the Kinect sensor placed on a static tripod. A picture of the setup is shown in Figure 6a.

Stations have been implanted at 0.8 m range from the reference wall, and then from a 1 m up to 6 m range every 25 cm (Figure 6b). In this way, the sensor is progressively moved away from the plane, while covering the whole operating range of the camera. To limit the influence of noise, at each sensor position, 50 successive depthmaps of the wall have been acquired with intervals of one second. Finally, to insure the parallelism between sensor and wall and thus to avoid the addition of a possible rotation effect, the distances have been surveyed by tachometry at each new position, using small prisms at the two sensor extremities.

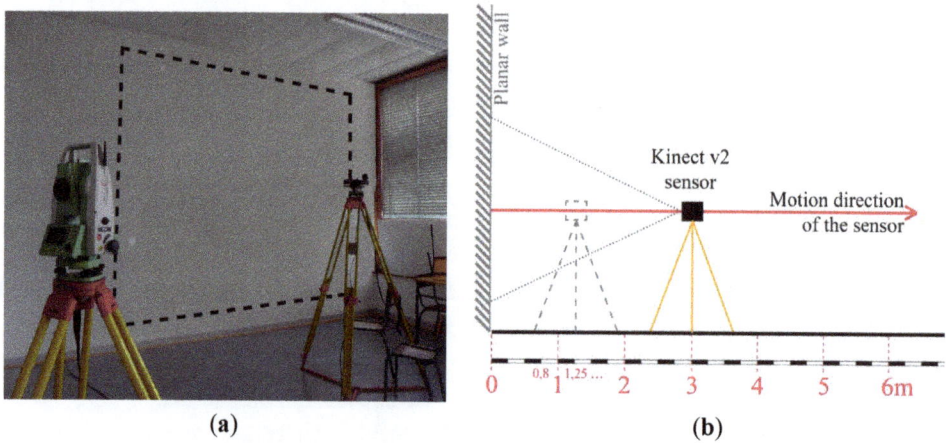

(a) (b)

Figure 6. (a) Picture of the experimental setup with the observed area in dotted lines; (b) Global schema of the acquisition protocol.

4.2.2. Data Analysis

The results obtained led us to consider two steps of depth calibration. First, deviations of measurements for a small central patch of the frames are assessed through B-spline fitting as suggested in [42]. Secondly, the deformations extended to the whole sensor array are investigated, in order to highlight the need of a potential per-pixel correction.

First of all, a central matrix of 10×10 pixels is considered in the input images acquired during the experiment. This enables to compute mean measured distances from the sensor for each position. Then, the deviations between real and measured distances are plotted on a graph as a function of the range. Each of the 50 deviations obtained from the 50 depthmaps acquired per station is represented as a point. As depicted in Figure 7a, a B-spline function is estimated within these values. Since the sensor was accurately placed on the tripod with respect to its fixing screw, a systematic offset occurs on raw measurements because the reference point for the measurement does not correspond to the optical center of the lens. The influence of this offset corresponding to the constant distance between fixing point and lens (approximately 2 cm) is removed on this graph. It appears that the distortions for the averaged central area vary from −1.5 cm to 7 mm, which is rather low regarding the technology investigated. At 4.5 m range, a substantial variation is observed. Under 4.5 m range, the deviations are rather included within an interval of variation of almost 1 cm (from −1.5 cm to 7 mm).

Since a set of 50 successive depthmaps is acquired for each position of the sensor, a standard deviation can also be computed over each sample. Figure 7b presents a separate graph showing the evolution of the computed standard deviations as a function of the range. As it can be seen, the standard deviation increases with the range. This means that the scattering of the measurements increases around the mean estimated distance when the sensor moves away from the scene. Moreover, for the nearest range (0.8 m), the standard deviation reported stands out among all other positions. As a matter of fact, measurements realized at the minimal announced range of 0.5 m would probably be still less precise. Since a clear degradation with depth is showed, it makes sense to bring a correction.

4.2.3. Survey of Local Deformations on the Whole Sensor Array

The second part of our depth calibration approach consists of extending the deformations analysis to the whole sensor array. For this purpose, only the first 10 datasets acquired on the planar wall corresponding to the first 10 positions from 0.8 to 3 m are considered. It appears that the corresponding point clouds are not planar, but rather exhibits a curved shape. On these 10 successive point clouds, planes have been fitted by means of least-squares adjustment. To realize a proper adjustment in the Kinect point clouds, some outlying points corresponding to null values in the depthmaps have to be previously removed. Then, the Euclidean distances between a point cloud and its corresponding plane are calculated. It provides a matrix of residuals, in which for each pixel the deviation of the actual measured distance with respect to the adjusted plane is stored.

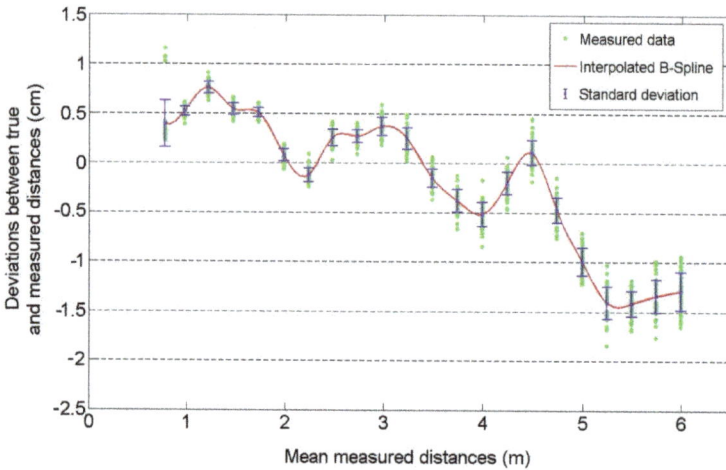

(a)

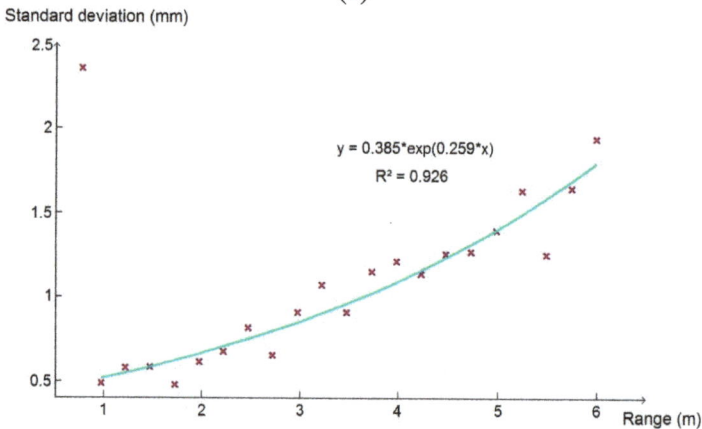

(b)

Figure 7. (a) Deviations (in cm) between true distances and measured distances, as a function of the range (m). (b) Evolution of the standard deviation (in mm) calculated over each sample of 50 measurements, as a function of the range (m).

Figure 8 shows the residuals with respect to the plane at 1 m range. The residuals have been computed and represented in the form of a point cloud (in red), in which a surface is interpolated (here color-coded). First, it appears that the residuals are more important at sensor boundaries and especially on the corners. As known from the literature, these radial effects increase with the distance between sensor and target.

454

In our case, the residuals can reach about 1 centimeter up to tens of centimeters in the corners at 3 m range.

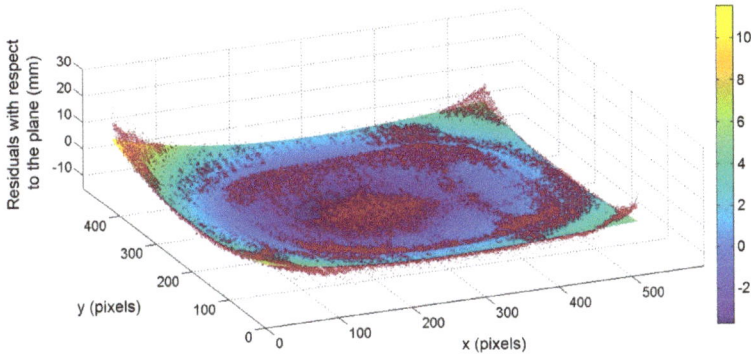

Figure 8. Point cloud of residuals (mm) in red, at 0.8 m, together with the surface (color-coded) fitting the residuals.

To define an adapted correction, the parameters of the interpolated surfaces previously mentioned are stored. The chosen surface model is polynomial, in order to keep a convenient computation time for the correction step. This post-processing step consists of delivering signed corrections interpolated from the residuals surfaces computed at each range. Thus, to correct each pixel of the depthmap, given the distance value it contains, a residual value will be interpolated considering the nearest ranges. In that way, a correction matrix is computed, that needs to be added to the initial depthmap as a per-pixel correction.

4.3. Influence of Corrections

All the corrections related to depth calibration are part of post-processing carried out on the depthmaps. In order to visualize the influence of the computed correction on the systematic distance measurement error, the dataset consisting of measurements realized on a planar wall at known ranges is used. To apply the calibration, an adjustment of the errors model in the form of a spline previously determined is required. Indeed, knowing the curve parameters and the distance measured allows inferring the correction values applied to the measurements. For the analysis, a central patch of 10×10 pixels on the corrected data is considered as before. The remaining deviations between real and corrected distances are shown in Figure 9a. It appears that the average deviations are reduced to zero through this calibration. These good results can also be explained by the fact that the dataset investigated is the one used to determine the correction parameters. However, it enables validating the model. At the same time, the standard deviations computed over the 50 corrected measurements per station are almost the same or have slightly decreased compared to the initial data. They confirm the increase of the scattering of the measurements when the range raises, but they look promising for the investigated device.

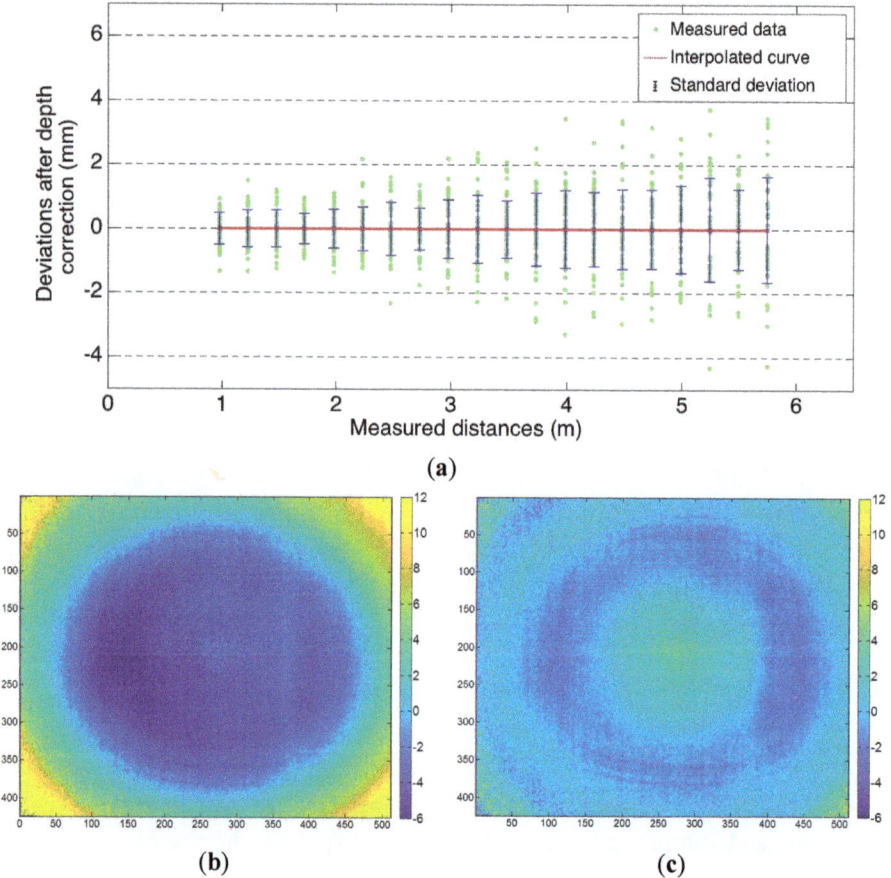

(a)

(b)   (c)

Figure 9. (a) Remaining deviations (in mm) between real and corrected distances, as a function of the range (m); Colorized representation of the residuals (in mm) with respect to a fitted plane (range: 1.25 m), (b) before correction and (c) after correction of local deformations.

Considering the correction extended to the whole sensor array, the approach used is based on surface approximation within the residual values calculated with respect to a plane. These residuals are visible on a colorized representation for the initial data (Figure 9b) and after correction of the local deformations (Figure 9c) with the method proposed beforehand. A 1.25 m range is considered in the figures.

Whereas before correction the residuals have negative values in a circular central area and can reach more than 1 cm on the corners, they are reduced to some millimeters after correction. The residuals on the corners are also much less significant, so that a positive impact of the second step of the calibration is noticed. For the 10 ranges considered for this correction, the residuals are always reduced after calibration. This is also true for the standard deviation calculated on plane fitting residuals. For the ranges considered, the standard deviations vary from about 2–5 mm before correction, and from about 1–3 mm after correction. Nevertheless, a less significant effect is observed at 2.25 m range. In addition, these corrections are time-consuming due to the local processing of each pixel.

Consequently, depending on the accuracy required by the applications, the need for a depth calibration must be discussed. Indeed, the user could decide to use only the measurements provided by a limited central area of the sensor, regarding the fact that fewer deformations affect the middle of the sensor array. However, the works presented in the following section of this paper have all been realized with calibrated data.

5. Experimental Approach for 3D Modeling with Kinect v2

To perform a three-dimensional reconstruction of an object or a scene with the Kinect sensor, several ways can be considered for the alignment and the merge of the raw data coming from the sensor. Among them, the SDK provided by Microsoft offers a tool called "Kinect Fusion" [32] to answer to a possible use of the Kinect sensor as a 3D scanning device. In the literature, many works have shown the potential of Kinect v1 for modeling purposes. In [30], a standard deviation of about 1 mm is reached for the 3D reconstruction of a small statue. A precision in the order of magnitude of some millimeters is also reported in [58] for spheres and planes adjustment using acquisitions from a Kinect v1 sensor. Because of the technical differences between first and second version of Kinect, an improvement in terms of precision can be expected.

Regarding the calibration results obtained in this article, a priori error lower than 5 mm can be assumed for acquisitions realized at ranges smaller than 4 m. Even if a sub-millimetric precision can be offered by laser scanner or photogrammetry techniques for 3D reconstruction of small objects, it is interesting to test the Kinect v2 sensor for similar applications. In this section, a reconstruction approach with this device is discussed after the presentation of the object under study. This will lead to a qualitative assessment of the models produced.

5.1. Object under Study and Reference Data

The object under study is a fragment of a sandstone balustrade of about 40×20 centimeters, coming from the Strasbourg Cathedral Notre-Dame (France). As shown in Figure 10a, it presents almost a symmetric geometry and flat faces, except on its extremities.

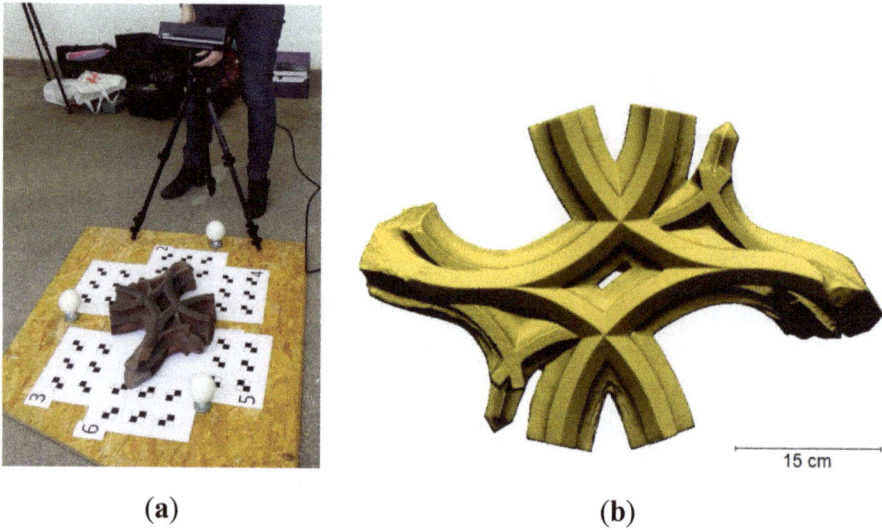

(a)   (b)

Figure 10. (a) Object under study (a sandstone balustrade fragment); and (b) reference mesh realized after acquisition with a measuring arm.

A measuring arm from FARO (FARO ScanArm) was used to generate a ground truth dataset of the fragment. It provides a very high metric quality thanks to its sub-millimetric accuracy (0.04 mm according to the specifications). Such a device is often used for metrological purposes, but because of its cost, its bad portability and its fragility, other devices are often preferred in many missions. In this study, the dataset obtained with the measuring arm constitutes a reference and will be used to assess the quality of the model created with the Kinect v2 sensor. The reference point cloud of the fragment contains about 5 million points after a spatial resampling of 0.1 mm. Based on this point cloud, a meshed model was created using the commercial software 3D Reshaper (Technodigit). The resulting mesh is composed of about 250,000 faces (Figure 10b).

5.2. Reconstruction Approach

In the following subsections, the experimental workflow used to reconstruct the object under study with a Kinect v2 sensor is described. Once the data have been acquired, the produced point clouds need to be registered before the final creation of a meshed model.

5.2.1. Data Acquisition and Registration

First of all, a set of data of the balustrade fragment has been acquired with the Kinect v2 sensor. The way used to carry out the acquisitions is rather inspired by the common workflow set up in image-based reconstruction methods. A circular network composed of acquisitions from eight different viewpoints was performed around the object under study. An overlap of about 80% on the object was reported between two successive range images. In order to overcome the deformations affecting the measurement on the borders of the sensor, the object was located in the central part of the sensor. As shown in Figure 10a, flat targets were also placed around the object during the acquisitions for the point clouds' registration to be done in a post processing step. The sensor was placed at about 1 m of the object. Indeed, the results of depth calibration presented in the previous section emphasize the fact that considering scene observation at a range from 1–4 m, the global depth-related deformations are smaller. At the chosen range, it is possible to capture the whole object within each range image, as well as a sufficient number of flat targets for the future registration. At this range, the related spatial resolution of the object is about 2–6 mm, depending on the shapes encountered.

The survey of the sensor specifications in Section 3.2 also shows that the frame averaging allows a reduction of the measurement noise inherent to the sensor. Thus, 10 successive acquisitions were realized at the framerate of 0.5 s for each viewpoint. Based on raw data, the averaging of the depth images acquired for each viewpoint as well as the computation of individual point clouds from these depth images are performed. Then, a target-based registration was performed by extracting the target centers manually. A spatial resolution between 2 and 4 mm is reported on the ground where the targets are located. This method based on target selection provides an overall registration error of about 2 cm, even with a refinement of the registration using the ICP method. This is high regarding the requirements for 3D modeling and the dimensions of the fragment, especially since the more adverse effect appears in the center of the frames where the object is located.

Thus, another method has been considered for point clouds registration. The workflow applied is summarized in Figure 11. The first step consists in the colorization of the point clouds with the infrared data provided by the sensor. This allows removing the ground under the object in the point clouds, through filtering. Moreover, noise reduction has been performed depending on the point of view. Then, homologous points have been selected directly on the object on pair-wise successive point clouds. It appears that this method is more reliable than the previous one. Indeed, after a registration refinement, the residual registration error between two successive point clouds varies from 2–4 mm. This operation has been repeated to register the eight individual point clouds which are counted to be between 13,000 and 14,000 points each. These better results are due to the fact that the common points chosen directly on the object are hence located close to the center of the point cloud. The use of targets directly on the object could be considered.

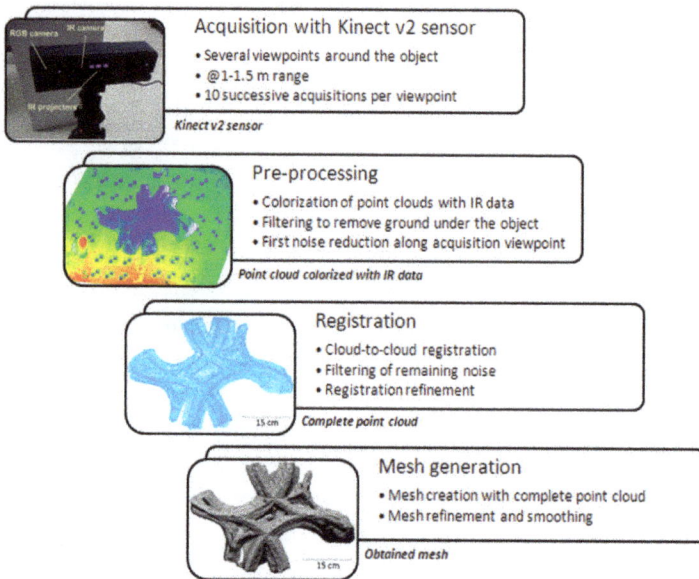

Figure 11. Processing chain applied in this study for 3D reconstruction based on Kinect, from the acquisition of point clouds to the mesh creation.

5.2.2. Creation of a Meshed Model

The complete point cloud of the balustrade fragment, as well as the mesh based on it and created through the 3D Reshaper software, are presented in Figure 11. Noise remains in the obtained complete point cloud and is hard to remove without losing geometric information. Consequently, the creation of the mesh is time-consuming. Indeed, several parameters can be empirically set during the mesh creation, which have a substantial influence on the resulting mesh. They can provide, for example, a good visual smoothed result but with a loss of details. Because of the low quality of the initial point cloud, even the processing steps' order might influence the resulting mesh obtained following this methodology.

As depicted in Figure 11, the geometry of the final mesh suffers from several faults compared to the real fragment. Its metric accuracy will be assessed in the next Subsection. There is obviously a correlation between the deformations observed on the final mesh and error sources related to the raw data themselves. First of all, border effects can be observed along the line of sight of the sensor and constitute noise. A second problem is related to deformations of the geometry linked to the viewpoint. Among all error sources, these two problems which were detailed in [33] obviously contribute to the low visual quality of the mesh performed.

5.3. Accuracy Assessment

In order to assess the accuracy of the Kinect-based model, the reference mesh previously presented was used in order to make a series of comparisons between them. The reference point cloud obtained by the measuring arm offers a very high density of points which is almost 15 times larger than the density of the point cloud obtained with the Kinect v2. This high point density variation may be an issue for the cloud-to-cloud comparison algorithms, resulting in a loss of efficiency in the distance computation. Additionally, considering meshes, a loss of geometric quality or a data simplification due to mesh creation can occur. Comparisons between different kinds of data were therefore carried out.

The complete point cloud of the balustrade fragment and its corresponding mesh derived from Kinect raw data were compared to the reference mesh. Figure 12a presents the results of the cloud-mesh comparison and Figure 12b presents the results of the comparison between meshes. The deviations between datasets are color-coded, and projected on the computed data. Despite a possible bias due to the triangulation method applied during the mesh creation, for both methods the results look similar.

In both comparisons, a mean deviation of about 2 mm is reported for almost 40% of the models, and a standard deviation of about 2.5 mm is observed. Besides, maximal deviations of about 1.5 cm are reached. They correspond to local deformations affecting the model and are particularly high on the borders of the object. These deformations of at least 1 cm provide to the object false shapes, especially for the edges on the upper part of the object, or on its extremities.

Regarding the calibration results (Figure 7a), for the used range an *a priori* error in the order of magnitude of 5 mm can be assumed. It appears that 90% of the deviations are lower than the *a priori* error. Thus, the final error of about 2 mm on a large part of the model is more than acceptable for the device under study. It looks promising since a lot of improvements in the processing chain can be considered. Nevertheless, the remaining deviations in the order of magnitude of one centimeter still represent a significant percentage of the global size of the object. Besides, the visual rendering suffers from many deformations and the model is too far from the real shape of the object. This might be detrimental to archeological interpretation.

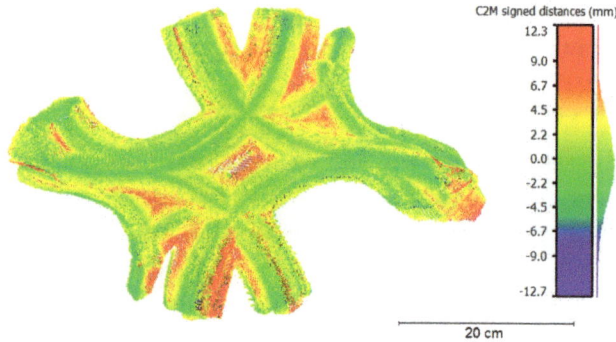

C2M signed distances (mm)

```
12.3
9.0
6.7
4.5
2.2
0.0
-2.2
-4.5
-6.7
-9.0
-12.7
```

20 cm

(a)

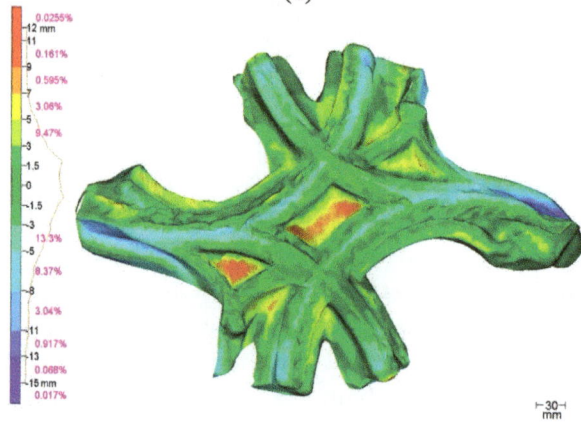

```
0.0255%
12 mm
11
  0.161%
9
  0.595%
7
  3.06%
5
  9.47%
3
1.5
0
1.5
3
  13.3%
5
  0.37%
8
  3.04%
11
  0.917%
13
  0.068%
15 mm
  0.017%
```

├─30─┤
mm

(b)

Figure 12. (a) Comparison between Kinect-based point cloud and reference mesh; and (b) comparison between mesh created from Kinect point cloud and reference mesh. The deviations are in millimeters.

5.4. Kinect Fusion Tool

A first version of Kinect Fusion tool was already proposed for the first version of the device, and it has been adapted and directly added to the SDK of Kinect v2. It allows creating automatically accurate meshes of objects with a visual rendering close to reality. It is mainly based on an Iterative Closest Point (ICP) algorithm [59], a well-known algorithm in the field of computer vision.

In [58], the benefits of using Kinect Fusion tool compared to individual scans of a Kinect v1 are exposed in terms of accuracy and precision. For example, at 2 m range a precision of about 1 mm on sphere estimation is reached thanks to Kinect Fusion, whereas the precision is larger than 7 mm with the sensor alone. Despite the common use of this 3D reconstruction tool for Kinect sensors due to their good results, the major drawback of this tool is the "black box" effect related to the use of the standalone version available in the SDK, for which the algorithms are not available. Indeed, the user can only change a few settings but do not really manage all processing steps that are carried out on the raw data. However, improvements of the workflow applied in Kinect Fusion are possible through open source implementations and libraries. This was, for example, already tested with Kinect v1 in [60], where they suggest adding calibration results and estimated exterior parameters to the implementation. The deviations in the models produced are reduced compared to the initial Kinect Fusion models.

In [33], the Kinect Fusion tool associated with the Kinect v2 sensor was used in its standalone version to scan another architectonic fragment. After a cloud-mesh comparison with a reference model, a standard deviation of 1.4 mm and a mean deviation of 1.1 mm were reported for the Kinect Fusion model. Since this fragment was more complex than the one investigated in this article, this led us to conclude about the satisfactory precision provided by Kinect Fusion. Moreover, the aim of this article was rather to analyze a more experimental methodology; that is why the Kinect Fusion tool will not be further investigated here.

6. Discussion

This section aims at analyzing the results obtained for the 3D model creation with the Kinect v2 sensor. Not only will the previously assessed accuracy be commented on, but rather the characteristics of the experimental reconstruction method will presented along with potential improvements. First of all, some differences between Kinect v1 and the v2 investigated in this article will be discussed.

6.1. Differences between Kinect v1 and v2

The Kinect v2 sensor presents several technical benefits compared to the first version. First of all, it offers a larger horizontal as well as vertical field of view. According to this criterion and to the change of depth sensing technology, the point clouds obtained present a better resolution. Moreover, the point clouds produced by Kinect v1 sensor show striped irregularities, which is not the case with the second version. Then, the resolution of color images are increased. To summarize these technical features, specifications of both Kinect versions can be found in the form of tables in previous contributions as for example [24] or [37]. Finally, as depicted in Section 3.3.2, outdoor acquisitions can be envisaged since Kinect v2 seems to be much less sensitive to daylight.

Considering now the accuracy of the data captured by both versions of the sensor, a metrological comparison between Kinect v1 and Kinect v2 carried out in [37] is examined. The assessment of data acquired on a same artefact shows that accuracy and precision are better with Kinect v2. There is also less influence of the range on the decrease in precision for Kinect v2 measurements. Data performed for other applications as 3D reconstruction also confirm the benefits of the second version compared to the first one. In the field of people tracking, for example, it is shown in [61] that the tracking is 20% more accurate with Kinect v2 and it is also possible in outdoor environments. It appears that Kinect v2 is generally better in many cases. However, Kinect v1 and v2 both suffer from a limited accuracy on edges.

6.2. RGB-D Camera as an Alternative to Photogrammetry?

Since a 3D reconstruction method for close-range objects with Kinect v2 has been presented in this paper, it seems interesting to compare it to a more common approach. For this purpose, [33] presents models obtained on the same archaeological material through DSLR camera pictures acquisition and use of an image-based modeling solution. With a mean deviation smaller than 1 mm and very local small deviations, photogrammetry obviously provides a high level of detail on very accurate models. As expected, the results obtained in Section 5 of this paper are clearly still far from achieving such accuracy. Nevertheless, regarding the initial use of the Kinect v2 device, the accuracy cannot constitute a unique comparison point. To extend the comparison to further considerations, other characteristics should be investigated. Table 2 offers a comparative overview between both mentioned methodologies. Not only the accuracy reached has been taken into account, but also criteria such as acquisition and computation times, or cost, have been considered in order to underline the benefits and drawbacks of each strategy.

Table 2. Comparative overview of some criteria for the image-based and for the Kinect-based 3D reconstruction methods.

	Photogrammetry (DSLR Camera)	Kinect v2 (Registration Method)
Cost	low to medium	low
Portability	good	good, but requires USB 3 and external electricity supply
Object size	from small objects to entire rooms or buildings	rather limited to objects
Indoor/outdoor	indoor + outdoor	mainly indoor
Ease-of-use	rather easy, needs care for photographs acquisition	needs some practice, for registration especially
Processing time	some hours all in all	long, because of manual treatments
Accuracy	very high, up to sub-millimetric (depends on the camera)	≈1 cm, to be improved

Whereas the cost is comparable for the digital camera and the Kinect sensor, it seems that the photogrammetric survey remains easier to apply than the method considered in Section 5. This is also true in terms of portability because of the external electricity supply still required by the Kinect sensor. Besides, the more experimental method consisting of registration of Kinect point clouds remains very time-consuming for a final lower accuracy. Indeed, it appeared that in the meshes presented in the previous section that a lot of details were not at all or badly represented, which is associated with a loss of geometrical information. These results tend to confirm the higher efficiency of image-based reconstruction methods when accurate 3D models are required, regardless of the object size.

6.3. Suggested Improvements

Regarding the results obtained for the mesh creation in Section 5.3, it is obvious that the Kinect sensor presents technical constraints which are actually problematic for 3D modeling applications. Nevertheless, improvements in the reconstruction method are necessary in order to enhance the metrical quality of the models produced. Since there is no unique approach for aligning and merging Kinect raw data in order to get a three-dimensional reconstruction of a scene, several solutions can be considered. An approach would be to use the colorimetric information produced by the camera [31]. In the methodology developed in the abovementioned contribution, SURF algorithm [62] was used for the localization and matching of keypoints on color images obtained from a Kinect v1 device. After that, a first alignment is provided, which can be refined by using the ICP principle. This idea of combining the RGB information as a complement to frame-to-frame alignment based on 3D point clouds helps overcoming the reduced accuracy of the depth data.

Other open issues related to the considered acquisition process and registration method have to be investigated. Indeed, it has been shown in [33] that some geometric deformations are directly related to the acquisition viewpoints. As a matter of fact, the choice of a network of eight circular acquisitions realized in the proposed experiment is questionable. The presence of almost one more zenithal view could perhaps influence the final result. Besides, depending on the geometry of the object observed, the idea of performing stripes rather than circular networks should also modify the result. Regarding the registration methodology carried out, its process was very long because of the manual tasks required. A higher automation of the registration could result in a reduction in processing time, but also a removal of potential user handling influences. Moreover, the pair-wise alignment of successive point clouds presents the problem of error propagation, so that a global adjustment could be considered to minimize the global error. Another approach to consider would be to apply weights to the points selected for the registration, as a function of the incident angle of the point clouds they belong to. This would confer more importance to more reliable measurements, probably enhancing at the same time the quality of the registration.

7. Conclusions

This article offers an overview of a set of experiments realized with a RGB-D camera, in order to assess its potential for close-range 3D modeling applications.

The sensor studied is a low-cost motion sensing device from Microsoft, namely the Kinect v2. The main idea was to investigate its measurement performance in terms of accuracy, as well as its capacity to be used for a particular metrological task. To do that, experiments for raw data enhancement and a calibration approach were presented. An experimental method for 3D reconstruction with Kinect was also described, with its related issues.

Several tests contributed to highlighting errors arising from the environment and the properties of the captured scene, as well as errors related to the sensor itself. As reported in many contributions dealing with RGB-D cameras, a pre-heating time has to be considered to obtain constant measurements. It has been estimated that almost 30 minutes is needed with the Kinect v2 sensor. Then, acquiring a tenth of successive depthmaps from a single viewpoint enables reducing the measurement noise through temporal averaging of the measured distances. Besides, it appeared that the color and reflectivity of the items have a great influence, both on the intensity images and on the distance measurements provided by the camera. Considering outdoor efficiency, the achieved results look promising.

Geometric as well as depth calibrations have been performed. Whereas the geometric lens distortions can be computed with a standard method also used for DSLR cameras, depth calibration requires further studies. A systematic depth-related error reported for range imaging devices was first reduced. The remaining deviations of some millimeters after calibration are very satisfactory. Finally, the local deformations which appear especially on the sensor array boundaries were reduced. This quite time-consuming step does not really enhance the measurements observed for the central area of the sensor array. Nevertheless, it was applied for the acquisitions dedicated to 3D modeling.

An experimental methodology for 3D reconstruction of an archaeological fragment was finally reported. Based on a mainly manual workflow from point cloud registration to mesh creation, it was proved to be very time-consuming. Besides, regarding the accuracy achieved while using the same sensor with the automatic tool Kinect Fusion, the models presented in Section 5 are obviously highly correlated to that explored in this paper. As a matter of fact, even if their accuracy in an order of magnitude of 1 cm is rather satisfactory, many improvements could be considered. For example, an acquisition process including more zenithal views could enable reducing the influence of deformations related to grazing angles. Adding colorimetric information is also a possible improvement, as well as more automation during the registration process.

Regarding the small price difference between RGB-D cameras and photogrammetric survey, as well as the wide range of robust software available for photogrammetry, as expected this second solution remains the easiest and most accurate method for 3D modeling issues. However, one should underline that the

466

use of a Kinect sensor for this purpose was a real challenge. However, the technology of RGB-D camera looks promising for many other applications, once the major limitations in terms of geometric performances have been addressed. This kind of devices seems to have a bright future, and this can be confirmed when considering, for example, the Project Tango [63] developed by Google. A smartphone or a tablet including a range imaging camera is used to reconstruct a scene in real-time, so that it looks attractive, easy and fast. Similarly, the Structure Sensor [64] offers a 3D scanning solution with a tablet, based on structured light projection and detection similarly to the first Kinect version. Since data quality assessment is still an open issue in such projects, acquisitions performed with such innovative techniques will no doubt be the focus of future contributions.

Acknowledgments: The authors would like to thank the Oeuvre Notre-Dame foundation and the City of Strasbourg for the provision of the Cathedral's balustrade fragment. They also want to thank Marie-Anne Mittet for her advices particularly concerning the setting up of the device under study.

Author Contributions: Elise Lachat realized its Master thesis project entitled "Assessment of a 3D camera (Kinect v2) for measurement and reconstruction of close-range 3D models" in the Photogrammetry and Geomatics group of the INSA of Strasbourg, under the supervision of Tania Landes and Pierre Grussenmeyer. This article is based on the results obtained during this project. The experiments were realized with the help of Hélène Macher, who also contributed to the writing of this paper. Tania Landes and Pierre Grussenmeyer also participated to the writing of this paper.

Conflicts of Interest: The authors declare no conflict of interest.

References

1. Remondino, F. Heritage recording and 3D modeling with photogrammetry and 3D scanning. *Remote Sens.* **2011**, *3*, 1104–1138.
2. Remondino, F.; Rizzi, A.; Agugiaro, G.; Girardi, S.; De Amicis, R.; Magliocchetti, D.; Girardi, G.; Baratti, G. Geomatics and geoinformatics for digital 3D documentation, fruition and valorization of cultural heritage. In Proceedings of the EUROMED 2010 Workshop "Museum Futures: Emerging Technological and Social Paradigms", Lemessos, Cyprus, 8–13 November 2010.
3. Sansoni, G.; Trebeschi, M.; Docchio, F. State-of-the-art and applications of 3D imaging sensors in industry, cultural heritage, medicine, and criminal investigation. *Sensors* **2009**, *9*, 568–601.
4. Hullo, J.F.; Grussenmeyer, P.; Fares, S. Photogrammetry and dense stereo matching approach applied to the documentation of the cultural heritage site of Kilwa (Saudi Arabia). In Proceedings of CIPA Symposium, Kyoto, Japan, 10–15 October 2009.
5. Grussenmeyer, P.; Alby, E.; Landes, T.; Koehl, M.; Guillemin, S.; Hullo, J.F.; Smigiel, E. Recording approach of heritage sites based on merging point clouds from high resolution photogrammetry and terrestrial laser scanning. *Int. Arch. Photogramm. Remote Sens. Spat. Inf. Sci.* **2012**, *XXXIX-B5*, 553–558.

467

6. El-Hakim, S.F.; Beraldin, J.A.; Picard, M.; Godin, G. Detailed 3D reconstruction of large-scale heritage sites with integrated techniques. *IEEE Comput. Graphic. Appl.* **2004**, *24*, 21–29.

7. Henry, P.; Krainin, M.; Herbst, E.; Ren, X.; Fox, D. RGB-D mapping: Using Kinect-style depth cameras for dense 3D modeling of indoor environments. *Int. J. Robot. Res.* **2012**, *31*, 647–663.

8. Freedman, B.; Shpunt, A.; Machline, M.; Arieli, Y. Depth Mapping Using Projected Patterns. Patent US 2010/0118123, 13 May 2010.

9. Mittet, M.-A.; Grussenmeyer, P.; Landes, T.; Yang, Y.; Bernard, N. Mobile outdoor relative localization using calibrated RGB-D cameras. In Proceedings of the 8th International Symposium on Mobile Mapping Technology, Tainan, Taiwan, 1–3 May 2013; pp. 2–7.

10. Dal Mutto, C.; Zanuttigh, P.; Cortelazzo, G.M. *Time-of-Flight cameras and Microsoft Kinect™*; Springer Science & Business Media: New York, NY, USA, 2012.

11. Curless, B. Overview of active vision techniques. In Proceedings of SIGGRAPH 2000 Course on 3D Photography, New Orleans, LA, USA, 23–28 July 2000.

12. May, S.; Droeschel, D.; Holz, D.; Fuchs, S.; Malis, E.; Nüchter, A.; Hertzberg, J. Three-dimensional mapping with time-of-flight cameras. *J. Field Robot.* **2009**, *26*, 934–965.

13. Henry, P.; Krainin, M.; Herbst, E.; Ren, X.; Fox, D. RGB-D mapping: Using depth cameras for dense 3D modeling of indoor environments. In Proceedings of the 12th International Symposium on Experimental Robotics, Delhi, India, 18–21 December 2010.

14. Dong, H.; Figueroa, N.; El Saddik, A. Towards consistent reconstructions of indoor spaces based on 6D RGB-D odometry and KinectFusion. In Proceedings of the IEEE International Conference Intelligent Robots and Systems, Chicago, IL, USA, 14–18 September 2014; pp. 1796–1803.

15. Kahn, S.; Bockholt, U.; Kuijper, A.; Fellner, D.W. Towards precise real-time 3D difference detection for industrial applications. *Comput. Ind.* **2013**, *64*, 1115–1128.

16. Hussmann, S.; Ringbeck, T.; Hagebeuker, B. A performance review of 3D TOF vision systems in comparison to stereo vision systems. In *Stereo Vision*; Bhatti, A., Ed.; InTech: Vienna, Austria, 2008; pp. 103–120.

17. Cavagnini, G.; Sansoni, G.; Trebeschi, M. Using 3D range cameras for crime scene documentation and legal medicine. *Proc. SPIE* **2009**, *7239*.

18. Mittet, M.A.; Landes, T.; Grussenmeyer, P. Localization using RGB-D cameras orthoimages. *Int. Arch. Photogramm. Remote Sens. Spat. Inf. Sci.* **2014**, *XL-5*, 425–432.

19. Bakkay, M.C.; Arafa, M.; Zagrouba, E. Dense 3D SLAM in dynamic scenes using Kinect. In Proceedings of 7th Iberian Conference on Pattern Recognition and Image Analysis, Santiago de Compostela, Spain, 17–19 June 2015; pp. 121–129.

20. Shotton, J.; Sharp, T.; Kipman, A.; Fitzgibbon, A.; Finocchio, M.; Blake, A.; Moore, R. Real-time human pose recognition in parts from single depth images. *Commun. ACM* **2013**, *56*, 116–124.

21. Keskin, C.; Kıraç, F.; Kara, Y.E.; Akarun, L. Real time hand pose estimation using depth sensors. In *Consumer Depth Cameras for Computer Vision*; Fossati, A., Gall, J., Grabner, H., Ren, X., Konolige, K., Eds.; Springer: London, UK, 2013; pp. 119–137.

22. Kuznetsova, A.; Leal-Taixé, L.; Rosenhahn, B. Real-time sign language recognition using a consumer depth camera. In Proceedings of 2013 IEEE International Conference on Computer Vision Workshops (ICCVW), Sydney, Australia, 2–8 December 2013; pp. 83–90.

23. Hsieh, C.T. A new Kinect-based scanning system and its application. *Appl. Mech. Mater.* **2015**, *764*, 1375–1379.

24. Amon, C.; Fuhrmann, F.; Graf, F. Evaluation of the spatial resolution accuracy of the face tracking system for Kinect for windows v1 and v2. In Proceedings of the 6th Congress of the Alps-Adria Acoustics Association, Graz, Austria, 16–17 October 2014.

25. Butkiewicz, T. Low-cost coastal mapping using Kinect v2 time-of-flight cameras. In Proceedings of 2014 Oceans-St. John's, St. John's, NL, Canada, 14–19 September 2014; pp. 1–9.

26. Lahamy, H.; Lichti, D.; El-Badry, M.; Qi, X.; Detchev, I.; Steward, J.; Moravvej, M. Evaluating the capability of time-of-flight cameras for accurately imaging a cyclically loaded beam. *Proc. SPIE* **2015**, *9528*.

27. Guidi, G.; Remondino, F.; Morlando, G.; Del Mastio, A.; Uccheddu, F.; Pelagotti, A. Performances evaluation of a low cost active sensor for cultural heritage documentation. In Proceedings of the 8th Conference on Optical 3D Measurement Techniques, Zurich, Switzerland, 9–12 July 2007.

28. UCSD Researchers Modify Kinect Gaming Device to Scan in 3D. Available online: http://ucsdnews.ucsd.edu/archive/newsrel/general/2011_07arkinect.asp (accessed on 28 June 2015).

29. Richards-Rissetto, H.; Remondino, F.; Agugiaro, G.; Robertsson, J.; von Schwerin, J.; Girardi, G. Kinect and 3D GIS in archaeology. In Proceedings of 18th International Conference on Virtual Systems and Multimedia, Milan, Italy, 2–5 September 2012; pp. 331–337.

30. Menna, F.; Remondino, F.; Battisti, R.; Nocerino, E. Geometric investigation of a gaming active device. *Proc. SPIE* **2011**, *8085*.

31. Quintana, M.; Zvietcovich, F.; Castaneda, B. 3D reconstruction of archaeological walls using Kinect. In Proceedings of the EUROMED 2014 Conference, Lemessos, Cyprus, 3–8 November 2014; pp. 47–58.

32. Newcombe, R.A.; Izadi, S.; Hilliges, O.; Molyneaux, D.; Kim, D.; Davison, A.J.; Fitzgibbon, A. KinectFusion: Real-time dense surface mapping and tracking. In Proceedings of 10th IEEE International Symposium on Mixed and Augmented Reality, Basel, Switzerland, 26–29 October 2011; pp. 127–136.

33. Lachat, E.; Macher, H.; Landes, T.; Grussenmeyer, P. Assessment of the accuracy of 3D models obtained with DSLR camera and Kinect v2. *Proc. SPIE* **2015**, *8528*.

34. Foix, S.; Alenya, G.; Torras, C. Lock-in time-of-flight (ToF) cameras: A survey. *Sens. J.* **2011**, *11*, 1917–1926.

35. Lefloch, D.; Nair, R.; Lenzen, F.; Schäfer, H.; Streeter, L.; Cree, M.J.; Kolb, A. Technical foundation and calibration methods for time-of-flight cameras. In *Time-of-Flight and Depth Imaging. Sensors, Algorithms, and Applications*; Grzegorzek, M., Theobalt, C., Koch, R., Kolb, A., Eds.; Springer: Berlin/Heidelberg, Germany, 2013; pp. 3–24.

36. Khoshelham, K. Accuracy analysis of Kinect depth data. *Int. Arch. Photogramm. Remote Sens. Spat. Inf. Sci.* **2011**, *XXXVIII-5*, 133–138.

37. Gonzalez-Jorge, H.; Rodríguez-Gonzálvez, P.; Martínez-Sánchez, J.; González-Aguilera, D.; Arias, P.; Gesto, M.; Díaz-Vilariño, L. Metrological comparison between Kinect I and Kinect II sensors. *Measurement* **2015**, *70*, 21–26.

38. Lindner, M.; Schiller, I.; Kolb, A.; Koch, R. Time-of-flight sensor calibration for accurate range sensing. *Comput. Vis. Image Underst.* **2010**, *114*, 1318–1328.

39. Kahlmann, T.; Remondino, F.; Ingensand, H. Calibration for increased accuracy of the range imaging camera Swissranger^{TM}. *Int. Arch. Photogramm. Remote Sens. Spat. Inf. Sci.* **2006**, *XXXVI-5*, 136–141.

40. Schiller, I.; Beder, C.; Koch, R. Calibration of a PMD-camera using a planar calibration pattern together with a multi-camera setup. *Int. Arch. Photogramm. Remote Sens. Spat. Inf. Sci.* **2008**, *XXXVII-3a*, 297–302.

41. Lindner, M.; Kolb, A. Calibration of the intensity-related distance error of the PMD ToF-camera. *Proc. SPIE* **2007**, *6764*.

42. Lindner, M.; Kolb, A. Lateral and depth calibration of pmd-distance sensors. In *Advances in Visual Computing: Second International Symposium, ISVC 2006 Lake Tahoe, NV, USA, November 6-8, 2006. Proceedings, Part II*; Springer: Berlin/Heidelberg, Germany, 2006; pp. 524–533.

43. Lenzen, F.; Kim, K.I.; Schäfer, H.; Nair, R.; Meister, S.; Becker, F.; Theobalt, C. Denoising strategies for time-of-flight data. In *Time-of-Flight and Depth Imaging. Sensors, Algorithms, and Applications*; Grzegorzek, M., Theobalt, C., Koch, R., Kolb, A., Eds.; Springer: Berlin/Heidelberg, Germany, 2013; Volume 8200, pp. 25–45.

44. Guomundsson, S.A.; Aanæs, H.; Larsen, R. Environmental effects on measurement uncertainties of time-of-flight cameras. In Proceedings of IEEE International Symposium on Signals, Circuits and Systems (ISSCS), Iasi, Romania, 13–14 July 2007; Volume 1, pp. 1–4.

45. Chow, J.; Ang, K.; Lichti, D.; Teskey, W. Performance analysis of a low-cost triangulation-based 3D camera: Microsoft Kinect system. *Int. Arch. Photogramm. Remote Sens. Spat. Inf. Sci.* **2012**, *XXXIX-B5*, 175–180.

46. Kinect for Windows. Available online: https://www.microsoft.com/en-us/kinectforwindows/default.aspx (accessed on 28 June 2015).

47. Grzegorzek, M.; Theobalt, C.; Koch, R.; Kolb, A. *Time-of-Flight and Depth Imaging. Sensors, Algorithms and Applications*; Springer: Berlin/Heidelberg, Germany, 2013; Volume 8200.

48. Hansard, M.; Lee, S.; Choi, O.; Horaud, R.P. *Time-of-flight cameras: Principles, Methods and Applications*; Springer: London, UK, 2012.

49. Kolb, A.; Koch, R. *Dynamic 3D Imaging*; Springer: Berlin/Heidelberg, Germany, 2009; Volume 5742.

50. Remondino, F.; Stoppa, D. *TOF Range-Imaging Cameras*; Springer: Heidelberg, Germany, 2013.

51. Chiabrando, F.; Chiabrando, R.; Piatti, D.; Rinaudo, F. Sensors for 3D imaging: Metric evaluation and calibration of a CCD/CMOS time-of-flight camera. *Sensors* **2009**, *9*, 10080–10096.

52. Lachat, E.; Macher, H.; Mittet, M.A.; Landes, T.; Grussenmeyer, P. First experiences with Kinect v2 sensor for close range 3D modelling. *Int. Arch. Photogramm. Remote Sens. Spat. Inf. Sci.* **2015**, *XL-5/W4*, 93–100.

53. Herrera, C.; Kannala, J.; Heikkilä, J. Accurate and practical calibration of a depth and color camera pair. In Proceedings of 14th International Conference on Computer Analysis of Images and Patterns, Seville, Spain, 29–31 August 2011; pp. 437–445.

54. Kinect Calibration. Available online: http://nicolas.burrus.name/index.php/Research/Kinect (accessed on 28 June 2015).

55. Hansard, M.; Horaud, R.; Amat, M.; Evangelidis, G. Automatic detection of calibration grids in time-of-flight images. *Comput. Vis. Image Underst.* **2014**, *121*, 108–118.

56. Camera Calibration Toolbox for Matlab. Available online: http://www.vision.caltech.edu/bouguetj/calib_doc/ (accessed on 28 June 2015).

57. Kim, C.; Yun, S.; Jung, S.W.; Won, C.S. Color and depth image correspondence for Kinect v2. In *Advanced Multimedia and Ubiquitous Engineering*; Park, J., Chao, H.C., Arabnia, H., Yen, N.Y., Eds.; Springer: Berlin/Heidelberg, Germany, 2015; Volume 352, pp. 111–116.

58. Bueno, M.; Díaz-Vilariño, L.; Martínez-Sánchez, J.; González-Jorge, H.; Lorenzo, H.; Arias, P. Metrological evaluation of KinectFusion and its comparison with Microsoft Kinect sensor. *Measurement* **2015**, *73*, 137–145.

59. Besl, P.J.; McKay, N.D. Method for registration of 3-D shapes. *Proc. SPIE* **1992**, *1611*.

60. Pagliari, D.; Menna, F.; Roncella, R.; Remondino, F.; Pinto, L. Kinect fusion improvement using depth camera calibration. *Int. Arch. Photogramm. Remote Sens. Spat. Inf. Sci.* **2014**, *XL-5*, 479–485.

61. Zennaro, S.; Munaro, M.; Milani, S.; Zanuttigh, P.; Bernardi, A.; Ghidoni, S.; Menegatti, E. Performance evaluation of the 1st and 2nd generation Kinect for multimedia applications. In Proceedings of IEEE International Conference on Multimedia and Expo, Torino, Italy, 29 June–3 July 2015; pp. 1–6.

62. Bay, H.; Tuytelaars, T.; Van Gool, L. Surf: Speeded up robust features. In *Computer Vision–ECCV 2006*; Leonardis, A., Bischof, H., Pinz, A., Eds.; Springer: Berlin Heidelberg, Germany, 2006; Volume 3951, pp. 404–417.

63. Project Tango. Available online: https://www.google.com/atap/project-tango/ (accessed on 28 June 2015).

64. Structure Sensor. Available online: http://structure.io/ (accessed on 24 August 2015).

Automatic Geometry Generation from Point Clouds for BIM

Charles Thomson and Jan Boehm

Abstract: The need for better 3D documentation of the built environment has come to the fore in recent years, led primarily by city modelling at the large scale and Building Information Modelling (BIM) at the smaller scale. Automation is seen as desirable as it removes the time-consuming and therefore costly amount of human intervention in the process of model generation. BIM is the focus of this paper as not only is there a commercial need, as will be shown by the number of commercial solutions, but also wide research interest due to the aspiration of automated 3D models from both Geomatics and Computer Science communities. The aim is to go beyond the current labour-intensive tracing of the point cloud to an automated process that produces geometry that is both open and more verifiable. This work investigates what can be achieved today with automation through both literature review and by proposing a novel point cloud processing process. We present an automated workflow for the generation of BIM data from 3D point clouds. We also present quality indicators for reconstructed geometry elements and a framework in which to assess the quality of the reconstructed geometry against a reference.

Reprinted from *Remote Sens.* Cite as: Thomson, C.; Boehm, J. Automatic Geometry Generation from Point Clouds for BIM. *Remote Sens.* **2015**, *7*, 11753–11773.

1. Introduction

The main techniques for capturing 3D data about the built environment from terrestrial platforms are through laser scanning or total station measurements. The latter being the predominant method for building surveys where a predetermined set of point measurements are taken of features from which 2D CAD plans are produced. Terrestrial laser scanning has been the technology of choice for the 3D capture of complex structures that are not easily measured with the sparse but targeted point collection from a total station since the technology was commercialised around 2000. This includes architectural façades with very detailed elements and refineries or plant rooms where the nature of the environment to be measured makes traditional workflows inefficient. This is particularly exemplified in the increased use of freeform architecture by prominent architects such as Frank Gehry and Zaha Hadid [1] where laser scanning presents the most viable option for timely data capture of complex forms.

Building Information Modelling (BIM) is the digital data flow surrounding the lifecycle of an asset or element of the built environment, instigated to provide better information management to aid with decision making. As a process, BIM has been gaining global acceptance across the Architecture, Engineering, Construction, and Operations (AECO) community for improving information sharing about built assets. A key component of this is a data-rich object-based 3D parametric model that holds both geometric and semantic information. By creating a single accessible repository of data, then other tools can be utilised to extract useful information about the asset for various purposes.

Although BIM has been extensively studied from the new build process, it is in retrofit where it is likely to provide the greatest impact. In the UK alone, at least half of all construction by cost is on existing assets [2]. With the need to achieve international environmental targets and construction being one of the largest contributors of CO_2 emissions in the UK, sustainable retrofit is only going to become more relevant. This is supported by the estimate of the UK Green Building Council that of total building stock in the UK, the majority will still exist in 2050 [3]. Therefore, many existing buildings will need to be made more environmentally efficient if the Government is to reach its sustainability targets.

With the introduction of BIM and the data-rich 3D parametric object model at its heart, laser scanning has come to the fore as the primary means of data capture. This has been aided by both the US and UK Governments advising that laser scanning should be the capture method of choice for geometry [4,5]. However, little thought has been given about how to integrate this in to the BIM process due to the change in the nature of the information requirements of a BIM model and uncertainty over level of detail or information that should be provided by a Geomatic Land Surveyor. It has been proposed that a point cloud represents an important lowest level of detail base (stylised as LoD 0) from which more information rich abstractions can be generated representing higher levels of detail [6].

Traditional surveying with scanning currently does not result in a product that is optimal for the process of BIM due to the historical use of non-parametric CAD software to create survey plans. Therefore, a process shift is required in workflows and modelling procedures of the stakeholders who do this work to align themselves with this. The shorthand name given to the survey process of capture to model is Scan to BIM. Technically Scan to BIM as a phrase is wrongly formed as the end result is not BIM as usually understood, *i.e.* the process, but a 3D parametric object model that aids the process at its current level of development.

Even though, from a BIM perspective, creating parametric 3D building models from scan data appears new, it actually extends back to the early days of commercialised terrestrial laser scanning systems creating parameterised surface representations from segmented point clouds [7] and goes back further than this in the aerial domain for external parametric reconstruction [8,9]. A system of note from the close-range photogrammetry domain is Hazmap; originally developed to facilitate the capture and parametric modelling of complex nuclear plants in the 1990s [10]. Hazmap consisted of a panoramic imaging system using calibrated cameras attached to a robotic total station. This would capture 60 images per setup and use a full bundle adjustment together with total station measurements for scale to localise the sensor setup positions. After capture, the system made use of a plant design and management system (PDMS) interface that allowed the user to take measurements in the panoramic imagery and export them via a macro to the PDMS where the plant geometry could be modelled using a library of parametric elements.

One of the earliest pieces of research with a workflow that would be recognised today as scan to BIM is in [11]. Their key conclusion was to consider what tolerance is acceptable both in surveying and modelling as assuming orthogonality is rarely true in retrofit but may be desirable to simplify the modelling process.

1.1. Standards for Modelling

The survey accuracy requirement set by the Royal Institution of Chartered Surveyors (RICS) was for 4cm accuracy for building detail design at a drawing resolution of 20 cm [12]. However, in a 3D modelling context, drawing resolution now seems less relevant and so in 2014 an updated guidance note was published splitting accuracy into plan and height with measured building surveys banded between ±4–25 mm depending on job specification [13]. We choose the RICS over other guidelines such as those for the survey of historic buildings, as our work focuses more on the modelling of contemporary buildings where RICS is more relevant.

Before this updated guidance, a UK survey specification for a BIM context did not exist. Therefore, survey companies took it upon themselves to create in-house guides. The most comprehensive of these is by Plowman Craven who freely released their specification, focused around the parametric building modeller Revit, and documenting what they as a company will deliver in terms of the geometric model [14].

One of the immediate impressions of this document is the number of caveats that it contains with respect to the geometry and how the model deviates from reality. This is partially due to the reliance on Revit and the orthogonal design constraints that this encourages, meaning that representing unusual deviations that exist in as-built documentation have to be accounted for in this way; unless very time consuming (and therefore expensive) bespoke modelling is performed. This experience is borne out by literature where the tedium of modelling unique components [11] and the unsuitability of current BIM software to represent irregular geometry such as walls out of plumb [15] are recognised. However Plowman Craven, as outlined in their specification above, do make use of the availability of rich semantic detail to add quality information about deviations from the point cloud to the modelled elements. In the medium to long term, the establishment of the point cloud as a fundamental data model is likely to happen as models and the data they are derived from start to exist more extensively together in a BIM environment.

Larsen *et al.* [15] endorses this view and considers that the increased integration of point clouds into BIM software makes post processing redundant. However, they then contradict this by envisaging that a surveyor provides a "registered, cleaned and geo-referenced point cloud". By post processing, Larsen *et al.* appear to be referring to the modelling process and see the filtering and interpretation of data to be obsolete with the point cloud a "mould" that other professions in the lifecycle can make use of as necessary.

1.2. Current State of the Art

Automated modelling is seen as desirable commercially to reduce time and therefore cost and make scanning a more viable proposition for a range of tasks in the lifecycle, such as daily construction change detection [16,17].

Generally, digital modelling is carried out to provide a representation or simulation of an entity that does not exist in reality. However for existing buildings, the goal is to model entities as they exist in reality. Currently the process is very much a manual one and recognised by many as being time-consuming, tedious, subjective and requiring skill [11,18]. The general manual process as in creating 2D CAD plans, from point clouds requires the operator to use the cloud as a guide in a BIM tool to effectively trace around the geometry, requiring a high knowledge input to interpret the scene as well as add the rich semantic information that really makes BIM a valuable process.

The orthogonal constraints present in many BIM design tools limit the modelling that can be achieved without intense operator input. Depending on the type and use of the model this is not necessarily a disadvantage. In many cases a geometric representation is not required to have very tight tolerances [7]. This further emphasises the need to define fitness for purpose and we have given some examples of UK standards and specifications above.

1.2.1. Research

Both Computer Science and Geomatics are investigating the automated reconstruction of geometry from point clouds, especially as interior modelling has risen in prominence with the shift to BIM requiring rich parametric models. Geomatics has a track record in this with reconstruction from facades, pipework and from aerial LIDAR data as in [19,20]. An early review of methods is given by [21].

The ideas and approaches taken to aiding the problem of geometry reconstruction have mainly come from computer science. This community uses parametric modelling as a paradigm mainly implemented for invented or stylised representations of the externals of buildings using techniques such as procedural modelling and grammars, *i.e.* algorithms to generate the model [22]. This rule-based approach to automating the modelling process can fit well to the parametric models that are intrinsically rule driven; [23] presents this approach. Rules can also be represented by shape grammars as shown by [24].

The focus of computer science on BIM has mainly been on algorithms to speed the modelling of geometry from point clouds, as well as applying other vision techniques from robotics for scene understanding. This is all related to automating the understanding of the environment, which is an important prerequisite for providing robots with autonomy [25].

In terms of the reconstruction of building elements, the focus has been on computational geometry algorithms to extract the 3D representation of building elements through segmentation, including surface normal approaches [26], plane sweeping [27] and region growing [28]. Segmentation of range measurement data is a long established method (initially from computer vision for image processing) for classifying data with the same characteristics together. An example of this is Hoover *et al.* [29],which brought together the different approaches to this topic that were being pursued at the time and presented a method for evaluating these segmentation algorithms.

Existing work that has shown promising results towards automating the reconstruction process of geometry include [30,31], however these do not result in a parametric object-based model as used in BIM but in a 3D CAD model that needs to be remodelled manually, a point made by Volk *et. al.* [32] who provide an extensive review of the area.

However Nagel *et al.* [33] points out that the full automatic reconstruction of building models has been a topic of research for many groups over the last 25 years with little success to date. They suggest the problem is with the high reconstruction demands due to four issues: definition of a target structure that covers all variations of building, the complexity of input data, ambiguities and errors in the data, and the reduction of the search space during interpretation.

1.2.2. Commercial

Given the above statement about achieving full automation, there are a few commercial pieces of software that have emerged in recent years and could be described as semi-automated. To the best of the authors' knowledge all these tools rely on Autodesk Revit for the geometry generation. Below the prominent packages are summarised.

The first is by ClearEdge3D who provides solutions for plant and MEP object detection alongside a building-focused package called Edgewise Building. This classifies the point cloud into surfaces that share coplanar points, with the operator picking floor and ceiling planes to constrain the search for walls [34]. Once found, this geometry can be bought into Revit via a plugin to construct the parametric object-based geometry. In its wall detection, Edgewise uses the scan locations to aid geometric reasoning; a constraint it forces by only allowing file-per-scan point clouds for processing.

The other main solution is Scan to BIM from IMAGINiT Technologies, which is perhaps the most successful solution in terms of deliverable [35]. This is a plugin to Revit and therefore relies on much of the functionality of Revit to handle most tasks (including loading the point cloud and geometry library) and essentially just adds some detection and fitting algorithms along with a few other tools for scan handling. The main function is wall fitting whereby the user picks three points to define the wall plane from which a region growing algorithm detects the extents. The user then sets a tolerance and selects which parametric wall type element in the Revit model should be used. There is also the option of fitting a mass wall, which is a useful way of modelling a wall face that is not perfectly plumb and orthogonal. The downside to this plugin is that it only handles definition by one surface meaning that one side of a wall has to be relied upon to model the entire volume, unless one fitted a mass wall from each surface and did a Boolean function to merge the two solids appropriately.

Kubit, now owned by Faro, and Pointcab both provide tools that aid the manual process of tracing the points in Autodesk software but do not, as of this time, automate the geometry production [36,37]. Autodesk itself did trial its own Revit module for automated building element creation from scan data, which it shared with users through its Autodesk Labs preview portal. However, this module is no longer available and it remains to be seen if this will ever be integrated into future production editions of Revit.

2. Proposed Method

In the following, we describe our methods to automate the identification of geometric objects from point clouds and vice versa. We concentrate on the major room bounding entities, *i.e.*, walls. Other more detailed geometric objects such as windows and doors are not currently considered.

2.1. Overview

Two methods are presented in this paper, one to automatically reconstruct basic Industry Foundation Classes (IFC) geometry from point clouds and another to classify a point cloud given an existing IFC model. The former consists of three main components:

(1)   Reading the data into memory for processing
(2)   Segmentation of the dominant horizontal and vertical planes
(3)   Construct the IFC geometry

While the latter consists of two:

(1)   Reading the IFC objects bounding boxes from the file
(2)   Use the bounding boxes to segment the point cloud by object

By using E57 for the scans and IFC for the intelligent BIM geometry, this work is kept format agnostic, as these are widely accepted open interchange formats, unlike the commercial solutions which overly rely on Autodesk Revit for their geometry creation. IFC was developed out of the open CAD format STEP, and uses the EXPRESS schemata to form an interoperable format for information about buildings. This format is actively developed as a recognised open international standard for BIM data: ISO 16739 [38]. Within IFC, building components are stored as instances of objects that contain data about themselves. This data includes geometric descriptions (position relative to building, geometry of object) and semantic ones (description, type, relation with other objects) [39].

This work makes use of two open source libraries as a base from which the routine presented has been built up. The first is the Point Cloud Library (PCL) version 1.7.0, which provides a number of data handling and processing algorithms for point cloud data [40]. The second is the eXtensible Building Information Modelling (xBIM) toolkit version 2.4.1.28, which provides the ability to read, write and view IFC files compliant with the IFC2x3 TC1 standard [41].

2.2. Point Cloud to IFC

2.2.1. Reading In

Loading the point cloud data into memory is the first step in the process. To keep with the non-proprietary, interoperable nature of BIM the E57 format was decided as the input format of choice to support. The LIBE57 library version 1.1.312 provided the necessary reader to interpret the E57 file format [42] and some code was written to transfer the E57 data into the PCL point cloud data structure. In this case, only the geometry was needed so only the coordinates were taken into the structure.

2.2.2. Plane Model Segmentation

With the data loaded in, the processing can begin. Firstly the major horizontal planes are detected as these likely represent the floor and ceiling components and then the vertical planes which likely represent walls (Figure 1b,c). The plane detection for both cases is done with the PCL implementation of RANSAC (RANdom SAmple Consensus) [43] due to the speed and established nature. The algorithm is constrained to accept only planes whose normal coefficient is within a three degree deviation from parallel to the Z-axis (up) for horizontal planes and perpendicular to Z for vertical planes. Also a distance threshold was set for the maximum distance of the points to the plane to accept as part of the model. Choosing this value is related to the noise level of the data from the instrument that was used for capture. As a result, the stopping criteria for each RANSAC run is when a plane is found with 99% confidence consisting of the most inliers within tolerance. This is an opportunistic or "greedy" approach, based on the assumption that the largest amount of points that most probably fit a plane will be the building element. This approach can lead to errors especially where the plane is more ambiguous, but is fast and simple to implement. Recent developments could improve this such as Monszpart *et al.* [44], which provides a formulation that allows less dominant planes to not become lost in certain scenes.

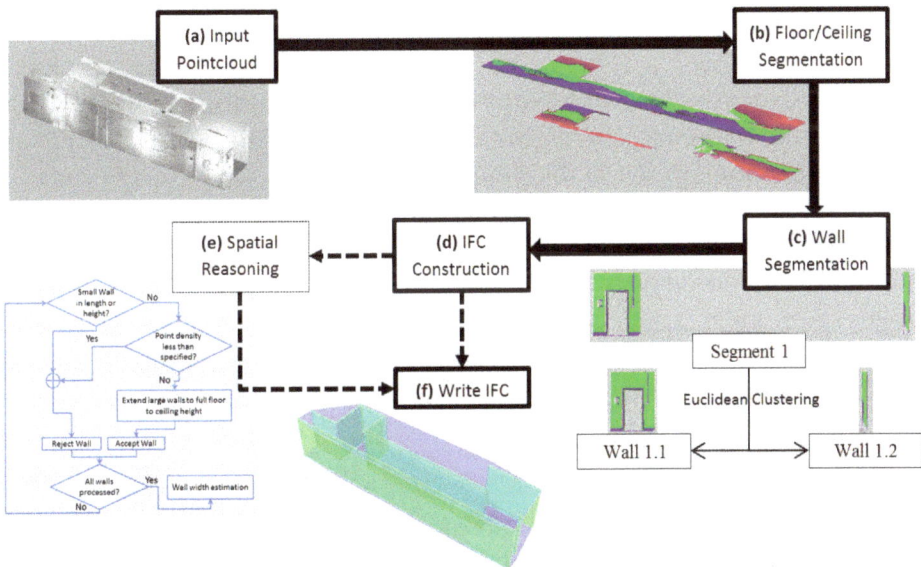

Figure 1. Flowchart of Point Cloud to IFC algorithm steps. (**a**) Load Point Cloud; (**b**) Segment the Floor and Ceiling Planes; (**c**) Segment the Walls and split them with Euclidean Clustering; (**d**) Build IFC Geometry from Point Cloud segments; (**e**) (Optional) Spatial reasoning to clean up erroneous geometry; (**f**) Write the IFC data to an IFC file.

The downside to extracting points that conform to a planar model using RANSAC alone is that the algorithm extracts all of the points within tolerance across the whole plane model, irrelevant of whether they form part of a contiguous plane. Therefore a Euclidean Clustering step [45] was introduced after the RANSAC to separate the contiguous elements out, as it could not be assumed that two building elements that shared planar coefficients but were separate in the point cloud represented the same element (Figure 1c). Euclidean Clustering separates the point data into sets based on their distance to each other up to a defined tolerance. Along with this, a constraint or condition was applied to preserve planarity, preventing points that lay on the plane but formed part of a differently oriented surface from being included. The constraint manifested as the dot product between the normal ranges of each point added to the clusters had to be close to parallel. Accepted clusters to extract are chosen based on the amount of points assigned to the cluster as a percentage of the total data set.

With the relevant points that represent dominant planes extracted, the requisite dimensions needed for IFC geometry construction could be measured. This meant extracting a boundary for the slabs and a length and height extent for the walls; the reasons for this are provided in the next section.

Once all clusters are extracted, further information for each is collected. For each slab's cluster, the points are projected onto the RANSAC-derived plane. The convex hull in the planar projection is calculated to give the coordinates of the boundary. The two corner points that describe the maximum extent of the cluster are computed to describe the wall (Figure 2). This finds the extent well along the X/Y direction (blue points of Figure 2), but the height of the wall cannot be guaranteed to be found by this method. This situation is illustrated in Figure 2, because parts of the wall can extend further than the places of maximum and minimum height. To counter this, the minimum and maximum Z coordinates (height) were sought by sorting all the coordinates and returning both the lowest and the highest values. By using these Z values, an overall wall extent can be defined (black points of Figure 2).

Figure 2. Example plot of minimum and maximum (black) and maximum segment (blue) coordinates on the points of a convex hull (red) calculated for a wall cluster point cloud (white).

2.2.3. IFC Generation

Each IFC object can be represented a few different ways (swept solid, brep, *etc.*) To create the IFC, the geometry of the elements needs to be constructing using certain dimensions. In this work the IFC object chosen for wall representation is IfcWallStandardCase, which handles all walls that are described by a vertically extruded footprint. The slab representation chosen is IfcSlab, defined similarly to the walls by extruding a 2D perimeter of coordinates vertically down by a value [46].

We start by creating an initializing an empty model into which the IFC objects can be added. Each object is created from the information extracted previously by the segmentation code detailed in the last section. First the slabs are added by providing a boundary, extrusion depth or thickness and the level in Z at which he slab is extruded from. The walls are represented by their object dimensions (length, width, height) and a placement coordinate and rotation (bearing) of that footprint in the global coordinate system. As the usefulness of BIM is as much about the semantic information alongside the geometry, mean and standard deviation information on the RANSAC plane fits is added as a set of properties to the geometry of walls and slabs.

2.2.4. Spatial Reasoning

An optional step of geometric reasoning can be added to the IFC generation process to clean up the reconstructed geometry (Figure 1e). This implements some rules or assumptions about the reconstructed elements to change or remove them from the model. The following rules have been implemented and can be customized by user-defined thresholds.

- Reject small walls: Planes that are too small in length or occupy too little of the height between floor and ceiling levels are removed from the model. For the experiments described here we chose 100 mm as the minimum length and 1/3 of the floor to ceiling distance as the minimum height.
- Extend large walls: Large planes which do not extend fully from floor to ceiling are automatically extended so their lowest height is at floor level and highest is at ceiling level. For the experiments we chose walls that are greater than 1m in length or occupy 2/3 of the height between floor and ceiling levels.
- Merge close planes with similar normal: Planes that have parallel normals and within an offset distance from each other are merged into one wall of overall thickness being the distance between the planes. If a plane pair is not found a default value of 100 mm thickness is used. For the experiments we chose a distance threshold of 300 mm. This currently relies on user input but a database of priors that would be expected in the building would streamline this.

Once successfully created with or without geometric reasoning applied the walls and slabs can be stored to the model and saved as an IFC file. Figure 3 shows an example of the resultant geometry that is obtained through both processes. This file can then be viewed in any IFC viewer or BIM design tools such as Autodesk Revit.

2.3. IFC to Point Cloud

The process described in the previous section of generating the geometry can, in effect, be reversed and the model used to extract elements out of the point cloud within a tolerance. In so doing, these segments of the point cloud can be classified and assessed for their quality of representation in the model based on the underlying point cloud data that occupies that geometric space. This process could easily be the start of a facilitation process towards 4D change detection as a project evolves on site against a scheduled design model for that time epoch.

Figure 3. Example of corridor reconstruction without (blue) and with (red) spatial reasoning applied.

2.3.1. Reading the IFC

We start by loading the IFC file into memory. Then, for each IFC object that is required, the coordinates of an object-oriented bounding box are extracted as the return geometry. A user specified tolerance can be added to take account of errors and generalisations made during modelling, thus enlarging the bounding boxes by that amount.

2.3.2. Classifying the Point Cloud

Once the bounding boxes are known, we can then extract the points within each box and colour them by type. This is performed with a 3D convex hull of the box which is then used to filter the dataset. Then, each segment can be written into the same E57 file as separate "scans" within the structure, each named after the IFC object they represent (Figure 4).

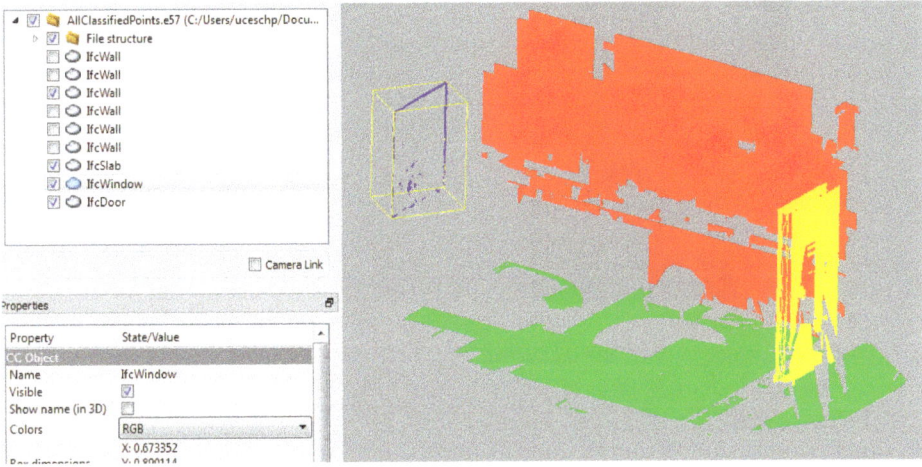

Figure 4. E57 point cloud with each element classified by IFC type (loaded in CloudCompare with four of nine elements visible).

3. Findings

3.1. Data

The following is a brief outline of the data used to test the methods developed in this paper. A fuller description for the dataset that forms a benchmark for indoor modelling can be found in [47]. The dataset is freely available to download at: http://indoor-bench.github.io/indoor-bench. The datasets used are both sections of the Chadwick Building at UCL, each captured with state of the art methods of static and mobile laser scanning and accompanied by a manually created IFC model. This represents a typical historical building in London that has had several retrofits over the years to provide various spaces for the changing nature of activities within the UCL department housed inside.

3.1.1. Basic Corridor

This first area is a long repetitive corridor section from the second floor of the building. It roughly measures 1.4 m wide by 13 m long with a floor to ceiling height of 3 m. The scene features doors off to offices at regular intervals and modern fluorescent strip lights standing proud of the ceiling as can be seen in Figure 5, along with the data captured with each instrument.

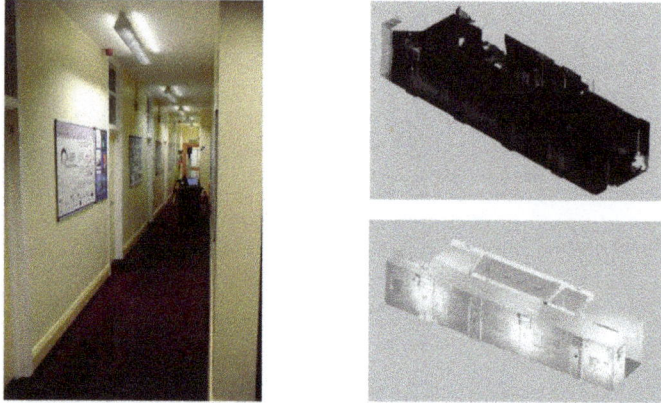

Figure 5. Image of corridor (**left**) and resulting point cloud data collected with a Viametris Indoor Mobile Mapping System (**top right**) and Faro Focus 3D S (**bottom right**).

3.1.2. Cluttered Office

The second indoor environment is a standard office from the modern retrofitted mezzanine floor of the Chadwick Building. It roughly measures 5 m by 3 m with floor to ceiling height of 2.8 m at its highest point. The environment contains many items of clutter, which occlude the structural geometry of the room including filing cabinets, air conditioning unit, shelving, chairs and desks as illustrated in Figure 6, along with the data captured with each instrument.

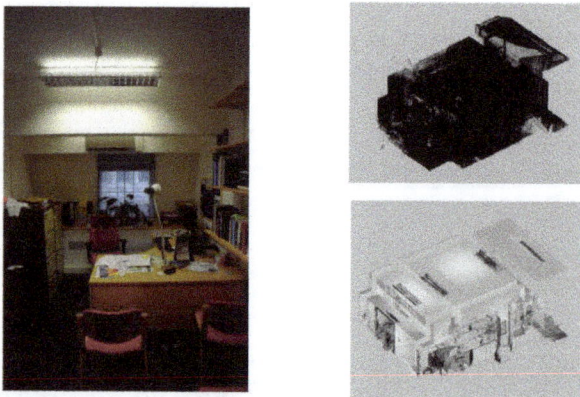

Figure 6. Image of office (**left**) and point cloud data collected with a Viametris Indoor Mobile Mapping System (**top right**) and Faro Focus 3D S (**bottom right**).

3.2. Results

The data outlined in the previous section was fed into the point cloud to IFC algorithm described earlier in this paper and the results are presented here. The results from this process are split into those that are qualitative and can be observed in relation to the reference human-made model and more quantitative results that put figures to the deviations seen in the form of a quality metric that acts as a discrepancy measure between reference and test datasets.

3.2.1. Quality Metric

To compare the reconstructed geometry to the reference model the following is performed. For each wall in the reference model, the reconstructed test wall with the nearest centroid in 2D is transformed into the local coordinate space (as plotted in Figure 7) of the reference that then can be measured for geometric deviation in relation to this coordinate space.

This quality metric is composed of three criteria:

- the Euclidean distance offset between reference (indicated by subscript 1) and automated (indicated by subscript 2) wall centroids

$$d = \sqrt{(x_1 - x_2)^2 + (y_1 - y_2)^2} \tag{1}$$

- the area formed by the absolute difference in magnitude of the wall in width and length

$$A = |length_1 - length_2| * |width_1 - width_2| \tag{2}$$

- and the Sine of the angular difference

$$\Delta = \sin(\alpha_1 - \alpha_2) \tag{3}$$

The distance d and the area A are normalised by dividing them by the maximum value. The quality metric is then computed by a weighted sum of the three normalised values

$$q = \sum w_1 * \bar{d} + w_2 * \bar{A} + w_3 * \Delta \tag{4}$$

In other words, these values from formulas 1, 2, and 3 represent the quality of the global placement, the object construction and the angular discrepancy, respectively. All of these values indicate a better quality detection and therefore success of the fit when they tend to 0. The metric is a weighted sum where we currently choose all the weights w_i to be $1/3$ to keep the metric within the range 0-1. A value under 0.1 of this measure can be generally considered as very good. An alternative measure that gives similar relative results to this quality metric is the Hausdorff Distance.

However, the ability to categorise the fit in our quality sum value by looking at the magnitude of the three components allows greater understanding and finer control through weighting than the measure provided by a Hausdorff Distance. The angular difference, for instance, could be more heavily weighted so that smaller angular changes have a greater impact on the final measure.

An example for the quality calculation using data comparing one of the corridor walls to the automatically extracted wall from static scan data is shown in Table 1 for a wall that is considered good by the quality metric and one that is considered almost ten times worse. To illustrate why these fits provide the quality values that they do, the wall geometries are plotted in Figure 7. The fit for wall 4, has been successfully recovered from both scan datasets looking at the plotted data, leading to good quality values, whereas wall 15 has a worse quality value by an order of magnitude as it is wrong in length by almost 2 m as illustrated in the plot. The value is low overall as its angular discrepancy is small against the reference, unlike wall 15 from the mobile data where the size and angle of the wall is wrong (Figure 7b) producing a larger quality value.

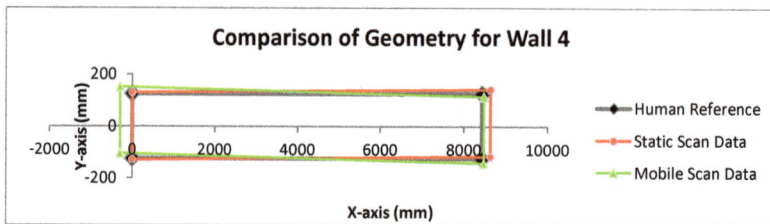

(a)

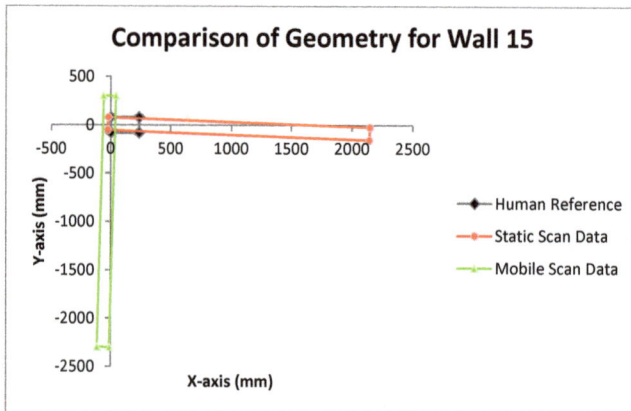

(b)

Figure 7. Plots of boundary placement of manual and automatically created geometry from the static scan data for (**a**) wall 4 and (**b**) wall 15 of the corridor dataset.

Table 1. Example of data from a wall comparison with quality metric for a well (wall 4) and poorly (wall 15) reconstructed case of the corridor dataset against the static scan derived geometry.

	Delta Centroid X (mm)	Delta Centroid Y (mm)	Angular Difference (Deg)	Delta Wall Length (mm)	Delta Wall Width (mm)	Quality Metric
Wall 4	102.9	5.8	179.9	209.6	249.6	0.029
Wall 15	944.6	38.3	177.3	1935.7	153.8	0.233

3.2.2. Point Cloud to IFC—Corridor

The majority of the space-bounding walls have been reconstructed as shown by Figure 8, with the central wall seeming to have the best construction from both static and mobile scan data. This is borne out by the quality measure (Figure 9) where this central wall (wall 4) scored the best. However, this wall does have a 200 mm over-extension in its reconstruction, representing an error of 2.3% of total length. This is due to extraneous noise outside the main wall affecting the maximum segment measurement for wall length extent as in Figure 10.

As edges are inferred as being the edge of the detected plane, then extraneous points from clutter or "split pixels" can affect the size of the recovered wall in the clustering stage as described above. Overall, the static scan data has allowed more walls to be recovered with greater confidence than the mobile data where the greater noise present in the data has created some ambiguity with multiple wall fits.

The spatial reasoning also appears to have been largely successful as all of the walls reconstructed from the static and a majority from the mobile data are actually walls in the reference with good approximations of width. The one wall from the mobile data that is incorrect runs along the Southeast edge of the plan view in Figure 8 and is caused by a planar façade to a long lectern that is geometrically close to a wall in planarity and point density. Figure 8 also shows that some of the walls detected from the static data have their normal pointing in the wrong direction leading to the wall lying on the wrong side of the wall surface found. This is a problem that is prevalent in almost every wall without a second side as there is no easy way to interpret the correct sidedness just from the geometry alone. This could be improved with geometric reasoning whereby the scan location is taken into account allowing ray casting to the points to attribute viewshed per location and therefore more robust normal prediction.

Table 2 provides some statistics about the walls that were well reconstructed from both sources of data. Length has generally been overestimated from both sets of data for the reasons described earlier with Figure 10. In the two components that are related to the walls' width (delta of the centroid in Y and wall width itself) there is little difference between each dataset. This is to be hoped for, as the width/Y placement in local coordinates is dependent on the RANSAC plane found from the data. The error in wall width stems from the inability to estimate the wall width for two of the walls as only one side was scanned.

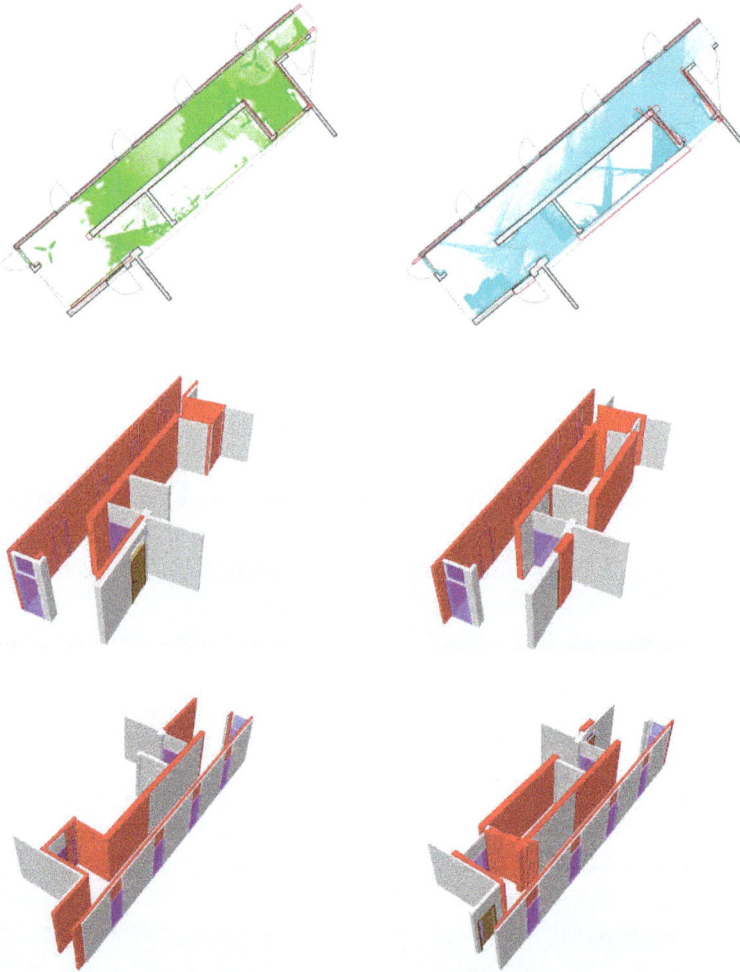

Figure 8. Extracted walls (red) of the corridor from the static (green) and mobile (blue) point cloud data overlaid on the human-generated model (grey/white).

Table 2. Statistics for the accuracy of the same reconstructed corridor walls considered good quality from both datasets (walls 3, 4 and 7) against the reference.

Statistics for Good Quality Corridor Walls	Delta Centroid in X RMS (mm)	Delta Centroid in Y RMS (mm)	Angular Difference RMS (Deg.)	Delta Wall Length RMS (mm)	Delta Wall Width RMS (mm)
Static Scan Data	180	43	0.20	371	186
Mobile Scan Data	256	45	0.50	401	186

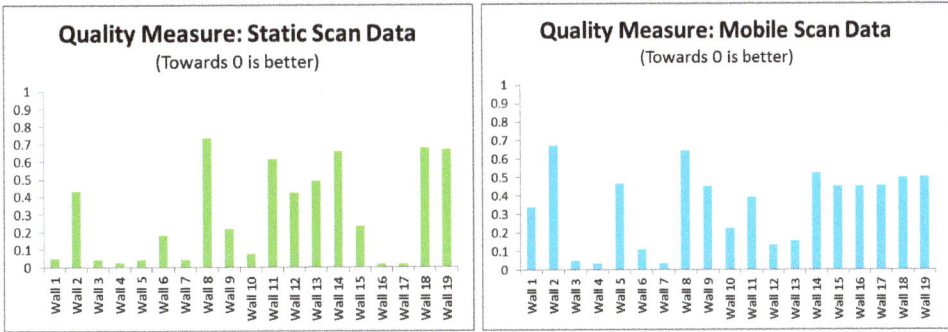

Figure 9. Charts of calculated reconstruction quality for each wall in the reference human-created model of the corridor against the automated geometry from the two scan datasets.

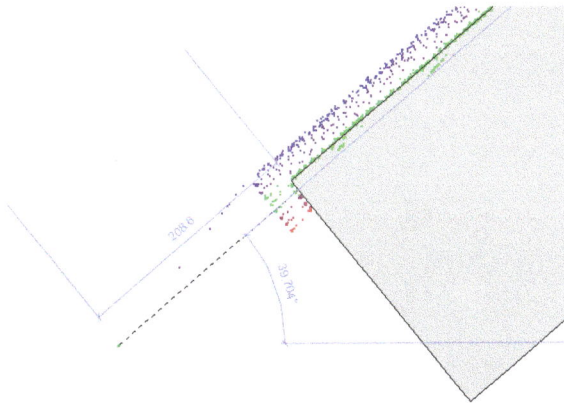

Figure 10. Noise that caused over-extension from the static corridor scan data shown against reference wall.

3.2.3. Point Cloud to IFC—Office

The main bounding walls of the office have been reconstructed with varying degrees of success. To aid with the description, starting at the wall with the door (Figure 11), the wall numbers going clockwise are 5, 4, 6, and 3. Walls 1 and 2 in the reference model are small height walls from the ceiling and floor respectively and can be seen at the end of the room above and below the window in Figure 6.

The geometry for walls 3 and 4, which have both sides captured in the static data, seems fairly good with an over-estimation of thickness due to the clutter present on those walls of the office. Wall 6 in the same data is well placed but with no backside scan data its width is incorrect, whereas in wall 5 the open door and chairs close to the outside face of the wall have affected the reconstruction. The extraneous walls from the clutter have survived due to their size being above the small wall limit, so the desk and filing cabinets are considered by the reasoning as partial walls, which is too simplistic, although correct for the first wall parallel to the window. A change to the algorithm to take the space of the room into account, e.g., maximum bounded area, may be a quick way to remove these surviving "walls" from clutter.

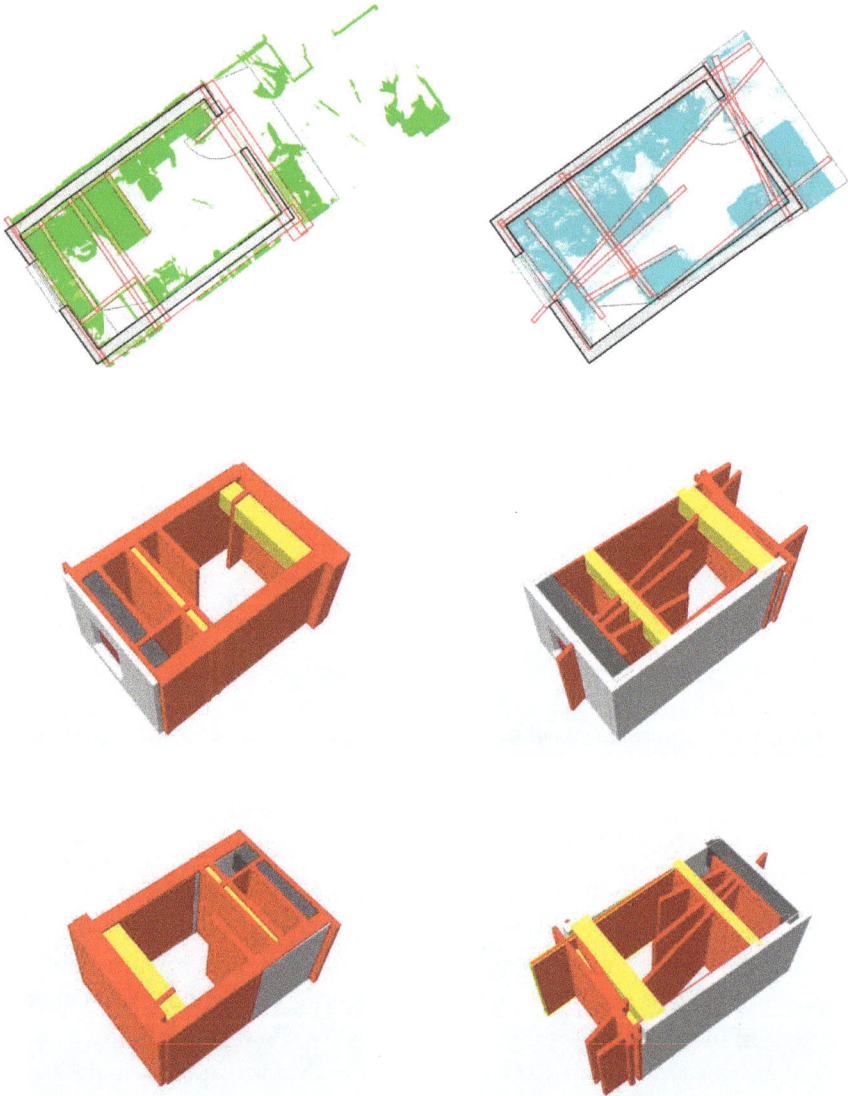

Figure 11. Extracted walls (red) of the office from the static (green) and mobile (blue) point cloud data overlaid on the human-generated model (grey/white).

The mobile scan data does not have the second side to most of the walls in the data meaning that the thicknesses could not be estimated; the exception being wall 5 where the same problems that affected the static data reconstruction appear here. However, fewer of the elements of clutter have been turned into extraneous walls, probably due to the lower resolution of the objects and higher noise level of the mobile point cloud. The fact that the door has led to a plane being successfully fitted through the whole office would suggest that the threshold values for the Euclidean Clustering were too high for data from this sensor.

The chart of the static quality scores shows a fairly even level of quality with wall 3 being "best"; Figure 11 shows it as the wall with the least incorrect estimation in both width and length so seems valid. By looking at the results of the quality metric in Figure 12, the patchiness of the reconstruction can be seen. The figure for the mobile data shows a gross error for wall 4, which has depressed the quality values for walls 5 and 6 more than their visual placement would imply would be acceptable as such a good quality fit.

Figure 12. Charts of calculated reconstruction quality for each wall in the reference human-created model of the office against the automated geometry from the two scan datasets.

Looking at the statistical data for the reconstruction of bounding walls that were not failed reconstructions as shown in Table 3, the trend is much the same as the data from the corridor test. The wall width RMS values are similar between datasets with the length having quite large RMS values from the mobile scan data. The width is for similar reasons highlighted in the same section for the corridor data, namely the inability to guess a wall thickness given one wall side captured. The length errors in this set appear mainly down to clutter and manifest more on one side of the wall from the mobile reconstruction as the centroid mean X coordinate is much larger. In addition, the higher noise from the mobile scanners tied with the greater clutter in the scene has increased the ambiguity in the detection leading to increased errors.

Table 3. Statistics for the accuracy of three of the main four bounding office walls from both datasets (walls 3, 5 and 6) against the reference model; wall 4 is ignored for the gross error from the mobile data detection as seen in Figure 12.

Statistics for Good Quality Office Walls	Delta Centroid in X RMS (mm)	Delta Centroid in Y RMS (mm)	Angular Difference RMS (Deg.)	Delta Wall Length RMS (mm)	Delta Wall Width RMS (mm)
Static Scan Data	159	101	0.47	539	204
Mobile Scan Data	378	100	0.45	602	206

4. Conclusions

The work presented in this paper has shown the applicability and limits to full automated reconstruction of object-based "intelligent" BIM geometry from point clouds in a format agnostic way. There has been partial success towards the aim of fully automatic reconstruction, especially where the environment is simple and not cluttered. Where both sides of a wall are present, the reconstruction has tended to be more reliable from both data sources. However, clutter, as is usual in the indoor environment, does have an effect. This was shown by the planar model presented here being supported by dense wall hosted or connected elements such as shelving or filing cabinets. A way to detect this data through more involved scene understanding would help. A method of doing this may be to take the intensity return or colour data present in the scans into account.

Clutter in the environment also has another effect and that is to hide or shadow the building features that need to be constructed. This is mitigated to a degree in the field by good survey design, but point clouds generated by imaging systems will, by their nature, suffer from occlusions somewhere in the dataset. This is something the routine presented here is affected by and unable to overcome in its current state e.g., if an alcove is hidden by an open door.

Computing power is an inherent problem when handling point cloud datasets and the approach presented here exacerbates it by relying on RAM for fast access to the whole dataset for processing. Downsampling is usually performed to keep the data manageable and could be applied in this case for very large datasets, however cloud computing presents a tangible opportunity to reduce the effect of this limitation by providing a scalable amount of computing power per process.

The data capture method does not seem to have a large bearing on the success of reconstruction except in the cluttered scene where the reduced resolution and more even sampling of the clutter objects in the mobile scan seems to prevent as many extraneous planes surviving the spatial reasoning step and affecting the geometry construction.

The question of what represents a good and bad reconstruction is crucial for comparisons and benchmarking. To provide a way of quantifying this, a quality measure has been developed based on the placement, size and angular discrepancies in relation to a reference model. This is not only useful for scoring automatically reconstructed geometry but could also prove useful when comparing an as-built model from scans against a base design model for verification throughout the construction phase of a project.

The logic required to make decisions about whether a certain geometry configuration is feasible already exists in commercial packages for design model verification (e.g., Solibri Model Checker [48]). It could easily be envisaged that this set of existing rules could be applied to automated models, as they are currently used to assess human-generated ones for building regulation infringements.

In terms of the scan-to-BIM workflow this work could be seen as a supplement to the semi-automatic construction of the bounding structural features of a building. For survey purposes the attribution of fit data is useful for quality assurance. Other stakeholders may require other semantic information that can be added to the IFC geometry created by this process or swapped out with more specific elements if required. Irregular and complex shapes require a more involved process and it will depend on fitness for purpose as to whether the recovery of an exact geometric description is worthwhile. Heritage and construction cases would tend to require this whereas simulation and operational management phases of BIM would accept a geometric generalisation.

Overall, the quality measure works well but the normalisation against a maximum value from the data means that the results from two different comparisons are not directly comparable, something that could be changed by setting the maximum limit to a threshold value after which values are set to 1.

Point clouds are now the basis for a large amount of 3D modelling of existing conditions and their importance has never been greater. However the complexity of effort to generate information for the BIM process is significant. Pure automation is currently not at a stage to be viable as the varied, cluttered nature of the indoor environment means that, currently, human intuition still triumphs over computerised methods alone. Ultimately the work presented in this paper is still best practically seen as an aid to a human user who can edit, accept or reject the geometry recovered by the algorithm in a semi-automated process.

Further Work

To continue this work, it would be beneficial to test the routine on larger datasets, such as one floor or many floors of a building as that would be a truer test of the algorithms main application. Further development of the process to take an IFC and classify a point cloud from it would help reverse the idea of a point cloud as being geometrically rich but information poor, bringing more value to the point cloud as a dataset.

Certain steps of the algorithm can also be refined. Generally refining the algorithm so that it is not so influenced by clutter would be beneficial. Linked to that is adding topological links so that walls that are close together can be joined, then logical building rules could be better applied (e.g., about bounded room volumes) and it would help to mitigate any incorrect determination of length from noise. The quality measure could also be refined by investigating weighting the three components differently or thresholding the normalisation so that a quality value from one comparison is equal to the same value from another set of data.

Acknowledgments: The authors would like to thank the reviewers and copy editors, whose helpful comments led to a better paper overall.

Author Contributions: Charles Thomson designed, executed and performed the analysis of the work presented in this paper as well as wrote the article with the guidance of Jan Boehm throughout the process.

Conflicts of Interest: The authors declare no conflict of interest.

References

1. Pottmann, H. Geometry of architectural freeform structures. *Proceedings of the 2008 ACM Symposium on Solid and Physical Modeling—SPM '08*; ACM Press: New York, NY, USA, 2008; Volume 209, p. 9. Available online: http://portal.acm.org/citation.cfm?doid=1364901.1364903 (accessed on 2 July 2015).

2. Cabinet Office. *Government Construction Strategy*; London, UK, 2011. Available online: https://www.gov.uk/government/uploads/system/uploads/attachment_data/file/61152/Government-Construction-Strategy_0.pdf (accessed on 10 April 2012).

3. UK Green Building Council. Retrofit. Available online: http://www.ukgbc.org/content/retrofit (accessed on 20 April 2013).

4. U.S. General Services Administration. *GSA BIM Guide for 3D Imaging*; Washington, DC, USA, 2009. Available online: http://www.gsa.gov/graphics/pbs/GSA_BIM_Guide_Series_03.pdf (accessed on 10 April 2012).

5. BIM Task Group. Work-streams and Work-packages. Available online: http://www.bimtaskgroup.org/work-streams-wps/ (accessed on 25 October 2012).

6. Li, S.; Isele, J.; Bretthauer, G. Proposed methodology for generation of building information model with laserscanning. *Tsinghua Sci. Technol.* **2008**, *13*, 138–144.

7. Runne, H.; Niemeier, W.; Kern, F. Application of laser scanners to determine the geometry of buildings. In *Optical 3D Measurement Techniques V*; Grün, A., Kahmen, H., Eds.; Department of Applied and Engineering Geodesy, Institute of Geodesy and Geophysics, Vienna University of Technology: Vienna, Austria, 2001; pp. 41–48.

8. Haala, N.; Anders, K. Fusion Of 2D-GIS and image data for 3D building reconstruction. *Int. Arch. Photogram. Remote Sens.* **1996**, *31*, 285–290.

9. Suveg, I.; Vosselman, G. Reconstruction of 3D building models from aerial images and maps. *ISPRS J. Photogramm. Remote Sens.* **2004**, *58*, 202–224.

10. Chapman, D.P.; Deacon, A.T. D.; Hamid, A. HAZMAP: A remote digital measurement system for work in hazardous environments. *Photogramm. Rec.* **1994**, *14*, 747–758.

11. Rajala, M.; Penttilä, H. Testing 3D building modelling framework in building renovation. *Communicating space(s), Proceedings of the 24th Education and Research in Computer Aided Architectural Design in Europe Conference*; Bourdakis, V., Charitos, D., Eds.; University of Thessaly: Volos, Greece, 2006; pp. 268–275. Available online: http://www.mittaviiva.fi/hannu/studies/2006_rajal_penttila.pdf (accessed on 3 June 2013).

12. Royal Institution of Chartered Surveyors. *Scale*; London, UK, 2010. Available online: http://www.rics.org/uk/knowledge/more-services/guides-advice/rics-geomatics-client-guide-series/ (accessed on 24 April 2014).

13. Royal Institution of Chartered Surveyors. *Measured Surveys of Land, Buildings and Utilities*; London, UK, 2014, 3rd ed. Available online: http://www.rics.org/uk/knowledge/professional-guidance/guidance-notes/measured-surveys-of-land-buildings-and-utilities-3rd-edition/ (accessed on 24 April 2014).

14. Plowman Craven Limited. *PCL BIM Survey Specification v2.0.3*; Harpenden, UK, 2012. Available online: http://www.plowmancraven.co.uk/bim-survey-specification/ (accessed on 26 April 2014).

15. Larsen, K.E.; Lattke, F.; Ott, S.; Winter, S. Surveying and digital workflow in energy performance retrofit projects using prefabricated elements. *Autom. Constr.* **2011**, *20*, 999–1011.

16. Day, M. Scan-to-BIM: A reality check. Available online: http://aecmag.com/index.php?option=content&task=view&id=498 (accessed on 25 May 2013).

17. Eastman, C.M.; Teicholz, P.; Sacks, R.; Liston, K. *BIM Handbook: A Guide to Building Information Modeling for Owners, Managers, Designers, Engineers and Contractors*, 2nd ed.; John Wiley and Sons: Hoboken, NJ, USA, 2011.

18. Tang, P.; Huber, D.; Akinci, B.; Lipman, R.; Lytle, A. Automatic reconstruction of as-built building information models from laser-scanned point clouds: A review of related techniques. *Autom. Constr.* **2010**, *19*, 829–843.

19. Pu, S.; Vosselman, G. Automatic extraction of building features from terrestrial laser scanning. *Int. Arch. Photogramm. Remote Sens. Spat. Inf. Sci.* **2006**, *36*, 25–27.

20. Tao, C.V. 3D data acquision and object reconstruction for AEC/CAD. In *Large-scale 3D Data Integration: Challenges and Opportunities*; Zlatanova, S., Prosperi, D., Eds.; CRC Press: Boca Raton, FL, USA, 2005; pp. 39–56.

21. Hichri, N.; Stefani, C.; De Luca, L.; Veron, P. Review of the "As-Built BIM" approaches. *Int. Arch. Photogramm. Remote Sens. Spat. Inf. Sci.* **2013**, *XL-5/W1*, 107–112.

22. Müller, P.; Wonka, P.; Haegler, S.; Ulmer, A.; Van Gool, L. *Procedural Modeling of Buildings*; ACM Press: New York, NY, USA, 2006.

23. Kelly, G.; McCabe, H. A survey of procedural techniques for city generation. *ITB J.* **2006**, *14*, 87–130.

24. Khoshelham, K.; Díaz-Vilariño, L. 3D Modelling of interior spaces: Learning the language of indoor architecture. *ISPRS Int. Arch. Photogramm. Remote Sens. Spat. Inf. Sci.* **2014**, *XL-5*, 321–326.

25. Besl, P.J.; Jain, R.C. Three-dimensional object recognition. *ACM Comput. Surv.* **1985**, *17*, 75–145.

26. Barnea, S.; Filin, S. Segmentation of terrestrial laser scanning data using geometry and image information. *ISPRS J. Photogramm. Remote Sens.* **2013**, *76*, 33–48.

27. Budroni, A.; Boehm, J. Automated 3D reconstruction of interiors from point clouds. *Int. J. Archit. Comput.* **2010**, *8*, 55–73.

28. Adan, A.; Huber, D. 3D reconstruction of interior wall surfaces under occlusion and clutter. In Proceedings of the IEEE International Conference on 3D Imaging, Modeling, Processing, Visualization and Transmission (3DIMPVT), Hangzhou, China, 16–19 May 2011; pp. 275–281.

29. Hoover, A.; Jean-Baptiste, G.; Jiang, X.; Flynn, P.J.; Bunke, H.; Goldgof, D.B.; Bowyer, K.; Eggert, D.W.; Fitzgibbon, A.; Fisher, R.B. An experimental comparison of range image segmentation algorithms. *IEEE Trans. Pattern Anal. Mach. Intell.* **1996**, *18*, 673–689.

30. Xiong, X.; Adan, A.; Akinci, B.; Huber, D. Automatic creation of semantically rich 3D building models from laser scanner data. *Autom. Constr.* **2013**, *31*, 325–337.

31. Jung, J.; Hong, S.; Jeong, S.; Kim, S.; Cho, H.; Hong, S.; Heo, J. Productive modeling for development of as-built BIM of existing indoor structures. *Autom. Constr.* **2014**, *42*, 68–77.

32. Volk, R.; Stengel, J.; Schultmann, F. Building Information Modeling (BIM) for existing buildings—Literature review and future needs. *Autom. Constr.* **2014**, *38*, 109–127.

33. Nagel, C.; Stadler, A.; Kolbe, T.H. Conceptual requirements for the automatic reconstruction of building information from uninterpreted 3D models. In *GeoWeb 2009 Academic Track - Cityscapes*; Kolbe, T.H., Zhang, H., Zlatanova, S., Eds.; ISPRS: Vancouver, BC, Canada, 2009; Volume 38, pp. 46–53.

34. ClearEdge3D. Edgewise Building. Available online: http://www.clearedge3d.com/products/edgewise-building/ (accessed on 9 June 2015).

35. IMAGINiT Technologies. Scan to BIM. Available online: http://www.imaginit.com/software/imaginit-utilities-other-products/scan-to-bim (accessed on 19 June 2015).

36. Faro 3D Software GmBH. Software for Surveying and As-Built Documentation. Available online: http://faro-3d-software.com/index.php (accessed on 9 June 2015).

37. Pointcab. Point Cloud Processing Software. Available online: http://www.pointcab-software.com/en/ (accessed on 9 June 2015).

38. BuildingSMART International. IFC Overview Summary. Available online: http://www.buildingsmart-tech.org/specifications/ifc-overview/ifc-overview-summary (accessed on 4 December 2012).

39. Bonsma, P.P. Basic IFC knowledge. Available online: http://www.buildingsmart-tech.org/implementation/ifc4-implementation/basic-ifc-knowledge (accessed on 7 Feburary 2013).

40. Rusu, R.B.; Cousins, S. 3D is here: Point Cloud Library (PCL). In Proceedings of the IEEE International Conference on Robotics and Automation, Shanghai, China, 9–13 May 2011; pp. 1–4.

41. Ward, A.; Benghi, C.; Ee, S.; Lockley, S. The eXtensible Building Information Modelling (xBIM) Toolkit. Available online: http://nrl.northumbria.ac.uk/13123/ (accessed on 20 Feburary 2015).

42. E57.04 3D Imaging System File Format Committee. libE57: Software Tools for Managing E57 files (ASTM E2807 standard). Available online: http://www.libe57.org/ (accessed on 20 Feburary 2015).

43. Fischler, M.A.; Bolles, R.C. Random sample consensus: A paradigm for model fitting with applications to image analysis and automated cartography. *Commun. ACM* **1981**, *24*, 381–395.

44. Monszpart, A.; Mellado, N.; Brostow, G.J.; Mitra, N.J. RAPTER: Rebuilding Man-made Scenes with regular arrangements of planes: SIGGRAPH 2015. Available online: http://geometry.cs.ucl.ac.uk/projects/2015/regular-arrangements-of-planes/ (accessed on 18 June 2015).

45. Rusu, R.B. Semantic 3D Object Maps for everyday manipulation in human living environments. *KI—Künstliche Intell.* **2010**, *24*, 345–348.

46. Liebich, T.; Adachi, Y.; Forester, J.; Hyvarinen, J.; Karstila, K.; Reed, K.; Richter, S.; Wix, J. IFC2x Edition 3 Technical Corrigendum 1. 2007. Available online: http://www.buildingsmart-tech.org/ifc/IFC2x3/TC1/html/index.htm (accessed on 2 March 2015).

47. Thomson, C.; Boehm, J. Indoor modelling benchmark for 3D geometry extraction. *ISPRS Int. Arch. Photogramm. Remote Sens. Spat. Inf. Sci.* **2014**, *XL-5*, 581–587.

48. Solibri. Model Checker. Available online: http://www.solibri.com/products/solibri-model-checker/ (accessed on 20 June 2015).

A Survey of Algorithmic Shapes

Ulrich Krispel, Christoph Schinko and Torsten Ullrich

Abstract: In the context of computer-aided design, computer graphics and geometry processing, the idea of generative modeling is to allow the generation of highly complex objects based on a set of formal construction rules. Using these construction rules, a shape is described by a sequence of processing steps, rather than just by the result of all applied operations: shape design becomes rule design. Due to its very general nature, this approach can be applied to any domain and to any shape representation that provides a set of generating functions. The aim of this survey is to give an overview of the concepts and techniques of procedural and generative modeling, as well as their applications with a special focus on archeology and architecture.

Reprinted from *Remote Sens.* Cite as: Krispel, U.; Schinko, C.; Ullrich, T. A Survey of Algorithmic Shapes. *Remote Sens.* **2015**, *7*, 12763–12792.

1. Introduction

In the context of computer-aided design (CAD) and shape description, the digital creation of a shape is called modeling. The most common representation of a shape is a composition of elementary objects. However, a shape can also be described by its generating process. In this case, the description is called a generative model. A generative model does not describe a shape by the parts it consists of, but by the operations and steps needed to be performed in order to create it; *i.e.*, a generative model is an algorithm. Its implementation is an algorithmic description written in a programming language. Depending on the used software engineering paradigm, a generative model may also be called a procedural model or a functional model, if the algorithm is implemented in a procedural way, respectively functionally.

For many purposes in CAD, the mightiness of a Turing-complete programming language may lead to potential problems, such as the halting problem. In order to avoid these problems, CAD frameworks often offer a language that is not Turing-complete; *i.e.*, the set of language features is reduced to parametric modeling.

In generative modeling, the object is not just the end result of applied operations, as this paradigm describes a shape by a sequence of processing steps. The result is a paradigm shift from shape design to rule design. This general approach can be applied to many domains.

1.1. Ruler and Compass

Geometry from the days of the ancient Greeks placed great emphasis on problems of constructing various geometric figures using only a ruler without markings (to draw lines) and a compass (to draw circles). Ruler-and-compass constructions are based on EUCLID's axioms [1] using points, lines and circles that have already been constructed. The resulting geometric primitives together with the ruler-and-compass constructions are the first algorithmic descriptions of generative models. EUCLID's Elements is probably the most successful textbook ever written. It still influences modern curricula of mathematics [2]. As the history of geometry [3,4] is not within the scope of article, this article jumps directly to modern uses of generative modeling techniques.

The long history of geometric constructions [6] is also reflected in the history of civil engineering and architecture [7]. Gothic architecture, especially window tracery, exhibits a good example of these constructions. Their complexity is achieved by combining only a few basic geometric patterns. SVEN HAVEMANN and DIETER W. FELLNER show how constructions of prototypic Gothic windows can be formalized using generative modeling techniques [8]. By combining modular construction rules, it is possible that complex configurations can be obtained from elementary constructions. The different combinations of specific parametric features can be grouped together, leading to the concept of styles. A differentiation between basic shape and appearance allows, for example, the creation of ornamental decoration, as seen in Figure 1 [5]. This leads to an extremely compact representation for a whole class of shapes [9].

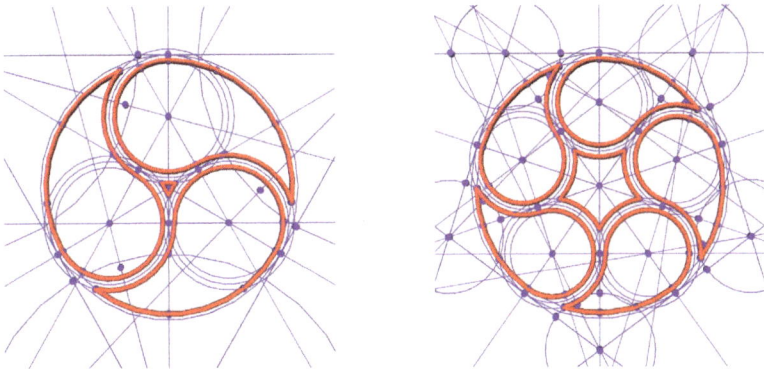

Figure 1. Compass-and-ruler operations have long been used in interactive procedural modeling. This Gothic window construction was created in the framework presented by WOLFGANG THALLER *et al.* using direct manipulation without any code or graph editing [5].

1.2. Natural Patterns

In today's procedural modeling systems, scripting languages and grammars are often used to create a set of rules to achieve a description of an object or pattern. Early systems based on grammars were Lindenmayer systems, short L-systems, named after ARISTID LINDENMAYER [10]. They were successfully used for modeling plants [11] or fractal structures [12]. Given a set of string rewriting rules, complex strings are created by applying these rules to simpler strings. Starting with an initial string, the predefined set of rules form a new, possibly larger string. In order to use L-systems to model geometry, an interpretation of the generated strings is necessary.

The modeling power of L-systems was limited to creating fractals and plant-like branching structures. This limitation led to the introduction of parametric L-systems. The idea is to associate numerical parameters with L-system symbols to address continuous phenomena, which were not covered satisfactorily by L-systems alone.

In combination with additional 3D modeling techniques, Lindenmayer systems can be used to generate complex geometry. ROBERT F. TOBLER et al. introduce a combination of subdivision surfaces, fractal surfaces and parametrized L-systems to create models of natural phenomena [13,14]. Different combinations can be used at each level of resolution. Since the whole description of such multi-resolution models is procedural, their representation is very compact and can be exploited by level-of-detai renderers.

This trade-off between data storage and computation time can be found in various fields of computer graphics, e.g., the tessellation of curved surfaces specified by a few control points directly on the GPU. The result is low storage costs, allowing the generation of complex models only when needed, while also reducing memory transfer overheads. Although L-systems are parallel rewriting systems, derivation through rewriting leads to very uneven workloads. Since the interpretation of an L-system is an inherently serial process, they are not straightforwardly applicable to parallel processing. In 2010, MARKUS LIPP et al. presented a solution to this algorithmic challenge [15].

2. Languages and Grammars

Scripting languages have been designed for a special purpose, e.g., for client-side scripting in a web browser. Nowadays, scripting languages are used for many different applications. JavaScript, for example, is used to animate 2D and 3D graphics in the Virtual Reality Modeling Language (VRML) [16] and Extensible 3D (X3D) [17] files. It checks user forms in PDF files [18], controls game engines [19], configures applications, defines 3D shapes [20] and performs many more tasks. According to JOHN K. OUSTERHOUT, scripting languages use a higher level of abstraction compared to system programming languages, as they are often typeless and interpreted to emphasize the rapid application development purpose [21]. System programming languages, on the other hand, are designed for creating algorithms and data structures based on low-level data types and memory operations. Consequently, graphics libraries [22], shaders [23] and scene graph systems [24,25] are usually written in C/C++ dialects [26], whereas procedural modeling frameworks incorporate scripting languages, such as Lua, JavaScript, *etc.*

2.1. Language Processing and Compiler Construction

For the evaluation of procedural descriptions, typically techniques used for the description of formal languages and compiler construction are used [27]. There is a wide range of different concepts of languages to describe a shape, including all kinds of linguistic concepts [28]. The main categories to describe a shape are:

- rule-based: using substitutions and substitution rules to generate complex structures out of simple starting structures [29–32];
- imperative and scripting-based: using a scripting engine and techniques from predominant programming languages [20,33,34] or;
- GUI and dataflow-based: using new graphical user interfaces (GUI) and intelligent GUIs to detect structures in modeling tasks, which can be mapped onto formal descriptions [35,36].

The general principles of formal descriptions and compiler construction are the same in all cases: independent of ahead-of-time compilation, just-in-time compilation or interpretation [37]. In the first stage of the compilation process, the input source code is passed to the lexer and parser. The first step here is to convert a sequence of characters into a sequence of tokens, which is done by special grammar rules forming the lexical analysis. Typically, only a limited number of characters is allowed for an identifier: all characters A–Z, a–z, digits 0–9 and the underscore _ are allowed with the condition that an identifier must not begin with a digit or an underscore. The lexer rules are embedded in another set of rules: the parser rules. They evaluate the resulting sequence of tokens to determine their grammatical structure. The complete grammar is of a hierarchical structure and consists of rules for analyzing all possible statements and expressions that can be formed in the language, thus forming the syntactic analysis.

For each available language construct, a set of rules ensures syntactic correctness and incorporates mechanisms to report possible syntactic errors and warnings. These rules are also used to create the intermediate abstract syntax tree (AST) structure that is a representation of the input source code to be used for the next stage: semantic analysis. Once all statements and expressions of the input source code are collected in the AST, a tree walker checks their semantic relationships for errors and warnings. After performing all compile-time checks, a translator uses the AST to generate platform-specific files, possibly involving other intermediate structures.

As mentioned in the Introduction, the first procedural modeling systems were L-systems. Later on, L-systems were used in combination with shape grammars to model cities. YOGI PARISH and PASCAL MÜLLER presented a system that generates a street map enriched with geometry for buildings using a number of image maps as input [38]. The resulting framework called CityEngine is a modeling environment for the shape grammar Computational Geometry Algorithm (CGA) Shape. MARKUS LIPP et al. presented another modeling approach based on CGA Shape following the notation of PASCAL MÜLLER [31,35]. It enables more direct local control of the underlying grammar by introducing visual editing. Principles of semantic and geometric selection are combined, as well as functionality to store local changes persistently over global modifications.

SVEN HAVEMANN takes a different approach to generative modeling. He proposes a stack-based language called the Generative Modeling Language (GML) [33]. The postfix notation of the language is very similar to that of Adobe Postscript. High-level shape operations are created by using low-level shape functionality. A number of applications are based on the GML platform, because it is easily extensible and offers an integrated visualization engine. Current efforts in the context of the GML are devoted to directly creating interactive generative visualizations for the web.

Generative modeling inherits the methodologies of 3D modeling and programming [39], which lead to drawbacks in usability and productivity. The need to learn and use a programming language is a significant inhibition threshold, especially for non-computer scientists. The choice of the scripting language has a huge influence on the usability and effectiveness of procedural modeling. Processing is a good example of how an interactive, easy to use, yet powerful development environment can open up new user groups. It had been initially created to serve as a software sketchbook and to teach students fundamentals of computer programming. It quickly developed into a tool that is used for creating visual arts [40].

Processing is a Java-like interpreter offering new graphics and utility functions together with some usability simplifications. The large community behind the tool produced libraries to facilitate computer vision, data visualization, music, networking and electronics. The success of Processing is based on two factors: the simplicity of the programming language and the interactive experience. Instant feedback of the scripting environments allows the user to program via "trial and error".

2.2. Scripting Languages for Generative Modeling

There are many different programming paradigms in software development that are also used in the field of generative modeling, where some paradigms emerged to be useful for specific domains.

Imperative: Many generative models are described using classical programming paradigms: a programming language is used to generate a specific object, possibly using a library that utilizes some sort of geometry representation and operations to perform changes. Any modeling software that is scriptable by an imperative language or provides some sort of API falls into this category.

Dataflow based: A generative description can be represented by a directed graph of the data flowing between operations. This graph representation also allows for a graphical representation; visual programming languages (VPLs) allow one to create a program by linking and modifying visual elements. Many VPLs are based on the dataflow paradigm. Examples in the domain of generative modeling are the Grasshopper3D (online) plug-in for the Rhinoceros3D (online) modeling suite, or the work of GUSTOVA PATOW *et al.* built on top of the procedural modeler Houdini (online) [41].

Rule-based systems: Another different representation for generative modeling is rule-based systems. These systems provide a declarative description of the construction behavior of a model by a set of rules. An example are L-systems, as described in the Introduction. Furthermore, the seminal work of GEORGE STINY and JAMES GIPS introduced shape grammars, as a formal description of capturing the design of paintings and sculptures [42]. Similar to formal grammars, shape grammars are based on rule replacement.

2.3. Shape Grammars

A shape grammar consists of shape rules and a generation engine that selects and processes rules. A shape rule defines how an existing shape can be transformed. The work of PETER WONKA *et al.* applied the concepts of shape grammars to derive a system for generative modeling of architectural models [43]. This system uses a combination of a spatial grammar system (split grammar) to control the spatial design and a control grammar, which distributes the design ideas spatially (e.g., set different attributes for the first floor of a building). Both of these grammars consist of rules with attributes that steer the derivation process. The grammar consists of two types of rules: split and convert. The split rule is a partition operation, which replaces a shape by an arrangement of smaller shapes that fit in the boundary of the original shape. The convert rule replaces a shape by a different shape that also fits in the boundary of the original shape.

This system has further been extended by the work of PASCAL MÜLLER *et al.*, which introduced a component split to extend the split paradigm to arbitrary 3d meshes, as well as occlusion queries and snap lines to model non-local influences of rules [31]. For example, two wall segments that intersect each other should not produce windows, such that the window of one wall coincides with the other wall; therefore, occlusion queries are used to decide if a window should be placed or not.

JEAN-EUDES MARVIE *et al.* have shown that the derivation of a split grammar, starting from an initial shape, yields a tree structure, which suggests that the derivation can be sped up by a parallel implementation [44]. Parallel generation is especially useful in an urban context, with scenes with high complexity and detail. The work of LARS KRECKLAU *et al.* used GPU-accelerated generation in the context of generating and rendering highly detailed building facades [45]; the work of ZHENGZHENG KUANG *et al.* proposes a memory-efficient procedural representation of urban buildings for real-time visualization [46].

With more advanced shape grammar systems, the non-local influences are a problem because they introduce dependencies between arbitrary nodes of the derivation tree. Recent work by MARKUS STEINBERGER *et al.* shows how to overcome this problem in a GPU implementation [47]. Furthermore, the same authors presented methods to interactively generate and render only the visible part of a procedural scene using procedural occlusion culling and the level of detail [48].

2.4. Tools and Environments

A selection of commonly-used tools and programming environments for generative modeling is listed in the Tables 1 and 2.

Table 1. Overview of generative/procedural 3D modeling tools and approaches (Part 1).

Tool Name	Application Domain	Programming Category	Environment
Blender Scripting	general purpose modeling	python scripting	open source modeling software blender
CGAL, The Computational Geometry Algorithms Library[49]	general purpose modeling	C++	CGAL open source project
CityEngine [31]	urban modeling	CGA shape	commercial integrated development environment CityEngine
Generalized Grammar G^2 [30]	scientific	python scripting	commercial modeling software Houdini
Generative Modeling Language (GML) [33]	CAD	postscript dialect	proprietary, integrated development environment for polygonal and subdivision modeling
Grasshopper 3D	visual arts, rapid prototyping, architecture	visual programming based on dataflow graphs, Microsoft .NET family of languages	commercial modeling software Rhinoceros3D
HyperFun [50]	scientific	specialized high-level programming language	proprietary geometry kernel FRep (Function Representation)
Maya Scripting	general purpose modeling	Maya Embedded Language (MEL) and python scripting	commercial modeling software Autodesk Maya
OpenSCAD	CAD	OpenSCAD language	open source, based on CGAL geometry kernel
PLaSM	scientific	python scripting, Function Level scripting	integrated development environment Xplode
Processing	visual arts, rapid prototyping	Java dialect	open source, integrated development environment Processing

Table 2. Overview on generative/procedural 3D modeling tools and approaches (Part 2).

Tool Name	Application Domain	Programming Category	Environment
PythonOCC	general purpose modeling and CAD	python scripting	Open CASCADE Technology
Revit Scripting	architecture	Microsoft .NET family of languages	commercial modeling software Autodesk Revit
siteplan [51]	rapid prototyping, architecture	interactive GUI-based modeler	open source, integrated development environment siteplan
SketchUp Scripting	architecture, urban modeling and CAD	Ruby scripting	commercial modeling software SketchUp
Skyline Engine [41]	urban modeling	visual programming based on dataflow graphs, python scripting	commercial modeling software Houdini
speedtree	plants/trees	interactive GUI-based modeler, SDK for C++	standalone modeler and integration into various game engines
Terragen	landscape modeling	interactive GUI-based modeler	free and commercial, integrated development environment Terragen
XFrog [11]	plants/trees	interactive GUI-based modeler	integrated development environment, standalone and plugins for Maya and Cinema4D

3. Modeling by Programming

3D objects consisting of organized structures and repetitive forms are well suited for procedural descriptions, e.g., by the combination of building blocks or by using shape grammars.

3.1. Building Blocks and Elementary Data Structures

Creating shapes with elementary data structures requires the definition of modeling operations. Depending on the underlying representation, certain modeling operations are difficult or impossible to implement. The selection of operations for these data structures are manifold and can be grouped as follows:

- Instantiations are operations for creating new shapes.
- Binary creations are operations involving two shapes, such as constructive solid geometry (CSG) operations.
- Deformations and manipulations stand for all deforming and modifying operations, like morphing or displacing.

Building blocks can also be regarded as modeling operations. When creating an algorithmic description of a shape, an important task is to identify inherent properties and repetitive forms. These properties must be accounted for in the structure of the

505

description. Identified subparts or repetitive forms are best mapped to functions in order to be reusable. However, the true power of an algorithmic description becomes obvious when parameters are introduced for these functions. Even if only used to position a subpart at a different location. From that point on, the algorithmic description no longer stands for a single object, but for a whole object family.

3.2. Architectural Modeling with Procedural Extrusions

This method utilizes the paradigm of footprint extrusion to automatically derive geometry from a coarse description. The inputs to this system are polygons whose segments can be associated with an extrusion profile polygon. The system utilizes the weighted straight skeleton method [52] to calculate the resulting geometry.

The growing demand for new building models for virtual worlds, games and movies, makes the easy and fast creation of modifiable models more and more important [53]. Nevertheless, 3D modeling of buildings can be a tedious task due to their sometimes complex geometry [54]. For historic buildings, especially the roofs can be challenging. JOHANNES EDELSBRUNNER *et al.* present a new method of combining simple building solids to form more complex buildings and give emphasis to the blending of roof faces [55]. Their method can be integrated in common pipelines for procedural modeling of buildings and extends their expressiveness compared to existing methods.

3.3. Deformation-Aware Shape Grammars

Generative models based on shape and split grammar systems often exhibit planar structures. This is the case because these systems are based on planar primitives and planar splits. There are many geometric tools available in modeling software to transform planar objects into curved ones, e.g., free-form deformation [56]. Applying such a transformation as a post-processing step might yield undesirable results. For example, if a planar facade of a building is bent into a curved shape, the windows inside the facade will have a curved surface, as well. Another possibly unwanted property arises when an object is deformed by scaling: the windows on a facade would have different appearances.

RENÉ ZMUGG *et al.* introduced deformation-aware shape grammars, which integrate deformation information into grammar rules [57]. The system still uses established methods utilizing planar primitives and splits; however, measurements that determine the available space for rules are performed in deformed space. In this way, deformed splits can be carried out, and the deformation can be baked at any point to allow for straight splits in deformed geometry.

3.4. Procedural Shape Modeling

The effectiveness of procedural shape modeling can be demonstrated with the mass customization of consumer products [58]. A generative description composed of a few well-defined procedures can generate a large variety of shapes. Furthermore, it covers most of the design space defined by an existing collection of designs; in this case, wedding rings.

The basic shape of most rings can be defined using a profile polygon, the angular step size defined by the number of supporting profiles to be placed around the ring's center, the radius and a vertex transformation function. A ring's design variations are decomposed into a set of transformation functions. Each function transforms selected parts of the profile in a certain way. Effects can be combined by calling a sequence of different transformations. The creation of the basic shape is separated from optional steps to create engravings, change materials or add gems. Engravings are implemented as per-vertex displacements (to maintain the option for 3D-printing) and can be applied on quadrilateral parts of the ring's mesh using half-edges to specify position and spatial extent.

Materials, like gold, silver and platinum, are used for wedding rings. Their surfaces can be treated with various finishing techniques, like polishing, brushing or hammering. In order to account for these effects, a per-pixel shading model is used featuring anisotropic highlights. By using a cube map, visually appealing reflections are created, and predefined surface finishes can be applied using normal mapping techniques. Procedural gem instances can also be placed on the ring.

Figure 2. The presented generative description is able to produce a large variety of wedding rings. Features, like engravings, recesses, different materials, unusual forms and gems, can be created and customized.

The presented approach is used in a hardware-accelerated server-side rendering framework [59], which has been included in an online system called REx by Johann Kaiser. It offers an intuitive web interface for configuring and visualizing wedding rings.

This work demonstrates the efficiency of procedural shape modeling for the mass customization of wedding rings. The presented generative description is able to produce a large variety of wedding rings. Figure 2 shows a few results of the parametric toolkit.

3.5. Variance Analysis

The analysis and the visualization of the differences of similar objects is important in many research areas: scan alignment, nominal/actual value comparison and surface reconstruction, to name a few. In computer graphics, for example, differences of surfaces are used to validate reconstruction and fitting results of laser scanned surfaces. Scanned representations are used for documentation, as well as analysis of ancient objects, revealing the smallest changes and damage. Analyzing and documentation tasks are also important in the context of engineering and manufacturing to check the quality of productions.

CHRISTOPH SCHINKO *et al.* contribute a comparison of a reference/nominal surface with an actual, laser-scanned dataset [60]. The reference surface is a procedural model whose accuracy and systematics describe the semantic properties of an object, whereas the laser-scanned object is a real-world dataset without any additional semantic information. The first step of the process is to register a generative model (including its free parameters) to a laser scan. Then, the difference between the generative model and the laser scan is stored in a texture, which can be applied to all instances of the same shape family.

A generative model represents an ideal object rather than a real one. The combination of noisy 3D data with an ideal description enhances the range of potential applications. This bridge between both the generative and the explicit geometry description is very important: it combines the accuracy and systematics of generative models with the realism and the irregularity of real-world data, as pointed out by DAVID ARNOLD [61]. Once the procedural description is registered to a real-world artifact, we can use the fitted procedural model to modify a 3D shape. In this way, we can design both low-level details and high-level shape parameters at the same time.

3.6. Semantic Modeling

In the context of digital libraries, semantic metadata play an important role. They provide semantic information that is vital for digital library services: indexing, archival and retrieval. Depending on the field of application, metadata can be classified according to the following criteria [62]:

Data type: The data type of the object can be of any elementary data structure (e.g. polygons, non-uniform rational b-splines (NURBS), subdivision surfaces, *etc.*).
Scale of semantic information: This property describes whether metadata are added for the entire dataset or only for a sub-part of the object.
Type of semantic information: The type of metadata can be descriptive (describing the content), administrative (providing information regarding creation, storing, provenance, *etc.*) or structural (describing the hierarchical structure).

Type of creation: The creation of the semantic information for an object can be done manually (by a domain expert) or automatically (e.g., using a generative description).

Data organization: The two basic concepts of storing metadata are storing the information within the original object (e.g., Exchangeable Image File Format (Exif) data for images) or storing it separately (e.g., using a database).

Information comprehensiveness: The comprehensiveness of the semantic information can be declared varying from low to high in any gradation.

Many concepts for encoding semantic information can be applied to 3D data. Despite the large number of 3D data formats, only a few are standardized, non-proprietary and support semantic markup [63]:

Collada The XML-based Collada format is an ISO standard and allows storing metadata, like title, author, revision, *etc.*, not only on a global scale, but also for parts of the scene [64]. This file format can be found in Google Warehouse where metadata are, for example, used for geo-referencing objects.

IGES Initial Graphics Exchange Specification (IGES), an American National Standards Institute (ANSI) standard since 1980, allows the definition of annotations, including dimensioning data, as well as labels and notes [65]. This file format is used as a vendor-neutral exchange format among CAD systems.

JT The Jupiter Tesselation (JT) file format has been an ISO standard since 2012 and is used for product visualization and data exchange in CAD systems [66]. Annotations in the form of attributes and properties, as well as filters are supported by this format. It is accompanied by the XML-based format for product lifecycle management (PLMXML) to represent product structure hierarchy.

PDF 3D PDF 3D is an ISO standard and allows one to store annotations separated from the 3D data, even allowing annotating of the annotations [67]. An advantage is that the viewer application is widely spread, and PDF documents are the quasi standard for textual documents.

STEP The standard for the exchange of product model data (STEP) has been an ISO standard since 1994 divided into different parts, data models and environments [68]. The current Application Protocol 242 supports product data and non-geometrical metadata.

X3D The X3D file format is an XML-based ISO standard for representing 3D computer graphics [69]. It supports a number of different metadata nodes, providing arrays of strongly typed data.

While a standard has advantages for accessibility, long-term archival and many other aspects, it does not solve inherent problems; *i.e.*, due to the persistent naming

509

problem, a modification of the 3D model can break the integrity of the semantic information. Any change of the geometry can cause the referenced part of the model to no longer exist or be changed. Nevertheless, there are many examples for semantic modeling in various contexts [70–75].

4. Inverse Modeling

The full potential of the generative techniques is revealed when the inverse problem is solved; *i.e.*, what is the best generative description of one or several given instances of an object class? This problem can be interpreted in different ways. The simplest way is to create a generative model out of a given 3D object and to store it in a geometry definition file format. Obviously, this is not the desired result, as the generative model can only represent a single object, not a family of objects.

4.1. Parsing Shape Grammars

Shape grammars can be used to describe the design space of a class of buildings/facades. An interesting question in this context is: given a set of rules and measurements of a building, typically photographs or range scans, which application of rules yields the measurements? Here, the applied rules can also be seen as a parse tree of a given input.

The work of HAYKO RIEMENSCHNEIDER *et al.* [76] utilizes shape grammars to enhance the results of a machine learning classifier that is pre-trained to classify pixels of an orthophoto of a facade into categories, like windows, walls, doors and sky. The system applies techniques from formal language parsing to parse a two-dimensional split grammar consisting of horizontal and vertical splits, as well as repetition and symmetry operations. For the reduction of the search space, an irregular grid is derived from the classifications, and the parsing algorithm is applied to yield the most probable application of rules that yields a classification label per grid cell. These parse trees can easily be converted into procedural models.

FUZHANG WU *et al.* also address the problem of how to generate a meaningful split grammar explaining a given facade layout [77]. Given a segmented facade image, the system uses an approximate dynamic programming framework to evaluate if a grammar is a meaningful description. However, the work does not contribute to the problem of facade image segmentation.

4.2. Model Synthesis

PAUL MERELL and DINESH MANOCHA present an approach that, given an object (*i.e.*, a mesh) and constraints, derives a locally-similar object [78,79]. This method is related to texture synthesis. It computes a set of acceptable states, according to several types of constraints, and constructs parallel planes that correspond to the face orientations of the input model. The intersections of these planes yield possible vertex positions in the output model. Acceptable states are assigned to a vertex, while incompatible states are removed in its neighborhood. The system terminates if every vertex has been assigned a state.

4.3. Inverse Procedural Modeling of Trees

The method proposed by ONDREJ STAVA *et al.* estimates the parameters of a stochastic tree model, given polygonal input tree models [80]. This is done in such a way that the stochastic model produces trees similar to the input. The parameters are estimated using Markov chain Monte Carlo (MCMC) optimization techniques. A statistical growth model consisting of 24 geometrical and environmental parameters is used. The authors propose a similarity measure between the statistical model and a given input mesh that consists of three parts: shape distance, measuring the overall shape discrepancy, geometric distance, reflecting the statistics of the geometry of its branches, and structural distance, encoding the cost of transforming a graph representation of the statistical tree model into a graph representation of the input tree model. The MCMC method has also been applied by other methods to find parameters of a statistical generative model [81–83].

4.4. Parameter Fitting and Shape Recognition

TORSTEN ULLRICH and DIETER W. FELLNER presented an approach that uses generative modeling techniques to describe a class of objects and to identify objects in real-world data, e.g., laser scans [84]. A point cloud P and a generative model M are the input datasets of the algorithm. It answers the questions:

1. can the point cloud be described by the generative model, and if so,
2. what are the input parameters x_0, such that $M(x_0)$ is a good description of P.

A hierarchical optimization routine based on fuzzy geometry and a differentiating compiler are used. The complete generative model description $M(x_1, \ldots, x_k)$ (including all possibly called subroutines) is differentiated with respect to the input parameters. This differentiating compiler offers the possibility to use gradient-based optimization routines in the first place. Without partial derivatives, many numerical optimization routines cannot be used at all or in a limited way.

5. Architecture, Engineering and Design

5.1. Generative Architectural Design

The usage of generative modeling techniques in architecture is not limited to buildings of the past [85,86]. Over the last few decades, architects have used a new class of design tools that support generative design. Generative modeling software extends the design abilities of architects and may even help to reduce costs by harnessing computing power in new ways. Computers, of course, have long been used to capture and implement the design ideas of architects by means of CAD and 3D modeling. Generative design actually helps architects design by using computers to extend human abilities [87].

An impressive example is the Helix Bridge in Singapore (see Figure 3). The 280-m bridge is made up of three 65-m spans and two 45-m end spans. The major and minor helices, which spiral in opposite directions, have overall diameters of 10.8 m and 9.4 m, respectively. The outer helix is formed from six tubes, which are set equidistant from one another, whereas the inner helix consists of five tubes. The bridge design is the product of inseparable collaboration between architects (Cox Architecture and Architects 61) and civil engineers (Arup Consultant). For its 280-m length, the dual helix structure of the bridge utilizes five-times less steel than a conventional box girder bridge. This fact enabled the client to direct the structure to be constructed entirely of stainless steel for its longevity.

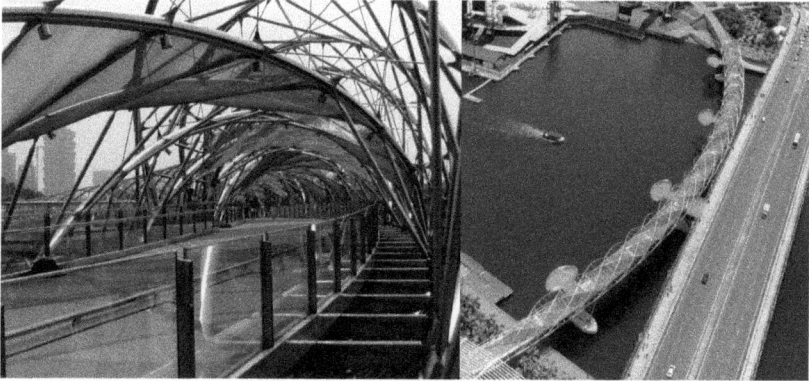

Figure 3. The Helix Bridge is a pedestrian bridge in the Marina Bay area in Singapore. Its generative design has been optimized numerically. Furthermore, the bridge was fully modeled in order to visualize its form and geometrical compatibility, as well as to visualize the pedestrian experience on the bridge.

Another example of generative, architectural design has been presented by TORSTEN ULLRICH et al. [88]. They interpret a generative script as a function, which is nested into an objective function. Thus, the script's parameters can be optimized according to an objective. They demonstrate this approach using architectural examples: each generative script creates a building with several free parameters. The objective function is an energy efficiency simulation that approximates a building's annual energy consumption. Consequently, the nested objective function reads a set of building parameters and returns the energy needs for the corresponding building. This nested function is passed to a minimization and optimization process.

The outcome is the best building (within the family of buildings described by its script) concerning energy efficiency. The contribution is a new way of modeling: the generative approach separates design and engineering. The complete design is encoded in a script, and the script ensures that all parameter combinations (within a fixed range) generate a valid design. Then, the design can be optimized numerically.

The adjustment of architectural forms to local and specific conditions is a fundamental study. When discussing energy consumption and solar power harnessing in buildings, important aspects have to be taken into account, e.g., the relation between a building form and its energy behavior and the local weather conditions on an all-year basis. Several studies were published so far trying to answer these questions. "Form follows energy" has become an omnipresent dogma in architecture, but its realization is difficult. The manual analysis of the various relations between form, volume and energy consumption has to face many, not only numerical, problems.

The new approach by TORSTEN ULLRICH et al. [88] for architectural design is opening the door to new possibilities for the user. It relieves the user from additional, interdisciplinary burdens: the designer can concentrate on the design, while the civil engineer can focus on engineering aspects. This new approach based on procedural modeling can be used in many different fields of product design.

5.2. Engineering Design

The research area of computational design synthesis (CDS) is concerned with the automation of synthesis activities in design [89]. Computer systems are used to generate design candidates for a specific task. For example, the work of MICHAEL J. PUGLIESE et al. [90] investigated the possibility to capture brand specifics using a shape grammar. A more recent synthesis approach presented by FOREST FLAGER et al. is concerned with the sizing optimization of steel structures [91]. For an extensive overview of this topic, the authors refer to the work of AMARESH CHAKRABARTI et al. [92].

Each design process that involves repetitive tasks is perfectly suited for a generative approach. Engineering processes can be classified into repetitive and creative processes. In contrast to creative processes, repetitive ones consist of nearly identical tasks and are therefore independent of creative decisions. This is a precondition for modeling them in a system of rules, as is shown in this practical example [93]: Liebherr manufactures and sells an extensive range of products, including different kinds of cranes. Each crane has to be partially or fully engineered to the needs of the customer. Nevertheless, the design process of ascent assemblies is based on repetitive tasks that are described by a set of invariant rules. These rules have been modeled and stored by Liebherr. The integration into the existing CAD pipeline now allows a construction engineer to create ascent assemblies only by determining the defining parameters and filling out the corresponding input fields in a user interface. Using the procedural approach, the efforts of engineering ascent assemblies have been reduced to 10%.

5.3. Urban Modeling

In the context of urban modeling, procedural systems can be used to cover different levels of detail, as has been shown in a survey for urban reconstruction by PRZEMYSLAW MUSIALKSI et al. [94]. On a coarse scale, the procedural paradigm is applicable to the generation of terrain, e.g., using methods based on hydrology, as presented by JEAN-DAVID GÉNEVAUX [95], or for the inexpensive reconstruction of the landscape surrounding a road, as presented by CARLOS ANDÚJAR et al. [96].

Such systems have also been used for the generation of roads. ERIC GALIN et al. present an algorithm that generates roads on terrain [97]. JAN BENES et al. present a model for growing procedural road networks in and close to cities [98]. The work of MARKUS LIPP et al. is concerned with interactive modeling of entire city layouts [99] using procedural methods. An overview of modeling the appearance and behavior of urban spaces is given by CARLOS A. VANEGAS et al. [100].

Within the scale of a building, PAUL MERELL et al. present a method for automatic creation of residential building layouts; FAN BAO et al. formulate a constrained optimization to characterize good building layouts and a method to let a user explore the space of good building layouts [101]. Procedural systems are also used in the context of facade modeling: PRZEMYSLAW MUSIALSKI et al. present an interactive framework for modeling building facades from images [102]. FAN BAO et al. show a technique to create procedural facade variations from a single layout [103]. MICHAEL SCHWARZ et al. present an approach for designing exterior lighting for buildings with complex constraints [104]. When it comes down to the interior of a building, PAUL MERELL et al. present an automatic method for furniture placement following interior design guidelines [105].

5.4. Building Information Modeling

Procedural modeling can also be helpful in the context of building information modeling (BIM), the new paradigm of today's building industry [106]. The American National Building Information Model Standard (NBIMS-US) project committee defines BIM as "a digital representation of physical and functional characteristics of a facility. A BIM is a shared knowledge resource for information about a facility forming a reliable basis for decisions during its life cycle; defined as existing from earliest conception to demolition" [107]. Other definitions are summarized in a literature review by ABBASNEJAD and MOUD, who conclude that a generally accepted comprehensive definition of BIM has not been established yet, and different stakeholders (architects, builders, owners, etc.) have mixed expectations towards

BIM [108]. In contrast to established computer-aided design (CAD), a building information model does not just store the geometry of a building, but includes semantic data about the functions of the buildings and its elements. Furthermore, BIM is intended to be used throughout the building's life cycle, containing information for planning, design, construction, operation and maintenance. That is, a model is not only used by architects, contractors and suppliers, but by all kinds other users, e.g., government agencies, owners, real estate agents, facility managers, *etc.*

EASTMAN *et al.* help to understand BIM by describing examples that are not BIM technology. As already mentioned, models without object attributes, but only 3D data, are not considered BIM. Furthermore, models composed of multiple 2D drawings that have to be combined or models that do not automatically reflect changes made in one view in other views are not building information models. Moreover, EASTMAN *et al.* consider parametric object capabilities as essential for BIM. Parametric objects in BIM can include rules to automatically modify associated objects (e.g., a wall is changed when a door is placed in it) and for ensuring feasibility (e.g., regarding size and manufacturability) [109]. Such intelligent objects are similar to the idea of generative modeling, where a 3D object is described by the operations necessary to generate the object, rather than the result of these operations [110,111].

One use-case is documenting a building "as-built BIM" [112], to aid, amongst others, in the application scenarios of restoration, documentation and maintenance. Such a model is built from measured data, which is typically acquired by terrestrial laser scanning (TLS) or image-based approaches (photogrammetry or structure from motion techniques), which yields point positions in 3D. From these point clouds, a mesh can be created using 3D surface reconstruction techniques, e.g., Poisson surface reconstruction [113]. Furthermore, the surface appearance has to be acquired [114]. In the case of building interiors, the generation of orthographic images can be used for surface color representation and the retrieval of additional semantic information [115]. Such semantic relationships have to be acquired and represented within the model [116]; see also the foregoing section about semantic enrichment. A recent example of the usage of parametric elements for the reconstruction and documentation of complex architecture is the case of a reactor building, as shown by JEAN-FRANÇOIS HULLO *et al.* [117].

For historic building information modeling (HBIM), procedural methods have been used to aid the reconstruction and documentation process. CONOR DORE *et al.* applied a shape grammar approach to model classical building facades for HBIM [118] and reconstructed the Four Courts, a historic classical building in Dublin City [119] using rule-based modeling in ArchiCAD. Another recent example of creating a HBIM model with rich semantics from terrestrial laser scanning data has been shown by RAMONA QUATTRINI *et al.* in the case study of the Church of Santa Maria at Portonovo [120] using Autodesk Revit.

In the context of functional building information modeling (FBIM), generative techniques can be used to semantically filter a CAD dataset of a building. A major future challenge in the building industry is to reduce primary energy use of buildings. Hence, energy performance simulation becomes an increasingly important topic. Accurate, yet efficient simulation depends on simple building models. Most of the required data can be found in BIMs. However, typical BIM data contains a lot of irrelevant data, in particular geometric representations, which are too detailed for energy performance simulation. Using generative modeling techniques, DANIEL LADENHAUF *et al.* [121,122] show an approach of geometry simplification subject to semantic and functional groups. These simplified models are sufficiently accurate for energy calculations and small enough so that they do not flood simulation software with unnecessary details. As these semantically-filtered models are generated automatically, they simplify the design process significantly and offer an energy calculation, even at early design stages.

6. Archeology and Cultural Heritage

The increasing number of (3D) documents makes digital library services become more and more important. A digital library provides markup, indexing and retrieval services based on available metadata. In a simple case, metadata are of the Dublin Core type: title, creator/author, time of creation, *etc* [123]. This is insufficient for large collections of 3D objects, because of their versatility and rich structure.

6.1. Semantic Enrichment

Scanned models are used in raw data collections, for documentation archival, virtual reconstruction, historical data analysis and for high-quality visualization for dissemination purposes [124]. Navigation and browsing through the geometric models should be possible on the semantic level; this requires higher level semantic information. The need for semantic information becomes immediately clear in the context of electronic data exchange, storage and retrieval [125,126]. The problem of 3D semantic enrichment is closely related to the shape description problem [127]:

How to describe a shape and its structure
on a higher, more abstract level?

The traditional way of classifying objects, pursued both in mathematics and, in a less formal manner, in dictionaries, is to define a class of objects by listing their distinctive properties. This approach is hardly realizable, because of the fact that definitions cannot be self-contained. They depend on other definitions, which leads to circular dependencies that cannot be resolved automatically by strict reasoning, but rely on intuitive understanding at some point.

An alternative, non-recursive approach for describing shape uses examples. Each entry in a picture dictionary is illustrated with a photo or a drawing. This approach is widely used, for example, in biology for plant taxonomy. It avoids listing an exhaustive list of required properties for each entry. However, it requires some notion of similarity, simply because the decision of whether object x belongs to class A or B requires measuring the closeness of x to the exemplars $a \in A$, respectively $b \in B$. This decision can be reached by a classifier using statistics and machine learning [128,129]. A survey on content-based 3D object retrieval is provided by BENJAMIN BUSTOS et al. [130]. Statistical approaches clearly have their strength in discriminating object classes. However, feature-based object detection, e.g., of rectangular shapes, does not yield object parameters: the width and height of a detected rectangle must typically be computed separately.

To describe a shape and its construction process, its inner structure must be known. Structural decomposition is well in line with human perception. In general, shapes are recognized and coded mentally in terms of relevant parts and their spatial configuration or structure [131]. One idea to operationalize this concept was proposed, among others, by MASAKI HILAGA et al., who introduce the Multiresolution Reeb Graph, to represent the skeletal and topological structure of a 3D shape at various levels of resolution [132]. Structure recognition is a very active branch in the field of geometry processing. The detection of shape regularities [133], self-similarities [134] and symmetries [135,136] is important to understand a 3D shape. To summarize, structural decomposition proceeds by postulating that a certain type of general regularity or structure exists in a class of shapes. This approach clearly comes to its limits when very specific structures are to be detected, i.e., complicated constructions with many parameter interdependencies.

A possibility to describe a shape is realized by the generative modeling paradigm [29,137]. The key idea is to encode a shape with a sequence of shape-generating operations and not just with a list of low-level geometric primitives. In its practical consequence, every shape needs to be represented by a program, i.e., encoded in some form of programming language, shape grammar [31], modeling language [33] or modeling script [138].

The implementation of the "definition by algorithm" approach is based on a scripting language [84]: Each class of objects is represented by one algorithm M. Furthermore, each described object is a set of high-level parameters x, which reproduces the object, if an interpreter evaluates $M(x)$. As this kind of modeling resembles programming rather than "designing", it is obvious to use software engineering techniques, such as versioning and annotations. In this way, model M may contain a human-readable description of the object class it represents.

In contrast to other related techniques using fitting algorithms, such as "Creating Generative Models from Range Images" by RAVI RAMAMOORTHI and JAMES ARVO, the approach by TORSTEN ULLRICH can classify data semantically [139,140]. Although RAVI RAMAMOORTHI and JAMES ARVO also use generative models to fit point clouds, they modify the generative description during the fitting process. As a consequence, the optimization can be performed locally with a computational complexity that is significantly reduced. However, starting with the same generative description to fit a spoon as well as a banana does not allow one to generate or preserve semantic data.

An example illustrates this process. The generative model to describe a vase takes 13 parameters: $R(r_x, r_y, r_z)$ is the base reference point of the vase in 3D and $T(t_x, t_y, t_z)$ is its top-most point. The points R and T define an axis of rotational symmetry. The remaining seven parameters define the distances $d_0, \ldots, d_6$ of equally-distributed Bézier vertices to the axis of rotation (see Figure 4). The resulting 2D Bézier curve defines a surface of revolution: the generative vase.

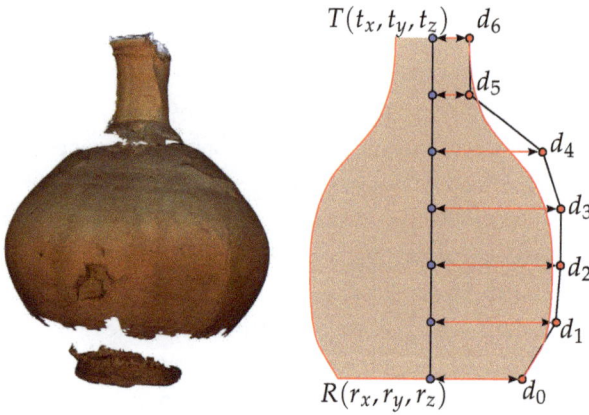

Figure 4. The vase on the left-hand side is a digitized artifact of the "Museum Eggenberg" collection. It consists of 364,774 vertices and 727,898 triangles. The example of a procedural shape on the right-hand side takes two points R and T in 3D and distance values, which define the control vertices of a Bézier curve.

6.2. Cultural Heritage

The huge volume of cultural objects is a challenge, even for the most ambitious plans for digitization campaigns [141]. The fact that probably 90 percent of museum collections are in storage and not accessible to the public is almost demanding for digitization and public accessibility. However, the digitization alone is only part of a larger process that begins at a field excavation and does not end with the presentation in museum exhibitions. Secondary exploitation, database access and sustainable long-time archiving of digitized artifacts is also part of the process [142]. A very important aspect is the choice of the 3D format used during this process [143,144]. However, the availability of large quantities of cultural heritage data will enable new methods for analysis and new applications [145].

The presented modeling system by CHRISTOPH SCHINKO et al. is restricted to techniques to meet sustainability conditions. By using JavaScript, the inhibition threshold to use a programming language is reduced, resulting in a beginner-friendly tool with a high degree of usability [111]. RENÉ BERNDT et al. present a system for the production of three-dimensional interactive illustrations in the domain of medieval castles [146]. A special focus is on creating generic modeling tools that increase the usability with a unified 3D user interface.

One of the advantages of procedural modeling techniques is the included expert knowledge within an object description [84]. Classification schemes used in architecture, archeology and other domains can be mapped to procedures [147]. When a procedural object description is available, only the type and instantiation parameters have to be identified in order to create an object [148] (see Figure 5). It is then also possible to use the fitted procedural model to modify existing 3D shapes [149].

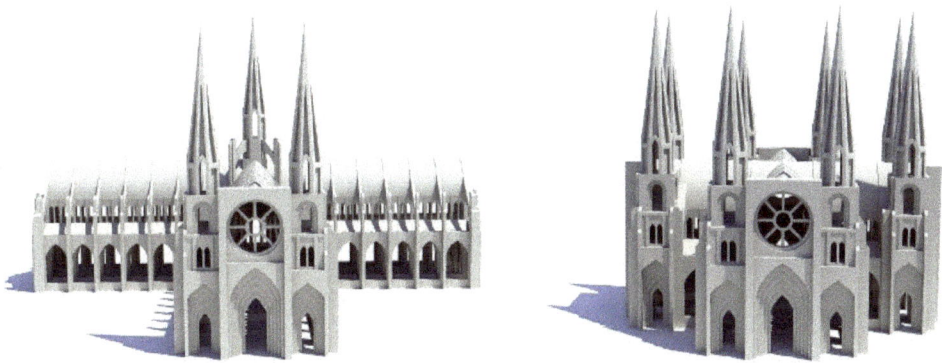

Figure 5. Gothic architecture is defined by strict rules with its characteristics. The generative description of Gothic cathedrals encodes these building blocks and the rules on how to combine them. These building blocks have been created by MICHAEL CURRY, http://www.thingiverse.com/thing:2030.

Another use-case is the creation of several building hypotheses in the context of historic analysis, as shown by ERICA CALOGERO et al. [150] in a case study that investigated different hypothesis for parts of the Louvre. Furthermore, MARIE SALDAÑA et al. carried out a similar approach for parts of the city of Rome [151]. Both works were carried out using the Esri CityEngine.

7. Open Research Questions

According to DIETER W. FELLNER and SVEN HAVEMANN, several research challenges have to be met: from the classification of shape representations via generic, stable and detailed 3D markup to 3D query operations [125,152–154].

A particularly important problem occurs in the context of internal structure organization and interfaces. Within a composition of modeling functions, where each function is attached via its parameters to topological entities defined in previous states of the model, referenced entities must be named in a persistent way in order to be able to reevaluate the model in a consistent manner. In particular, when a reevaluation leads to topological modifications, references between entities used during the design process are frequently reevaluated in an erroneous way, giving results different from those expected. This problem is known as the "persistent naming problem" [155].

Acknowledgments: The authors gratefully acknowledge the support of the European Commission's Seventh Framework Programme (FP7/2007-2013) under Grant Agreement No. 600908 (DURAARK: Durable Architectural Knowledge), founded by the program, ICT-2011-4.3-Digital Preservation, as well as the generous support of the Austrian Research Promotion Agency, the Forschungsförderungsgesellschaft (FFG), for the research project GINGER (Graphical Energy-Efficiency-Visualization in Architecture), Grant Number 840190, and the research project ProFitS (Procedural Fitting Service), Grant Number 840254. Furthermore, the authors would like to thank Daniel Ladenhauf for his research on building information modeling, as well as Wolfgang Thaller and Volker Settgast for permitting use of their images in Figures 1 and 3, respectively.

Author Contributions: This survey has been conducted by the stated authors, with Ulrich Krispel and Christoph Schinko contributing mainly to sections 2–5, and Torsten Ullrich mainly contributing to sections 1, 6–7.

Conflicts of Interest: The authors declare no conflict of interest.

References

1. Heiberg, J. *Euclid's Elements of Geometry*; Fitzpatrick Richard: Austin, TX, USA, 2007.
2. Hartshorne, R. Teaching geometry according to Euclid. *Not. AMS* **2000**, *47*, 460–465.
3. Maxfield, J.E.; Coolidge, J.L. *A History of Geometrical Methods*, 1st ed.; Dover Publications: New York, NY, USA, 2003.
4. Scriba, C.J.; Schreiber, P. *5000 Jahre Geometrie: Geschichte, Kulturen, Menschen (English: 5000 Years of Geometry: History, Cultures, Men)*; Springer: Berlin, Germany, 2004.
5. Thaller, W.; Krispel, U.; Zmugg, R.; Havemann, S.; Fellner, D.W. A Graph-Based Language for Direct Manipulation of Procedural Models. *Int. J. Adv. Softw* **2013**, *6*, 225–236.
6. Martin, G.E. *Geometric Constructions*; Springer: Berlin, Germany, 1998.
7. Mitchell, W.J. *The Logic of Architecture: Design, Computation, and Cognition*; MIT Press: Cambridge, UK, 1990.

8. Havemann, S.; Fellner, D.W. Generative parametric design of gothic window tracery. *IEEE Proc. Shape Model. Appl.* **2004**, doi:10.1109/SMI.2004.1314525.

9. Berndt, R.; Fellner, D.W.; Havemann, S. Generative 3D models: A key to more information within less bandwidth at higher quality. In Proceedings of the 10th International Conference on 3D Web Technology, Gwynedd, UK, 29 March–1 April 2005.

10. Prusinkiewicz, P.; Lindenmayer, A. *The Algorithmic Beauty of Plants*; Springer: Berlin, Germany, 1990.

11. Deussen, O.; Lintermann, B. *Digital Design of Nature: Computer Generated Plants and Organics*; Springer: Berlin, Germany, 2005.

12. Mandelbrot, B.B. *The Fractal Geometry of Nature*; W. H. Freeman and Co.: New York, NY, USA, 1982.

13. Tobler, R.F.; Maierhofer, S.; Wilkie, A. A multiresolution mesh generation approach for procedural definition of complex geometry. *Proc. Shape Model. Int.* **2002**, *6*, 35–44.

14. Tobler, R.F.; Maierhofer, S.; Wilkie, A. Mesh-based parametrized L-systems and generalized subdivision for generating complex geometry. *Int. J. Shape Model.* **2002**, *8*, 173–191.

15. Lipp, M.; Wonka, P.; Wimmer, M. Parallel generation of multiple L-systems. *Comput. Graph.* **2010**, *34*, 585–593.

16. Brutzman, D. The virtual reality modeling language and Java. *Commun. ACM* **1998**, *41*, 57–64.

17. Behr, J.; Dähne, P.; Jung, Y.; Webel, S. Beyond the web browser–X3D and immersive VR. *IEEE Virtual Real. Tutor. Workshop Proc.* **2007**, *28*, 5–9.

18. Breuel, F.; Bernd, R.; Ullrich, T.; Eggeling, E.; Fellner, D.W. Mate in 3D–publishing interactive content in PDF3D. In *Digital Publishing and Mobile Technologies*, Proceedings of the 15th International Conference on Electronic Publishing, İstanbul, Turke, 22–24 June 2011.

19. Di Benedetto, M.; Ponchio, F.; Ganovelli, F.; Scopigno, R. SpiderGL: A JavaScript 3D graphics library for next-generation WWW. In Proceedings of the 15th International Conference on Web 3D Technology, Los Angeles, CA, USA, 24–25 July 2010.

20. Schinko, C.; Strobl, M.; Ullrich, T.; Fellner, D.W. Scripting technology for generative modeling. *Int. J. Adv. Softw.* **2011**, *4*, 308–326.

21. Ousterhout, J.K. Scripting: Higher level pogramming for the 21st century. *IEEE Comput. Mag.* **1998**, *31*, 23–30.

22. OpenGL Architecture, R.B. *OpenGL Reference Manual*; Addison-Wesley: Boston, MA, USA, 1993.

23. NVidia. CUDA C Programming Guide. Available online: https://docs.nvidia.com/cuda/cuda-c-programming-guide/ (accessed on 29 June 2015).

24. Reiners, D.; Voss, G.; Behr, J. OpenSG: Basic concepts. *OpenSG Symp.* **2002**, *1*, 1–7.

25. Voß, G.; Behr, J.; Reiners, D.; Roth, M. A multi-thread safe foundation for scene graphs and its extension to clusters. *EGPGV* **2002**, *4*, 33–37.

26. Eckel, B. *Thinking in C++: Introduction to Standard C++, Practical Programming*; Prentice Hall: New Jersey, NJ, USA, 2003.

27. Parr, T. *Language Implementation Patterns: Create Your Own Domain-Specific and General Programming Languages*; Pragmatic Bookshelf: North Carolina, NC, USA, 2010.
28. Chomsky, N. Three models for the description of language. *IRE Trans. Inf. Theory* **1956**, 2, 113–124.
29. Özkar, M.; Kotsopoulos, S. Introduction to shape grammars. In Proceedings of the International Conference on Computer Graphics and Interactive Techniques, Los Angeles, CA, USA, 11–15 August 2008.
30. Krecklau, L.; Pavic, D.; Kobbelt, L. Generalized use of non-terminal symbols for procedural modeling. *Comput. Graph. Forum* **2010**, 29, 2291–2303.
31. Müller, P.; Wonka, P.; Haegler, S.; Andreas, U.; van Gool, L. Procedural modeling of buildings. *ACM Trans. Graph.* **2006**, 25, 614–623.
32. Snyder, J.M.; Kajiya, J.T. Generative modeling: A symbolic system for geometric modeling. *ACM SIGGRAPH Comput. Graph.* **1992**, 26, 369–378.
33. Havemann, S. Generative Mesh Modeling. Ph.D. Thesis, Technische Universit, Braunschweig, Germany, 2005.
34. Krecklau, L.; Kobbelt, L. Procedural modeling of interconnected structures. *Comput. Graph. Forum* **2011**, 30, 335–344.
35. Lipp, M.; Wonka, P.; Wimmer, M. Interactive visual editing of grammars for procedural architecture. *ACM Trans. Graph.* **2008**, 27, 1–10.
36. Thaller, W.; Krispel, U.; Havemann, S.; Fellner, D. Implicit nested repetition in dataflow for procedural modeling. In Proceedings of the International Conference on Computational Logics, Algebras, Programming, Tools, and Benchmarking (Computation Tools), Nice, France, 22–27 July 2012; pp. 45–50.
37. Schinko, C.; Ullrich, T.; Fellner, D.W. Minimally invasive interpreter construction—How to reuse a compiler to build an interpreter. In Proceedings of the International Conference on Computational Logics, Algebras, Programming, Tools, and Benchmarking (Computation Tools), Nice, France, 22–27 July 2012; pp. 38–44.
38. Parish, Y.; Müller, P. Procedural modeling of cities. In Proceedings of the 28th annual conference on Computer graphics and interactive techniques, Los Angeles, CA, USA, 1 August 2001; pp. 301–308.
39. Ullrich, T.; Krispel, U.; Fellner, D.W. Compilation of procedural models. In Proceeding of the 13th International Conference on 3D Web Technology, Los Angeles, CA, USA, 9–10 August 2008; pp. 75–81.
40. Reas, C.; Fry, B.; Maeda, J. *Processing: A Programming Handbook for Visual Designers and Artists*; The MIT Press: Cambridge, MA, USA, 2007.
41. Patow, G. User-friendly graph editing for procedural modeling of buildings. *IEEE Comput. Graph. Appl.* **2012**, 32, 66–75.
42. Stiny, G.; Gips, J. Shape grammars and the generative specification of painting and sculpture. *IFIP Congr.* **1971**, 2, 125–135.
43. Wonka, P.; Wimmer, M.; Sillion, F.; Ribarsky, W. Instant architecture. *IACM Trans. Graph.* **2003**, 22, 669–677.

44. Marvie, J.E.; Buron, C.; Gautron, P.; Hirtzlin, P.; Sourimant, G. GPU Shape grammars. *Comput. Graph. Forum* **2012**, *31*, 2087–2095.

45. Krecklau, L.; Born, J.; Kobbelt, L. View-dependent realtime rendering of procedural facades with high geometric detail. *Comput. Graph. Forum* **2013**, *32*, 479–488.

46. Kuang, Z.; Chan, B.; Yu, Y.; Wang, W. A compact random-access representation for urban modeling and rendering. *ACM Trans. Graph.* **2013**, doi:10.1145/2508363.2508424.

47. Steinberger, M.; Kenzel, M.; Kainz, B.; Müller, J.; Peter, W.; Schmalstieg, D. Parallel generation of architecture on the GPU. *Comput. Graph. Forum* **2014**, *33*, 73–82.

48. Steinberger, M.; Kenzel, M.; Kainz, B.; Wonka, P.; Schmalstieg, D. On-the-fly generation and rendering of infinite cities on the GPU. *Comput. Graph. Forum* **2014**, *33*, 105–114.

49. The CGAL Project CGAL User and Reference Manual. Available online: http://doc.cgal.org/latest/Manual/ (accessed on 29 June 2015).

50. Pasko, A.; Adzhiev, V. Function-based shape modeling: Mathematical framework and specialized language. *Lect. Notes Comput. Sci.* **2004**, *2930*, 132–160.

51. Kelly, T.; Wonka, P. Interactive architectural modeling with procedural extrusions. *ACM Trans. Graph.* **2011**, doi:10.1145/1944846.1944854.

52. Aurenhammer, F. Weighted skeletons and fixed-share decomposition. *Comput. Geom.* **2008**, *40*, 93–101.

53. Watson, B.; Wonka, P. Procedural methods for urban modeling. *IEEE Comput. Graph. Appl.* **2008**, *28*, 16–17.

54. Whiting, E.; Ochsendorf, J.; Durand, F. Procedural modeling of structurally-sound masonry buildings. *ACM Trans. Graph.* **2009**, doi:10.1145/1618452.1618458.

55. Edelsbrunner, J.; Krispel, U.; Havemann, S.; Sourin, A.; Fellner, D.W. Constructive roof geometry. In Proceedings of the 2014 International Conference on Cyberworlds, Santander, Spain, 6–8 October 2014.

56. Sederberg, T.W.; Parry, S.R. Free-form deformation of solid geometric models. *ACM SIGGRAPH Comput. Graph.* **1986**, *13*, 151–160.

57. Zmugg, R.; Thaller, W.; Krispel, U.; Edelsbrunner, J.; Havemann, S.; Fellner, D.W. Procedural architecture using deformation-aware split grammars. *Visual Comput.* **2013**, *12*, 1–11.

58. Berndt, R.; Schinko, C.; Krispel, U.; Settgast, V.; Havemann, S.; Eggeling, E.; Fellner, D.W. Ring's anatomy—parametric design of wedding rings. *Content* **2012**, *4*, 72–78.

59. Schinko, C.; Berndt, R.; Eggeling, E.; Fellner, D. A scalable rendering framework for generative 3D content. In Proceedings of the 19th International ACM Conference on 3D Web Technologies, Vancouver, BC, Canada, 8–10 August 2014.

60. Schinko, C.; Ullrich, T.; Schiffer, T.; Fellner, D.W. Variance analysis and comparison in computer-aided design. In Proceedings of the International Workshop on 3D Virtual Reconstruction and Visualization of Complex Architectures, rento, Italy, 2–4 March 2011.

61. Arnold, D. Procedural methods for 3D reconstruction. *Rec. Model. Vis. Cult. Herit.* **2006**, *1*, 355–359.

62. Ullrich, T.; Settgast, V.; Berndt, R. Semantic enrichment for 3D documents: Techniques and open problems. In *the Networked World: Transforming the Nature of Communication*, Proceedings of the International Conference on Electronic Publishing, Helsinki, Finland, 16–18 June 2010.

63. Settgast, V. Processing Semantically Enriched Content for Interactive 3D Visualizations. Ph.D. Thesis, Technische Universität, Graz, Austria, 2013.

64. International Organization for Standardization (ISO) / Publicly Available Specification (PAS) 17506:2012 (Industrial Automation Systems and Integration–COLLADA Ddigital Asset Schema Specification for 3D Visualization of Industrial Data). Available online: http://www.iso.org/iso/catalogue_detail.htm?csnumber=59902 (accessed on 29 June 2015).

65. U.S. Product Data Association (US PRO), Formerly ANS US PRO/IPO-100-1996 (Initial Graphics Exchange Specification IGES 5.3). Available online: http://webstore.ansi.org/RecordDetail.aspx?sku=SAE+J+1881-2001+(SAE+J1881-2001) (accessed on 29 June 2015).

66. International Organization for Standardization (ISO) 14306:2012 (Industrial Automation Systems and Integratio–JT File Format Specification for 3D Visualization). Available online: http://www.iso.org/iso/catalogue_detail.htm?csnumber=60572 (accessed on 29 June 2015).

67. International Organization for Standardization (ISO) 32000-1:2008 (Document Management–Portable Document Format–Part 1: PDF 1.7). Available online: http://www.iso.org/iso/catalogue_detail.htm?csnumber=51502 (accessed on 29 June 2015).

68. International Organization for Standardization (ISO) 10303-1:1994 (Industrial Automation Systems and Integration–Product Data Representation and Exchange–Part 1: Overview and Fundamental Principles). Available online: http://www.iso.org/iso/catalogue_detail?csnumber=20579 (accessed on 29 June 2015).

69. International Organization for Standardization (ISO) / International Electrotechnical Commission (IEC) 19775-1:2013 (Information technology–Computer Graphics, Image Processing and Environmental Data Representation–Extensible 3D (X3D)–Part 1: Architecture and Base Components). Available online: http://www.iso.org/iso/catalogue_detail?csnumber=60760 (accessed on 29 June 2015).

70. Boulch, A.; Houllier, S.; Marlet, R.; Tournaire, O. Semantizing complex 3D scenes using constrained attribute grammars. *Proc. Eur. Symp. Geom. Proc.* **2013**, *32*, 33–42.

71. Haegeler, S.; Müller, P.; Van Gool, L. Procedural modeling for digital cultural heritage. *J. Image Video Process.* **2009**, *9*, 1–11.

72. Mendez, E.; Schall, G.; Havemann, S.; Fellner, D.W.; Schmalstieg, D.; Junghanns, S. Generating semantic 3D models of underground infrastructure. *IEEE Comput. Graph. Appl.* **2008**, *28*, 48–57.

73. Thaller, W.; Zmugg, R.; Krispel, U.; Posch, M.; Havemann, S.; Fellner Dieter, W. Creating procedural windowbuilding blocks using the generative fact labeling method. *Proc. ISPRS Int. Workshop 3D-ARCH* **2013**, *5*, 235–242.

74. Van Gool, L.; Martinovic, A.; Mathias, M. Towards semantic city models. *Proc. Photogramm. Week* **2013**, *1*, 217–232.

75. Yong, L.; Mingmin, Z.; Yunliang, J.; Haiying, Z. Improving procedural modeling with semantics in digital architectural heritage. *Comput. Graph.* **2012**, *36*, 178–184.

76. Riemenschneider, H.; Krispel, U.; Thaller, W.; Donoser, M.; Havemann, S.; Fellner, D.W.; Bischof, H. Irregular lattices for complex shape grammar facade parsing. In Proceedings of the IEEE Conference on Computer Vision and Pattern Recognition (CVPR), Providence, RI, USA, 16–21 June 2012.

77. Wu, F.; Yan, D.M.; Dong, W.; Zhang, X.; Wonka, P. Inverse procedural modeling of facade layouts. *ACM Trans. Graph.* **2014**, doi:10.1145/2601097.2601162.

78. Merrell, P.; Manocha, D. Continuous model synthesis. *ACM Trans. Graph.* **2008**, doi:10.1145/1409060.1409111.

79. Merrell, P.; Manocha, D. Model Synthesis: A general procedural modeling algorithm. *IEEE Trans. Vis. Comput. Graph.* **2010**, *17*, 715–728.

80. Stava, O.; Pirk, S.; Kratt, J.; Chen, B.; Měch, R.; Deussen, O.; Benes, B. Inverse procedural modelling of trees. *Comput. Graph. Forum*, **2014**, *33*, 118–131.

81. Talton, J.O.; Lou, Y.; Lesser, S.; Duke, J.; Mech, R.; Koltun, V. Metropolis procedural modeling. *ACM Trans. Graph.* **2011**, doi:10.1145/1944846.1944851.

82. Vanegas, C.A.; Garcia-Dorado, I.; Aliaga, D.G.; Benes, B.; Waddell, P. Inverse design of urban procedural models. *ACM Trans. Graph.* **2012**, doi:10.1145/2366145.2366187.

83. Yu, L.F.; Yeung, S.K.; Tang, C.K.; Terzopoulos, D.; Chan, T.F.; Osher, S. Make it home: Automatic optimization of furniture arrangement. *ACM Trans. Graph.* **2011**, doi:10.1145/1964921.1964981.

84. Ullrich, T.; Fellner, D.W. Generative object definition and semantic recognition. In Proceedings of the Eurographics Workshop on 3D Object Retrieval, Llandudno, UK, 10 April 2011.

85. Müller, P.; Vereenooghe, T.; Ulmer, A.; van Gool, L. Automatic reconstruction of Roman housing architecture. *Rec. Model. Vis. Cult. Heritage* **2006**, *1*, 287–298.

86. Müller, P.; Vereenooghe, T.; Wonka, P.; Paap, I.; Van Gool, L. Procedural 3D reconstruction of Puuc buildings in Xkipche. *Proc. Eur. Symp. Virtual Real. Archaeol. Cult. Heritage (VAST)* **2006**, *1*, 139–146.

87. Hohmann, B.; Krispel, U.; Havemann, S.; Fellner, D.W. Cityfit: High-quality urban reconstructions by fitting shape grammars to images and derived textured point clouds. *Proc. ISPRS Int. Workshop 3D-ARCH* **2009**, *3*, 61–68.

88. Ullrich, T.; Silva, N.; Eggeling, E.; Fellner, D.W. Generative Modeling and Numerical Optimization for Energy Efficient Buildings. In Proceedings of the IECON 2013-39th Annual Conference of the IEEE Industrial Electronics Society, Vienna, Austria, 10–13 November 2013.

89. Campbell, M.I.; Shea, K. Guest editorial: Computational design synthesis. *AI EDAM* **2014**, *28*, 207–208.

90. Pugliese, M.; Cagan, J. Capturing a rebel: Modeling the Harley-Davidson brand through a motorcycle shape grammar. *Res. Eng. Design* **2002**, *13*, 139–156.

91. Flager, F.; Soremekun, G.; Adya, A.; Shea, K.; Haymaker, J.; Fischer, M. Fully Constrained Design: A general and scalable method for discrete member sizing optimization of steel truss structures. *Comput. Struct.* **2014**, *140*, 55–65.

92. Chakrabarti, A.; Shea, K.; Stone, R.; Cagan, J.; Campbell, M.; Vargas-Hernandez, N.; Wood, K.L. Computer-based design synthesis research: An overview. *J. Comput. Inf. Sci. Eng.* **2011**, doi:10.1115/1.3593409.

93. Frank, G.; Hillbrand, C. Automatic support of standardization processes in design models. In Proceedings of the 2012 IEEE International Conference on Intelligent Engineering Systems (INES), Lisbon, Portugal, 13–15 June 2012.

94. Musialski, P.; Wonka, P.; Aliaga, D.G.; Wimmer, M.; van Gool, L.; Purgathofer, W. A survey of urban reconstruction. *Comput. Graph. Forum* **2012**, *31*, 1–28.

95. Génevaux, J.D.; Galin, E.; Guérin, E.; Peytavie, A.; Beneš, B. Terrain generation using procedural models based on hydrology. *ACM Trans. Graph.* **2013**, doi:10.1145/2461912.2461996.

96. Andújar, C.; Chica, A.; Vico, M.A.; Moya, S.; Brunet, P. Inexpensive reconstruction and rendering of realistic roadside landscapes. *Comput. Graph. Forum* **2014**, *33*, 101–117.

97. Galin, E.; Peytavie, A.; Marechal, N.; Guerin, E. Procedural generation of roads. *Comput. Graph. Forum* **2010**, *29*, 429–438.

98. Benes, J.; Wilkie, A.; Krivanek, J. Procedural modelling of urban road networks. *Comput. Graph. Forum* **2014**, *33*, 132–142.

99. Lipp, M.; Scherzer, D.; Wonka, P.; Wimmer, M. Interactive modeling of city layouts using layers of procedural content. *Comput. Graph. Forum* **2011**, *30*, 345–354.

100. Vanegas, C.A.; Aliaga, D.G.; Wonka, P.; Müller, P.; Waddell, P.; Watson, B. Modelling the appearance and behaviour of urban spaces. *Comput. Graph. Forum* **2010**, *29*, 25–42.

101. Bao, F.; Yan, D.M.; Mitra, N.J.; Wonka, P. Generating and exploring good building layouts. *ACM Trans. Graph.* **2013**, doi:10.1145/2461912.2461977.

102. Musialski, P.; Wimmer, M.; Wonka, P. Interactive coherence-based facade modeling. *Comput. Graph. Forum* **2012**, *31*, 661–670.

103. Bao, F.; Schwarz, M.; Wonka, P. Procedural facade variations from a single layout. *ACM Trans. Graph.* **2013**, doi:10.1145/2421636.2421644.

104. Schwarz, M.; Wonka, P. Procedural design of exterior lighting for buildings with complex constraints. *ACM Trans. Graph.* **2014**, doi:10.1145/2629573.

105. Merrell, P.; Schkufza, E.; Li, Z.; Agrawala, M.; Koltun, V. Interactive furniture layout using interior design guidelines. *ACM Trans. Graph.* **2011**, doi:10.1145/2010324.1964982.

106. Abrishami, S.; Goulding, J.S.; Rahimian, F.P.; Ganah, A. Integration of BIM and Generative Design to Exploit AEC Conceptual Design Innovation. Available online: http://clok.uclan.ac.uk/11420/ (accessed on 29 June 2015).

107. National Institute of Building Sciences. Frequently Asked Questions About the National BIM Standard. Available online: https://www.nationalbimstandard.org/faqs (accessed on 29 June 2015).

108. Abbasnejad, B.; Moud, H.I. BIM and basic challenges associated with its definitions, interpretations and expectations. *Int. J. Eng. Res. Appl.* **2013**, *3*, 287–294.

109. Eastman, C.; Teicholz, P.; Sacks, R.; Liston, K. *BIM Handbook*, 2nd ed.; John Wiley & Sons: New Jersey, NJ, USA, 2011.

110. Krispel, U.; Schinko, C.; Ullrich, T. The rules behind—Tutorial on generative modeling. *Proc. Symp. Geom. Process.* **2014**, *12*, 1–49.

111. Schinko, C.; Strobl, M.; Ullrich, T.; Fellner, D.W. Modeling procedural knowledge—A generative modeler for cultural heritage. In *Digital Heritage*; Springer: Berlin, Germany, 2010.

112. Hichri, N.; Stefani, C.; De Luca, L.; Veron, P. Review of the "as-built BIM" approaches. *ISPRSI Arch. Photogramm. Remote Sens. Spatial Inf. Sci.* **2013**, *XL-5/W1*, 107–112.

113. Kazhdan, M.; Bolitho, M.; Hoppe, H. Poisson surface reconstruction. In Proceedings of the Fourth Eurographics Symposium on Geometry Processing, Cagliari, Sardinia, 26–28 June 2006; pp. 61–70.

114. Weyrich, T.; Lawrence, J.; Lensch, H.P.A.; Rusinkiewicz, S.; Zickler, T. Principles of appearance acquisition and representation. *Found. Trends. Comput. Graph. Vis.* **2009**, *4*, 75–191.

115. Krispel, U.; Evers, H.L.; Tamke, M.; Viehauser, R.; Fellner, D.W. Automatic texture and orthophoto generation from registered panoramic views. *Int. Arch. Photogramm. Remote Sens. Spat. Inf. Sci.* **2015**, *XL-5/W4*, 131–137.

116. Tamke, M.; Blümel, I.; Ochmann, S.; Vock, R.; Wessel, R. From point clouds to definitions of architectural space-Potentials of automated extraction of semantic information from point clouds for the building profession. In Proceedings of the 32nd eCAADe Conference, Northumbria, UK, 10–12 September 2014; pp. 557–566.

117. Hullo, J.F.; Thibault, G.; Boucheny, C. Advances in Multi-sensor scanning and visualization of complex plants: The utmost case of a reactor building. *Int. Arch. Photogramm. Remote Sens. Spat. Inf. Sci.* **2015**, *XL-5/W4*, 163–169.

118. Dore, C.; Murphy, M. Semi-automatic modelling of building Faç with shape grammars using historic building information modelling. *Int. Arch. Photogramm. Remote Sens. Spat. Inf. Sci.* **2013**, *XL-5/W1*, 57–64.

119. Dore, C.; Murphy, M.; McCarthy, S.; Brechin, F.; Casidy, C.; Dirix, E. Structural simulations and conservation analysis-Historic building information model (HBIM). *Int. Arch. Photogramm. Remote Sens. Spat. Inf. Sci.* **2015**, *XL-5/W4*, 351–357.

120. Quattrini, R.; Malinverni, E.S.; Clini, P.; Nespeca, R.; Orlietti, E. From TLS to HBIM. High quality semantically-aware 3D modeling of complex architecture. *Int. Arch. Photogramm. Remote Sens. Spat. Inf. Sci.* **2015**, *XL-5/W4*, 367–374.

121. Ladenhauf, D.; Berndt, R.; Eggeling, E.; Ullrich, T.; Battisti, K.; Gratzl-Michlmair, M. From building information models to simplified geometries for energy performance simulation. In Proceeding of the First International Academic Conference on Places and Technologies, Belgrade, Yugoslavia, 3–4 April 2014.

122. Ladenhauf, D.; Berndt, R.; Krispel, U.; Eggeling, E.; Ullrich, T.; Battisti, K.; Gratzl-Michlmair, M. Geometry simplification according to semantic constraints. *Comput. Sci. Res. Dev.* **2014**, *11*, 1–7.

123. Initiative, D.C.M. Dublin Core Metadata Initiative. Available online: http://dublincore. org/ (accessed on 29 June 2015).

124. Settgast, V.; Ullrich, T.; Fellner, D.W. Information technology for cultural heritage. *IEEE Potentials* **2007**, *26*, 38–43.

125. Fellner, D.W. Graphics content in digital libraries: Old problems, recent solutions, future demands. *J. Univers. Comput. Sci.* **2001**, *7*, 400–409.

126. Fellner, D.W.; Saupe, D.; Krottmaier, H. 3D documents. *IEEE Comput. Graph. Appl.* **2007**, *27*, 20–21.

127. Maybury, M.T. *Multimedia Information Extraction*; John Wiley & Sons: New Jersey, NJ, USA, 2012.

128. Bishop, C.M. *Pattern Recognition and Machine Learning*; Springer: Berlin, Germany, 2007.

129. Ulusoy, I.; Bishop, C.W. Generative versus discriminative methods for object recognition. In Proceedings of the 2005 IEEE Computer Society Conference on Computer Vision and Pattern Recognition, San Diego, CA, USA, 20–26 June 2015.

130. Bustos, B.; Keim, D.; Saupe, D.; Schreck, T. Content-based 3D object retrieval. *IEEE Comput. Graph. Appl.* **2007**, *27*, 22–27.

131. King, B.D.; Wertheimer, M. *Max Wertheimer & Gestalt Theory*; Transaction Publishers: New Jersey, NJ, USA, 2005.

132. Hilaga, M.; Shinagawa, Y.; Kohmura, T.; Kunii, T.L. Topology matching for fully automatic similarity estimation of 3D shapes. In Proceedings of the 28th Annual Conference on Computer Graphics and Interactive Techniques, Los Angeles, CA, USA, 12–17 August 2011; pp. 203–212.

133. Pauly, M.; Mitra, N.J.; Wallner, J.; Pottmann, H.; Guibas, L.J. Discovering structural regularity in 3D geometry. *ACM Trans. Graph.* **2008**, *27*, 1–11.

134. Bokeloh, M.; Wand, M.; Seidel, H.P. A connection between partial symmetry and inverse procedural modeling. In Proceedings of the ACM SIGGRAPH 2010, Los Angeles, CA, USA, 27–29 July 2010.

135. Mitra, N.J.; Guibas, L.J.; Pauly, M. Partial and approximate symmetry detection for 3D geometry. *ACM Trans. Graph.* **2006**, *25*, 560–568.

136. Mitra, N.J.; Guibas, L.J.; Pauly, M. Symmetrization. *Int. Conf. Comput. Graph. Interact. Tech.* **2007**, *26*, 1–8.

137. Ullrich, T.; Schinko, C.; Fellner, D.W. Procedural modeling in theory and practice. In Proceedings of the 18th WSCG International Conference on Computer Graphics, Visualization and Computer Vision, Plzen, Czech Republic, 27 March 2010.

138. Autodesk. Autodesk Maya API. Available online: http://docs.autodesk.com/MAYAUL/ 2014/ENU/Maya-API-Documentation/index.html (accessed on 29 June 2015).

139. Ramamoorthi, R.; Arvo, J. Creating generative models from range images. *Proc. ACM SIGGRAPH* **1999**, *1*, 195–204.

140. Ullrich, T. Reconstructive Geometry. Ph.D. Thesis, Technische Universität, Graz, Austria, 2011.

141. Arnold, D. Computer graphics and cultural heritage: From one-way inspiration to symbiosis. *Comput. Graph. Appl.* **2014**, *34*, 76–86.

142. Havemann, S.; Settgast, V.; Krottmaier, H.; Fellner, D.W. On the integration of 3D models into digital cultural heritage libraries. In Proceedings of the 7th International Symposium on Virtual Reality, Archaeology and Cultural Heritage (VAST), Nicosia, Cyprus, 30 October–4 November 2006.

143. Niccolucci, F. XML and the future of humanities computing. *ACM SIGAPP Appl. Comput. Rev.* **2002**, *10*, 43–47.

144. Niccolucci, F.; D'Andrea, A. An ontology for 3D cultural objects. In Proceedings of the 7th International Symposium on Virtual Reality, Archaeology and Cultural Heritage (VAST), Nicosia, Cyprus, 30 October–4 November 2006.

145. Arnold, D. Computer graphics and cultural heritage: Continuing inspiration for future tools. *Comput. Graph. Appl.* **2014**, *34*, 70–79.

146. Berndt, R.; Gerth, B.; Havemann, S.; Fellner, D.W. 3D modeling for non-expert users with the castle construction kit v0.5. In Proceedings of the 6th International Symposium on Virtual Reality, Archaeology and Cultural Heritage (VAST), Pisa, Italy, 8–11 Novermber 2005.

147. Ullrich, T.; Settgast, V.; Fellner, D.W. Semantic fitting and reconstruction. *J. Comput. Cult. Heritage* **2008**, *1*, 1201–1220.

148. Ullrich, T.; Schinko, C.; Schiffer, T.; Fellner, D.W. Procedural descriptions for analyzing digitized artifacts. *Appl. Geomat.* **2013**, *5*, 185–192.

149. Schinko, C.; Ullrich, T.; Fellner, D.W. Modeling with high-level descriptions and low-level details. In Proceeding of the International Conference on Computer Graphics, Visualization, Computer Vision and Image Processing, Lisbon, Portugal, 15–19 July 2014.

150. Calogero, E.; Arnold, D. Generating alternative proposals for the louvre using procedural modeling. In Proceedings of the 4th ISPRS International Workshop 3D-ARCH, Trento, Italy, 2–4 March 2011.

151. Saldana, M.; Johanson, C. Procedural modeling for rapid-prototyping of multiple building phases. *Int. Arch. Photogramm. Remote Sens. Spat. Inf. Sci.*, **2013**, *1*, 205–210.

152. Fellner, D.W.; Havemann, S. Striving for an adequate vocabulary: Next generation metadata. In Proceedings of the 29th Annual Conference of the German Classification Society, Magdeburg, Germany, 9–11 March 2005.

153. Havemann, S.; Fellner, D.W. Seven research challenges of generalized 3D documents. *IEEE Comput. Graph. Appl.* **2007**, *3*, 70–76.

154. Havemann, S.; Ullrich, T.; Fellner, D.W. The meaning of shape and some techniques to extract it. *Multimed. Inf. Extr.* **2012**, *1*, 81–98.

155. Marcheix, D.; Pierra, G. A survey of the persistent naming problem. In Proceedings of the ACM Symposium on Solid Modeling and Applications, Saarbrucken, Germany, 17–21 June 2002.

Chapter 5:
Structural Analysis

Photogrammetric, Geometrical, and Numerical Strategies to Evaluate Initial and Current Conditions in Historical Constructions: A Test Case in the Church of San Lorenzo (Zamora, Spain)

Luis Javier Sánchez-Aparicio, Alberto Villarino, Jesús García-Gago and Diego González-Aguilera

Abstract: Identifying and quantifying the potential causes of damages to a construction and evaluating its current stability have become an imperative task in today's world. However, the existence of variables, unknown conditions and a complex geometry hinder such work, by hampering the numerical results that simulate its behavior. Of the mentioned variables, the following can be highlighted: (i) the lack of historical information; (ii) the mechanical properties of the material; (iii) the initial geometry and (iv) the interaction with other structures. Within the field of remote sensors, the laser scanner and photogrammetric systems have become especially valuable for construction analysis. Such sensors are capable of providing highly accurate and dense geometrical data with which to assess a building's condition. It is also remarkable, that the latter provide valuable radiometric data with which to identify the properties of the materials, and also evaluate and monitor crack patterns. Motivated by this, the present article investigates the potential offered by the combined use of photogrammetric techniques (DIC and SfM), as well as geometrical (NURBs and Hausdorff distance) and numerical strategies (FEM) to assess the origin of the damage (through an estimation of the initial conditions) and give an evaluation of the current stability (considering the deformation and the damage).

Reprinted from *Remote Sens.* Cite as: Sánchez-Aparicio, L.J.; Villarino, A.; García-Gago, J.; González-Aguilera, D. Photogrammetric, Geometrical, and Numerical Strategies to Evaluate Initial and Current Conditions in Historical Constructions: A Test Case in the Church of San Lorenzo (Zamora, Spain). *Remote Sens.* **2016**, *8*, 60.

1. Introduction

The conservation of built heritage is today considered a fundamental aspect of modern society. Their artistic, cultural, and intrinsic value make these constructions extremely important. Complementary to this, the lack of the building's own mechanical values and the characteristic behavior of its masonry, the complex interaction between components, and the lack of documentation, make the analysis of such constructions remarkably difficult. Currently, and derived from these considerations, numerous regulations propose the integration of different approaches among which are [1]: (i) the study of the construction's history; (ii) inspection; (iii) monitoring; and (iv) structural analysis.

Regarding the numerical calculations, the static graphic [2] and limit analyses [3] traditionally provided the necessary tools to study the stability and bearing capacity of historical structures [4]. However, such numerical strategies have among their drawbacks the difficulty to evaluate damages [1].

In contrast with these models, the Finite Element Method (FEM) has been widely used for the evaluation of historical buildings at different levels; from complex and large constructions through macromodelling techniques [5], to the use of micromodelling strategies [6], where the units are independently discretized, or homogenized [7]. However, the large number of involved variables, as well as interaction with other structures, conditions the results.

It is in the field of built constructions where remote sensors and especially photogrammetric and laser scanner systems have proven great worth for their analysis [3,8–10]. These sensors are able to provide accurate and dense geometric and radiometric values with which to assess these buildings, as well as obtaining the data through non-intrusive means. Despite this, the data they provide (in form of dense and accurate point clouds) is largely untapped, since it is only used for the construction of simplified CAD models [10].

On one hand, the present article introduces two novel robustness parameters (based on geometrical components) in order to increase their applicability, obtained from the symmetrical Hausdorff distance [11]. These parameters, called Global Hausdorff metric (GHm_s) and Local Hausdorff metric (LHm_s), help ascertain whether the variables or simulated conditions improve or worsen the numerical results, in comparison with the real deformation provided by the photogrammetric and laser scanner systems.

On the other hand, the article introduces a methodology based on a Non-Uniform Rational B-Splines (NURBs) modelling strategy, with the purpose of providing an accurate geometrical model (with the current deformation and damage) for the evaluation of the current stability of the construction. This strategy is able to take advantage of the properties provided by the Structure from Motion products: (i) density; (ii) accuracy; and (iii) photorealistic texture, within a numerical environment.

In order to confirm the feasibility of the proposed geometrical strategies (GHm_s, LHm_s and NURBs modelling), they are applied to a case study: the dome of the church of San Lorenzo in Sejas de Aliste (Zamora, Spain). This construction, built in brick masonry, has suffered severe structural damages, shown through significant deformation, cracking and plastic hinges that reduce its bearing capacity.

It seems necessary to perform a structural evaluation in order to design efficient restoration actions.

The article is organized as follows: Section 1 consists of an introduction and brief state of the art, Section 2 describes the different image-based techniques that were employed; Section 3 is made up of the description of the construction, the current deformation, damage, and the numerical aspect through the FEM; Section 4 describes two robustness indices based on geometrical discrepancies, a manual calibration of the model and a complementary strategy to evaluate the current stability of the construction (considering the complex geometry and the presented cracking); and finally, Section 5 shows the conclusions.

2. Image Based Approaches: Digital Image Correlation and Structure from Motion

The great diversity of approaches today, along with their flexibility, place image-based procedures as a suitable solution for the analysis of constructions [3,9,12], materials [13,14], and pathologies [8].

The different methodologies that comprise this approach, particularly in the field of numerical evaluation of constructions, highlight: (i) Digital Image Correlation (DIC); and (ii) image-based modelling procedures. While the former provides mechanical data of materials and constructive solutions (in the form of displacement and strains), the latter allows the definition of a dense, accurate, and photorealistic geometrical model of the construction. Their combination provides relevant information for the numerical analysis of the structure.

2.1. 2D Digital Image Correlation

A wide variety of methodologies has been developed and used to study material and union behavior. Some of these are [14,15]: (i) Moiré interferometry; (ii) Holography interferometry; (iii) Shearography; and (iv) Digital Image Correlation.

These methodologies prove to have important advantages, compared to traditional methods based on strain gauges or LVDT's (Linear Variable Differential Transformer) such as their non-invasiveness and their full-field data information. In comparison, traditional methods provide only local information and require direct contact with the tested material. Within this wide range of techniques, the use of Digital Image Correlation (DIC) stands out.

To characterize the materials used in the dome, various compression tests were performed separately on each material (three in each material) during the experimental campaign that was carried out. Considering the procedure defined by [16], an extra specification, such as the mortar joint material (made by gympsum mortar), was considered.

In order to verify the flexibility and accuracy of the shown method, one standard sensor was used for both the DIC and the SfM: a digital reflex camera Canon 500D. However, in contrast to the image-based modelling strategy, DIC requires the preparation of the analyzed specimen, following the approach defined by [16]: (i) MIG (Mean Intensity Gradient) evaluation [17] of the speckle pattern; (ii) camera pose estimation [18]; and (iv) camera calibration [19].

Once the specimen has been correctly pre-processed, different images were captured during the test (Figure 1). Also, concerning the test setup, a large focal distance and working distance were used in order to minimize the geometrical distortion, out of plane displacements (approximation to a telecentric lens system) [20], depth of field, and light conditions.

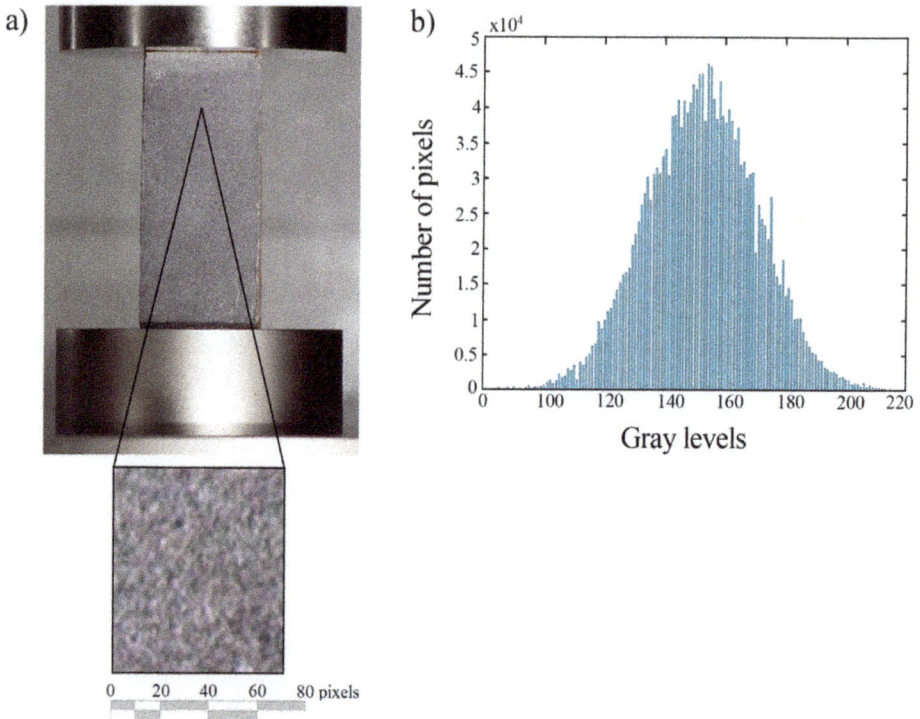

Figure 1. (a) Detail view of the brick and speckle pattern applied during the Digital Image Correlation (DIC) test; (b) Histogram of the speckle pattern.

The basic principle of DIC is the tracking (or matching) of the different areas of the images which were captured during the test (before and after deformation occurs), called subsets. As an initial approximation of this tracking, a correlation coefficient (generally the Zero Normalized Cross Correlation) [20] is used. Later, this initial approximation is optimized by the use of a non-linear strategy (such as the Inverse Compositional Gauss Newton method) [20] which allows the evaluation of the displacement suffered by the subset along the different captured images (Figure 2) [16].

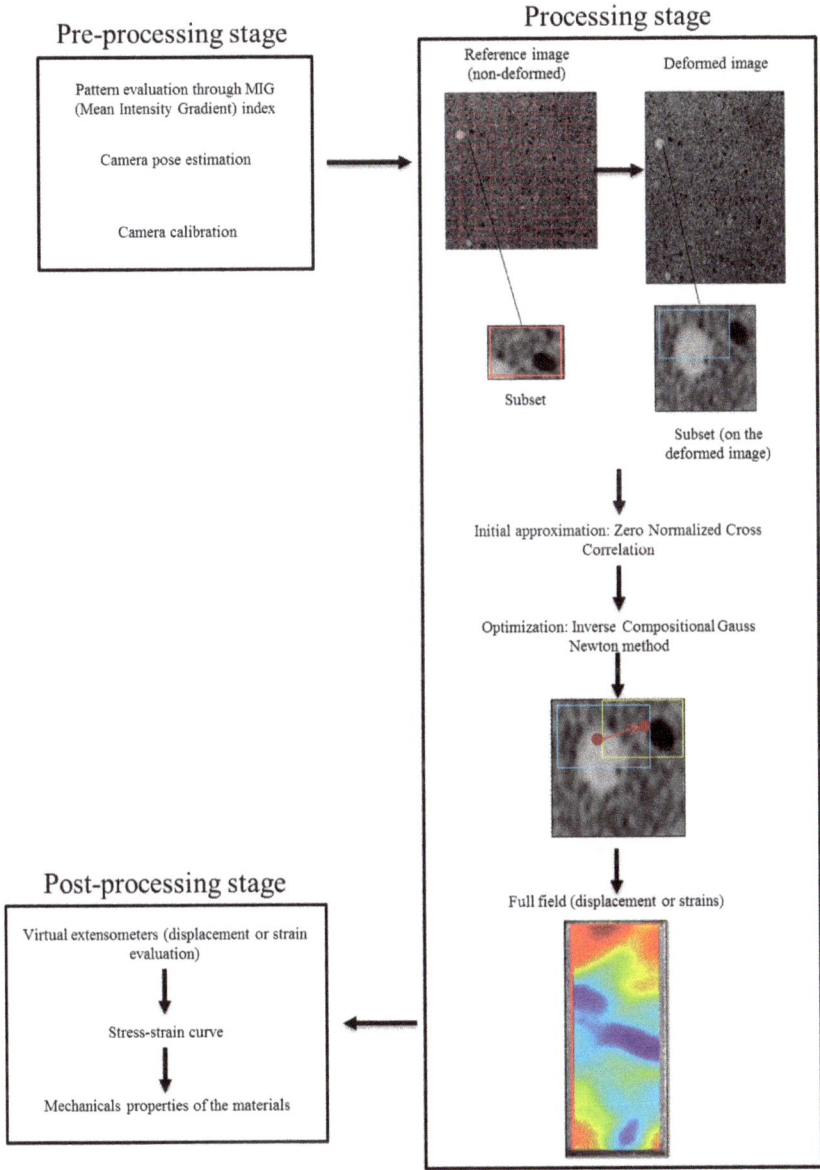

Figure 2. Digital Image Correlation general outline. In red the reference subset, in blue the initial seed, and in yellow, the final location of the subset.

Complemented to this optimization process an interpolation process (based on splines) is used with the aim of obtain sub-pixel accuracy [20]. Considering multiple subsets in the image, their analysis can provide a full-field displacement. Later, the strains suffered by the specimen during the test, which allow the evaluation of its mechanical properties, can be obtained by a direct relationship between the obtained displacement on the measurement point and the initial length of the virtual extensometers [16]. A total of three virtual extensometers were placed on the ROI: (i) A-A′ and B-B′ in the longitudinal direction; and (ii) C-C′ in the transversal direction.

Concerning accuracy, there are different studies [15,21,22] that endorse the DIC's precision for the assessment of the material's mechanical properties. For a standard test configuration, the accuracy may range from values of 0.01–0.04 pixels. Considering a conservative threshold at 0.1 pixels, and an acceptable accuracy for the test of 0.01 mm (from which the critical pixel size is set at 0.1 mm), the test's configuration is shown in (Table 1).

Table 1. Summary of the different properties set during the Digital Image Correlation (DIC) test carried out with a Canon EOS 500D and a macrolens system 70–300 mm.

Values Adopted during the Digital Image Correlation (DIC) Test	
Aperture	7.1
Focal length (mm)	200
Working distance (mm)	2700
Pixel size (mm)	0.063
Acquisition frequency (Hz)	0.33

Once the stress-strain curve has been obtained (by a relationship between the stress applied by the compression press and the strains obtained by DIC) (Figure 3b), it is possible to extract the mechanical properties of the materials as follows (Table 2): (i) the Young Modulus was considered as the ratio between one third of the maximum force achieved and the mean strain provided by the longitudinal extensometers (A-A′ and B-B′); (ii) for the Poisson ratio, the relationship between the strains provided by the longitudinal extensometers and those obtained by the transversal extensometer (C-C′) was taken into account; and (iii) the compression strength was considered as the maximum pressure supported by the specimen.

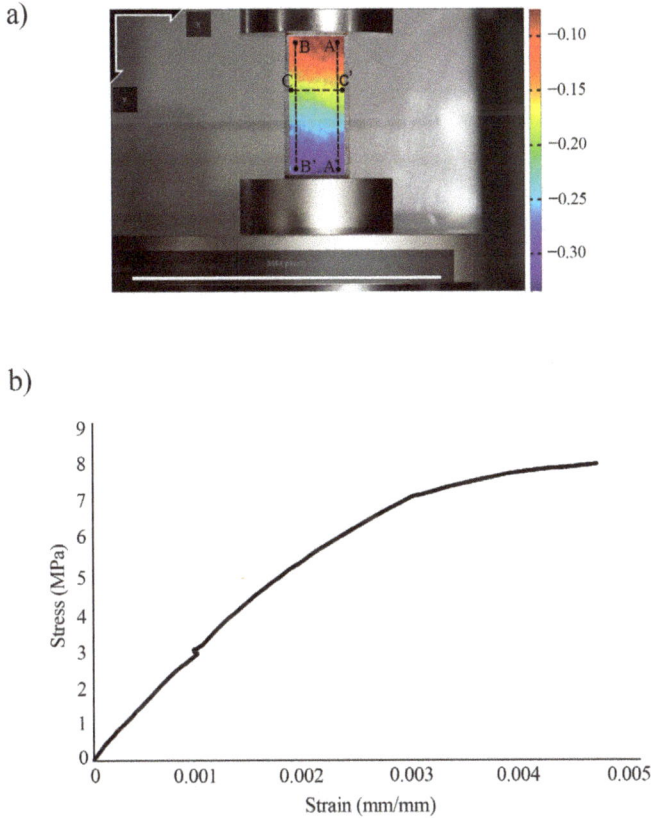

a)

b)

Figure 3. Results after the experimental campaign (2D DIC). (a) Deformation measurement, expressed in pixels, between two captures and positioning of the virtual extensometers; (b) Stress-strain curve obtained with the virtual extensometer A-A'.

Table 2. Mechanical properties obtained by the performed DIC test.

	Mechanical Properties Obtained by the DIC Test	
	Clay brick	Gypsum mortar
E (GPa)	3.10 ± 0.30	1.15 ± 0.06
ν (-)	0.22 ± 0.05	0.23 ± 0.02
fc (MPa)	7.80 ± 0.90	2.12 ± 0.10

To assess the accuracy of the previously mentioned procedure, a comparative study was carried out, between the strain rate applied by the compression press (with an average value of -1.77×10^{-6} s^{-1}), and the one obtained by the different performed DIC tests and the different virtual extensometers used (with an average value of -1.93×10^{-6} s^{-1}). The results obtained demonstrate the accuracy and suitability of the applied configuration and algorithms, with an estimated precision of 0.056 pixels (which correspond to an approximate value of 3.53 µm). This value proves to be lower than the previously shown critical value.

2.2. Image-Based Modelling: Structure from Motion

In recent years the image-based modelling strategy, called Structure from Motion (SfM), has positioned itself as an attractive alternative to laser scanning systems. Its flexibility—as it can be integrated into different types of platforms (e.g., UAV [9])—low-cost, and qualities of the point cloud (high density, photorealistic texture, and accuracy) place the solution at a vantage position in the evaluation of historical buildings [23].

This technique integrates within its operating structure the advantages of computer vision (automation and flexibility) and photogrammetry (accuracy and reliability) [23] to obtain high density three dimensional models whose accuracy can compete with those of the laser scanner system [24,25].

For this case study, a standard SfM strategy is applied, comprising the following stages: (i) automatic extraction and keypoint matching by applying the Affine-Scale Invariant Feature Transform (ASIFT) algorithm [26]; (ii) automatic hierarchical orientation of images; and (iii) dense model generation through the MicMac algorithm. For further details on this methodology see [12]. Concerning the photogrammetric network a convergent protocol was used, combining a total of 32 cameras with high overlap (around 90%) and throwing a mean GSD (Ground Sample Distance) of 1.61 mm. Complementary to these, different circular targets (along the lower part of the pendentive) were used to scale the model (this measurement were taken by a total station using a radiation approach).

As a result of the implementation of the above-mentioned methodology it is possible to obtain a dense and photorealistic texture point cloud (Figure 4a). Afterwards, applying CAD conversion techniques (meshing, surface parameterization, extrusions, revolutions, *etc.*), or even generating true-ortho-images, increase further the applicability of the obtained product. More precisely, they help to accurately build CAD models suitable for subsequent numerical simulations, as well as complementary products, which analyze patterns of deformation and cracking for the pathological characterization of the structure [9] (Figure 4b).

Concerning the total error, associated with this point cloud, a quadratic error propagation was used Equation (1). Into this approach, two sources were considered: (i) the error coming from the bundle adjustment of the photogrammetric network; and (ii) the error corresponding to the scaling process Equation (2).

$$\varepsilon_t = \sqrt{\varepsilon_p{}^2 + \varepsilon_s{}^2} \qquad (1)$$

$$\varepsilon_s = \sqrt{2\varepsilon_i^2 + \varepsilon_m^2} \tag{2}$$

where ε_t represents the total error; ε_s the scale error; ε_p the error associated with the photogrammetric network; ε_i the origin error stablished as $\sqrt{2} * (\dfrac{pixelsize}{\gamma})$, where γ is the subpixel accuracy of the target detection algorithm (estimated in 0.5); and ε_m the error associated with the total station.

As a result a budget error of ε_t = 4.38 mm was obtained (with values of ε_p = 3.22 mm, ε_i = 1.14 mm, ε_m = 2.50 mm, and ε_s = 2.97 mm).

a)   b)

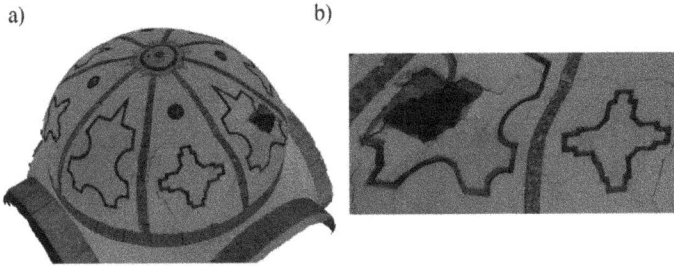

Figure 4. (**a**) 3D model obtained by the proposed methodology; (**b**) Detail view of the most damaged section through the texture model.

3. Structural Evaluation of San Lorenzo's Dome

3.1. San Lorenzo Church

The church is built with irregular masonry walls (slabs of slate) fixed with lime mortar and at the corners finished with granite masonry. The parish church of San Lorenzo in Sejas de Aliste is located in the region of Aliste, Zamora province (Spain), 32 m long and 17 m wide, it belongs to the family of temples with transept crossing, Latin cross-shaped floor plan, and transept and nave at different heights (Figure 5).

The transept crossing is the most representative element of this temple. Its importance in the building is highlighted in the interior through the semi-elliptical dome that shelters the whole crossing. Its eight ribs marked with bands stand out. The transept is highlighted in the outside as well, covering the dome with a hipped roof that rises above the nave and transept height. This roof is built with a pavilion-shaped chestnut-timber framing, with regularly placed rafters that lean on the main beams and bear the load of the roof, made up of curved tiles and wooden roof boards (Figure 6a). Overall stability is obtained by use of tie beams at the top of the bearing walls, which collect the loads of the rafters and the hip rafters or main beams. Angle-ties, placed at 45° in each corner under the hip rafters prevent the transversal deformation of the tie beams (Figure 6a).

Figure 5. San Lorenzo church: (**a**) Orthophoto of the main façade through the methodology proposed; (**b**) Orthophoto of the west façade (chancel) of the construction; (**c**) Floor plan-view of the church, red color indicates the damaged area of the dome.

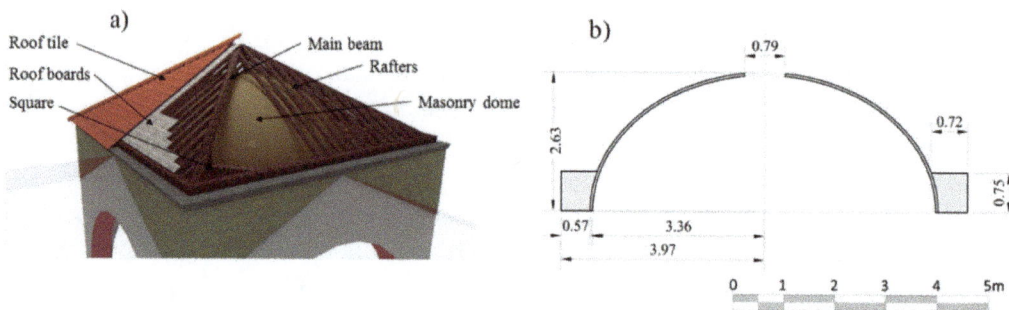

Figure 6. (**a**) Constructive section of the church's transept; (**b**) Transversal section of the dome geometry (initial state estimated by the Structure from Motion (SfM) point cloud) with dimensions in meters.

Concerning the dome, the construction has an estimated diameter of 6.72 m and a total height (measured from the pendentive) of 2.63 m. This structure was built with traditional tile brick and gypsum mortar, reaching a total thickness of 5.00 cm. It resembles, from a construction point of view, a Catalan vault (Figure 6b). It is also worth mentioning the presence of an infill (basically composed of a mixture of sand, clay and fragments of bricks with a medium compaction) at the support of the dome. This infill reaches a total height of 0.75 m and an average thickness of 0.65 m, and its presence contributes to the stability of the construction.

3.2. Present Damage and Deformation

The characterization of both, deformations and cracking patterns, is key to understanding the structure in terms of stability and safety. The high density, accuracy, and photorealistic texture of the point cloud obtained by the proposed methodology (Section 2.2) can address this task foregoing any need of physical contact with the structure. Through evaluation of the obtained product, it is possible to obtain a hypothesis for the origin of the damage.

It is worth noting that there is widespread damage in the area enclosed by three ribs (corresponding to the southern part of the dome). This area has two main cracks, in the parallel direction, which are interconnected through the presence of two plastic hinges. At its maximum, there is a deflection of 19.70 cm (compared to the initial estimated model) (Figure 7).

These structural pathologies seem to be related to the presence of assymetric loads acting on part of the dome's shell (Figure 7a). More specifically the current damages, which are located under the south wing, can be attributed to a failure of the timber structure.

On one hand, the evaluation of photogrammetric products (which are the result of the previously defined SfM strategy) allows an estimation of the possible causes of the dome's damage. However, it is required to have numerical strategies to verify these assumptions and assess the current state of the construction. For the present case study, and considering the hypothesis of failure of the timber structure several numerical analyses were performed: (i) numerical evaluation of the timber structure for the worst load case: snow; (ii) evaluation of the dome's stability under self-weight; and (iii) numerical evaluation of the interaction timber-dome as a result of a timber failure.

Previous investigations carried out by [16], verified the stability of the timber structure for the most adverse load case: the presence of snow. Yielding a maximum deflection of 2.35 cm, it proves to be insufficient in order to interact with the dome. Considering these results, it is possible to conclude that the interaction between the cover and the dome seems to be linked with the presence of pathological agents (mainly moisture and biological organisms) which reduce the bearing capacity of the timber structure until it fails and rests on the dome.

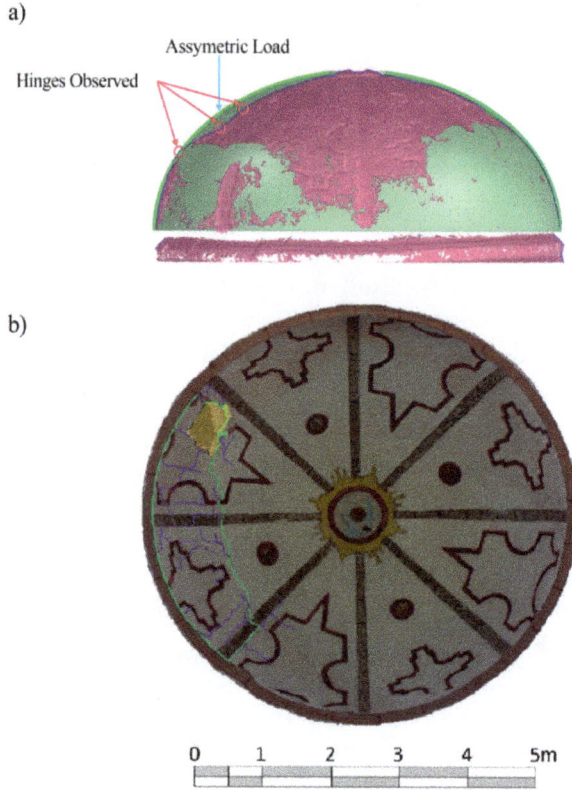

a)

b)

Figure 7. Results of the visual inspection over the different photogrammetric products: (**a**) Surface comparison between the initial proposed model and the most deformed one estimated by the SfM point cloud; (**b**) Damage inspection in the orthophoto, in green the main observed cracks, in blue the secondary cracks, in yellow the material removal.

3.3. Numerical Simulation of the Initial State of the Dome: Self-Weight and South Wing Support

Understanding the degradation mechanisms present in the construction requires a geometrical model of its initial state, a material characterization, as well as its boundary conditions, and load assessment. According to these, several numerical simulations (through non-linear static analysis) were performed in order to understand the causes and the construction's initial conditions. Several improvements, regarding the geometric and mechanical aspects, are introduced in comparison to the previous investigations performed by [16]: (i) consideration and modelization of the infill-dome interaction; and (ii) account of gypsum as union material.

Regarding the mechanical aspect, a macromodelling strategy of the masonry was followed. This technique blends the bricks, mortar joints as well as the brick-mortar interface into one continuum assuming homogeneous material properties (Table 3). Also, the recommendations exposed by reference [6,27] were considered. An initial estimation of the masonry's Young modulus was estimated using the formulas displayed in [28], setting the initial Young modulus at 2.54 GPa. However, further visual inspections showed the presence of an erratic masonry with low overlap between units. In accordance with this, a reduction of the initial Young modulus was considered for subsequent simulations (half of the initial estimated), yielding a final value of 1.22 GPa, analogous to those used in similar studies [7,29].

Table 3. Mechanical properties adopted for the macromodelling of the masonry.

Mechanical Properties for the Masonry Structure		
E_m (GPa)	Young Modulus	1.22
δ_m (kg/m^3)	Density	1800.00
ν_m (-)	Poisson coefficient	0.25
$f_{t,m}$ (MPa)	Tensile strength	0.16
$f_{c,m}$ (MPa)	Compressive strength	1.60
$d_{t,m}$ (mm)	Ductility index in tensile	0.093
$d_{c,m}$ (mm)	Ductility index in compression	1.6
$\beta_{c,m}$ (-)	Shear retention	0.2

For the numerical simulation of the infill, a Morh-Coulomb failure criterion was considered, with its mechanical properties set according to the visual inspection (medium compaction) and the recommendations shown by [7,29,30] (Table 4).

Table 4. Mechanical properties adopted for the infill simulation.

Infill Mechanical Properties		
E_i (GPa)	Infill Young Modulus	0.80
δ_i (kg/m^3)	Infill density	1800.00
G (GPa)	Infill shear modulus	$E_i/2$
f_i (MPa)	Cut-off tension	0.02
Φ_i (deg)	Infill friction angle	39
c_i (MPa)	Infill cohesion	$1 \times f_i$

Concerning the load (for the numerical evaluation of the interaction between timber structure and the dome), a value of 8000 N was considered, resulting from the combination of different loads: (i) 650 N/m^2 for the arabic tiles and wooden board; and (ii) 400 N/m^2 for the snow load. Finally, the numerical model (for both simulations), had a total of 46,181 high order solid elements (CTE30) [31] (Figure 8a).

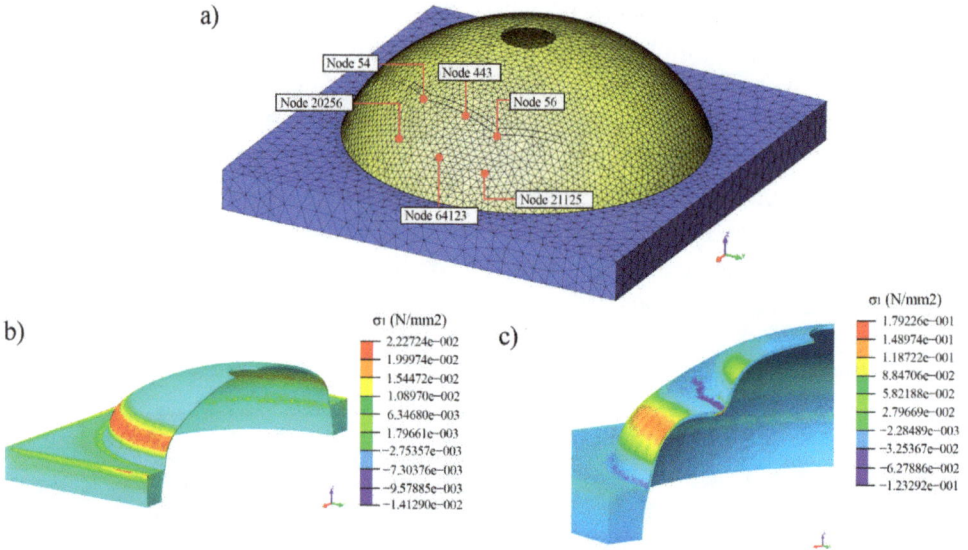

Figure 8. (a) Isometric view of the mesh and the control points (nodes) used for the numerical simulations; (b) First principal stress distribution, expressed in N/mm^2 for the self-weight case; (c) First principal stress distribution, expressed in N/mm^2, for the numerical model which considers the asymmetric load.

It is possible to observe that in the absence of external loads acting on the dome, the structure seems to be stable under its own weight (Figure 8b). In spite of this, considering the support of the south wing (roof tile, boards, and rafters), the dome begins to present damage (cracking) and its deformation tendency (Figure 8c) seems to be similar to the one shown in the photogrammetric model (Figure 7a).

However, in terms of deformation, considering for this purpose six control nodes along the damage area (Figure 8a), the model exhibits high rigidity. This suggests that the initial mechanical conditions are inadequate to reproduce the damage and deformation presented in the dome (Table 5).

The high discrepancies shown in the previous numerical simulation suggest the need for an optimization of the mechanical properties. However, performing such an optimization requires inevitably having robustness indices to quantify the level of improvement/worsening introduced by the different variable's variations.

Table 5. Comparison between the obtained and expected displacement of the control nodes in the numerical model (initial considerations).

Control Node	Displacement Obtained (mm)	Displacement Expected (mm)
54	1.95	148.00
20256	0.48	46.00
56	1.57	198.00
21125	0.54	52.00
443	2.12	196.00
64123	0.57	25.00

Exploiting the advantages offered by the SfM or laser scanner systems, two novel robustness parameters (based on geometric discrepancies) are proposed: (i) a global parameter, based on the similarity between the numerical and real model; and (ii) a local index which provides data about the geometrical variations introduced by the new variables considered in different areas of the construction.

4. SfM, NURBS Modelling, Global and Local Hausdorff Metrics: Geometrical Strategies to Improve the Knowledge about the Initial and Current State of the Constructions

4.1. Global and Local Hausdorff Metrics as Geometric Accuracy Indices

The Hausdorff distance or Hausdorff metric is used in a wide range of fields, such as point cloud [32] and meshes [33] comparison, object recognition [34], and image comparison and matching [35]. This metric proves to be a robust strategy for the similarity evaluation of two compact and non-empty sub-sets within a metric space. It is formulated as follows Equations (3) and (4):

$$d(y, X) = \min_{x \in X} \|y - x\|_2 \tag{3}$$

$$d_H(Y, X) = \max_{y \in Y} d(y, X) \tag{4}$$

where $\|.\|_2$ stands for the Euclidean norm; min the minimum value (distance); max the maximum distance; X and Y are the two compact sub-sets defined by the numerical and photogrammetric nodes; and x and y the considered points inside these sub-sets.

It is worth mentioning that, considering the previously defined concept of Hausdorff distance, the value of the norm does not have a symmetrical nature; it is therefore different in each direction ($d_H(X,Y) \neq d_H(Y,X)$). For that reason, the symmetrical Hausdorff distance d_{SH} Equation (5) is used as metric comparison to avoid potential errors of geometrical similarity. This way a more robust solution is provided for geometry comparison.

$$d_{SHi} = \max \{d_H(y, X), d_H(x, Y)\} \tag{5}$$

where d_{SHi} is the symmetrical Hausdorff distance; of sub-set i, between models (numerical and photogrammetric); and x and y are two points that respectively belong to sub-sets X and Y.

On the other hand, understanding the global structural behavior of the analyzed construction inevitably requires several numerical analyses in order to adapt the simulated behavior to the real one. It is necessary to take into account the consideration that new conditions or new values of variables may worsen or improve the global and/or local result of the structure. It is therefore possible to define, out of the previously shown comparison metric Equation (3), two novel geometrical indices of robustness that represent improvements or worsening in the new numerical simulations in comparison to a reference model, considering the different variations of the variables or conditions: Global Hausdorff metric Equation (6); and Local Hausdorff metric Equation (7).

$$GHm_s = \left(\frac{\sum\limits_{i=1}^{n} d_{SH}(i) - \sum\limits_{i=1}^{n} d_{SH_b}(i)}{\sum\limits_{i=1}^{n} d_{SH_b}(i)} \right) \times 100 \tag{6}$$

$$LHm_s(i) = \frac{d_{SH}(i)}{d_{SH_{ref}}(i)} \tag{7}$$

where GHm_s represent the Global Hausdorff metric index and LHm_s the Local Hausdorff metric index, $d_{SH(i)}$ the symmetrical Hausdorff distance to cluster i considered for the model; $d_{SHb}(i)$ the symmetrical Hausdorff distance for cluster i of the base model (the model that results from the geometrical discrepancies between the initial model and the photogrammetric one); and $d_{SHref}(i)$ the symmetrical Hausdorff distance from cluster i to the reference one (which may be the base model).

On one hand, GHm_s is able to provide a global value, expressed in percentage, for the improvement/worsening of the numerical simulation model in comparison to the model that was considered as base model. On the other hand, LHm_s provides a comparison of the variations between the numerical model and the reference model at a local level (values lower than one indicates a local improvement and values higher than one, a worsening).

For this case study, the reference model was considered to be the base model, obtained by the application of Equation (3) between the photogrammetric model and the non-deformed numerical model (Figure 9).

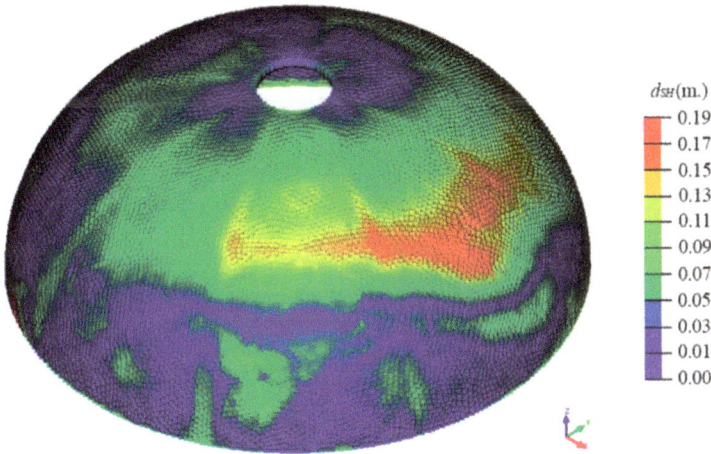

Figure 9. Graphical distribution of the different considered symmetrical Hausdorff distance (d_{SH}) (expressed in m) for the base model.

Finally, and considering GHm_s and LHm_s as the robustness indexes, a manual calibration was carried out, according to the established lower and upper bounds (Table 6) (Figure 10). According to [36], which provides a range of mechanical properties for historical masonry constructions, the upper and lower bounds were established with a safety factor of 1.35, since nowadays only visual inspection and geometrical survey are available (without an extensive experimental campaign).

Table 6. Parameters and variables considered during the manual calibration stage.

	Variable	Initial Value	Upper Bound	Lower Bound	Update Value
$f_{t,j}$ (MPa)	Masonry tensile strength	0.16	0.20	0.05	0.13
E_i (GPa)	Infill Young Modulus	0.80	1.00	0.05	0.50
E_m (GPa)	Masonry Young Modulus	1.22	0.89	1.33	0.90
β (-)	Shear retention factor	0.20	0.01	0.20	0.15

Noteworthy is the presence of a red area (Figure 10). The said phenomenon is associated with the presence of an offset in the spatial distribution of the plastic hinge in comparison to the photogrammetric one. Considering the results provided by the GHm_s and LHm_s indices (Figure 10b) (Table 7), a mild improvement in the geometrical similarity between the photogrammetric and numerical model (Figure 9b) is observable compared with the initial conditions (Figure 10a) and previous studies carried out on the dome [16] (presence of an infill, independent oculus, and manual calibration of the mechanical properties).

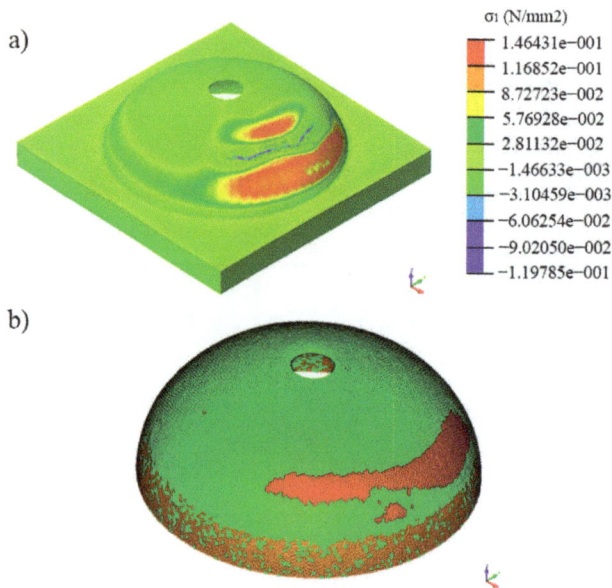

Figure 10. (a) First principal stress distribution, expressed in N/mm2 of the updated model; (b) Geometrical accuracy, in terms of Local Hausdorff metric (LHm_s) of the updated model; in green, values where the geometrical model improves the results, in orange values where no improvements are carried out and in red, areas where the updated numerical model displays worse behavior.

On one hand, the obtained numerical results, with a value of GHm_s of 7.40%, are insufficient to study the current stability of the dome based on an initial state model. The discrepancies, derived from the large number of currently unknown variables, call for the use of additional sensors as well as additional experimental campaigns (in laboratory and in field tests).

On the other hand, the causes of the current damage and deformation correspond to the initial one: a local failure of the timber structure (south wing) could be the cause of pathological agents acting on the wood (moisture and biological agents).

In order to understand the current stability of the construction it is required to evaluate it with the actual deformation and damage (cracks). Motivated by this, and given the geometrical and radiometric properties provided by the SfM systems, a geometric strategy is defined below.

Table 7. Comparison between expected and predicted displacement of the considered control nodes.

Control Node	Displacement Obtained (mm)	Displacement Expected (mm)
54	16.56	148.00
20256	22.80	46.00
56	26.10	198.00
21125	31.52	52.00
443	46.19	196.00
64123	38.92	25.00

4.2. Analysis of the Current Stability of the Construction Based on a SfM and NURBs Approach

It should be stressed that the structural evaluation of historical constructions not only implies the assessment of the damage's causes, but also requires a thorough understanding of the current stability (considering the actual deformation and damage), in order to take efficient restoration actions on the construction and to predict its integrity in case of different events (e.g., earthquakes). With the aim of improving the knowledge of the current stability, with respect to previous studies (Section 3.3 and [16]), a new approach is needed.

Although the point clouds obtained by the previously defined SfM approach, rich in geometric (density and accuracy) and radiometric (photorealistic texture) features, accurately represent the actual state of the construction, it is required to have additional strategies capable of exporting these properties into a numerical environment. The resulting mesh (triangulation of the SfM point cloud) has significant shortcomings to be considered as a suitable CAD/CAM model. Among its deficiencies, the following stand out [37]: (i) High density/resolution, which implies a large number of triangles and (ii) inadequate shapes.

Under the said framework, a methodology able to exploit these features based on the Non-Uniform Rational B-Splines (NURBs) and enhanced by the integration of structural pathologies (such as cracks and lack of material) is proposed. It follows the workflow shown below (Figure 11).

Considering the point cloud as the starting point, this product is firstly meshed by a standard Delaunay triangulation. Usually, these meshes present a non-manifold structure, which implies a low quality product with non-natural triangles which hinder the NURBs' generation. In order to minimize this drawback we use a topological reconstruction, which generates a maniflod mesh, based on the approach defined by [38].

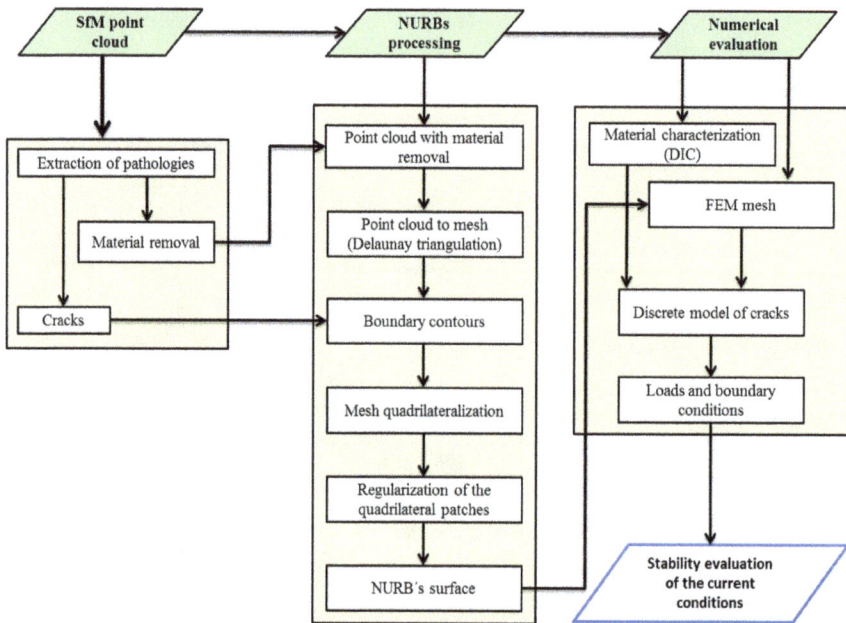

Figure 11. Proposed workflow for the study of the current stability of the construction.

Once the mesh has been correctly defined, a region clustering (boundary contours) was carried out, comprising two critical stages. In a first stage the boundary vertex (limits of the construction, lack of material and the absence of an oculus piece) of the mesh are extracted. Later, in a second stage, the cracks are integrated into these boundary contours, through a projection of the observed crack path on the SfM point cloud, as established [9].

After that, a correct representation of these regions by means of NURBS patches is required. For this purpose and in order to build a regular base on which to correctly estimate the parameters of the different regions, a quadrilaterization of the mesh is carried out. This procedure is based on the combination of Morse theory and Spectral mesh analysis according to [37]. This methodology guarantees a complete quadrilateral description of the mesh, with a C^1 (tangential continuity) between neighboring patches, ensuring a continuity along the edges.

Since the construction's surface needs to be fitted using different NURBs regions (quadrilaterial patches), a regularization process of these regions is necessary. This procedure comprises several steps [37]: (i) selection of one random border in the considered path and its opposite; (ii) border's regularization using B-Splines with a lambda density; and (iii) matching between points by means of the Fast Marching Method.

Finally, the points obtained by the regularization procedure are used as control points to fit each quadrilateral path to a NURBs' surface. It is worth mentioning, that, in construction elements such as arches, vaults or domes, the acquisition of its extrados (as a point cloud) is not possible in most of the cases, due to the presence of several setbacks (e.g., presence of infill, poor lighting conditions or lack of accessibility). Therefore, this previously shown strategy restricted the analysis of such constructions with the membrane theory (Mindlin-Reissner or Kirchhoff-Love theories). These theories limit the geometry of the numerical model's different elements to a minimum recommended size of ten times the construction's thickness. It implies as well the waste of the geometrical potentialities offered by the previously shown methodology.

Based on what is remarked above, a complementary strategy is proposed; able to estimate the construction's extrados based on its intrados geometry. This methodology is made up of the following stages: (i) decorative elements removal; (ii) normal estimation of the points by means of eigenvalue analysis of the covariance matrix [39]; (iii) translation of each point along the normal direction (with a value equivalent to the construction's thickness); (iv) point cloud meshing based on the Poisson approach and (v) projection of the cracks along its orthogonal direction. As a result, an accurate geometrical model of the construction is obtained with which to evaluate its actual stability (Figure 12).

Regarding the numerical aspect, and for the present case study, an incremental static non-linear FEM was carried out [40]. The material properties and the modelling strategy remain the same as those estimated for the initial model (considering the most appropriate ones) in Section 3.3, including a discrete model of the cracks.

For the present case study, this cracking is modelled considering the residual transversal stiffness (shear strength) through Equation (8). Concerning the normal stiffness, only a contribution in compression was considered, dismissing any contribution to the tensile regime according to Equation (9) (Table 8).

$$K_{t,c} = \frac{G_b G_m}{h_m (G_b - G_m)} \beta_{crack} \tag{8}$$

$$K_{n,c} = \frac{E_b E_m}{h_m (E_b - E_m)} \tag{9}$$

where $K_{t,c}$ and $K_{n,c}$ represent the tangential and normal stiffness respectively; G_b and G_m the shear modulus of brick and mortar, respectively; E_b and E_m the Young modulus of brick and mortar, respectively; h_m the mortar thickness; and β_{crack} the shear retention factor.

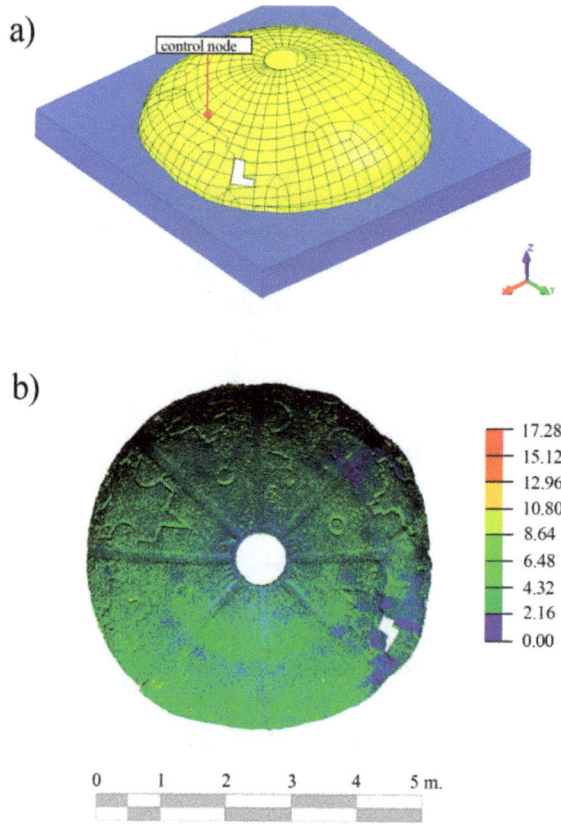

Figure 12. (a) Isometric view of the considered mesh model; (b) Discrepancies, expressed in mm, between the Non-Uniform Rational B-Splines (NURBs) and the photogrammetric models.

Table 8. Mechanical properties considered for the interaction between macroblocks (cracks).

Mechanical Properties of the Cracks		
h_m (mm)	Mortar thickness	15.00
G_b (N/mm^2)	Brick's shear modulus	1.27
G_m (N/mm^2)	Mortar's shear modulus	0.47
β_{crack}	Shear retention factor	$\beta_{c,m}$
$K_{t,c}$ (N/mm^3)	Tangential stiffness	121.88
$K_{n,c}$ (N/mm^3)	Normal stiffness (compression)	49.74

Finally a mesh for the numerical simulation is provided, with a total of 45,350 elements, clustered in: 45,196 high order solid elements and 154 high order interface elements.

For the stability analysis, all the loads acting on the dome (self-weight, infill pressure and asymmetric load) were considered. Afterwards, the estimated safety factor was established as the ratio between the current load and the collapse load obtained in the numerical simulation (Figure 13).

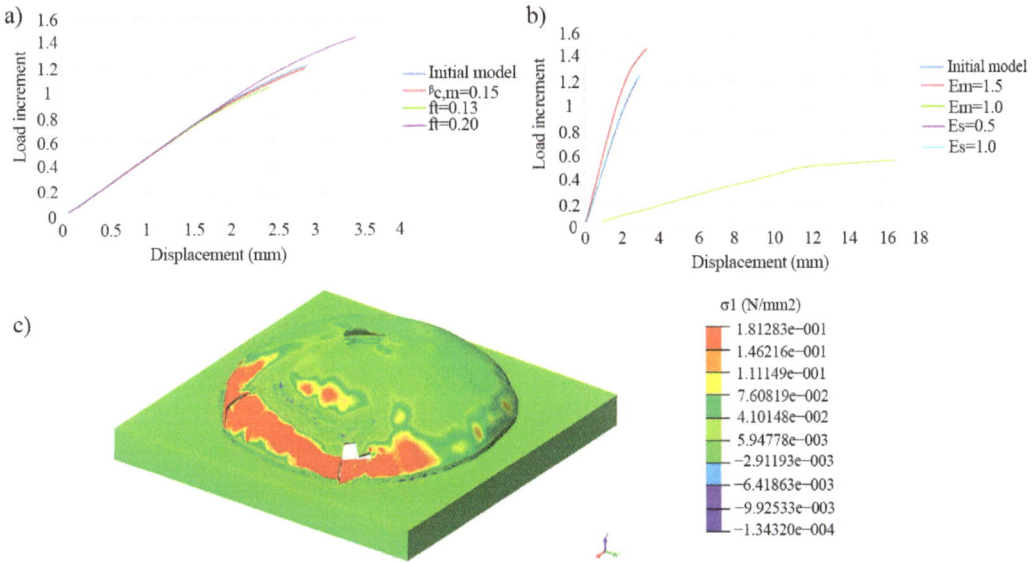

Figure 13. (a) Parametric analysis of different tensile strengths and shear retention factors; (b) Parametric analysis of different masonry and infill's Young modulus; (c) Maximum principal stress (σ_1), expressed in N/mm^2, at collapse of the initial considered model.

According to the study carried out in the Section 4.1 and the inspection of the SfM point cloud, the collapse mechanisms are mainly due to the formation of plastic hinges in the tensile regime.

The complexity of the model and the uncertainties associated with the variables (e.g., soil properties) require the study of the influence of different mechanical variables in the global stability of the construction, through parametrics analysis.

For these analyses, only the most important mechanical properties (to the tensile regime) were considered, namely: (i) Young modulus; (ii) tensile strength; and (iii) shear retention factor.

It can be observed, that the stability of the dome is mainly conditioned by the mechanical properties of the masonry, rather than the mechanical properties of the infill. Therefore, a safety factor (considering the initial mechanical properties) of 1.23 was established (Figure 13c).

However, it is worth mentioning that only the most important cracks were taken into account. By following a discrete strategy, minor and diffuse cracks were not considered. For this reason further investigation, integrating complementary approaches is necessary in order to obtain a better estimation of the actual stability.

5. Conclusions

Based in the already established photogrammetric techniques of Digital Image Correlation (DIC) and image-based modelling (SfM), and complemented with geometrical (NURBs modelling and Hausdorff distance) and numerical methodologies (FEM), the strategies defined and used in the article allow the needs of structural evaluation of historical constructions to be met.

On one hand, two novel geometric quality indices are introduced and defined, called Global Hausdorff metric or GHm_s and Local Hausdorff metric or LHm_s. They allow to assess globally (GHm_s) and locally (LHm_s) the robustness, in geometric terms, of the obtained numerical model in comparison to the point cloud (deformed shape) of the construction. These indices can calibrate the different variables, based on the geometrical similarity between models acting on the numerical simulation.

On the other hand, with the aim of evaluating the actual stability of the construction and exploiting the geometrical and radiometric components of the obtained products (SfM point clouds), a modelling strategy based on NURBs is proposed. This strategy is able to profit from these properties to obtain an accurate geometrical model (with the actual deformation and damage), that serves as a basis for subsequent numerical analysis.

In order to validate these parameters and modelling methodology, it was applied to a real case study: the dome of the church of San Lorenzo in Sejas de Aliste (Zamora, Spain). Several simulations were carried out to understand the degradation process between the initial and the current state, and to corroborate the viability of the defined robustness parameters with a value of 7.40%, for the GHm_s. When studying the current construction's stability, through the modelling strategy defined in the article, the results reveal a damaged construction with an estimate safety factor of 1.23.

However, the complexity of the model, the initial state, the absence of comprehensive knowledge of the different construction stages and the need of more experimental campaigns hinder the numerical results and the correct estimation of the safety factor. Taking this into account, further research will focus on the following aspects: (i) dynamical tests; and (ii) a robust calibration procedure (e.g., Non-Linear Square Minimization) based on the geometrical indices defined to enhance the numerical simulation of the dome.

Complementary to this, concerning the used image-based procedures, their potential includes: (i) flexibility (these may be used in the evaluation of mechanical properties of materials as well as geometrical models); (ii) wide range of applications, its use may be extended to other types of constructions such as tunnels or bridges; (iii) non-contact and non-destructive techniques; (iv) low associated cost; and (v) abundance of geometric and radiometric data. However, the methodology has some limitations: (i) the lack of geometrical information in non-visible areas, requiring complementary sensors such as electric tomography or ground penetration radar; and (ii) the model's accuracy, with several millimeters of error, restricting this strategy to constructions with large deformations.

Acknowledgments: Authors would like to thank Remote Sensing and the anonymous reviewers. Also the assistance provided by the University of Vigo and Minho. Authors would also like to thank architect Jose Luis Bordell for his valuable comments.

Author Contributions: All of the authors conceived and designed the study. Luis Javier Sánchez-Aparicio implemented the geometrical methodologies defined in the article. He also evaluated the structural behavior of the construction and carried out the experimental campaigns with the support of Alberto Villarino. Jesús García-Gago investigated the construction, its materials and constructive solutions, and also contributed to the experimental campaign. Luis Javier Sánchez-Aparicio, Alberto Villarino, Jesús García-Gago and Diego González-Aguilera wrote the manuscript.

Conflicts of Interest: The authors declare no conflict of interest.

References

1. Saloustros, S.; Pelà, L.; Roca, P.; Portal, J. Numerical analysis of structural damage in the church of the poblet monastery. *Eng. Fail. Anal.* **2015**, *48*, 41–61.

2. Huerta, S. The analysis of masonry architecture: A historical approach: To the memory of professor Henry J. Cowan. *Arch. Sci. Rev.* **2008**, *51*, 297–328.

3. Riveiro, B.; Solla, M.; de Arteaga, I.; Arias, P.; Morer, P. A novel approach to evaluate masonry arch stability on the basis of limit analysis theory and non-destructive geometric characterization. *Autom. Constr.* **2013**, *31*, 140–148.

4. Heyman, J. *The Stone Skeleton: Structural Engineering of Masonry Architecture*; Cambridge University Press: Cambridge, UK, 1997.

5. Ramos, L.F.; Aguilar, R.; Lourenço, P.B.; Moreira, S. Dynamic structural health monitoring of Saint Torcato Church. *Mech. Syst. Signal Process.* **2013**, *35*, 1–15.

6. Lourenço, P.B. Recent advances in masonry modelling: Micromodelling and homogenisation. *Multiscale Model. Solid Mech. Comput. Approaches* **2009**, *3*, 251–294.

7. Milani, G.; Simoni, M.; Tralli, A. Advanced numerical models for the analysis of masonry cross vaults: A case-study in italy. *Eng. Struct.* **2014**, *76*, 339–358.

8. Del Pozo, S.; Herrero-Pascual, J.; Felipe-García, B.; Hernández-López, D.; Rodríguez-Gonzálvez, P.; González-Aguilera, D. Multi-sensor radiometric study to detect pathologies in historical buildings. *Int. Arch. Photogramm. Remote Sens. Spat. Inf. Sci.* **2015**.

9. Sánchez-Aparicio, L.J.; Riveiro, B.; González-Aguilera, D.; Ramos, L.F. The combination of geomatic approaches and operational modal analysis to improve calibration of finite element models: A case of study in Saint Torcato Church (Guimarães, Portugal). *Constr. Build. Mater.* **2014**, *70*, 118–129.

10. Villarino, A.; Riveiro, B.; Gonzalez-Aguilera, D.; Sánchez-Aparicio, L. The integration of geotechnologies in the evaluation of a wine cellar structure through the finite element method. *Remote Sens.* **2014**, *6*, 11107–11126.

11. Hausdorff, F. *Felix Hausdorff-gesammelte Werke Band III: Mengenlehre (1927, 1935) Deskripte Mengenlehre und Topologie*; Springer-Verlag: Berlin, Germany, 2008; Volume 3.

12. García-Gago, J.; González-Aguilera, D.; Gómez-Lahoz, J.; san José-Alonso, J.I. A photogrammetric and computer vision-based approach for automated 3D architectural modeling and its typological analysis. *Remote Sens.* **2014**, *6*, 5671–5691.

13. Rodríguez-Martín, M.; Lagüela, S.; González-Aguilera, D.; Rodríguez-Gonzálvez, P. Procedure for quality inspection of welds based on macro-photogrammetric three-dimensional reconstruction. *Opt. Laser Technol.* **2015**, *73*, 54–62.

14. Ghorbani, R.; Matta, F.; Sutton, M. Full-field displacement measurement and crack mapping on masonry walls using digital image correlation. In *Advancement of Optical Methods in Experimental Mechanics*; Jin, H., Sciammarella, C., Yoshida, S., Lamberti, L., Eds.; Springer International Publishing: New York, NY, USA, 2014; Volume 3, pp. 187–196.

15. Salmanpour, A.; Mojsilovic, N. Application of digital image correlation for strain measurements of large masonry walls. In Proceedings of the 5th Asia Pacific Congress on Computational Mechanics, Queens Town, Singapore, 11–14 December 2013; pp. 11–14.

16. Sánchez-Aparicio, L.; Villarino, A.; García-Gago, J.; González-Aguilera, D. Non-contact photogrammetric methodology to evaluate the structural health of historical constructions. *Int. Arch. Photogramm. Remote Sens. Spat. Inf. Sci.* **2015**.

17. Pan, B.; Lu, Z.; Xie, H. Mean intensity gradient: An effective global parameter for quality assessment of the speckle patterns used in digital image correlation. *Opt. Lasers Eng.* **2010**, *48*, 469–477.

18. Schweighofer, G.; Pinz, A. Robust pose estimation from a planar target. *IEEE Trans. Pattern Anal. Mach. Intell.* **2006**, *28*, 2024–2030.

19. Bouguet, J.-Y. Camera Calibration Toolbox for Matlab. 2004. Available online: http://www.vision.caltech.edu/ bouguetj/calib_doc/ (accessed on 7 August 2015).

20. Pan, B.; Quian, K.; Xie, H.; Asundi, A. Two-dimensional digital image correlation for in-plane displacement and strain measurement: A review. *Meas. Sci. Technol.* **2009**, *20*, 062001.

21. Xavier, J.; Fernandes, J.R.A.; Frazão, O.; Morais, J.J.L. Measuring mode I cohesive law of wood bonded joints based on digital image correlation and fibre Bragg grating sensors. *Compos. Struct.* **2015**, *121*, 83–89.

22. Pan, B. Bias error reduction of digital image correlation using gaussian pre-filtering. *Opt. Lasers Eng.* **2013**, *51*, 1161–1167.

23. Barazzetti, L.; Binda, L.; Scaioni, M.; Taranto, P. Photogrammetric survey of complex geometries with low-cost software: Application to the "G1" temple in Myson, Vietnam. *J. C Herit.* **2011**, *12*, 253–262.

24. Pierrot-Deseilligny, M.; de Luca, L.; Remondino, F. Automated image-based procedures for accurate artifacts 3D modeling and orthoimage generation. *Geoinform. FCE CTU* **2011**, *6*, 291–299.

25. Rodríguez-Gonzálvez, P.; Garcia-Gago, J.; Gomez-Lahoz, J.; González-Aguilera, D. Confronting passive and active sensors with non-gaussian statistics. *Sensors* **2014**, *14*, 13759–13777.

26. Morel, J.-M.; Yu, G. Asift: A new framework for fully affine invariant image comparison. *SIAM J. Imaging Sci.* **2009**, *2*, 438–469.

27. Selby, R.G.; Vecchio, F. *Three-dimensional Constitutive Relations for Reinforced Concrete*; Department of Civil Engineering, University of Toronto: Toronto, ON, Canada, 1993.

28. Freeda Christy, C.; Tensing, D.; Mercy Shanthi, R. Experimental study on axial compressive strength and elastic modulus of the clay and fly ash brick masonry. *J. Civil Eng.* **2013**, *4*, 134–141.

29. Atamturktur, S.; Li, T.; Ramage, M.H.; Farajpour, I. Load carrying capacity assessment of a scaled masonry dome: Simulations validated with non-destructive and destructive measurements. *Constr. Build. Mater.* **2012**, *34*, 418–429.

30. Recommendations, Maritime Works. *Geotechnical Recommendations for the Design of Maritime and Harbour Works (ROM 0.5-94)*; Puertos del Estado: Madrid, Spain, 1995.

31. Manie, J.; Kikstra, W.P. *Finite Element Analysis User's Manual-Release 9.4.4*; TNO DIANA BV: Delft, The Netherlands, 2011.

32. Girardeau-Montaut, D.; Roux, M.; Marc, R.; Thibault, G. Change detection on points cloud data acquired with a ground laser scanner. *Int. Arch. Photogramm. Remote Sens. Spat. Inf. Sci.* **2005**, *36*, W19.

33. Aspert, N.; Santa Cruz, D.; Ebrahimi, T. MESH: Measuring errors between surfaces using the Hausdorff distance. In Proceedings of the IEEE International Conference in Multimedia and Expo (ICME), Lausanne, Switzerland, 26–29 August 2002; Volume 1, pp. 705–708.

34. Alexandre, L.A. Set distance functions for 3D object recognition. In *Progress in Pattern Recognition, Image Analysis, Computer Vision, and Applications*; Ruiz-Shulcloper, J., di Baja, G.S., Eds.; Springer: Berlin, Germany, 2013; pp. 57–64.

35. Wu, J.-M.; Jing, Z.; Wu, Z.; Feng, Y.; Xiao, G. Study on an improved Hausdorff distance for multi-sensor image matching. *Commun. Nonlinear Sci. Numer. Simul.* **2012**, *17*, 513–520.

36. Ministerio delle Infrastructure. *Ntc (Nuove Norme Tecniche per le Construzioni)*; Ministerio delle Infrastructure: Rome, Italy, 2008.

37. Branch, J.W.; Prieto, F.; Boulanger, P. Automatic extraction of quadrilateral patches from triangulated surfaces using morse theory. In Proceedings of the 16th International Meshing Roundtable; Brewer, M.L., Marcum, D., Eds.; Springer: Berlin, Germany, 2008; pp. 199–212.

38. Attene, M. A lightweight approach to repairing digitized polygon meshes. *Vis. Comput.* **2010**, *26*, 1393–1406.
39. Schaer, P.; Skaloud, J.; Landtwing, S.; Legat, K. Accuracy estimation for laser point cloud including scanning geometry. In Proceedings of the 5th International Symposium on Mobile Mapping Technology, Padova, Italy, 29–31 May 2007.
40. Milani, G.; Valente, M. Comparative pushover and limit analyses on seven masonry churches damaged by the 2012 Emilia-Romagna (Italy) seismic events: Possibilities of non-linear finite elements compared with pre-assigned failure mechanisms. *Eng. Fail. Anal.* **2015**, *47*, 129–161.

Geometrical Issues on the Structural Analysis of Transmission Electricity Towers Thanks to Laser Scanning Technology and Finite Element Method

Borja Conde, Alberto Villarino, Manuel Cabaleiro and Diego Gonzalez-Aguilera

Abstract: This paper presents a multidisciplinary approach to reverse engineering and structural analysis of electricity transmission tower structures through the combination of laser scanning systems and finite element methodology. The use of laser scanning technology allows the development of both drawings and highly accurate three-dimensional geometric models that reliably reproduce geometric reality of towers structures, detecting imperfections, and particularities of their assembly. Due to this, it is possible to analyze and quantify the effect of these imperfections in their structural behavior, taking into account the actual geometry obtained, different structural models, and load hypotheses proposed. The method has been applied in three different types of metal electricity transmission towers with high voltage lines located in Guadalajara (Spain) in order to analyze its structural viability to accommodate future increased loads with respect that which are currently subjected.

Reprinted from *Remote Sens.* Cite as: Conde, B.; Villarino, A.; Cabaleiro, M.; Gonzalez-Aguilera, D. Geometrical Issues on the Structural Analysis of Transmission Electricity Towers Thanks to Laser Scanning Technology and Finite Element Method. *Remote Sens.* **2015**, *7*, 11551–11567.

1. Introduction

Traditionally, building high-voltage power lines has had few obstacles during their construction phase. Currently, this type of infrastructure is facing a number of setbacks: it has a considerable impact on the environment, on economic activities, and on the expansion of cities, besides its economic cost, including inspections and maintenance. All of these problems have led the companies that use and maintain this infrastructure to consider making the best use possible of the existing lines before placing new lines. Old lines were designed according to the standards of the time in which they were built and they were designed to bear a certain load. In many cases, these towers were designed over forty years ago, so the increased loads that will be placed on them will be far greater than the one for which they were designed. In addition to this fact, the design and execution data of the towers has, in most cases, disappeared, and in other cases, building regulations did not even exist at the time.

Due to this, addressing the re-use of existing power lines requires a geometric and structural analysis of the towers to assess their current state.

Formerly, the towers' dimensional analysis was performed through expeditious and manual methods (through the use of a gauge and a measuring tape) that required direct contact with the structure and, therefore, meant high risks and high costs. Afterwards, in search of a remote non-invasive measuring method, classic topographic measuring allowed thorough, notably intense, field work taking angular measurements and determining singular points indirectly through angular intersections. More recently, the existence of reflectorless electromagnetic surveying equipment has allowed direct measurement of distances and angles from a single point, making field work easier and more efficient, although it solely focuses on extracting unique and specific measurements determined by the topographer [1]. This has meant great uncertainty upon the elements of the tower, since the data was only taken at the point where the measurement is performed. In order to have the full representation of the geometry of the structure, in the last years laser scanning has presented as an interesting solution [2–5], due to the fact that they generate dense real-time point-clouds of the tower's geometry from a distance [6]. However, one of the major limitations of these terrestrial geotechnologies is the overall height of the tower, impossible to cover completely from the ground, which has led to the use of robotic unmanned aerial systems that take aerial images and, through photogrammetric procedures, obtain dense three-dimensional models of this type of infrastructure [7].

As for structural assessment, this kind of structure has been analyzed from different points of view as presented in the literature: the effects of loading in the stability of the tower [8–11]; the effects of the stiffness of connections [12–14]; and causes of failure [15–17]. No previous work was performed in order to evaluate the effect of geometric imperfections such as misalignments of structural members at joints or assembly imperfections.

Therefore, in view of what is mentioned above, this paper presents a non-destructive multidisciplinary approach that is articulated in two stages and that analyzes geometrically and structurally electricity transmission tower structures:

1. The first stage will address a detailed geometric description of the structure (reverse engineering) using a terrestrial laser scanner system, performing an "as built" model that provides information on the structure's most relevant data, such as geometry, type, and dimensions of the metallic profiles and their assembly drawings.

2. Secondly, and taking the geometric model obtained by the laser system as a starting point, three different structural finite element models will be developed: one model will have an ideal geometry considering the nodes of the transmission tower to be pinned, supposing that this model was the one

that was used in the original design of the towers. A second model will be developed with the ideal geometry of the previous model, but considering certain continuous elements of the transmission tower to be rigidly connected at both ends. The third model will use the real geometry obtained from the laser scanner (taking into account all possible initial imperfections as, for example, misaligned structural members at joints) and will also consider the stiffness behavior of the continuous elements of the tower. For its analysis the current Spanish standard for the design of such towers will be taken into account [18]. The analysis will be carried out in a linear elastic regime with the software SAP2000 [19] to obtain data of the displacements and stresses in transmission tower members for each load case established by the current codes. It is expected to obtain conclusions about the performance of these different structural approaches and, therefore, conclude which is the most appropriate modeling strategy in structural assessment procedures when an increase in loads to accommodate new services is demanded.

The paper is outlined as follows: after the introduction, in Section 2, the characteristics of the towers' considered structural models are described. Section 3 details: the equipment used, the production of the geometric models through computer aided design (CAD) software, and the data used in structural analysis. In section 4 the results are analyzed and end with the reached conclusions.

2. Structural Modeling of Truss Structures

A transmission tower could be considered as a three dimensional truss structure [20], which is comprised of a reticular structure formed by discrete elements (bar or rods), joined together at their ends by means of connections without friction and designed to withstand the external forces by means of axial internal forces in their members. The idealized structural model used for the study of this kind of structure is usually based on the following assumptions [21,22]:

1. Individual elements or rods are joined together at their ends by means of connections with no friction, which means the nodes transmit forces but do not transmit moments.
2. The centroidal axis of each member is straight and matches the line that joins each end of the member. The axis of all of the members that end in the same node is cut at a single point; otherwise moments will appear in these members so that the node could be at static equilibrium.
3. Whenever external loads are applied in the nodes, all the elements that comprise the structure are subjected to tensile and compressive forces, since there is no friction at the connections. This means that the self-weight of the elements is

replaced with loads applied at their ends. According to [23] bending caused by direct wind loads on the structural elements can be omitted.

4. The cross section of each member has a negligible area compared to its length.
5. The self-weight of the elements is negligible, since the loads supported by the structural elements are usually large in comparison to their weights.

Under the fulfillment of these assumptions, the structural elements are exclusively subjected to axial forces. For the case studies later reported in this paper, given the age of the towers and the virtual inexistence of structural calculation software at that time, it is logical to assume that they were either calculated manually, using the so-called Ritter method [21,22] or graphically, using the Cremona method [21,22].

However when looking at the analyzed tower's "as built" model we found out that assumption 1 and 2 are not true in the case studies herein presented:

1. In the case of members of the towers, their connections are bolted and thus they are not actually frictionless nodes. Additionally, some members, such as the truss chords, are continuous elements, which can evidently transmit moments from one side of a node to another [24].
Thus, these nodes are assumed to be rigidly connected behaving as elastic embedment (or simply rigid joints), which cause local bending moments due to rotations that take place as a consequence of the global deformation of the structure. Such effect causes the so-called secondary bending moments and consequently a secondary stress state, which are juxtaposed to the primary stresses derived from axial forces.
2. Furthermore, truss members that end in the same node do not always cut it at the same point, leading again to an apparition of secondary bending moments. Such bending moments depend upon both proportion of the misaligned or eccentricity and stiffness of the elements.

Both circumstances are observed in all towers herein analyzed. Figure 1 corresponds to an example of the aforementioned issues upon one of the case studies later described (Tower 1). Furthermore, as it can be appreciated, the ideal situation of perfectly-aligned elements assumed in the initial stage of structure design and calculation can suffer variations during its construction on site. This is shown in the "as built" model, obtained with the laser scanner, where small deviations in the horizontality of several truss members and even a small deviation in the vertical alignment of the main body of the tower can be appreciated (Figure 1).

All of these issues question the use of a model with ideal pinned connections, which was possibly the used method at the time of the structure's design and calculation, and whether it is a valid model to perform further calculations at the present time. Therefore, in view of the constraints and considerations discussed above, three different structural models of the towers are tested in order to compare the results with each other and draw conclusions about the deformations and the actual stress state of the tower's elements, all under different load hypotheses established by current standards [18].

1. Model with pinned joints and ideal geometry (Model 1). This model is supposed to be the one that was used for the design of the towers. A three-dimensional model with all members pin-jointed at nodes was developed that took into account all the previously-mentioned assumptions. Accordingly, the adoption of certain assumptions about the geometry provided by the laser scanner was needed in order to adapt the model to an ideal geometry. More specifically, it was assumed that all members concurring in one node were overlapping in a single common point since, in reality, these members were assembled more erratically. For this, the criterion used was to assume that the main vertical members of the tower and the horizontal members were fixed in their position, and the corrections changed the length and exact location of the diagonal braces that connect at each joint. Additionally, the geometry was simplified in certain areas of the horizontal sections of the towers, ignoring the bend and the inclination. Accordingly, vertical deviation of the main bodies of the towers was not taken into account. Finally, the obtained "as built" model measurements of certain modules that form the main body of the towers are simplified to theoretical values. For example, when certain lengths result in 1.99 m or 2.01 m, the value is simplified and considered to be 2 m. For clarity, Figure 2 presents a scheme with the principal structural parts of a transmission tower.

2. Model with rigid joints and ideal geometry (Model 2). According to the data proportioned from the "as built" model, assumption 1 is not valid because certain elements, such as truss chords, are continuous elements. Therefore, a second model was elaborated which considers an ideal geometry but where truss chords and horizontal members of the towers are considered rigidly connected at both ends and all other structural components (truss elements) are treated as pin jointed. In this way, the originated secondary bending moments and, consequently, secondary stresses can be taken into account and their impact in comparison to modeling the tower with all ideal pinned joints quantified.

3. Model with rigid joints and real geometry (Model 3). In contrast with the two previous models, this approach considers all of the real geometric singularities of the structure and includes them into the simulations; for example, it includes horizontal deviations of truss chords, the vertical deviation of the tower's main body, as well as assumptions 1–3 obsolescence. To create such a structural model, first a 3D wire-model was produced using the "as-built" model as a

dimensional foundation. Secondly, it was imported as a "dxf" file into the structure calculation software SAP2000 [19].

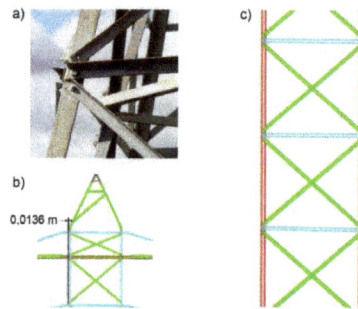

Figure 1. (**a**) Detail of the bolted connections with the inclusion of the eccentricity present at the linkage between profiles; (**b**) vertical deviation of the main body of tower; (**c**) horizontal deviation in the members of the main body of tower.

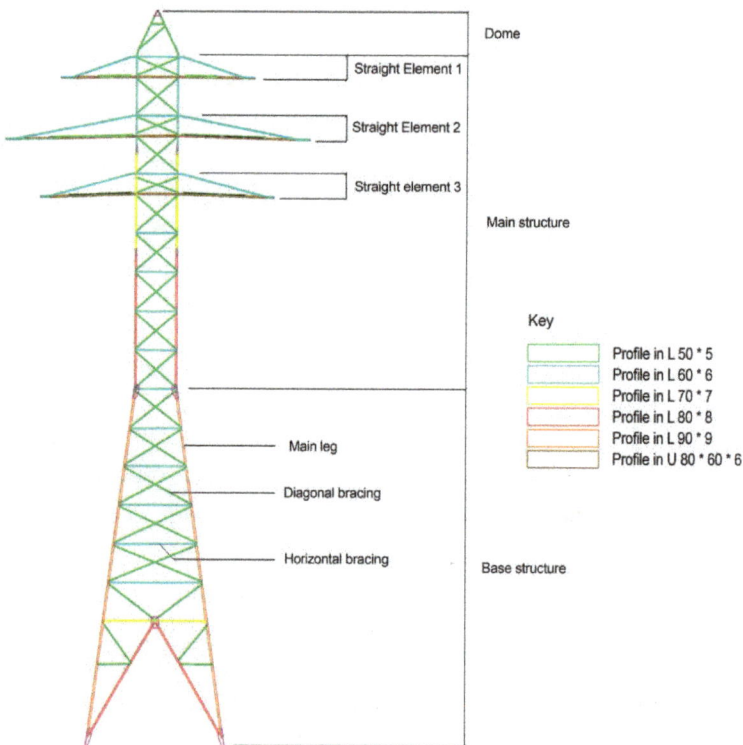

Figure 2. Scheme of main structural parts of transmission tower upon draws of one of the cases studies herein analyzed (Tower 1). Data was obtained from laser scanning technology according to the methodology presented in Section 3.

3. Methodology

3.1. Geometric Digitalization

A laser scanning survey was conducted in order to generate a 3D model accurately describing the structure of the transmission towers. A time-of-flight (ToF) terrestrial laser scanner (TLS), Leica ScanStation2, was used for recording external geometry. This scanner covers a field of view of 360° in the horizontal direction and 270° in the vertical direction, enabling the collection of full panoramic views. The distance measurement is obtained with a nominal accuracy of 4 mm at 25 m range. The horizontal and vertical angular accuracy is of 60 µrad (3.8 mgon). The diameter of the laser spot is 4 mm at 50 m. According to the technical specification of the instrument, it has a maximum acquisition rate of 50,000 points per second. The scanner incorporates a dual axis compensator, so the vertical Z-direction is perfectly defined during data acquisition. Due to the complexity of transmission towers and their heights, four scanner stations with a resolution of 5 mm at 25 m were required to enclose the whole structure. The resulting 3D point cloud (about 6 million points per tower) contains geometric data, normally given in Cartesian coordinates (XYZ), as well as the intensity values (I). This intensity measurement is referred to as the amount of reflected radiation with respect to the emitted radiation. Typically, this value is normalized in the range of 0–1 or 0–255, 8-bit format in our case. According to the principles of interaction between electromagnetic radiation and matter, the intensity values directly depend on the physical characteristics of the object surface, the wavelength of the incident radiation, and the distance between the laser and the object.

3.2. Geometric Modeling

The geometric modeling of the transmission towers was performed following four steps:

1. Cleaning and segmentation of point clouds in order to remove undesired data, such as reflections, noise or sensor artifacts. This step was performed manually.
2. Alignment of the point clouds from each scan under a common coordinate system. An automated registration method, iterative closest point (ICP) [25], was applied, supported by the identification of matching points and minimizing the Euclidean distance between corresponding point clouds. Initial approximations (three points) were manually identified by the user, trying to guarantee a good distribution around the area of interest and along the three main directions (X, Y, Z). A solid-rigid transformation based on the three points identified was executed. Afterwards, an automatic iterative process

to align the different scans was performed taking the Euclidean distance as a minimization criterion.

3. Generating cross-sections and technical drawings of the electrical towers, focusing on the steel profiles that make up each section of the tower and the arrangement of the connections used to define the linkage of these profiles. Different profiles were automatically generated along each main direction (X, Y, Z) in order to obtain vector information of the main sections of the towers and, thus, initial approximations to support the technical drawings and CAD model generation.

4. Obtaining a computer aided design (CAD) model. Since the structural analysis based on a FEM model does not cope with dense laser models, an important step which allows us to pass from the 3D point clouds to a solid geometric model was performed. This step consists in extruding the sections obtained in the step before along its normal direction. Manual interaction is required in this step in order to solve the different intersections between profiles and their connections. In addition, specific existing libraries based on standard steel profiles (*i.e.*, L-shaped and U channel profiles) were used for modeling the towers. Geometric modeling was done using Geomagic Spark, 2013 version.

3.3. Structural Analysis

The three towers are formed by angular steel profiles of different dimensions, and given the age of the towers and only for the purpose of the methodology developed in this article, we assume the lower specification for a steel material enabled by [26], type S-235.

This brings the following mechanical properties: Young's modulus of 2.1×10^8 kN/m^2, specific weight of 76.9729 kN/m^3, Poisson's coefficient of 0.3, and yield stress of 235 MPa.

Furthermore, the data of the power line, support type of the tower, and the mechanical characteristics of the electrical drivers are detailed below:

- High voltage power line with rated voltage of 132 kV and 50 Hz AC
- Two duplex-circuit line with alignment support.
- Span: 300 m between supports.
- Electrical driver of aluminum galvanized steel, type LA-280.

The boundary conditions of the three towers are assumed to be articulated supports in each of the legs that make up the outer frame of the towers, so that they have only limited movements according to the global axes (X, Y, Z). The constraints upon the members are explained in each of the structural models discussed above.

The different load conditions are obtained according to [18]. The following descriptions summarize the loads, always bearing in mind that the towers are located in the province of Guadalajara (Spain) and that they are power lines with alignment support and with suspension insulator strings [18].

- Permanent loads: The self-weight of the steel profiles that comprise the towers, electrical conductors, fittings, insulators, and the grounding wire.
- Wind load: It acts upon steel members of the towers, the insulators, and the suspension insulator strings.
- Imbalance of tensile forces: A longitudinal force equivalent to 25% of all unilateral tractions of electrical drivers and grounding wire. This tensile force will be applied at the point where the electrical conductors and the grounding wire are attached to the support, thus taking into account the torsion that these forces could create.
- Electrical conductor failure: A unilateral tensile force related to a single electrical conductor or a grounding cable's failure. The minimum admissible value for the failure is 50% of the broken cable's tension in the power lines that have two conductors per phase.

Taking into account aforementioned load patterns, the current standards [18] refer to certain calculation hypotheses that establish the load cases, shown in Table 1.

Table 1. Load cases considered in structural analysis of towers.

Tower Type	Force Direction	Hypothesis 1	Hypothesis 2	Hypothesis 3
Alignment support and suspension insulator strings	Vertical	Permanent loads, considering the electrical conductors and the grounding cables to withstand wind load according to a 120 km/h wind speed		
	Transversal	Wind load (120 km/h) on electrical conductors, cables, grounding cables and supports of tower	Not applicable	Not applicable
	Longitudinal	Not applicable	Imbalance of tensile stress	Electrical conductor and grounding cable failure

In order to clarify such load cases considered, in Figure 3 is detailed upon finite element model of one of the case studies herein analyzed (Tower 1), the arrangement of the loads in each of the three scenarios previously commented. Similarly, loads are arranged in Towers 2 and 3.

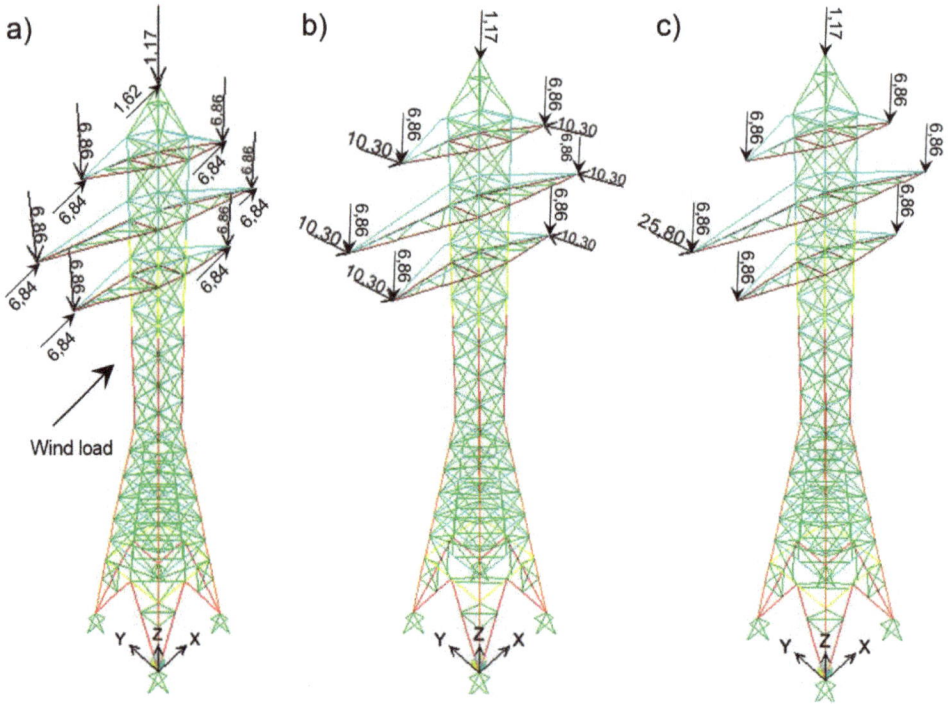

Figure 3. Loads cases considered in structural analysis. (**a**) Hypothesis 1: wind; (**b**) Hypothesis 2: Imbalance tractions; (**c**) Hypothesis 3: electrical conductor failure.

4. Experimental Results and Discussion

4.1. Case Studies

Since this study arises as a consequence of the need to analyze the structural viability of a series of electricity transmission towers that serve as support to an old power electricity line located between the cities of Guadalajara and Torija (Spain), three different cases studies were chosen for the development of the aforementioned methodology.

The electricity transmission towers chosen correspond to a type of tower known as "support alignment" which are disposed over different sections of electricity line. Additionally, in order to extend the study over different structural typologies, each tower corresponds to a different morphology.

The first tower is formed by both a main body support and another principal body (comprising horizontal bracings and diagonal bracings according to a St Andrew's disposition) and three horizontal symmetrical bodies for the support of the cables. The second tower only has a support body (formed by horizontal bracings and secondary diagonal bracings according to a St Andrew's disposition) and three asymmetric horizontal bodies. The third tower is similar to the second one, with exception in the diagonals forming the support body which are not arranged according to a St Andrew's disposition.

Figure 4 shows a photograph of the three towers that composes cases studies described above.

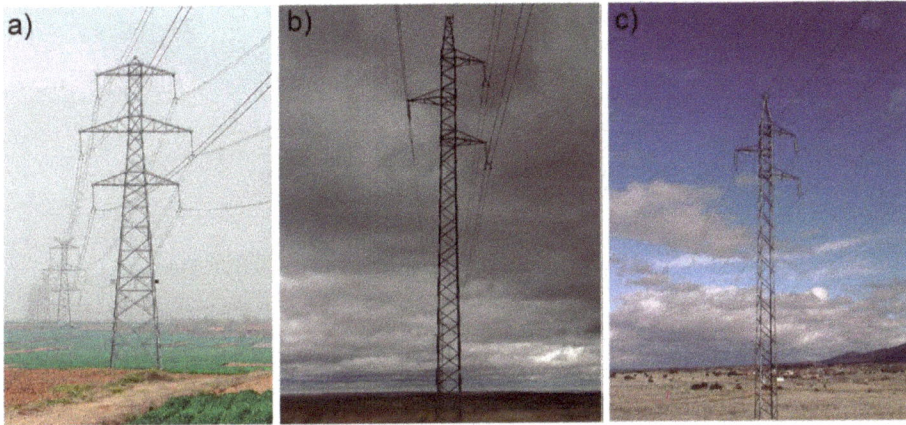

Figure 4. Transmission towers considered in this study: (**a**) tower 1; (**b**) tower 2; (**c**) tower 3.

4.2. Geometric Modeling

Following all steps described in Section 3.2, the point cloud data obtained as a result of a laser scanning survey was subsequently transformed into a CAD model valid for its implementation in the finite element software package SAP2000.

This is a key step required in this kind of reverse engineering process, since data obtained from laser scanning technology do not represent any valid information by itself for the purpose of finite element analysis without suitable data processing [27].

Therefore, taking this into account, CAD models for each one of towers analyzed together with drawings about its current disposition and assembly information were obtained.

Figure 5 shows CAD wire models obtained for each one of towers analyzed. Once such geometrical models were obtained, they were directly imported as a DXF file to SAP2000 software for the finite element analysis stage.

Main geometric data concerns to dimensions of the base, height of the tower, and length of horizontal bodies are displayed in Table 2. The length of the horizontal bodies of the towers is measured from the main body of the tower up to the farthest node. Tower 1 has three horizontal bodies with different dimensions, while in towers 2 and 3, all the horizontal bodies have similar dimensions.

Table 2. Main geometric data for transmission towers analyzed.

Tower	Base Dimensions (m)	Height (m)	Length of Horizontal Body (m)
1	7.25 × 7.25	37.25	3.95 6.25 5.0
2	1.5 × 1.5	18.50	2.0
3	1.5 × 1.5	18.50	2.15

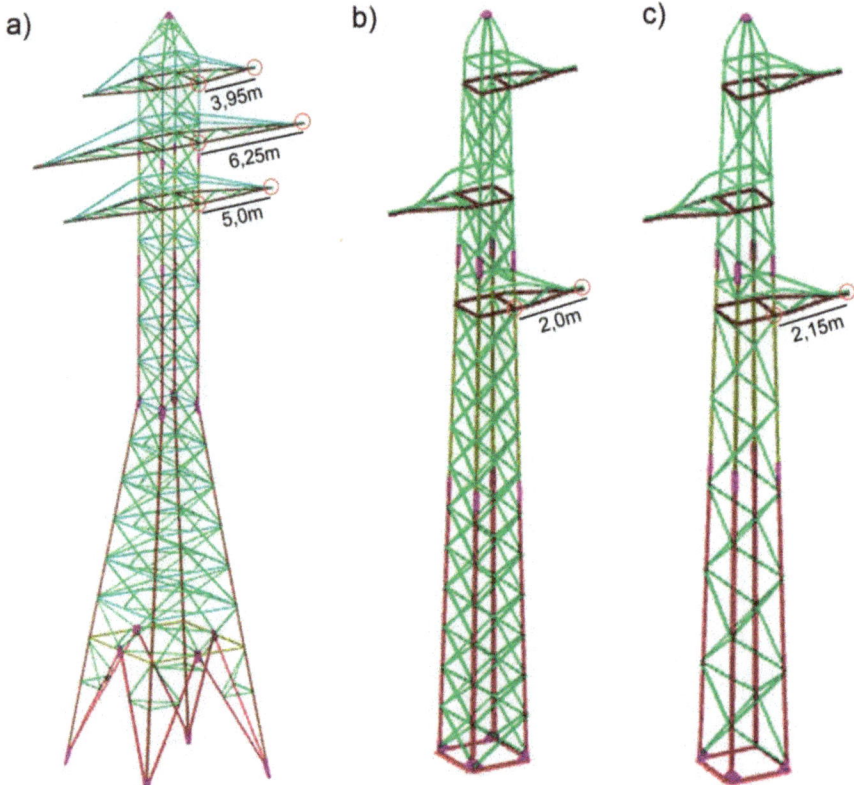

Figure 5. Geometrical CAD wireframe models of transmission towers analyzed: (a) tower 1; (b) tower 2; and (c) tower 3.

4.3. Structural Analysis

As was previously indicated, finite element models of transmission towers were analyzed in SAP2000 software. Within this package, frame elements were chosen so that stiffness against rotations could be considered in all of those nodes assumed to behave as rigid joints. For all other cases where moments will not be considered (as, for example, in diagonal members pin jointed to truss chords), releases end options in nodal degrees of freedom could be imposed for transforming frame elements to truss elements and, thus, only axial forces be considered. For all structural models herein developed, analysis was carried out considering linear elastic behavior.

Table 3 shows the number of frame elements and the number of degrees of freedom for each of the finite element model developed, taking into account the type of transmission tower structure and structural model approach.

Table 3. Number of frames elements and degrees of freedom for each one of the three different structural models considered upon cases studies analyzed.

Tower	Model 1		Model 2		Model 3	
-	Frame Elements	Degrees of Freedom	Frame Elements	Degrees of Freedom	Frame Elements	Degrees of Freedom
1	419	3320	419	3896	683	6772
2	241	1909	241	2348	321	3308
3	181	1448	181	2100	226	2408

4.3.1. Displacements

In this section, displacements [12,14,28] experimented by each tower under different load cases and structural models are analyzed and discussed.

To carry out the analysis and comparison, representative points of the three towers associated with the nodes of the horizontal bodies and the upper node of the dome were selected. Figure 6 shows an example of these nodes together with the displacements experimented for each tower under Loading Case 1 (wind load) and structural model considering real geometry for upper node of the dome (Node 1).

Table 4 reports both the values of maximum displacements obtained for Model 3 (real geometry) and the node where they take place (one of the aforementioned) for the different loading scenarios considered in each one of the transmission towers analyzed. Table 5 represents a comparison (in absolute value) between maximum displacements for Models 1 and 2 with respect to Model 3 (considered as the most accurate).

In both tables, the structural model considered is represented by "M" followed by a number indicating the corresponding model described in Section 2. Load cases considered are indicated by "H" denoting each one of the hypothesis described in Section 3.3 and Table 1 is applicable.

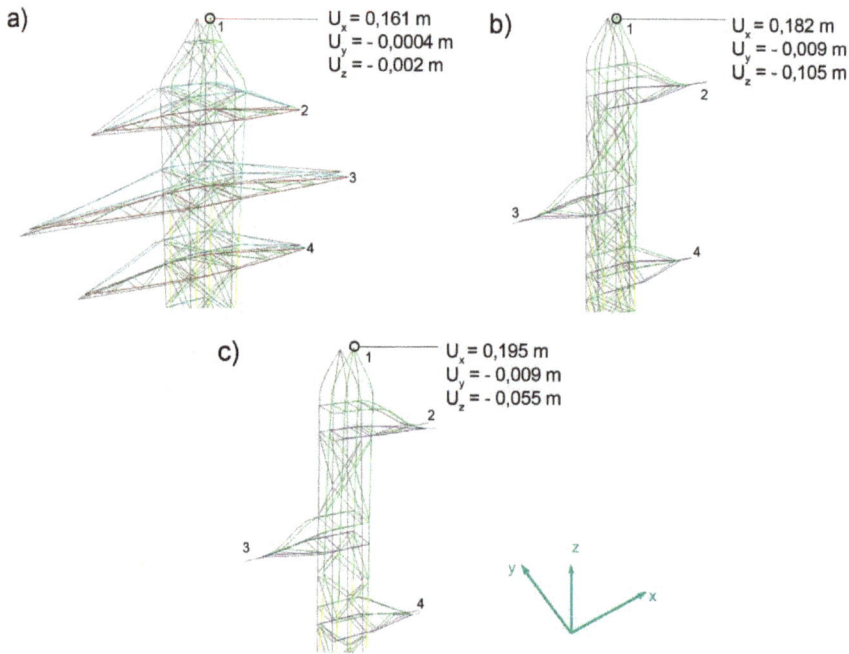

Figure 6. Nodes considered for analysis of displacements according to the global axis directions in each tower and structural model analyzed. (**a**) tower 1; (**b**) tower 2; (**c**)tower 3.

Table 4. Values of the maximum displacements experimented for Model 3 in all transmission towers and load cases considered with the indication of the node where they take place.

Tower	Model	Displacement Ux (mm)/Node			Displacement Uy (mm)/Node			Displacement Uz (mm)/Node		
		H1	H2	H3	H1	H2	H3	H1	H2	H3
1	M3	161/1	14/3	12/4	8/3	162/3	140/3	132/3	8/2	8/2
2	M3	182/1	44/2	11/2	9/1	88/1	125/3	123/2	11/2	15/2
3	M3	195/1	52/2	15/2	9/1	97/1	146/3	85/2	21/3	18/2

Table 5. Comparison between displacements experimented by Models 1 and 2 with respect to Model 3 in all transmission towers and load cases considered.

Tower	Models Compared	Maximum Deviation Ux (mm)/Node			Maximum Deviation Uy (mm)/Node			Maximum Deviation Uz (mm)/Node		
		H1	H2	H3	H1	H2	H3	H1	H2	H3
1	M3 front M1	93/1	6.3/3	4/4	5/3	52/3	62/3	27/3	4/2	4.6/2
	M3 front M2	95/1	6.3/3	4/4	5/3	65/3	78/3	29/3	4/2	5/2
2	M3 front M1	66/1	8.3/2	7/2	4/1	44/1	39/3	35/2	4/2	9/2
	M3 front M2	80/1	9/1	8/2	6/1	43/2	48/3	38/2	5/2	10/2
3	M3 front M1	75/1	9/2	6/2	7/1	38/1	37/3	38/2	3/3	5/2
	M3 front M2	65/3	7/1	7/2	4/2	32/2	42/3	43/3	5/3	6/3

The results displayed in Table 5 show that principal differences always take place in all towers under Hypothesis 1 (wind load case). The differences found between Model 3 and Model 1, reaching 95 mm in the global X direction is remarkable.

There are also significant differences for all towers, although less than aforementioned for displacements of nodes in the global Y direction under Hypotheses 2 and 3 (Imbalance tractions and cable break). The reasons that can explain these differences may be due to the following factors:

1. Consideration of the initial small vertical deviations of the towers' main bodies along with the application of wind forces in Hypothesis 1, precisely in that direction, may have accentuated the displacements in global X direction obtained in Model 3.

2. The fact of considering the true connections of the profiles onto the nodes (misalignment) has a drastic consequence upon the behavior of the structure since it directly affects to its stiffness.

In Models 1 and 2 "ideal nodes" connecting various elements mobilize stiffness of all concurrent elements (truss chords, diagonal, and horizontal members): axial stiffness in Model 1, and axial and bending stiffness in Model 2; in Model 3, however, it does not occur equally.

When considering real geometry, the existence of "intermediate" nodes inserted into the truss chords cause local bending moments and, thereby, additional rotations that accentuate local deformations of the structures. Therefore, we can state that the improper execution of the connections leads to a less stiff structure and may be the main cause of the differences found between the displacements obtained in the three towers for global X direction under Hypothesis 1 and for global Y direction under Hypotheses 2 and 3.

A representative example of this behavior can be seen in Figure 7, which shows the deformed shape of tower 1 for Models 2 and 3 under Load Case 3 (electrical conductor break), which subjects the body of the tower to a torque.

4.3.2. Stresses in Structural Elements

Finally, a comparison regarding the stresses in structural elements [12,14–16,29] is also established. Table 6 shows, based on the finite element results obtained, the maximum normal stresses for each transmission tower and its respective structural model. Particularly, it details for each tower, and for each model, the maximum normal stress in which member it occurs, and under which loading case is developed.

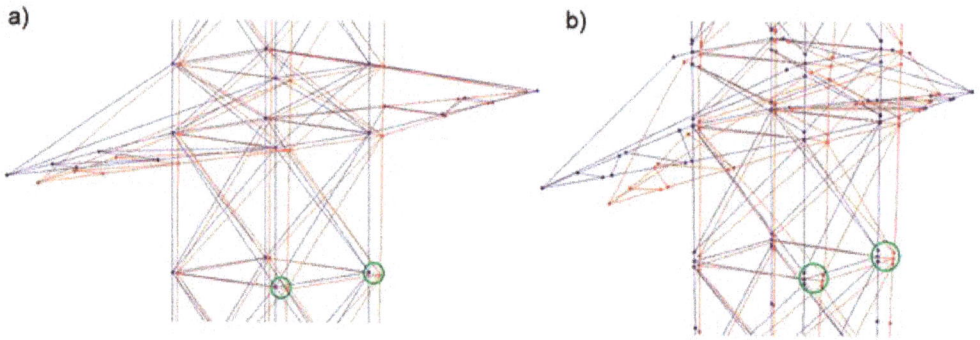

Figure 7. Detail of the different structural behavior of the Tower 1 under Load Case 3 in Model 2 (**a**) and Model 3 (**b**). Undeformed shape is shown in blue and deformade shape in red.

Table 6. Structural elements with maximum normal stresses for all transmission towers analyzed and structural models considered.

Tower	Model	Maximum Stress (Mpa)/Frame Element		
		Hypothesis 1	**Hypothesis 2**	**Hypothesis 3**
	1	215.15/90	165.26/401	195.56/468
1	2	225.36/90	170.23/401	205.21/468
	3	252.30/696	185.96/433	230.20/696
	1	218.96/380	155.50/442	221.32/405
2	2	222.56/380	163.89/442	232.45/405
	3	234.21/178	181.78/439	245.63/189
	1	222.25/272	167.25/353	221.45/107
3	2	232.63/272	175.63/353	226.12/107
	3	252.58/272	195.56/353	233.16/104

Based on Table 6, it can be observed that the differences in maximum normal stresses between Model 1 and Model 2 for all towers are quite small and the maximum of these values are always produced in the same members for both models. On the contrary, significant differences appear when compared to Model 3. Consider the case of Tower 1 and under Load Case 1 where the differences is approximately 37 MPa.

Moreover, it could be observed that the maximum stresses in Model 3 no longer occur at the same members that for Models 1 and 2. This is due to the singular arrangement of nodes in Model 3 (members do not intersect at a single point) which leads to a different discretization of frame elements compared to Models 1 and 2.

Rising stresses in Model 3 are mainly due to secondary stresses caused by the additional bending moments derived from small geometric eccentricities at diagonal and horizontal member's connections upon truss chords which are neglected in Model 1 and Model 2.

The higher the eccentricity is at the connections, the greater the induced bending moments will be. Likewise, the level of induced secondary stresses will be influenced by the stiffness of the truss chords; the greater is their stiffness the greater bending moments will be induced.

This is perhaps the cause of the observed differences for Tower 1 under Load Case 1, where the stiffness of the truss chords relative to the overall stiffness of the whole structure is greater than in the other two towers, thus providing greater increased stresses (37 Mpa) with respect to the theoretical Model 1.

It should be also noted that for Model 3, under certain load cases in all towers, some elements exceed the yield point of the material. This circumstance is highlighted in Figure 8, where for tower 1 and under Load Case 1, the element with maximum normal stress reaches 252 MPa, exceeding the elastic limit of the material in 17 MPa. Figure 8 also shows a representative image of the above discussed; the effects of improper execution of nodes upon truss chords and, consequently, the different discretization of frame elements in the model with ideal geometry and those based on real geometry.

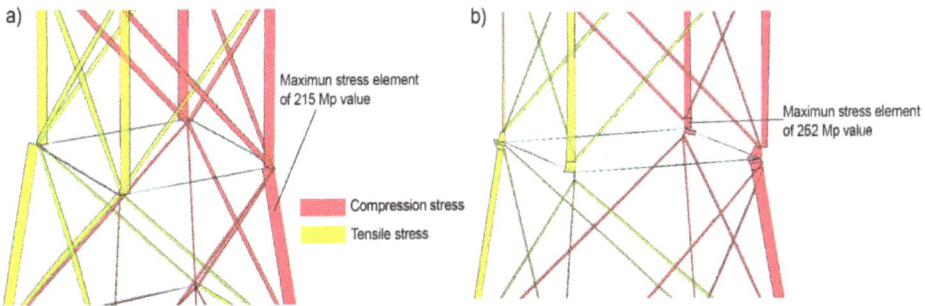

Figure 8. Normal stresses distribution onto Tower 1 for different structural models. (**a**) Model 1, with ideal representation of the nodes and maximum stress below the yield limit of steel; (**b**) Model 3, including the improper executions of the nodes and maximum stress exceeding the yield limit of steel.

As for results involving values of maximum normal stresses, they should be analyzed carefully. Due to the age of the structures, assumed material properties have been chosen according to the minimum value specified by the current regulatory codes; however, these values do not have to match the actual properties of the structure. Accordingly, appropriate experimental tests should be carried out in order to improve their characterization and, thus, derive proper conclusions about the real current safety state of the structure. Note, also, that the structural analysis of transmission towers was carried out assuming a linear elastic behavior.

Nonetheless, the results herein obtained could be considered as acceptable, bearing in mind that the present work is focused in defining an overall methodology able to detect the geometric imperfections present in electricity transmission tower structures by means of precise laser scanning systems and the procedures to incorporate them in structural models based on finite element methods.

5. Conclusions

Terrestrial laser scanning enabled performing a non-invasive remote survey of different transmission tower structures; it should be noted that these are objects of great complexity, not only for their size, but also by their geometry and their high heights. This technology allowed detecting significant imperfections in terms of connections between the members at the nodes, loss of verticality of the towers and the lack of horizontality in the horizontal bracings. The aforementioned imperfections motivated the consideration of different structural models for the towers, in order to analyze how this affects their structural behavior. To that purpose, three different models have been carried out: first, a model with an ideal geometry and considering perfectly pin jointed nodes (supposed model in original calculations of the structures); second, a model with ideal geometry, but taking into account real existing continuity in some profiles by means of rigid connections and pin jointed connections of all others elements upon them; and finally, a third model similar to that previous, but with a real geometry incorporating all imperfections obtained from laser scanning data.

When analyzing the results obtained in terms of displacements and stresses yielded by the different structural models considered, significant differences were observed. At those nodes considered for the comparison of displacements, differences between models with ideal geometry, and models with real geometry reached several centimeters, becoming the highest value (for Node 1 in Tower 1 under wind loading) 9.5 cm respect to the X global direction.

The study of stresses brings some other conclusions. Considering models with ideal geometry and considering pinned joints (Model 1) or rigid joints (Model 2) no significant differences were found. Indeed, maximum normal stresses in elements always take place in the same elements for both models. On the contrary, the model which accounts for real geometry (Model 3) presented notable differences when compared with their respective idealized models. Notorious increases in stresses were detected under certain loading conditions, even reaching the elastic limit of the steel in some occasions.

Differences observed in displacements, stresses in elements and, thereby, whole structural behavior of towers suggest that a detailed survey and conscientious structural analysis has to be carried out when these type of structures will be required for future uses as, for example, new communication services that increase their service loads.

Further studies could contemplate performing nonlinear analysis to extend and improve the results herein obtained, either by considering geometric nonlinearity effects such as P-Delta effects and plastic behavior of steel material. Moreover, due to the nature of the structures (quite slender, and with very slender members) the issue of structural stability should also be addressed. Therefore, it is expected that the combination in the use of the information already available, with the procedure herein developed, together with the consideration of more advanced topics related to strength and structural stability evaluation, will bring a deep insight in the behavior of the towers.

Acknowledgments: This study was funded in part by the Universities of Salamanca and Vigo. Authors would like to thank Remote Sensing and the anonymous reviewers which provided numerous comments and suggestions that contributed to improve significantly an earlier version of the manuscript.

Author Contributions: All authors contributed extensively to the work presented in this paper.

Conflicts of Interest: The authors declare no conflict of interest.

References

1. Ghilani, C.D.; Wolf, P.R. *Adjustment Computations: Spatial Data Analysis*, 4th ed.; John Wiley & Sons: Hoboken, NJ, USA, 2006.
2. Villarino, A.; Riveiro, B.; Gonzalez-Aguilera, D.; Sanchez-Aparicio, L. The integration of geotechnologies in the evaluation of a wine cellar structure through the Finite Element Method. *Remote Sens.* **2014**, *6*, 11107–11126.
3. Costanzo, A.; Minasi, M.; Casula, G.; Musacchio, M.; Buongiorno, M.F. Combined use of terrestrial laser scanning and IR thermography applied to a historical building. *Sensors* **2014**, *15*, 194–213.
4. Fregonese, L.; Barbieri, G.; Biolzi, L.; Bocciarelli, M.; Frigeri, A.; Taffurelli, L. Surveying and monitoring for vulnerability assessment of an ancient building. *Sensors* **2013**, *13*, 9747–9773.
5. Choi, S.W.; Kim, B.R.; Lee, H.M.; Kim, Y.; Park, H.S. A deformed shape monitoring model for building structures based on a 2D laser scanner. *Sensors* **2013**, *13*, 6746–6758.
6. Ioannidis, C.; Valani, A.; Georgopoulos, A.; Tsiligiris, E. 3D model generation for deformation analysis using laser scanning data of a Cooling Tower. In Proceedings of the 3rd IAG/12th FIG Symposium, Baden, Switzerland, 22–24 May 2006.
7. González-Aguilera, D.; del Pozo, S.; Lopez, G.; Rodriguez-Gonzálvez, P. From point cloud to CAD models: Laser and optics geotechnology for the design of electrical substations. *Opt. Laser Technol.* **2012**, *44*, 1384–1392.

8.   Shu, Q.; Yuan, G.; Zhang, Y.; Guo, G. Research on anti-foundation-displacement performance and reliability assessment of 500 KV transmission tower in mining subsidence area. *Open Civ. Eng. J.* **2011**, *5*, 87–92.

9.   Yang, F.; Yang, J.; Han, J.; Zhang, Z. Study on the limited values of foundation deformation for a typical UHV transmission tower. *IEEE Trans. Power Deliv.* **2010**, *2*, 2752–2758.

10. Yuan, G.L.; Li, S.M.; Xu, G.A.; Si, W.; Zhang, Y.F.; Shu, Q.J. The anti-deformation performance of composite foundation of transmission tower in mining subsidence area. *Proc. Earth Plan. Sci.* **2009**, *1*, 571–576.

11. Shu, Q.J.; Yuan, G.L.; Guo, G.L.; Zhang, Y.F. Limits to foundation displacement of an extra high voltage transmission tower in a mining subsidence area. *Int. J. Min. Sci. Technol.* **2012**, *22*, 13–18.

12. Yang, F.; Li, Q.; Yang, J.; Zhu, B. Assessment on the stress state and the maintenance schemes of the transmission tower above goaf of coal mine. *Eng. Fail. Anal.* **2013**, *31*, 236–247.

13. Prasad Rao, N.; Kalyanaraman, V. Non-linear behaviour of lattice panel of angle towers. *J. Constr. Steel Res.* **2001**, *57*, 1337–1357.

14. Da Silva, J.G.S.; da S. Vellasco, P.C.G.; de Andrade, S.A.L.; de Oliveira, M.I.R. Structural assessment of current steel design models for transmission and telecommunication towers. *J. Constr. Steel Res.* **2005**, *61*, 1108–1134.

15. Prasad Rao, N.; Samuel Knight, G.M.; Mohan, S.J.; Lakshmanan, N. Studies on failure of transmission line towers in testing. *Eng. Struct.* **2012**, *35*, 55–70.

16. Klinger, C.; Mehdianpour, M.; Klingbeil, D.; Bettge, D.; Häcker, R.; Baer, W. Failure analysis on collapsed towers of overhead electrical lines in the region Münsterland (Germany) 2005. *Eng. Fail. Anal.* **2011**, *18*, 1873–1883.

17. Zhuge, Y.; Mills, J.E.; Ma, X. Modelling of steel lattice tower angle legs reinforced for increased load capacity. *Eng. Struct.* **2012**, *43*, 160–168.

18. Órgano MINISTERIO DE INDUSTRIA, TURISMO Y COMERCIO. *Real Decreto 223/2008 Reglamento sobre Condiciones Técnicas y Garantías de Seguridad en Líneas Eléctricas de Alta Tensión y sus Instrucciones Técnicas Complementarias ITC-LAT 01 A 09*, 1st ed.; Boletín Oficial del Estado: Madrid, Spain, 2008.

19. Wilson, E. *Integrated Finite Element Analysis and Design of Structures*, 7th ed.; Computers and Structures, Inc.: Berkeley, CA, USA, 1998.

20. Eurocode 3: Design of steel structures—Part 1-1: General Rules and Rules for Buildings EN 1993-1-1: 2005. Available online: https://law.resource.org/pub/eur/ibr/en.1993.1.1.2005.pdf (accessed on 4 July 2015).

21. Beer, F.; Johnston, E.R., Jr.; Mazurek, D. *Vector Mechanics for Engineers: Statics*, 10th ed.; McGraw-Hill Science: London, UK, 2012.

22. Hibbeler, R.C. *Mecánica Vectorial Para Ingenieros: Estática*, 10th ed.; Pearson Educación: Mexico D.F., México, 2004.

23. Eurocode 1: Actions on structures Part 1-1: General Actions—Densities, Self-Weight, Imposed loads for buildings. UNE-EN 1991-1-1: 2003. Available online: http://infostore. saiglobal.com/emea/details.aspx?ProductID=400059 (accessed on 4 July 2015).

24. Argüelles, R.; Arriaga, F.; Atienza, J.R.; Martinez, J.J. *Estructuras de Acero II: Uniones y Sistemas Estructurales*, 1st ed.; Bellisco: Madrid, Spain, 2001.

25. Besl, P.; McKay, N. A method for registration of 3D shapes. *IEEE Pattern Anal. Machine Intell.* **1992**, *14*, 239–256.

26. Eurocode 0—Basis of Structural Design UNE-EN 1990: 2003. Available online: http:// infostore.saiglobal.com/store/details.aspx?ProductID=400051 (accessed on 4 July 2015).

27. Laefer, D.F.; Truong-Hong, L.; Fitztgerald, M. Processing of terrestrial laser scanning point cloud data for computational modelling of building facades. *Recent Pat. Comput. Sci.* **2011**, *44*, 16–29.

28. Liu, Y.C.; Liu, Z.D. Study on stabilization and rectification technology for inclined transmission tower. *Rock Soil Mech.* **2008**, *29*, 173–176.

29. Villarino, A.; Riveiro, B.; Martínez-Sánchez, J.; Gonzalez-Aguilera, D. Successful applications of geotechnologies for the evaluation of road infrastructures. *Remote Sens.* **2014**, *6*, 7800–7818.

MDPI AG

St. Alban-Anlage 66

4052 Basel, Switzerland

Tel. +41 61 683 77 34

Fax +41 61 302 89 18

http://www.mdpi.com

Remote Sensing Editorial Office

E-mail: remotesensing@mdpi.com

http://www.mdpi.com/journal/remotesensing